[redacted]

April '92

TREATMENT AND USE OF SEWAGE SLUDGE AND LIQUID AGRICULTURAL WASTES

Proceedings of a symposium held in Athens, Greece, from 1 to 4 October 1990 and organised jointly by

• the Commission of the European Communities, Directorate-General Science, Research and Development

and

• the National Technical University of Athens, Division of Water Resources

within the framework of the Concerted Action 'Treatment and Use of Organic Sludge and Liquid Agricultural Wastes' (COST 681).

TREATMENT AND USE OF SEWAGE SLUDGE AND LIQUID AGRICULTURAL WASTES

Edited by

P. L'HERMITE

Commission of the European Communities, Brussels, Belgium

ELSEVIER APPLIED SCIENCE
LONDON and NEW YORK

ELSEVIER SCIENCE PUBLISHERS LTD
Crown House, Linton Road, Barking, Essex IG11 8JU, England

Sole Distributor in the USA and Canada
ELSEVIER SCIENCE PUBLISHING CO., INC.
655 Avenue of the Americas, New York, NY 10010, USA

WITH 150 TABLES AND 113 ILLUSTRATIONS

British Library Cataloguing in Publication Data

Treatment and use of sewage sludge and liquid agricultural wastes.
I. L'Hermite, P. (Pierre), *1936–*
628.3

ISBN 1-85166-682-6

Library of Congress CIP data applied for

Publication arrangements by Commission of the European Communities, Directorate-General Telecommunications, Information Industries and Innovation, Scientific and Technical Communication Unit, Luxembourg

EUR 13411 EN

Printed in Northern Ireland by The Universities Press (Belfast) Ltd.

TABLE OF CONTENTS

SESSION V – IMPLEMENTATION OF REGULATIONS, PRACTICAL EXPERIENCES AND PUBLIC PERCEPTION

POSTER SESSION

Opening Session

CONCERTED ACTION - TREATMENT AND USE OF ORGANIC SLUDGE AND LIQUID AGRICULTURAL WASTES (COST 681)

R C RAMSAY
Water Directorate, Department of the Environment
Romney House, 43 Marsham Street
LONDON SW1P 3PY

COST PROJECTS

COST is the acronym for "Cooperation and Coordination in the Field of Scientific and Technical Research". COST projects are initiated by the Commission which negotiates agreements with Non-Member States to coordinate research in the participating countries in particular fields of interest, and submits the proposals to the Council of Ministers for adoption.

The Concertation Agreements bind the Community and the other participating States to provide the necessary funds to finance the administration, travel and subsistence costs of a Concertation Committee to coordinate research work being carried out in the various countries participating in the agreements. These funds do not extend to financing the actual research work.

The Concertation Committee is made up of delegates from all participating States with a secretariat provided by the Commission. Delegates may be assisted by experts as required. Details of the States participating in the various phases of the COST 68 - COST 681 projects are set out in Table I.

DEVELOPMENT OF COST 681
STAGE I: COST 68

The first project concerned with sewage sludge was COST 68, which ran from 1972 to 1974 and had as its objective:

> "the assessment of methods of processing and disposing of sewage sludge by the comparative evaluation, using standardized criteria, of existing plants of industrial scale in different countries."

The Concertation Committee submitted its report to the Commission in 1975, concluding that such international collaboration offered considerable advantages to the individual participating States in respect of the sharing of scientific and technical information. In addition the project showed that small, specialised, multinational teams of experts could ensure more efficient direction of research in the field. The Committee recommended that the work be extended with continuing participation of non-community countries. In consequence COST project 68 bis was initiated by the Commission in 1977.

STAGE II: COST 68 bis and 68 ter

Cost 68 bis , entitled Treatment and Use of Sewage Sludge, ran from 1977 to 1980 and covered the following research areas:

(a). **Sludge stabilization and odour problems:**

- definition and determination of degree of stability and relation to odour nuisance,

- comparative evaluation of stabilization procedures.

(b) **Problems related to sludge dewatering:**

- research on water binding forces,

- development and standardization of methods for the assessment of dewatering properties,

- problems related to the use of flocculants,

- comparative evaluation of thickening and dewatering equipment.

(c) **Analytical problems related to sludge treatment and use:**

- characterization of pathogens and evaluation of disinfection procedures,

- characterization and determination of pollutants (heavy metals, persistent organic compounds) in sludge and development of standardized analytical methods.

(d) **Environmental problems related to sludge use:**

- special processing of sludge for agricultural use (e.g. composting) including the improvement of disinfection procedures and pollutant removal.

- transfer of pollutants to plants and harmful effects on vegetation.

- effects of long range sludge application on soil quality and ground water,

- optimum land use of sludge, including sludge from dephosphatation plants.

The Concertation Committee submitted its final report to the Commission in May 1981, and as a result of its conclusions and recommendations its work was continued as COST project 68 ter, under the same title, from 1981 to 1983. Its final report recommended continuation of the work to include more involvement of farmers and in addition assessment of use of agricultural manures. As a result the current COST Project 681 was initiated by the Commission.

STAGE III: COST 681

COST 681, entitled Treatment and Use of Organic Sludge and Liquid Agricultural Wastes has been running since 1984. Two interim reports on this project have been submitted by the Concertation Committee, covering respectively the years 1984/1985 and 1986 - 1988. COST 681 covers the following research areas:

(a) **Treatment of sludge and agricultural wastes:**

- further improvement of conventional treatment methods, mainly with regard to their economic aspects, and of processes for fuel production from sludges and manures,

- study of technologies specifically applicable to small plants and of processes to eliminate heavy metals at their source;

(b) **Analysis of sludges and residues:**

- development and standardization of economic multi-element methods for the analysis of trace elements in sludges, soils and plants, and for the analysis of organic pollutants;

(c) **Hygienic aspects of treatment and use of sludges:**

- elaboration and improvement of methods for the detection and identification of bacteria, viruses and other pathogens, and study of their survival and contamination potential,

- investigations of the efficiency of hygienization processes, definitions of 'indicator organisms';

(d) **Nuisances:**

- odour characterization and emission control;

(e) **Environmental effects of spreading of sludges and manure:**

- long term field experiments on the accumulation of heavy metals, their availability to crops and on transfers of pollutants via soil to plants, and assessment of various application methods with regard to ground and surface water pollution;

(f) **Improvement of land-use of sludge and manure:**

- long-term field experiments on fertilizing value and soil-improvement properties of sludges and manures.

- improvement of treatment process and spreading equipment with regard to optimum land use,

- study of the agricultural value of residues from treatment processes,

- use of sludges and derived products for land reclamation and specific crops (e.g. biomass production).

HISTORY OF THE WORK

The more detailed projects began with COST 68 bis. Its wide range of objectives required the setting up of five specialised working groups so that all aspects of the project could be coordinated efficiently. Essentially these working groups have generally retained their separate identities through into COST 681. The five groups for COST 68 bis were defined as:

Working Party 1: Sludge Processing

Technical aspects of sludge processing (stabilization, dewatering etc.) and definition of parameters relevant to the processes (including analytical methods for their characterisation);

Working Party 2: Chemical pollution of sludge

Analytical aspects of the determination of heavy metals and organic pollutants, characterisation and measurement of odours, chemical state of pollutants, decontamination, etc;

Working Party 3: Biological pollution of sludge

Characterisation of bacteria, viruses, parasites, evaluation of pathogen reduction in sludge treatment processes and assessment of possible hazards for human health.

Working Party 4: Valorisation of sludge

Fertilizing properties and other positive aspects (e.g. improvement of soil structure), optimum application, improvement of product quality, etc.;

Working Party 5: Environmental effects of sludge

Potential constraints to agricultural use, due to heavy metals and organic pollutants; contamination of soil, ground and surface water; transfer of pollutants to crop.

Most of the detailed coordination and exchange of information was carried out through the working parties which met at intervals to discuss progress of the research programme in their field. In addition during the term of COST 68 bis two successful symposia were held to enable the results of a wide range of research projects to be presented to multinational audiences of experts. The first, at Cadarache in France, in February 1979, heard 49 separate techinical and scientific papers covering the work of all five working parties. The proceedings were published by the Commission in 1980. The second was held in Vienna in October 1980 and the proceedings were published in 1981. About 70 papers were presented, again covering the responsibilities of all the working parties.

During the course of COST 68 bis, the Concertation Committee was regularly informed about preparatory work by the Commission for a Directive on the agricultural use of sewage sludge - a paper was presented on this subject by the Commission at the Vienna symposium. In addition a group composed of the coordinators of the working parties reviewed the scientific basis for the draft proposal and attempted to reach a consensus on the recommended scientific criteria. The results of this work can be said to have influenced to a large extent the elements of the draft Directive as finally submitted to the Council of Ministers in 1981.

For the COST 68 ter programme, the terms of reference of the five working parties were defined in more detail as follows:

Working Party 1: Sludge Processing: New Techniques, Evaluation and Optimisation of Costs

- Technical aspects of sludge processing: stabilisation, elimination of malodours, physical behaviour of sludges during their handling and storage, improvement of the processes for the use of sludge in agricultural land, elimination of heavy metals, study of processes other than those aiming at agricultural use and of the use of biotechnology for energy production and valorisation of sludge.

- Standardisation of the methods for the characterisation of sludge and study and assessment of sludge production rates.

- Study of the effectiveness, cost and reliability of treatment techniques and of the economics of sludge processing units.

Working Party 2: Chemical Contamination and Analysis of Sludge and Soils

- Interlaboratory comparison experiments on analytical methods and study of the problems of sampling for analysis. Detection of the presence of organic substances and organic micropollutants. Comparison of costs of analysis.

- Analytical studies of the problems of mobility of heavy metals and of the conditions for their extraction.

- Study of the sources of heavy metals contamination.

- Problems of analysis and elimination of malodours.

Working Party 3: Hygienic Aspects related to Treatment and Use of Sewage Sludge

- Study of the significance of and the methodology to be used for indicators of pollution and of the efficiency of the treatment.

- Definition of the time-interval between spreading sludge and allowing animals to graze.

- Identification, extraction and numbering of viruses and parasitic cysts in sludges and soils; study of plasmid mediated antibiotic resistance.

- Comparison of disinfection procedures with regard to pathogen destruction, reliability and economic aspects.

- Design of epidemiological studies.

Working Party 4: Agricultural Value of Sludge

- Study of the availability of organic nitrogen in different types of sludges and of predictive methods for their nitrogen value. Study of the effects of season and time of application on the efficiency of nitrogen uptake by crops and grass.

- Study of phosphorous availability for crops, of the effects of large phosphorous accumulation in soils, of the interrelationships between phosphorus and heavy metals and of the problems related to the use of sludge orginated from urban plants using dephosphatation processes.

- Study of sludge ability to produce "soil organic matter" and of its effects on the physico-chemical properties of soils and on soil biology; use of sludge as substrate for specialised uses (horticulture, recreation areas, land reclamation).

- Basic studies of nitrogen and phosphorus behaviour in soils.

- Assessment of the "cost-benefit" from the farmer's point of view.

Working Party 5: Environmental Effects from Sewage Sludge

- Studies of crop uptake of metals from sludge-treated soil including cadmium, zinc, nickel, copper, lead, chromium, cobalt and mercury. Mass balance studies of cadmium in agricultural systems. Determination of the highest tolerable application rate of the various pollutants.

- Simulation of the availability and absorption of heavy metals by plants and evaluation of the potential toxicity of sludge with global methods.

- Study of the need for measures to be taken to prevent the pollution of ground water.

- Determination of possible harmful effects of organic compounds.

- Study of steps to be taken to avoid nuisance from smell, appearance and spray.

The working parties continued to meet periodically to exchange information on research in their particular areas, and a third international symposium was held in Brighton (UK) in September 1983. Some 32 separate papers were presented and additionally another 32 poster presentations were available for inspection and discussion during the symposium. The proceedings were published in 1984.

The first stage of COST 681 began in 1984, with five working parties having the following terms of reference;

Working Party 1: Processing of Organic Sludge and Liquid Agricultural Wastes

Technical aspects of sludge and manure processing aiming at further improvement of conventional methods (stabilisation, dewatering, composting, etc.) mainly with regard to their economic aspects.

- study of technologies, especially applicable to small plants processing sludge, and study of processes to eliminate heavy metals at their source,
- improvement of treatment processes (e.g. separation techniques for manures) and spreading equipment (e.g. injection devices) with regard to optimum land use and minimum nuisance.

Other types of processing of sludges, manures and agricultural residues with the aim to use certain residues as animal feed, raw materials for the production of chemicals, etc. as well as the use of sludges for the production of waste derived fuels...

For "odour measurement and control", an "ad hoc" expert group will evaluate the problems and provide synoptic documents on odour control and on the problems of standardisation of organoleptic measurements for assessing odour emission or immission.

For improvement of processes for biogas production from sludges and manures a close collaboration will be established with the contact coordination groups "Anaerobic Digestion and Carbohydrate Hydrolysis" and "Composting" of the R & D Programme on Recycling of Urban and Industrial Wastes.

Working Party 2: Chemical Contamination of Organic Sludge and Soils

Development and standardisation of economic multi-element methods for the analysis of trace elements in sludges, soils and plants.

- Performance of interlaboratory studies aimed at standardising analytical methods and study of the qualitative determination of organic pollutants and their standardisation.

Working Party 3: Hygienic Aspects Related to Treatment and Use of Organic Sludge

Essentially this working party will continue the work set out in its terms of reference for COST 68 ter.

Following a decision of General Directorate VI: (Agriculture) of the Commission of the European Communities to keep within the framework of its activities from 1984 and up to 1988 research on Communicable diseases resulting from storage, handling, transport and landspreading of manure, Working Party 3 will restrict its new activities to sewage sludge, the sanitary aspects of spreading of slurries and manures being considered within the research activities of General Directorate VI.
Among its new activities, Working Party 3 should pay attention to toxicology of trace elements related to nutrition practices.

Working Party 4: "Agricultural Value of Organic Sludge and Liquid Agricultural Wastes"

Long-term field experiments on the fertilising (N,P) and soils improvement properties of sludges and manures, study of the effects of lime-treated sludges, evaluation of residues from treatment process (e.g. anaerobic digestion of manure and food industry wastes) with regard to their use in agriculture, and mechanisms studies on N, P and organic matter behaviour in soils in connection with these practical problems.

Use of sludges and derived products for land reclamation and specific crops.

Working Party 5: Environmental Effects from Organic Sludge and Liquid Agricultural Wastes

Long-term field experiments on the accumulation of heavy metals in soils, their availability and their transfer from soil to plants.

Evaluation of the influence of sludge on flora and fauna in order to predict potential hazards, and comparisons with the effects of effluents from livestocks in order to assess their biological activity.

Assessment of various application methods (e.g. soil injection) with regard to ground and surface water pollution.

Study of the influence of other nutrients than sludge, i.e. slurries, fertilisers and/or other sources of effects of heavy metals, i.e. atmosphere, slurries, etc. on the environmental effects attributed to sludge.

Although the extent of their studies was extended to cover aspects of handling and disposing of agricultural slurries as well as sewage sludge, there was no permanent representation on the working parties of agricultural experts from participating States. It was agreed that such experts would be co-opted as required for discussion of particular problems, This facilitated continuation of the coordinating system set up under COST 68 bis and carried through COST 68 ter.

During the COST 681 programme two further international symposia have been held to discuss progress in research. Symposium number four was held in Rome in October 1985, when 25 primary papers were presented and a further 29 displayed as posters for inspection during the period of the symposium. The proceedings were published in 1986.

The fifth symposium was held in Amsterdam in September 1988 when 28 primary papers were presented and another 27 poster papers displayed for inspection. The proceedings were published in 1989.

The current symposium will be the last held under the COST 681 programme which is scheduled to end this year. At present there is no formal proposal for continuation of the coordination of European research into the treatment and use of sewage sludge, but the content of the papers to be presented here indicates to me that there are still several problems to be solved in this field. Perhaps the Commission will be able to arrange for some ad hoc coordination to be initiated for this important work.

Acknowledgements

The COST 68 - 681 programme has been a most successful example of how the Commission can organise and help direct multinational cooperation in research to useful effect. Its success has been due in no small measure to the energy and enthusiasm of Dr Heinz Ott and our indefatigable Secretary, Dr Pierre L'Hermite, of the Commission's DGXII (Environment and Waste Recycling), who have been associated with the programme from its very early days.

I must not forget the army of experts - about 100 strong - who form the cadres of the specialised working parties in COST 681 and have given of their time and knowledge to ensure that the Concertation Committee has been able efficiently to carry out the tasks set by the Commission. Nearly half of these representatives have been associated with the detailed work since the beginning of COST 68 bis, and many are here today to take part in our final symposium. I wish them and all other delegates an interesting and successful week here in Athens.

TABLE I

COUNTRIES PARTICIPATING IN COST 68 AND COST 681 PROJECTS

	COST 68	COST 68 bis	COST 68 ter	COST 681
	1972-1974	1977-1980	1981-1983	1984-1990
Austria	-	x	x	x
Belgium	x	x	x	x
Denmark	x	x	x	x
FRG	x	x	x	x
Finland	x	x	x	x
France	x	x	x	x
Greece	-	-	x	x
Ireland	-	x	x	x
Italy	x	x	x	x
Netherlands	x	x	x	x
Norway	x	x	x	x
Portugal*	-	-	-	x
Spain*	-	-	-	x
Sweden	x	x	x	x
Switzerland	x	x	x	x
Turkey	x	-	-	-
UK	x	x	x	x
Yugoslavia	x	-	-	-
Canada (guest)	-	x	x	x

* These countries have participated in the work of COST 681 since their entry into the EEC.

Session I

TECHNICAL AND ECONOMIC ASPECTS OF TREATMENT AND DISPOSAL OF SEWAGE SLUDGE AND SLURRIES

Chairman : F. COLIN (F)

Rapporteur : J.H. VOORBURG (NL)

SEWAGE SLUDGE PROCESSING : PROGRESS AND PROBLEMS

A M BRUCE

Water Research Centre
Swindon, Wilts, SN5 8YR
United Kingdom

Summary

As part of the EC COST 68/681 activity, Working Party 1 was established to co-ordinate and disseminate information concerning research in Member Countries on various aspects of sewage sludge processing. Most of the methods of treating sludge were studied or reviewed including thickening, disinfection, stabilisation, conditioning and dewatering, composting, thermal treatment, and transportation. The economics and energy aspects of sludge processing were also examined. Special consideration has been given to methods of characterisation of sludges and of harmonising methods. Progress has been made in our understanding of most processes and some important new developments in sludge processing have been assessed. Problems still remain, however, especially in the dewatering of certain types of sludge and in some aspects of stabilisation and incineration technology.

1. INTRODUCTION

When COST-Project 68 'Sewage Sludge Processing' was initiated in 1972, the objectives set were quite limited. There was a task of harmonising and developing methods of characterisation of sewage sludge and a task of undertaking a co-operative study of co-incineration of sewage sludge and domestic refuse. The Project was successful (1). This lead to further phases of COST 68 (bis and ter),and later to COST 681, and the formation of five separate Working Parties each with responsibility for a different aspect of sludge treatment and disposal. Working Party 1 (WP1) was given the broad task of co-ordinating and developing research on 'sludge processing'.

Within this framework, the tasks of WP1 became increasingly wide until virtually all aspects of sewage sludge treatment were included in its terms of reference. These terms were then extended to include ' liquid animal manures' so that the range of topics to be covered eventually became very extensive indeed.

After some 18 years of activity, involving some 15 meetings, 6 Workshops, the production of over 100 papers and involvement in 4 major Conferences, it is appropriate to examine in broad terms what progress has made on 'sludge processing' as identified or promoted through WP1 activity and to highlight some problems that remain to be resolved.

In making this review, it is necessary to confine attention to progress associated, at least loosely, with WP1 whilst recognising that much development in sludge processing has gone on outside the group, particularly in the commercial sector and in some countries not involved in COST, especially the USA. Developments in sludge processing technology proceed relatively slowly and a period of 10 years or more is not unusual for wide application of a new technology to be fully implemented.

Early in the life of WP1, it was agreed by the members that the essential object of all sewage sludge processing plants was to " render the sludge **suitable** for disposal at **minimum** cost". It was recognised that the definition of 'suitable' could vary considerably according to the method of disposal, and also from site to site and from country to country, so that a wide range of end products could be required. But it was considered that the second objective of "minimum cost" (subject to suitability of end product and environmental objectives) should apply everywhere. These objectives have always remained at the forefront of the considerations of WP1.

2. SLUDGE PRODUCTION

Information about sludge production rates is the starting point for the succesful design of a sludge processing plant and WP1 made this a high priority task. We now have reasonably good information about sludge production rates from different types of sewage treatment process and about the effects on production of operating factors (ie sewage type, temperature, plant loading, industrial waste component etc.). A review of data has been published by Puolanne(2).

For some forms of activated sludge treatment, production rates may be very high with as much as 70 per cent of the initial sewage polluting load being converted into sludge solids. At the other extreme (eg extended aeration treatment) production might be only 30 per cent but the resultant sludge is then often very difficult to dewater. Differences in production rates according to the type of sewage treatment process employed explain to some extent the large differences in reported 'per capita' production rates for different countries (ie 30 to 124 g TS / head.day.). But the large range remains a problem to explain fully(3). A protocol for measurement of sludge production has been suggested(4).

It has to be accepted that a 'sludgeless' sewage treatment system is an impossibility (unless all surplus solids are released with the effluent!). Even after incineration, or wet oxidation, 30-40 per cent of the original sludge dry solids remain for disposal, though the original volume is much reduced. But the incentive remains to develop sewage treatment processes which produce less sludge in a cost efective manner remains. The LOSLUJ process has some promise in this regard(5). Essentially, the process involves the biological filtration of finely-screened unsettled sewage so that much of the suspended matter normally removed by primary sedimentation passes on to the filter to be partially destroyed by biological action. Sludge production may be 30 per cent less than from conventional biological filter plants.

3. SLUDGE CHARACTERISATION

It is well known that sludges from different sources can exhibit wide differences in their physical, chemical and biological characteristics. In practice, this means that the designer of sludge treatment plant requires information about certain properties of the sludge in order to produce the best design specification. Characterisation tests are therefore necessary and the availability of laboratory or pilot equipment to conduct them is equally important.

WP1 has been particularly concerned with the development of tests to determine key physical characteristics relating to dewaterability, physical stability, and rheology. Altogether, some 28 different parameters for characterising sludges have been identified in a Compendium of Methods(6). Less than half the methods are Standard Methods, about one third are still tentative methods only and 12 methods are still at the exploratory stage. Thus, there is some way to go before a full suite of standardised and harmonised methods of testing sludges are available to the design engineer. A review of existing methods has been published(7).

In its early days, the COST 68 group produced a recommended method for determining the conditionability of sludges based on the use of the Capillary Suction Time apparatus and a standard stirrer. This approach has been further developed. The determination of particle size distribution, on which work was also done initially, has, however, not had much practical application.

Various useful studies have been encouraged by WP1 including Colin's work on physical state(8),and the studies by Spinosa et al on sludge centrifugability(9). Frost's work(10) to identify inter-relationships between various sludge properties, with the object of identifying a single parameter that would best quantify a range of physical properties, has also been of importance and should bear fruit in due course.

4.FUNDAMENTAL STUDIES

The behaviour of sewage sludges during consolidation, dewatering and stabilisation is complex and still not absolutely predictable. Fundamental studies to elucidate behaviour have continued to a limited extent as part of WP1 activity. Advances have been made in our fundamental knowledge of sludge dewatering(11,12,13), anaerobic digestion(14) and thermophilic aerobic digestion(15).

5.SLUDGE PROCESSING OPTIONS

One of the first outputs from WP1 was a sludge processing network diagram which included all feasible stages of treatment with the options available for each stage(16). This confirmed the commonality of methods in the different member countries. During the 15 years since it was first produced the network has shown little need for change. There is a revised version which now includes the new options of 'oil production' and 'wet oxidation'. These are the only significant additions, though neither is yet in full-scale use as a method of processing sewage sludge.

6.CONSOLIDATION

Sludge consolidation is often the most cost-effective process in sludge treatment. Sludge volumes can be more than halved with corresponding down-stream reductions in process plant size and in transport costs.

Gravity thickening has had a varied reputation over the years but more recent fundamental studies to model the behaviour of sludge during consolidation has lead to an improved design method(17) which has been succesfully implemented in the UK and in FR Germany. The key to success in this case has been the ability to measure 5 key parameters in samples of sludge by carrying out tests using a laboratory centrifuge and settling column. From the results of the tests, accurate predictions can be made of consolidation behaviour on the full scale and the optimum design of consolidation tank can be identified.

There are still problems in the consolidation of some types of sludge however, and further work is indicated to determine if operating methods used in the sewage treatment plant can be used to influence consolidation behaviour of the resultant sludges.

7.LIQUID SLUDGE STABILISATION

The stabilisation of liquid sludge to reduce its offensiveness, to provide some degree of disinfection and to make it more amenable to disposal, has been a major subject of study by WP1. Stabilisation has become more important since the implementation of the EC directive on use of sewage sludge in agriculture because that requires that sludge 'fermentabilty' is reduced before spreading on farm land. In the UK, for example, a Code of Practice identifies 7 alternative process 'routes' which can be used to meet the requirements of the directive; 4 of these involve liquid stabilistion with or without prior pasteurisation(18).

The options for liquid stabilisation are, essentially, anaerobic digestion (hot or cold), aerobic digestion (hot or cold) and lime stabilisation.

Anaerobic digestion

Several studies by WP1 have shown that heated anaerobic digestion is the most widely-used method of stabilisation in Europe (and North America) although the extent of use varies from country to country(19). The process was well developed over 60 years ago and is generally very successful. Progress in recent years has been related mainly to improved mixing of digester contents, improved heating systems and a reduction in design retention period for the sludge. The use of prefabricated steel digesters has also been a notable development in some areas(20). The mixing requirements for digesters have been reviewed(21) and a novel 'see-saw' design of digester has been proposed by Casey(22) with the expectation that it would show improved mixing characteristics with no increase in energy requirements. In a separate activity, the EC has funded considerable research on hydrolysis and anaerobic digestion of other wastes and some of these more fundamental studies have application to sewage sludge. There is now a good understanding of the kinetics and microbiology of anaerobic digestion.

The use of digester gas as an energy source has also received considerable attention. Wuhrmann(23) showed that in terms of energy efficiency, anaerobic digestion with a gas engine producing electricity represented the most attractive process and Loll showed a similar result in his study(24). The cost-effectiveness of using gas for power generation will depend on energy prices and in view of recent developments it might be predicted that there will be renewed interest in this approach even for smaller works.

Some problems still remain with anaerobic digestion. Foaming of digesters is an unpredictable occurence which can be very troublesome and is difficult to cure in some cases. Inhibition as a result of toxic industrial wastes still makes anerobic digestion vulnerable as a process in some situations.

Aerobic digestion

The development of thermophilic aerobic digestion(TAD) as a practical method of stabilisation and disinfection of sludge has been perhaps the most notable development in sludge processing during the period of activity of WP1. The success of the process owes much to Loll(25) who was responsible for early studies in Europe and for designing the first full-scale units in FR Germany. Some notable developments of the process also occurred later in the UK using different aeration methods(26,27). During the same period, the use of thermophilic aerobic treatment as a preliminary stage before anaerobic digestion was demonstrated in Switzerland in the form of the 'Aerotherm' system which is now finding wider application(28). The use of TAD as a biological pre-heating system for anaerobic digestion is another possibility which has received attention.

TAD is unique because it can operate autothermally at temperatures of

60degC and above, thus providing a high degree of disinfection during the stabilisation process. The major disadvantage of TAD is its relatively high operating cost though this is of less importance at small works (ie PE <20 000). The main challenge now is to develop more efficient aeration systems to reduce operating costs particularly for large plants.

The fundamental biochemistry and microbiology of TAD has been examined(15) and has given us a better understanding of the kinetics of the process but more studies need to be done to assist design and operation. Foaming may be an operational problem unless the design includes an effective system of control (eg foam cutters). In some cases, an odour control system is required to treat the off gas from a TAD plant.

Lime Stabilisation

The use of lime to raise the pH of sludge to above 12 is a relatively simple and effective method of acheiving temporary stabilisation and a high degree of disinfection. There has not been any significant development in this form of treatment and it is practiced more in some countries than in others. Much depends on the local cost of lime. The use of quicklime to provide a temperature increase in the sludge and partial drying is more rarely used but can be appropriate in some circumstances(29). The use of lime to cover sludge on landfills is another development(30).

8.DISINFECTION

A workshop on methods of sludge disinfection was one of the earliest activities of WP1(31). Later, an excellent review by Huber et al(32) presented the state of the art on the various methods. There is a trend towards positive disinfection but only Switzerland and FR Germany have legal requirements relating to the microbiological quality of sludges for use in agriculture. The EC Directive does not require positive disinfection to be employed nor is it seen to be necessary for public health reasons in most situations providing disposal practice is in accordance with the Directive.

In some situations, disinfection may provide a greater degree of security of disposal outlet for operators and the availability of cost-effective and reliable methods of disinfection may lead to its wider use in the coming years.

One problem in this area is that there is no simple standard test to determine the effectiveness of disinfection. Designers still await this development from the microbiologists. It appears that most microbiologists consider that a test for salmonella destruction must be included. The provision of 4 hours at 70degC would seem to satisfy most criteria of effectiveness.

Huber et al(32) separated existing methods of disinfection into three categories:

Category 1. Highly reliable; no re-infection problem.
(Pre pasteurisation folowed by anaerobic digestion; Thermophilic aerobic digestion followed by anaerobic digestion; Thermal drying to <15% moisture

Category 2. Processes with some technical problems or heavily dependent on calibre of staff.
(Composting in windrows; lime dewatered sludge; TAD alone)

Category 3. Short track-record or laboratory findings only
(Chemical treatment; beta-irradiation)

Since that review, much development work has been done on TAD and it now seems appropriate to elevate it to Category 1.

9. SLUDGE DEWATERING

Although there has been a trend in some areas towards the use of liquid sludge in agriculture, dewatering to form a solid cake remains a major processing procedure in all countries and it is usually essential where sludge is to be disposed of to landfill. Indeed, there is an increasing reqirement for dryer (and more physically-stable) cakes for landfilling with consequent demand for dewatering machinery that can meet higher solids specifications.

WP1 has maintained an active interest both in chemical conditioning techniques and in new developments in dewatering systems.

Chemical conditioning

The development of laboratory tests to select the most effective type and dose of polyelectrolyte for conditioning sludges has featured prominently in the activities of WP1(33,34). The original method recommended by COST 68 still has some application but other test procedures more directly related to the type of dewatering machine to be used have been proposed. The use of a standard polyelectrolyte (Praestol 444k) was originally required for tests but it is now seen as more appropriate to use the type which is available and likely to be used on the full scale. Many tests still depend on the CST apparatus and standard stirrer but other equipment (MFT test; PFT test) is also in use.

Dewatering equipment

There have been significant developments in the design and efficiency of all three major methods of dewatering(plate press, belt press and centrifuge) over the past 15 years. Most developments hve originated from the manufactures of equipment but some new ideas have come from independent research(35). There has been a sharp decline in the use of vacuum filters mainly because of their high cost, low efficiency and inability to produce high solids cakes.

A major review of dewatering methods by De Pinto et al (36) showed that routes involving thickening, anaerobic digestion and dewatering are usually the most economic.

10. INCINERATION

Co-incineration of sewage sludge with domestic refuse was one of the first topics of investigation for the COST 68 Group but this form of combined treatment of the different types of waste has not proved popular. A recent survey by Spinosa(37) has shown that incineration of sewage sludge alone does not currently play a major role in the treatment of sewage sludge in Europe though it is much more important in some countries (eg Belgium) than in others. Six European countries have no sludge incinerators. However, it is clear that with the increasing pressures on disposal outlets for sludge there will be a trend towards incineration to minimise the quantities of solid matter for disposal. In general, incinerator ash itself does not present a difficult disposal problem.

There has been considerable improvement in incinerator design in the last decade, especially in fluidised-bed technology which is now the 'state of the art' technique(38). Increasingly stringent standards for gaseous emmissions from incinerators has been reflected in improved systems for gas cleaning and it is now possible to meet the very strict controls applied in some countries (eg the TA Luft standards in FR Germany). There has also been an improvement in the energy efficiency of incineration plant, largely due to improved heat recovery systems and more effective dewatering and drying before burning to achieve autothermicity.

Costs of incineration show a very wide range according to local conditions but there is little doubt that under the best conditions this form of treatment can be competetive with other methods of treatment.

It has been shown in Switzerland that incineration of sewage sludge (as a fuel substitute) in modern rotary cement kilns can be succesful and will satisfy enviromnmental requirements providing the mercury content of the sludge is low(39).

11.COMPOSTING

The idea of producing a compost from sewage sludge is a very attractive one. Unfortunately, experience has shown that often there is only a very limited market for such a product though the situation could change in the future. Consequently, composting finds very limited use in Europe (40)

A great deal of research has been done on the process, much of it sponsored by the EC and its greatest potential seems to be in treating animal wastes. There are still problems however in relation to determining the maturity of compost and in maintaining quality control.

12.REMOVAL OF CONTAMINANTS

Throughout the period of activity of WP1 there has been regular consideration of the possibilities of improving the quality of sludges by removing undesirable constituents, particularly heavy metals, by special processing at sewage works. However, although various methods have been examined, the economics of extraction are always unfavourable and the conclusion is that the best way of reducing the problem of contamination is to control the source, ie trade waste control.

13.SPECIAL PROCESSES

Two special methods of processing sludge have come to the fore during the last 10 years and both have been considered by WP1.

Low Temperature pyrolysis to produce oil

The production of oil from sewage sludge by low temperature pyrolysis has been succesfully demonstrated at pilot scale and there is a good prospect of a full-scale system being installed within the next few years (41). This process is one which WP1 has been particularly interested in over the years of its evolution and it is one of the truly new developments in sludge processing in the last decade. Obviously, the economic attraction of the process increases with world oil prices and for this reason it may become of incresing interest in the years ahead. But the process also has merits in its own right.

Deep-hole wet oxidation

Wet oxidation of sludge by 'deep hole' technology is another new development in the last decade and, again, there seems a reasonable chance of a full-scale plant in the near future(42). This form of'emission-less' incineration has attractions but there is still along way to go to demonstrate its economic viability. Clearly, it is not appropriate for use in some geological conditions.

14.ENERGY

The importance of energy efficiency in sludge treatment has varied considerably in the last 20 years and is largely related to world energy prices. The study by Wuhrmann(23) to analyse completely the energy required for different form of treatment including the energy required to construct the plant has probably not greatly influenced planning decisions but it is a reminder that all aspects of energy use should be taken into consideration.

15.ECONOMICS

Bearing in mind the objective of 'minimum cost', special attention was given by WP1 to comparing costs of different processes and different process routes and a number of studieds have been done(43). There is general agreement as to which process routes are 'low cost ' and which are 'high cost' but it became clear early on that it would be extremely difficult to make close comparisons between countries because of different practices and currencies. There is no immediate solution to this problem but it may be possible in due course to use the ECU as the basis of comparison!

16.ANIMAL WASTES

While, in recent years, the terms of reference for WP1 included 'liquid animal wastes' it has been found extremely difficult to combine technical discussion on these wastes with that on sewage sludges. The problem is that the 'sewage industry' and the 'agricultural industry' are fundamentally different in most ways and this affects economic considerations and technological solutions. Only in the area of stabilisation, and to some extent in dewatering, is there common ground. Nielsen(44) has produced a review of the problems asociated with animal wastes and this confirms that diferent approaches are required for the two types of waste.

17.ODOURS

Sewage sludge handling and treatment always involves a risk of causing odour nuisance unless positive measures are taken to prevent it. Much is now known about how to contain odours and how to treat odorous air from treatment units(45). It is often important to be able to measure 'odour intensity' accurately using olfactometric techniques and a sub-group of WP1 under Voorburg has done excellent work in producing improved protocol for measuring odours by olfactometery(46).

18.ALTERNATIVE USES FOR SLUDGE

Perhaps the biggest challenge to WP1 was to identify practical alternatives for using sludges compared to the conventional uses. A review showed that there are many posibilities including use as a chemical feedstock ,animal feedstuff, source of metals, building material, and as an energy source but most do not have sufficient attraction to be used in practice. Probably only the production of oil (see above) has a realistic prospect for the near future.

17.CONCLUSIONS

WP1 has achieved a significant degree of success in terms of co-ordinating research in sludge processing in maintaining a watching brief over technological developments and encouraging dissemination of information among participating countries. Co-operation has beenexcellent and much work has been published through the European Commission. Problems remain but they will probably be the subject of consideration by the successors to WP1 sometime in the future!

REFERENCES

(1) COMMISSION OF THE EUROPEAN COMMUNITIES (1975) COST 68 Sewage Sludge Processing. Final Report of the Management Committee.

(2) PUOLANNE,J (1984) Sludge production rates. In 'Processing and Use of Sewage Sludge'(Eds P L'Hermite and H. Ott) D Reidel Publishing Co. Dordrecht.

(3) NEWMAN,P J.,BOWDEN,A V, and BRUCE,A M.(1989) Production, Treatment and Handling of Sewage Sludge. In'Sewage Sludge Treatment and Use'(Eds A H Dirkzwager and P L'Hermite) Elsevier Applied Science. London.

(4) BRUCE,A M.(1981) A note on the measurement of sewage sludge production. In 'Treatment and Use of Sewage Sludge'. COST 68bis III. Technical Annexes. Commission of the European Communities. Brussels.

(5) HOYLAND,G and RONALD,D.(1984) Biological Filtration of Finely-Screened Sewage. Water Research Report TR198. Water Research Centre,Medmenham.

(6) BRUCE,A M.(1983) Compendium of Methods Used for Characterisation of Sewage Sludges in Relation to Design and Operation of Sludge Treatment Processes. In'Final Report(II Scientific Report) Concerted Action. Treatment and Use of Sewage Sludge'. Commission of the European Communities. Brussels.

(7) HAUGAN,B E and MININNI,G.(1981) Characterisation of Sewage Sludges.In Characterization, Treatment and Use of Sewage Sludge' (Eds P L'Hermite and H Ott).D Reidel Publishing Co. Dordrecht.

(8) COLIN,F.(1984) Characterisation of the Physical State of Sludge. In 'Methods of Characterization of Sewage Sludge'(Eds T J Casey, P L'Hermite and P J Newman). D Reidel Publishing Co. Dordrecht.

(9) SPINOSA,L and MININNI,G.(1984) Assessment of Sludge Centrifugability. In'Methods of Characterization of Sewage Sludge'(Eds T J Casey, P L'Hermite and P J Newman). D Reidel Publishing Co. Dordrecht.

(10) FROST,R C.(1984) Inter-relation between Sludge Characteristics'. In'Methods of Characterization of Sewage Sludge'(Eds T J Casey. P L'Hermite and P J Newman). D Reidel Publishing Co. Dordrecht.

(11) CASEY,T J.(1981) Dewatering of Activated Sludge. In'Characterization, Treatment and Use of Sewage Sludge'(Eds P L'Hermite and H Ott) D Reidel Publishing Co. Dordrecht.

(12) SPINOSA,L and LOTITO,V.(1986) Optimisation of Processing, Managing and Disposing of Sewage Sludge 1. Treatment Process Models. In'Processing and Use of Organic Sludge and Liquid Agricultural Wastes(Ed. P L'Hermite) D Reidel Publishing Co. Dordrecht.

(13) RIBI,J.(1980) Untersuchungen zur Sclamm Filtration. In' Treatment and use of Sewage Sludge' Proceedings of the First European Symposium. Cadarache. Commission of the European Communities. Brussels.

(14) MARTY,B.(1986) Microbiology of Anaerobic Digestion. In'Anaerobic Digestion of Sewage Sludge and Organic Agricultural Wastes' (Eds A M Bruce, A Kouzeli-Katsiri and P J Newman). Elsevier Applied Science Publishers. London

(15) HAMER,G.(1989) Fundamental aspects of Aerobic Thermophilic Digestion. In'Treatment of Sewage Sludge: Themophilic Aerobic Digestion and Processing Requirements for Landfilling (Eds A M Bruce,F Colin and P Newman) Elsevier Applied Science. London.

(16) COMMISSION OF THE EUROPEAN COMMUNITIES (1981) Treatment and Use of Sewage Sludge. COST68 bis Final Report II Scientific Report. CEC Brussels.

(17) HOYLAND,G and DAY,M.(1984) Design Procedure for Batch Consolidation of Sewage Sludges. In'Methods of Characterization of Sewage Sludge (Eds T J Casey, P L'Hermite and P J Newman) D Reidel Publishing Co Dordrecht.

(18) DEPARTMENT OF THE ENVIRONMENT(1989) Code of Practice for Agricultural Use of Sewage Sludge. Department of the Environment. London

(19) COMMISSION OF THE EUROPEAN COMMUNITIES.(1981) Treatment and Use of Sewage Sludge. COST Project 68 bis. Final Report III Technical Annexes. CEC Brussels.

(20) NOONE,G P and BOYD,A K.(1981) Prefabricated Systems for low Cost Anaerobic Digestion. In'Characterization, Treatment and Use of Sewage Sludge'(Eds P L'Hermite and H Ott) D Reidel Publishing Co. Dordrecht.

(21) CASEY,T J.(1986) Requirements and Methods for mixing in Anaerobic Digesters. In' Anaerobic Digestion of Sewage Sludge and Organic Agricultural Wastes'(Eds A M Bruce, A Kouzeli-Katsiri and P J Newman) Elsevier Applied Science Publishers. London.

(22) CASEY,T J.(1984) Anaerobic Digestion of Sewage Sludge using a "See-Saw" Digester. In'Processing and Use of Sewage Sludge.' (Eds P L'Hermite and H Ott) D Reidel Publishing Co. Dordrecht

(23) WUHRMANN,K A (1981) Sludge Treatment and Energy. In'Treatment and Use of sewage Sludge'. COST Project68 bis. Final ReportIII. Technical Annexes. Commission of the European Communities.

(24) LOLL,U. (1984) Sludge Technology in Energy Production. In'Processing and Use of Sewage Sludge.(Eds P L'Hermite and H Ott) D Reidel Publishing Co Dordrecht.

(25) LOLL,U (1989) Combined aerobic thermophilic and anaerobic digestion of Sewage Sludge. In'Treatment of Sewage Sludge: Thermophilic Aerobic Digestion and processing Requirements for Landfilling' (Eds A M Bruce F Colin and P J Newman) Elsevier Applied Science. London

(26) MORGAN,S F. and GUNSON, H G.(1989) The Development of Aerobic Thermophilic Digestion Systems in the UK. In'Treatment and Use of Sewage Sludge: Thermophilic Aerobic Digestion and Requirements for Landfilling' (Eds A M Bruce, F Colin and P J Newman) Elsevier Apllied Science. London.

(27) BRUCE,A M.(1989). Other Investigations of Thermophilic Aerobic Digestion in the UK. In'Treatment and Use of Sewage Sludge: Thermophilic Aerobic Digestion and Requirements for Landfilling' (Eds A M Bruce, F Colin and P J Newman) Elsevier Applied Science.

(28) ZWIEFELHOFER, H P.(1983) Experiences with Aerobic Thermophilic Disinfection of Sewage Sludge. In' Disinfection of Sewage Sludge: Technical, Economic and Microbiological Aspects' (Eds A M Bruce, A H Havelaar and P L'Hermite) D Reidel Publishing Co. Dordrecht.

(29) TULLANDER,V.(1983) Quicklime Treatment of Sludge. In' Disinfection of Sewage Sludge:Technical,Economic and Microbiological Aspects'(Eds A M Bruce, A H Havelaar and P L'Hermite) D Reidel Publishing Co. Dordrecht.

(30) BALMER,P(1989) Experiences with limed sludge as the top cover for landfills. In' Treatment and Use of Sewage Sludge: Thermophilic Aerobic Digestion and Processing Requirements for Landfilling. Elsevier Applied Science. London

(31) DISINFECTION OF SEWAGE SLUDGE:TECHNICAL, ECONOMIC AND MICROBIOLOGICAL ASPECTS. (Eds A M Bruce, A H Havelaar and P L'Hermite) D Reidel Publishing Co. Dordrecht.

(32) HUBER,J. SIEGEL,O and BRUNNER,P H. Survey of Sewage Sludge Disinfection Processes.(Eds A M Bruce and P L'Hermite).Commission of the European Communities. Brussels

(33) HEIDE,B A and SPINOSA, L.(1984) Optimum use of Chemicals in Mechanical Sludge Dewatering. In'Processing and Use of Sewage Sludge' D Reidel Publishing Co. Dordrecht

(34) SPINOSA et al.(1986) Influence of Polyelectrolyte Characteristics on sludge conditioning:Lab Evaluations. In'New Developments in Processing of Sludges and Slurries' (Eds A M Bruce, P L'Hermite and P J Newman) Elsevier Applied Science Publishers. London

(35) LOLL,U.(1986) The CHP Continuous Filter Press. In'New Developments in the Processing of Sludges and Slurries' (Eds A M Bruce, P L'Hermite and P J Newman) Elsevier Apllied Science Publishers.London.

(36) DI PINTO,A C, MININNI, G and SANTORI,M.(1984) Economics of Sludge Processing. In' Processing and Use of Sewage Sludge.(ed P L'Hermite and H Ott) D Reidel Publishing Co. Dordrecht.

(37) SPINOSA,L.(1990) Sewage Sludge Incineration in Europe: Practices and Problems. In' Thermal Treatment of Sludges'(Eds F Colin, P J Newman and L Spinosa) E.Guyot SA Brussels.

(38) FROST,R C.(1990) Developments in Sludge Incineration. In'Thermal Treatment of Sludges'(Eds F Colin, P J Newman and L Spinosa) E Guyot SA Brusels.

(39) LANG,T A and OBRIST,A J.(1990) Incineration of Sludge in Cement Rotary Kilns. In'Thermal Treatment of Sludges'(Eds F Colin, P J Newman and L Spinosa) E Guyot SA Brussels.

(40) DUVOORT-VAN ENGERS,L F.and COPPOLA,S.(1986) State of the Art on Sludge Composting.In'Processing and Use of Organic Sludge and Liquid Agricultural Wastes.(Ed P L'Hermite) D Reidel Publishing Co. Dordrecht.

(41) CAMPBELL,H W.(1989) Status Report on Oil from Sludge Technology. In' Sewage Sludge Treatment and Use: New Developments.Technological Aspects and Environmental Effects'(Eds A H Dirkzwager and P L'Hermite). Elsevier Applied Science. London.

(42) DE BEKKER,P and VAN DEN BERG,J.(1990) Wet Oxidation as the Alternative for Sludge Treatment. In'Thermal Treatment of Sludges' (Eds F Colin, P J Newman and L Spinosa) E.Guyot. Brussels.

(43) DI PINTO,A C. MININNI,G and SANTORI,M.(1984) Economics of Sludge processing.In'Processing and Use of Sewage Sludge' (Eds P L'Hermite and H Ott) D Reidel Publishing Co. Dordrecht.

(44) NIELSEN,V C.(1986) Treatment Processes for Farm Wastes. In'Processing and Use of Organic Sludge and Liquid Agricultural Waste. (Ed P L'Hermite) D Reidel Publishing Co. Dordrecht

(45) EIKUM,A S.(1984). Studies of Processes in view of Preventing Odours. In'Processing and Use of Sewage Sludge' (Eds P L'Hermite and H Ott) D Reidel Publishing Co. Dordrecht.

(46) VOORBURG,J H.and VAN DEN BERG J J.(1989) Odour Problems with Sewage Sludge. In'Sewage Sludge Treatment and Use' (Eds A H Dirkzwager and P L'Hermite) Elsevier Applied Science. London.

(47) FROST,R C and CAMPBELL,H W.(1986) Alternative Uses of Sewage Sludge. In'Processing and Use of Organic Sludge and Liquid Agricultural Wastes' (Ed P L'Hermite) D Reidel Publishing Co. Dordrecht.

SLUDGE: WHAT IS THIS STUFF ANYWAY?

P. AARNE VESILIND

Department of Civil and Environmental Engineering
Duke University
Durham, North Carolina, USA

Summary

The COST group on sludge management has for almost twenty years conducted high quality and fruitful research, but this work has all been within the existing sludge research paradigm. In this paper, it is suggested that new thinking is needed in order to elevate the quality of future research on to a new paradigm. New ways of describing sludge are needed, and sludge disposal may be facilitated by appealing for public cooperation and by initiating changes in citizen behaviour.

1. INTRODUCTION

Sludge is the ubiquitous residual of human civilization. As that, it contains everything which we wish to discard, but cannot place in the trash container. As a result, sludge characteristics are specific to the country, city, part of a city and even the time of year. Further, each wastewater treatment plant treats water in order to meet effluent standards, and the selection of the treatment processes seldom if ever include a consideration of the characteristics of the sludge that is produced during the treatment. Sludge engineers thus have the greatest challenge of any branch of engineering: characterize, treat and dispose of a raw material over which we have almost no control.

What is most impressive and admirable of the COST sludge programs is that the participating countries recognized that even though they could not develop a "standard sludge", it is worth the time and money to help each other with the problem of sludge disposal.

And indeed, a lot of money and a lot of technical expertise has been expended in the COST process. The number and quality of the publications are simply outstanding, and will for many years represent the best research in the field.

That said, I want to chide all of us, me included, about our own myopia. You all have been undertaking the most concerted and imaginative research in sludge management, yet I wonder if we have gotten stuck in a rut, and if this may not be the time to ask some questions that may help liberate us from the ideas of the past, and direct our efforts to the future.

Thomas Kuhn, in his 1962 book The Structure of Scientific Revolutions proposes that science is done in paradigms, sets of specific rules and understandings which underlie a specific body of scientific work.(1) For example, the behaviourists in social science make basic assumptions about how the human mind acts, and build an entire research structure on these assumptions. Sigmund Freud, on the other hand, created a completely different social paradigm based on the psychology of human development.

Kuhn observed that it is very difficult for scientists in one paradigm to discuss matters of substance with those in another paradigm. In fact, even the rationality of the scientists in a competing paradigm is questioned. At any given time, a paradigm provides the network for scientific interaction and understanding among scientists who are doing what Kuhn called "puzzle solving":

> "Few people who are not actually practitioners of a mature science realize how much mop-up work of this sort a paradigm leaves to be done or quite how fascinating such work can prove in the execution...Mopping-up operations are what engage most scientists throughout their careers. They constitute what I am here calling normal science. Closely examined, whether historically or in the contemporary laboratory, that enterprise seems an attempt to force nature into the preformed and relatively inflexible box that the paradigm supplies."

Progress is made within the paradigm, but science does not take significant leaps forward. Whatever difficulties and anomalies scientists encounter are considered "noise" or laboratory artifacts. Eventually, the noise level becomes too great, and a revolutionary scientist is inspired to produce an entirely new paradigm, developing with it a set of procedures, assumptions and definitions. The new paradigm competes with the old one, and Kuhn calls this a "scientific revolution".

It seems to me that we in the sludge business have been working for too long in one paradigm. We seem to be "mopping-up" for the most part, and what we need are some revolutionary and visionary scientists who can carry us to the next paradigm.

In this paper, I do not pretend to show how we will get to that new way of doing sludge science, but I would like to suggest a few areas which seem to be ripe for such exploitation. Specifically, I ask what is it that we mean by the word "sludge"; how do we characterize it, and how do we manage it so that it can best be disposed of in the environment. Finally, I share with you some of the thinking in the USA concerning sludge disposal, not because it is exemplary, but simply as another means of facilitating discussion.

2. SLUDGE CHARACTERISTICS

There is an old story of the biochemist, when asked to explain what a steam locomotive is, ground the engine into fine dust and after analysis proclaimed that a steam locomotive is 95% steel, 2% copper, 2% water and 1% other materials. For the biochemist, working in his paradigm, this is an entirely proper characterization of a steam locomotive.

How would we in the sludge business respond to the question of what a sludge is? We would, I am sure, make use of various well known and understood characteristics: solids concentration, density, odour, colour, pathogen density, metals concentration, etc.

Such parameters are in fact based on Aristotelian objectivism. It was Aristotle who first assured us that there are some objective notions which really exist in nature, and that we can use these for our scientific purposes. These objective ideals are out "there", and our rational thinking creates symbols by which we can describe this ideal other world. Once we have developed a set of symbols for this purpose, these can be manipulated and categorized, and we can build entire scientific disciplines on our basic objectivist assumptions.

Through the past 2 millennia, we have seldom questioned Aristotle. Yet, how do we know he was right?

Take density, for example. This is a purely human concept. No material has an innate natural property called density. We have devised this property by dividing the mass by the volume, and then continued to think that this is an objective means for describing everything in nature. But is density a proper characteristic? What would happen if we question its validity? Can there be other parameters that will be more descriptive, honest, and useful?

We are locked into our objectivist notions of what is meant by sludge characteristics, and are unwilling or unable to question if these are indeed the proper parameters which will best solve our problems. Instead of asking what the best symbols for describing sludge may be, we continue to use the old characterizations in research after research, unable to question their validity.

There is an old story of a wise old American Indian medicine man who one morning was seen on his hands and knees in front of his hut. A friend came by and asked

"What are you doing groping around on your hands and knees?"

"I lost my buttons" replied the wise man, continuing to search.

"Where did you lose your buttons?" asked his friend.

"In the hut." was the reply.

"Why then are you looking out here?" asked the befuddled friend.

"Because the light is so much better here" replied the old man.

I would like to suggest that much of our sludge research has been done in the bright light, because it is easy to look there, but that the real buttons can be found in the hut, where the light is dim and the search is difficult. I illustrate this suggestion using three examples: Water content and distribution, particle size, and physical sludge characteristics.

Water in Sludge

One of our long standing symbols for defining sludge has been that sludge is made up of liquid and solids and that the solids are suspended in the liquid. Water is understood to be the liquid phase, and the solid phase is commonly divided into two types of solids, suspended and dissolved. Solids concentration is defined by evaporating off the liquid, and weighing the residual. Suspended solids are those which are caught on a filter and the remainder of the solids are called dissolved solids.

But the liquid phase in wastewater sludges cannot be only water. The liquid surrounding sludge particles must be composed of uncountable chemicals, all of which were either flushed down the drain, or created during the treatment process. The analytical test for solids, during which "water" is to be evaporated at 105 degrees C must also result in the loss of many liquids other than water. This becomes obvious when we attempt to get a sludge total solids test to stabilize with time. As time goes on, more chemicals volatilize, and the weight of the dried solids reaches equilibrium only after a long time. Similarly, running the test at different temperatures yields different solids concentrations. As the temperatures increase, more compounds volatilize, steadily decreasing the calculated solids concentration. What then is "water" in sludge?

Secondly, water in sludge is often considered to be independent of the solids. The assumption is that water simply surrounds the solids and does not interact with them. But it is understood that there are different kinds of water in sludge, and these behave in very different and important ways.(2,3)

A recent study in our laboratories involved the drying of sludge, with the weight of the sample measured continuously on a balance.(4) If we assume that the liquid phase is water, these experiments clearly show that there are at least four distinctly defined types of water in sludge. We call these free water, interstitial water, surface water and bound water. These are defined schematically in Figure 1. Free water is not associated with the particles; interstitial water is caught within the sludge flocs and travels with the sludge particles; while surface water surrounds the particles and behaves as part of the solids. The fourth form of water, bound water, is chemically combined with the particles.

Perhaps the most interesting type of water is surface water. The water molecules on the surface of the particle are oriented and layered, resembling "soft ice" at normal temperatures (5). These layers of oriented water molecules are tightly attached and travel with the particle, and can extend out to as much as 20 Angstroms. Farther out, water molecules may still be oriented, but are not attached to the particle.

These waters can be driven off the sludge solids by first evaporating the free water, then interstitial water, and finally surface water. Bound water is removed only during chemical dehydration. The sequential removal of these waters can be demonstrated by the curve shown in Figure 2. As the time advances, the change in mass decreases at a constant rate as the free water is driven off, and then decreases in two stages as the interstitial and surface water is removed sequentially.

Corroborating evidence for the existence of different types of water in sludge comes from freezing experiments. (6) As the water surrounding the sludge particles freezes, the ice crystals extract water first from the free water by sending ice needles into the sludge. Free water freezes first since more energy is needed to extract the interstitial water, and even more energy is needed for the surface water to freeze. Thus the freezing exotherm shows distinct jumps as each type of water is frozen.(7)

It thus is not adequate in many cases to speak of just "water" as the liquid phase of sludge. What we call water may be other liquids, and water exists in at least four different forms. The symbols we have constructed are too vague and the tool too blunt. It may be time to refine our way of thinking about water in sludge.

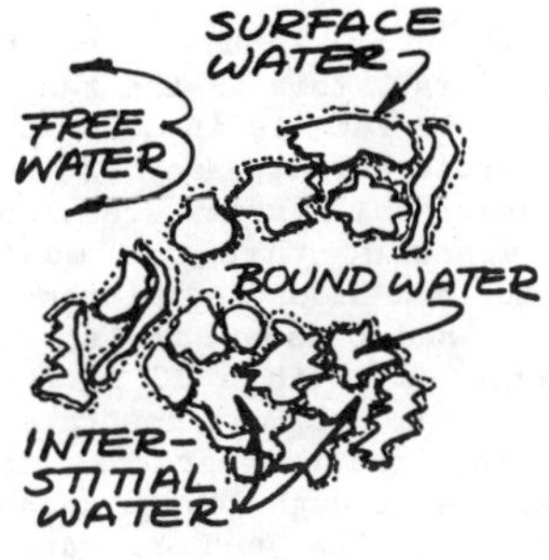

Figure 1: Definition of four types of water in sludge (3)

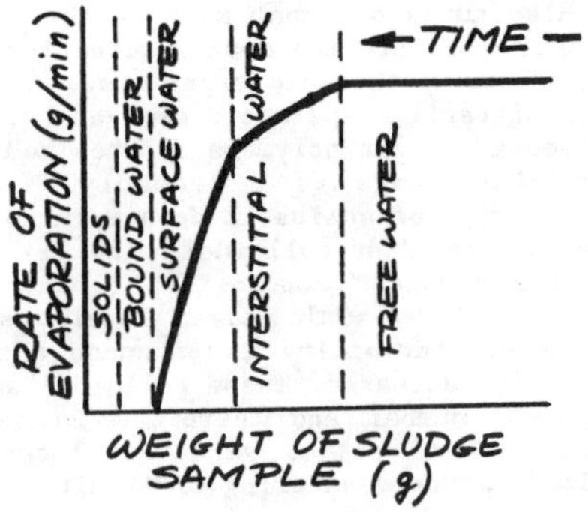

Figure 2: Drying curve, showing the rate of weight change as water is driven off (4)

Solids in Sludge

Solids are traditionally divided into suspended and dissolved fractions , using simple filtration and evaporation. These symbols of sludge have been codified in numerous standard tests and have produced a dichotomy; a solid is either suspended or dissolved. The dividing line between the two is defined by a piece of filter paper.

Yet all of us who have struggled to run suspended solids for sludge using the standard filtration test have given up, and either assumed that the value for suspended solids is almost the same as total solids, or resorted to a centrifugal method for measuring suspended solids concentration. With the latter test, the centrate is invariably cloudy, and try as we might, we seldom are able to obtain a clear centrate. So we assume that the centrate contains very little of the solids which would have been filtered out, and throw the centrate away, safe in the objective structured paradigm of defining sludge solids by the suspended/dissolved dichotomy.

There is mounting evidence, however, that the most important solids in sludge are precisely the ones we cannot easily categorize, and thus tend to ignore and discard as being troublesome. Even given the imprecise nature of particle size measurement techniques, it is evident that sludge particles describe a full particle size distribution curve.(8) Particle size distribution can of course by expressed in terms of the number of particles, the surface area of the particles, or their volume. Each one seems to be significant in some application. For example, the surface area of particles is the main variable governing polymer conditioning. In fact, we have found that the effect of polymers can be enhanced by violently agitating the sludge prior to conditioning.(9)

It also turns out that many of the solids which are in the range of 0.1 to 10 microns are the most interesting ones. First, this is the range for many of the pathogenic organisms, and thus their fate is in question when we arbitrarily separate dissolved from suspended solids. Many of the high molecular weight polymers synthesized by biological systems are also in this size range (5). Secondly, it is apparent that the most troublesome type of solids in dewatering operation are around 10 microns, slightly larger than colloidal. (10,11) Roberts and Olsson have shown that if the sludge "macrosolids" (solid filtered out with a 10 micron filter) are diluted with clean water instead of the original centrate, the sludge dewaterability is enhanced and the required polymer dose is significantly decreased. These colloidal solids, even though low in mass, are large in number and therefore in surface area. The surface water surrounding these solids, and the interstitial water captured within their flocs, makes dewatering difficult.

It seems therefore that to cavalierly cut the particle size distribution curve in half, calling one side suspended and the other dissolved solids, greatly and unjustly simplifies the actual situation.

We should retire with honor the concepts of suspended and dissolved solids, and invent new symbols which would more accurately define the solids portion of sludges. These categories may or may not even be based on size, since the shape of solids is as much of a concern in many cases as the volume or mass of the solids.

Physical Properties

Finally, sludge has been defined by a number of physical techniques borrowed from other fields. Perhaps the most interesting are rheological characteristics, a classical case of human-created symbols. We have defined something called viscosity, adopting the chemical engineering definition of viscosity as the slope of the stress/strain curve obtained by a viscometer, and now pretend that this is some basic property of material in the "other world" of Aristotle.

One problem in using this definition is, as many understand, that the rheogram for the sludge is not a straight line, and we cannot simply define a slope and call this the sludge viscosity. Viscosity for fluids such as sludge depends on the shear placed upon the sludge, and a further complication is that viscosity thus measured is time dependent. Although sludge flow characteristics have been put to good practical use, it nevertheless seems that viscosity is simply an inappropriate symbol for defining sludge fluid flow. Yet we persist in using it as if it had some meaningful relationship to sludge behaviour.

An even more interesting problem is the presence of the solids in the measurement of flow. Solids do not, of course, flow. They move with the liquid if the motion is slow enough and they are not moved out of the flow lines by inertia. Their presence clearly changes the way a two-phase fluid like sludge flows, but what exactly do they contribute to this change? Forster has shown that the surface components of sludge particles significantly influence the rheological characteristics.(13) It would seem reasonable that the particle shape, adhesive properties, or flexibility have an effect on how the flow properties are changed. Forster showed that water associated with the particles (capillary and interstitial water by our definitions above) can significantly influence sludge rheology, and that this could be modified by the action of metal ions.(13)

I believe there is a world of opportunity in defining new paradigms for sludge characterization using imaginative new rheological concepts.

3. SLUDGE DISPOSAL

At the present time, we in the United States are undertaking a comprehensive evaluation of sludge disposal. Briefly, the Environmental Protection Agency has devised a plan for the setting of national guidelines for sludge disposal. The Agency initially proposed preliminary guidelines, and these were open for public comment. At the present time, the EPA is reviewing the comments and will eventually promulgate the guidelines which will be enforced by the states.

The process for the development of these guidelines started with a review of the risks involved with sludge disposal. Six different disposal methods were identified:

1. Placement in a municipal solid waste landfill
2. Placement in a dedicated landfill
3. Incineration
4. Land spreading
5. Sale or give-away
6. Ocean disposal

Models were developed for estimating the effect of each of these disposal techniques on human health. The primary disease causing agents were identified as heavy metals and pathogens. The effect of disposal on the environment or on other specie were not considered.

The decision to use human health as a criterion is questionable. In EPA's own epidemiological studies, they could not demonstrate the adverse effect of sludge disposal on human health. Jakubowski admitted that "the preponderence of data was negative" from numerous epidemiological studies.(14) Farrell likewise admits that "We have not been able to find incidents of disease caused by land application of sludge when good farming practice was followed and a well stabilized sludge was used." (15) The situation with metal concentrations is of course even more problematical, since the effect of metals requires a very long time to be manifested, and there would be great difficulty in determining what the effect of a metal from sludge disposal would have on the total body burden.

During the initial screening, the sixth alternative, ocean disposal, invariably turned out to be the most safe. Yet the EPA has had a long-standing policy against ocean disposal, and the U.S. Congress has in fact made such disposal illegal, except in exceptional circumstances. Using this ban as its argument, the EPA simply dropped ocean disposal from its list of alternatives. Instead of recognizing that ocean disposal has adverse effect on the environment and that there may in fact be very good reasons for banning ocean disposal, it solved its dilemma by simply eliminating it, and considering the remaining five options on the basis of human health only.

The proposed guidelines divide sludge up into various groups, depending on its ultimate disposal. For example, only sludges low in pathogens and metals can be disposed of by sale or give-away. Pathogen densities obviously do not affect incineration.

Environmental effect are totally ignored. In incineration, for example, the effect of cadmium or mercury on downwind residents is calculated (a 70 kg person living for 70 years 100 m directly downwind of the stack), but the effect of sludge incineration on long term global problems is not considered.

If a municipality wants to develop a disposal scheme, it must produces a sludge which meets the appropriate limits. There are two ways of doing this: install treatment systems which will achieve the guidelines as shown, or produce a sludge which will pass the requirements without further treatment.

Over the years, many techniques have been developed for reducing pathogens in sludge, and this does not seem to represent a problem. Of greater concern is the heavy metal concentrations, since there are presently no effective and economical ways of removing metals from sludges. The temptation is to organize a concerted effort and develop the technology necessary for metals removal, similar to the systems we presently use to destroy pathogens.

But is there a better way? Could we instead change the nature of the sludge by changing the wastewater coming to the treatment plant?

The answer is clearly yes. We have already achieved impressive reductions in metals by imposing industrial pre-treatment requirements. The metals concentrations in many cities have been reduced by as much as 50%. But so far we have ignored the people, the source of the other 50%.

Hansen and Tjell have demonstrated that the metals excreted in human feces represent a small fraction of the total metals in sludge(16). The remaining metals must therefore come from cleaners and other chemicals flushed down the drain.

Engineers have for years operated in their objectivist paradigm, often ignoring or underestimating the public. In solid waste engineering, for example, we have invariably thought that the individual citizen is not interested in the disposition of municipal solid waste, and assumed that all recycling campaigns were doomed to failure. Interestingly enough, the past five years have witnessed impressive gains in citizen participation in the recycling of glass, paper and aluminium. Now most cities in the USA are realistically expecting recycling to take care of 25% or its solids waste, still low by world standards, but a significant jump from zero.

Why have we never tried this technique with sludge? Could we not expect citizens to become just as concerned about what is flushed down the drain as they seem to be in caring for what gets thrown into the trash? What would happen if we identified all the household products that contained high concentrations of metals detrimental to sludge, and asked the citizens to stop using them? My guess is that a high percentage would comply, and that we could produce clean sludges, suitable for disposal in our gardens, with a minimum of cost and a maximum of safety.

4. CONCLUSION

Conferences such as this give us ample opportunity to not only share the results of our experiments, but also to share our questions. And in this instance, to reminisce about the outstanding research conducted and professional friendships made during the several COST programs. In retrospect, perhaps it is not so much what you all have accomplished, as what you have attempted and where you have been.

As Ulysses might have said on his return from the odyssey : "Be grateful to Ithaca for the journey."

References

1. Kuhn, T. S. The Structure of Scientific Revolutions University of Chicago Press, Chicago IL (1962)
2. Laubenberger, G. and L. Hartmann "Physical structure of activated sludge in aerobic stabilization" Water Research v 5 pp 335-341 (1971)
3. Vesilind, P. A. Treatment and Disposal of Wastewater Sludges Ann Arbor Science, Ann Arbor MI (1974)
4. Vesilind, P. A. and K. R. Tsang "Moisture distribution in sludge" Proceedings IAWPRC Specialty Conference on Sludge Management, Los Angeles CA (1990)
5. Dugan, P. R. Biochemical Ecology of Water Pollution Plenum Press, New York (1972)
6. Vesilind, P. A. and C. J. Martel "On the freezing of water and wastewater sludges" In press, Journal of the Environmental Engineering Division, ASCE
7. Anderson, D. M. and A. R. Tice "The unfrozen interfacial phase in frozen soil water systems" Ecological Studies, Analysis and Synthesis v. 4, A. Hadas (ed) Springer-Verlag, Berlin (1973)
8. Kavanaugh, M. C. et al "Use of particle size distribution measurements for selection and control of Solid/Liquid Separation Processes" in Particulates in Water. M. C. Kavanaugh and J. O. Leckie (ed) Advances in Chemistry Series 189, American Chemical Society, Washington, DC (1980).
9. Wang, S-H. Unpublished. Duke University, Department of Civil and Environmental Engineering, Durham NC (1990)
10. Karr, P. R. and T. M. Keinath "Influence of particle size on sludge dewaterability" Journal, Water Pollution Control Federation, v 50 p 1911 (1978)
11. Knocke, W. R. and T. L. Zentkovich "Effects of mean cell residence time and particle size distribution on activated sludge vacuum dewatering characteristics" Journal, Water Pollution Control Federation, v 58 n 12 pp 1118-1123 (1986)
12. Roberts, K. and O. Olsson "Influence of colloidal particles on dewatering of activated sludge with polyelectrolyte" Environmental Science and Technology v 9 p 945-948 (1975)
13. Forster, C. F. "Bound water in sewage sludges and its relationship to sludge surfaces and sludge viscosities" Journal of Chemical Technology and Biotechnology v 33B pp76-84 (1983)
14. Jakubowski, W. "US EPA sponsored epidemiological studies of health effects associated with the treatment and disposal of wastewater and sewage sludges" in Epidemiological Studies of Risk Associated with the Agricultural Use of Sewage Sludge: Knowledge and Needs J . Block, A. H. Havelaar and P. L'Hermite (ed) Elsevier, London (1985)
15. Farrell, J. B. "Technical basis of pathogen and vector attraction reduction requirements in EPA's proposed sludge regulations" Unpublished, EPA Cincinnati OH (1990)
16. Hansen, J. Aa. and J. C. Tjell "Human Excretion of Heavy Metals and Other Elements" Unpublished, Lyngby, Denmark (1979).

STRATEGIC STUDIES ON SLUDGE

Harry A.A.M. WEBERS and Menno VISSER

Witteveen+Bos Consulting engineers
P.O. Box 233
7400 AE Deventer (NL)

Summary

The growing population results in an increasing number of sewage treatment plants and a growing amount of sewage sludge. Treatment and disposal of sewage sludge seems to become an urgent problem with increasing environmental problems.
In the Netherlands a systematic approach for strategic sludge management has been developed to calculate the most suitable and best possibilities for sludge treatment and disposal in the short and medium term.Such a strategic study comprises 6 phases:
inventory, survey of treatment methods,costs modelling,evaluation of treatment methods,development of scenarios and sensitivity analysis, and final conclusions and recommendations. Special attention is given to the possibilities for new treatment methods, environmental and energy aspects and future developments e.g. dephosphatation.
On basis of applicable treatment techniques, the investment and operation costs are calculated by a computer model and converted into annual expenses for treatment, transportation and disposal of the end product. As an example the results of a study for a province in the Netherlands with 23 sewage treatment plants and a sludge production of about 30,000 tds/year are given.
Several studies have proved that the systematic approach described is a useful management tool for sludge treatment and disposal.

1. INTRODUCTION

The growing population results in a large number of sewage treatment plants and growing amounts of sewage sludge, that have to be treated and disposed of : problems as a result of lacking treatment and disposal facilities are becoming more and more urgent. In the past almost all of the sludge could be utilized in agriculture, but at present other options need to be investigated. There are growing objections against the reuse in agriculture, due to high concentrations of organic pollutants and heavy metals that are a risk for contamination.

In the Netherlands and other countries severe guidelines have been developed for the use of sewage sludge as a fertilizer in agriculture and the production of black earth and compost. Governmental regulations and directives apply to the various sludge disposition methods.

Thus, the ultimate disposal of sludge is becoming more complex. Decision-makers and managers of water quality boards however should be capable of coming up with practical solutions for sludge problems. In designing new treatment plants or extending and rehabilitating existing plants, growing attention is paid to appropriate methods for sludge treatment and disposal .

In the Netherlands a systematic approach for sludge management has been developed for calculation of the most suitable possibilities for sludge treatment and disposal in the short and medium term. By means of this approach a strategic plan is set up indicating the costs of the most feasible technical solutions.

2. SLUDGE TREATMENT AND DISPOSAL IN THE EC AND THE NETHERLANDS

In general 2 methods of treatment and disposal can be distinguished: recycling or treatment and dumping of the end product. Recycling by means of land application reflects the utilization of sludge via agriculture for soil conditioning etc. Luxembourg, the Netherlands and Spain recycle the highest proportions of their sludge (1).

Spreading of sludge on agricultural land is more and more limited in several countries in the EC due to stricter quality standards for sewage sludge for land application. Especially components like heavy metals and organic toxic components are a matter of continuous concern at the moment because of the risks of contamination of the soil and groundwater and crops.

In previous studies data for sludge treatment and disposal in the EC were collected. In table 1 the results of a comparative survey for 1984 are summarized (1).

situation 1984		total EC	Netherlands
sludge production (tds)		5,564,000	202,000
number of stp		32,000	501
number stp > 10,000 p.e.		26,674	292
treatment	none	39 %	60 %
	anaerobic	56	44
	aerobic	16	35
conditioning	none	39 %	60 %
	Fe/Al/lime	17	20
	polyelect.	35	15
	heat	8	5
	other	1	0
dewatering	none	29 %	47 %
	air drying	6	8
	lagoons	2	7
	belt press	9	15
	plate press	10	20
	vacuum filt.	1	2
	centrifuge	8	1
	other	35	0
disposal	recycling	37 %	63 %
	landfill	44	27
	incineration	9	3
	sea	7	6
	other	3	1

Table 1. Sludge production, treatment and disposal in the EC and the Netherlands (1).

The most recent Dutch figures refer to 1987 (2). In 1987 there were about 490 (municipal) sewage treatment plants in the Netherlands with a capacity of 23.7 million population equivalents and a sludge production of 253,000 tons of dry solids after stabilisation. The applied treatment methods are summarized in table 2. Only a small amount is incinerated.

The main methods for sludge disposal in the Netherlands are land application and disposal to a landfill site (after dewatering). The most important reasons for the limitation of land application are :

a. strict criteria for sludge quality and product (black earth, compost)
b. bad reputation (risks of contamination)
c. competition by other fertilizers such as manure
d. varying demand in agriculture over the year.

natural dewatering (11.5 %) :	drying beds	9,800 tds
	lagoons	19,300 tds
mechanical dewatering (46.7 %):	belt presses	60,800 tds
	plate presses	46,600 tds
	vacuum filters	3,200 tds
	centrifuges	7,600 tds
no dewatering (41.8 %):	no dewatering	89,000 tds
	buffers	16,900 tds
further treatment	incineration	8,100 tds

Table 2. Sludge treatment in the Netherlands in 1987 (2).

The dumping of poorly dewatered sludge is also more limited because of:
a. growing lack of suitable landfill sites
b. bad consistency of the sludge (less than 35 % ds)
c. environmental problems (odour etc.).

The policy of our government and water authorities comes to (3):
a. prevention and reduction of sludge production by promoting sewage treatment methods with lower sludge yields
b. reuse of only clean sludge as a fertilizer in the year 2000
c. development of treatment methods as an alternative for incineration.

For the year 2000 the total production of municipal and industrial sludge in the Netherlands is estimated to be between 480,000 and 570,000 tons of dry solids, dephosphatation not included (4). The production of sewage sludge in 1994 is estimated at 275,000 tds (1). The increase in sludge production is a result of the growing population, new sewage treatment plants and further waste water treatment.

3. STRATEGIC STUDIES ON SLUDGE MANAGEMENT : APPROACH AND PHASES

It is expected that in the near future land application of sewage sludge will decrease to a negligible level in the Netherlands. In other countries sludge disposal to agriculture will also become more and more difficult, thus making alternative treatment and disposal necessary such as dewatering, drying or incineration before dumping.

Strategic and management studies on sludge treatment and disposal have been especially developed in order to evaluate possibilities for the treatment and disposal of sludge. Most studies for several regions in the Netherlands concern a sludge production of about 12,000 to 30,000 tds.

Finally a total strategic plan emerges with technical solutions and costs for the short and medium term and conclusions and recommendations for the sludge management. Each study consists of the following phases :
1. inventory of the present and future situation in a region:
   - general information, conditions and starting points
   - sludge production, quality and characteristics
   - sludge treatment and disposal methods and present costs
   - future developments in the region
2. survey of methods for treatment ("state of the art" technology)
   - design criteria for treatment methods
   - survey of techniques and methods

- costs calculations (investment, operational and annual costs)
- new developments

3. modelling : cost model for various treatment and disposal methods
   - calculation of annual treatment costs versus sludge production
   - transportation and dumping costs
   - selection figures
4. evaluation of treatment methods
5. drawing up of treatment scenarios up to the year 2000
   - influence of (partial) centralization
   - application of present facilities
   - back coupling between periods
   - sensitivity analysis (energy, chemicals, personnel)
6. conclusions and recommendations for sludge management and disposal.

4. INVENTORY AND SLUDGE TREATMENT METHODS

In the first phase the average sludge quality is compared to the new Dutch quality criteria e.g. for heavy metals (see table 3). In the near future criteria for organic pollutants will also be set.

year	until 1990	1991-1994	1995-2000	average conc. 1984	1987
mg/kg ds				(2)	(2)
Zinc	2,000	1,400	300	1,530	1,247
Copper	900	425	75	492	461
Lead	500	300	225	360	271
Chromium	500	350	75	129	94
Nickel	100	70	38	52	52
Arsenic	25	25	-	7	8
Cadmium	5	3.5	1.25	7	5
Mercury	5	3.5	0.75	2	2
org. matter	30 %	50 %	50 %	59 %	54 %

Table 3 : Concentration limits for sludge and weighted averages (2,5,8).

In 1986 the Dutch quality standards for the metals Cadmium and Mercury have been changed from 10 to 5 mg/kg ds. The average composition of the sewage sludge for 1984 and 1987 in the Netherlands is given in table 3. The concentrations of Zinc and Lead have increased with resp. 18.5 and 24.7% in de period 1984-1987 (2).As our limits will be tightened up in the future this will result in a reduction of sludge recycling and finally more dumping of the dewatered sludge.

In phase 2 treatment and disposal methods are investigated (6,8):

a. sludge treatment :
   - drying e.g. flash drying, rotary, Carver Greenfield etc.
   - composting e.g. windrow, tunnel etc.
   - incineration e.g fluidized bed, multiple hearth
   - high temperature process
   - other e.g. chemical fixation

b. sludge transportation by truck, vessel, pipeline or railroad

c. ultimate disposal :
   - land disposal (landfill)
   - reuse (land application, forestry, power generation)
   - other options (ocean disposal, lagoons etc).

In the Netherlands dewatering methods like air drying (sand beds and sludge drying lagoons), belt presses and plate (filter) presses are the most common to be selected and used for sludge dewatering.

But natural drying and belt presses generally produce end products which are too voluminous and too plastic:dumping of these products causes severe problems and moreover, the number of our disposal sites becomes more limited. Plate presses produce dry sludge cakes that can be disposed of without any problem but are more expensive (investment and chemicals).

In the long term it will be necessary for sewage treatment plant to have their own appropriate treatment unit or to make use of a central sludge dewatering plant in the area. Some developments are :

- for belt presses : pre-dewatering zone, high pressure zone
- for plate presses : membrane type press, polyelectrolytes, new press tissues and mixing technology.

Furthermore, necessary post-treatment methods are investigated, such as the production of black earth, composting, lime treatment, Vertech, Carver Greenfield, heat drying and incineration. In a study only the most feasible methods of sludge treatment and post-treatment are investigated.

Multiple effect drying (Carver Greenfield) is relatively successful in the United States especially in the food and beverage industry (6). Constraints are the high capital costs and the gaseous emissions (NH_3). Operation and maintenance costs are moderate and the end product can be used as a soil fertilizer.

An interesting example of the wet oxidation process is Vertech. The oxidation process is carried out in the liquid phase: in deep underground shafts high pressure and temperature are established for COD oxidation.

Multiple hearth and fluidized bed furnaces are most widely used for sludge incineration. Fluidized-bed incineration is among the best suited methods for incineration of sewage sludge and is economically feasible. Waste heat can be used as an energy source for drying sludge.The main pollution problem with sludge incineration arises from the disposal of the ashes (hazardous waste) and gaseous emissions (odours, SO_2, NO_x, HCL) (7). The recent Dutch quality criteria for incinerator exhaust gases are: 10 mg/Nm^3 dust, 50 mg/Nm^3 SO_2 and 0.01 mg/Nm^3 Cd and Hg. Reduction and removal of these substances can only be achieved by complex and expensive techniques such as after-burning, catalytic reduction and wet scrubbing. So far incineration has been considered as a valid final sludge treatment technique: the volume reduction is high (up to 95-99 %) and operational costs can be reduced considerably by energy recovery . The environmental problems and high capital costs however make incineration less feasible. In the Netherlands only a few incineration plants have been built (9,10).

Special attention is given to the following aspects :

a. quality and market possibilities for recycled sludge
b. possibilities for post-treatment of sludge dewatered by belt pressing, plate pressing or centrifugation (sludge characteristics),
c. comparison of various conditioning chemicals
d. possibilities for new treatment methods
f. environmental and energy aspects : pollution control measures are becoming more and more important e.g. for large scale treatment plants
g. future developments e.g. dephosphatation, which will increase the sludge production by some 30 to 40 % and reduce the combustion value
h. process complexibility and proven experience
i. effects of sludge stabilised by anaerobic and aerobic digestion.

5. COSTS MODELLING AND EVALUATION

On basis of applicable treatment techniques the investment and operation costs are calculated by a computer model and converted into annual expenses totalized for treatment, transportation and disposal of the end product (phase 3). Present costs of recycling methods are: 80 ECU/tds for land application and 215 ECU/tds for black earth.

Investment and operational costs for natural and mechanical sludge dewatering, lime treatment and composting in the range from 500 to 6,000 tds/year are shown in the figures 1 and 2. On basis of these costs annual costs (ECU/tds.year) have been calculated (figure 3).

For dewatering installations with a capacity of more than 6,000 tds/year, annual costs will hardly decrease with increasing capacity. At the moment plate press dewatering (after lime conditioning) and dumping is attractive: membrane presses are also suitable for conditioning with polyelectrolyte and produce a less voluminous end product that is suitable for post treatment.

Investment, operational and annual costs for (post)treatment methods on a large scale like heat drying (after filter press and belt press), Vertech, Carver Greenfield and incineration are shown in the figures 4,5 and 6. Of course costs levels may differ depending on the situation in a country and the applied design parameters.

Beside costs, other aspects of each method are determinant for its ultimate applicability. In phase 4 each method is evaluated , which is necessary to determine the technological and financial feasibility of the methods mentioned. Regarding the discussions in the Netherlands at present on the application of either sludge incineration or drying as a long term post treatment method, advantages and disadvantages of each method are to be evaluated. A comparison is made in table 4.

method after dewatering	heat drying	incineration
1.scale	+ (small/large)	- (large)
2.plant characteristics:		
flexibility	+ (day)	- (continuous)
sludge composition	- (homogenuous)	- (homogenuous)
air treatment	- (ammonia/odour)	- (dust/Hg/SO_2/dioxines)
3.energy:		
supply	100 %	30 %
co incineration	-	+ (RDF liquid fraction)
4.Residues:		
useful application	+ (landfill)	-
end product	- (odour)	+ (inert)
hazardous waste	0 %	15-55 %
5.Sludge characteristics:		
fresh	- (odour, NH_3)	+ (combustion value)
stabilized	+	+
P-sludge	+	- (ash content)
lime sludge	- (odour, NH_3)	+ (SO_2 reduction)

Table 4. Comparison of heat drying and incineration (9,10)

Drying seems to be more attractive than incineration as pollution control measures for drying are much simpler and no hazardous waste is disposed. However energy costs for drying are higher. It is expected that

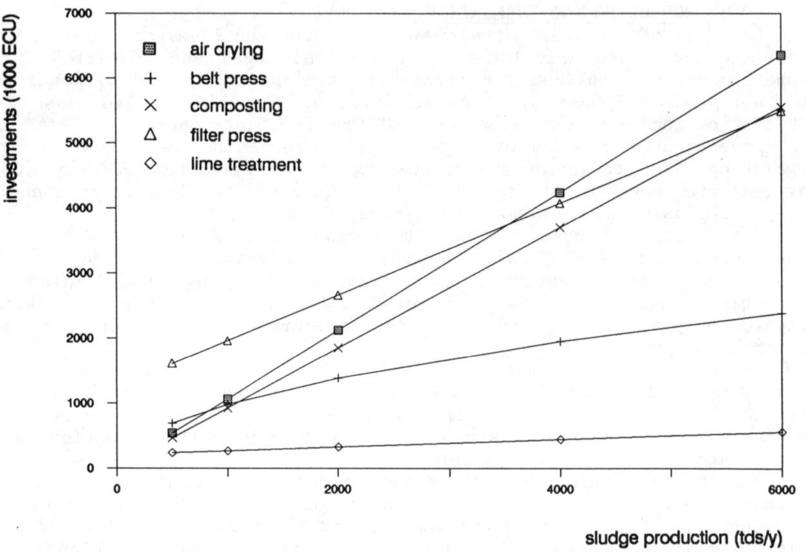

Figure 1: Investment costs for dewatering (ECU) vs. production.

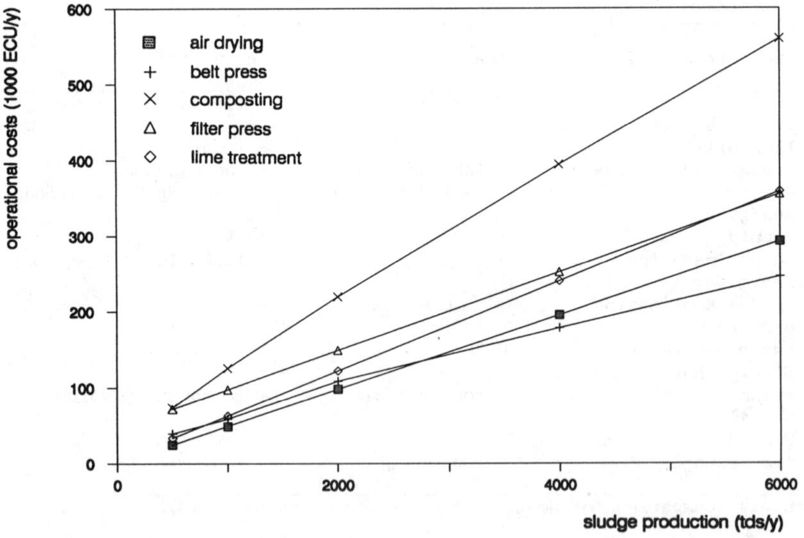

Figure 2: Operational costs for dewatering (ECU/y) vs. production.

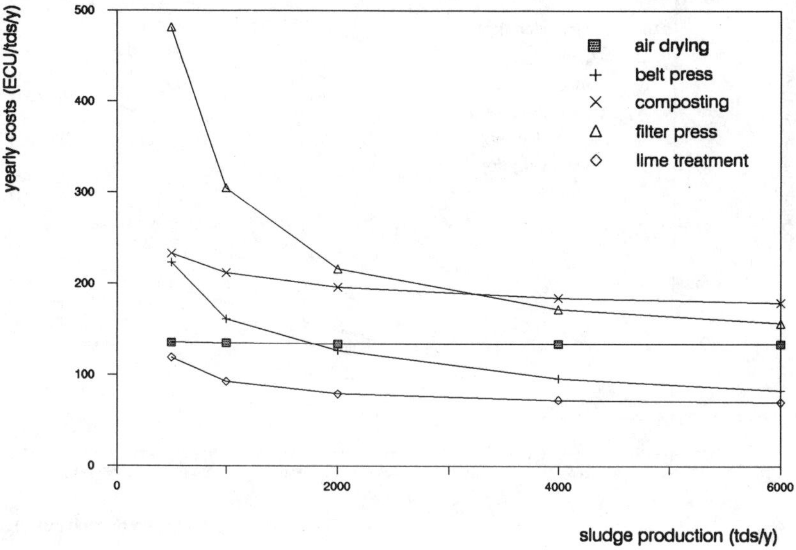

Figure 3: Annual costs for dewatering (ECU/tds.y) vs. production.

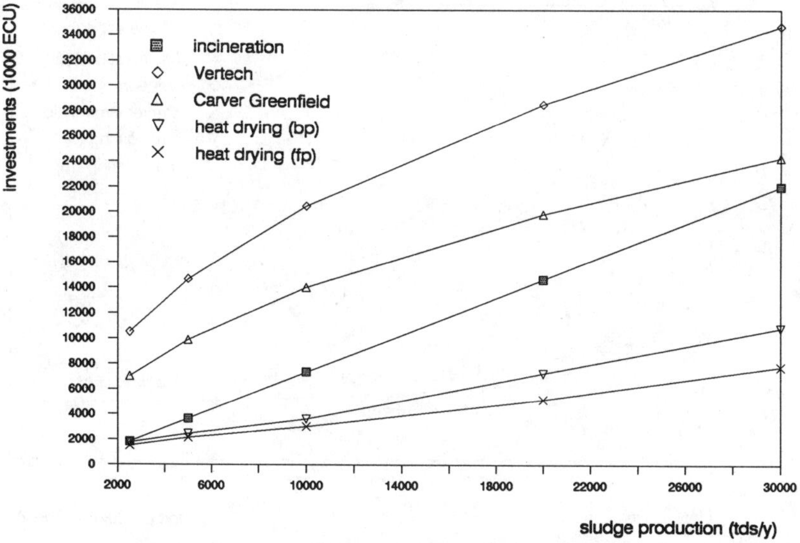

Figure 4: Investment costs for post-treatment (ECU) vs. production.

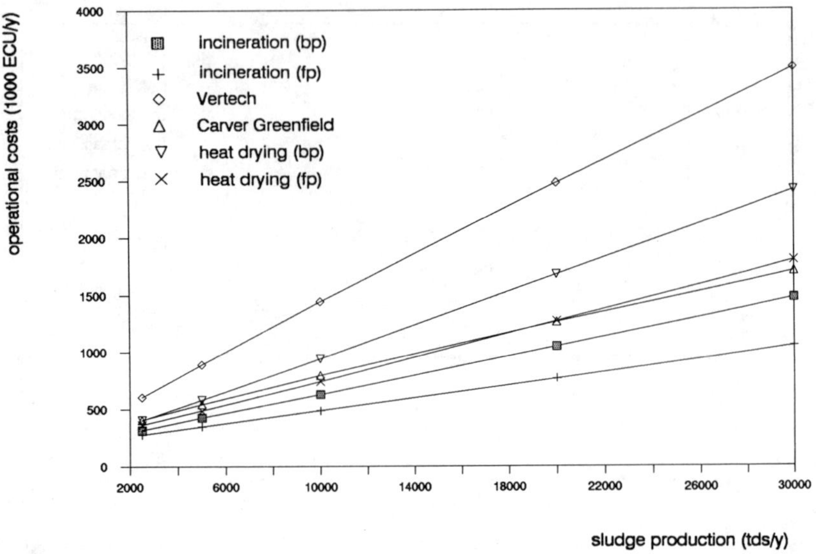

Figure 5: Operational costs for post-treatment (ECU/y) vs. production.

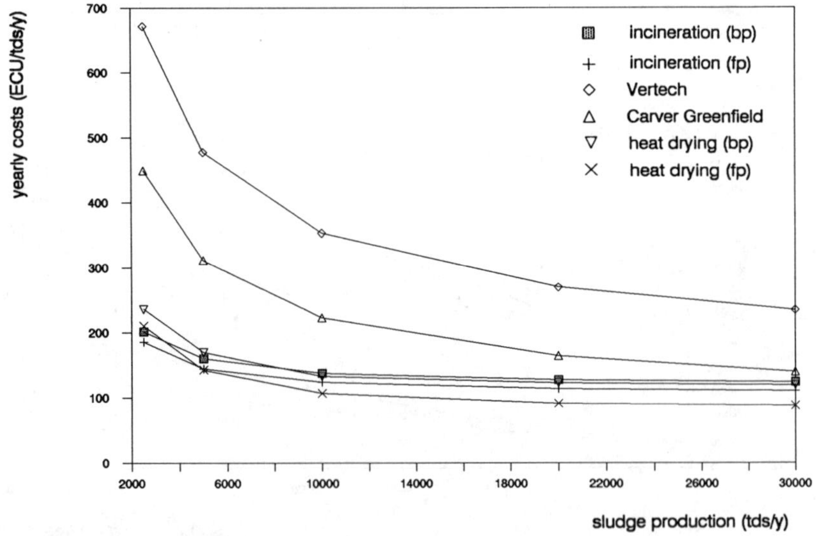

Figure 6: Annual costs for post-treatment (ECU/tds.y) vs. production.

large scale incineration and drying of sludge on a large scale will be applied in the Netherlands within 5 to 10 years.Small drying plants are a little more expensive than incinerators,but at higher capacities they are cheaper because of the handling the hazardous combustion ashes.

Finally a workable cost model and screening matrix can be drawn up. The degree of dewatering is very important for the final costs of post treatment methods such as drying (8). E.g. depending on the dry solids content, the specific drying costs vary from 90 ECU/tds for 30 % ds to 120 ECU/tds for 20 % ds for a drying capacity of 30,000 tds/year , using the exchange rate of july 1990 (see figure 6).

As an example the results are given from a study for a province in the Netherlands with 23 sewage treatment plants and a sludge production of about 30,000 tds/year (without dephosphatation). The annual costs are given in figure 7 : each scenario comprises transportation (liquid and cakes, ashes etc.), dewatering and post treatment and dumping costs.

In figure 7 the specified costs of the different scenarios have been calculated (transportation included):

1. belt press + landfill : 9.8 mln ECU/y
2. plate press + landfill : 9.4 mln ECU/y
3. belt press + drying + landfill : 8.6 mln ECU/y
4. plate press + drying : 9.5 mln ECU/y
5. belt press + incineration + landfill : 8.3 mln ECU/y
6. plate press + incineration + landfill : 9.6 mln ECU/y
7. belt press + Carver Greenfield + landfill : 9.1 mln ECU/y
8. Vertech + landfill : 9.7 mln ECU/y
9. composting : 8.6 mln ECU/y
10. composting + landfill : 9.8 mln ECU/y.

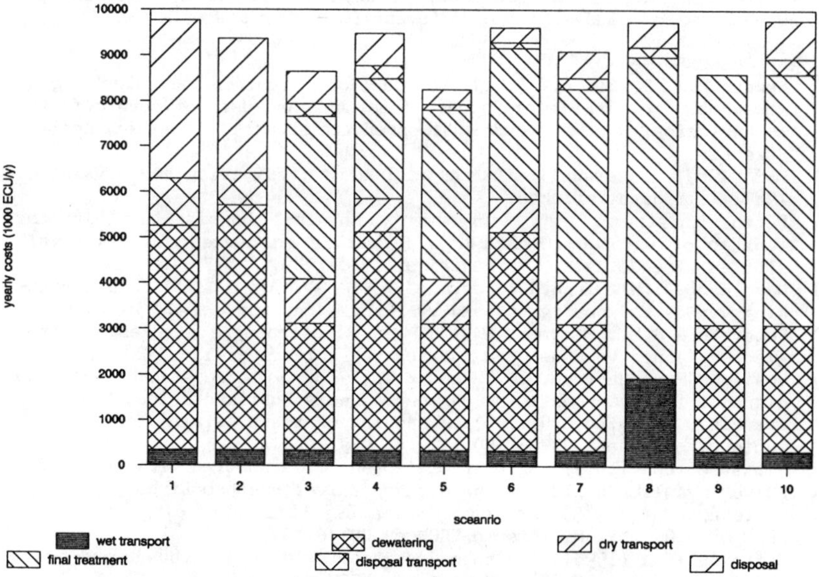

Figure 7: Different scenarios for sludge treatment and disposal for a region with 23 treatment plants and 30,000 tds/y (7 dewatering plants).

Depending on the degree of centralization of each scenario for sludge disposal, the ultimate total treatment and disposal costs at a capacity of 30,000 tds/year vary between 275 and 325 ECU/tds, including costs for transportation and dumping. The average transportation costs are 15 % but vary between 5 and 20 %.

Finally the cost-effects on energy, personnel, transportation, dumping costs and chemicals are calculated and evaluated by means of a sensitivity analysis. Dumping costs in the Netherlands may increase to 35 ECU/m^3 in the near future. The general objective is to render the sludge suitable for utilisation or disposal at minimum costs. Environmental aspects are evaluated on basis of a multicriteria analysis.

6. CONCLUSIONS AND RECOMMENDATIONS

1. Recycling of sludge in agriculture is becoming increasingly limited in the Netherlands, due to strict quality standards for sewage sludge for land application. Therefore other sludge treatment and disposal methods should be investigated.
2. In the Netherlands an approach has been developed for strategic sludge management. Several studies have proved that this systematic approach is a useful management tool for sludge treatment and disposal.
3. Natural dewatering and belt presses are relatively cheap, but in general they produce sludge that need to be post treated. Plate presses after lime conditioning produce hard sludge cakes ready for dumping.
4. Sludge drying is technically and economically attractive as a post treatment method, especially when waste heat is available.
5. It is expected that within 5 to 10 years drying and incineration of sludge will be applied on a large scale in the Netherlands .
6. Further studies will be necessary to obtain unit costs for treatment and disposal of sludge in the different EC member states.

7. REFERENCES

(1) Newman,P.J.,Bowden,A.V.(1989). Production, treatment and handling of sewage sludge. In: Sewage sludge treatment and use:new developments, technological aspects and environmental effects. Elsevier Applied Science.

(2) CBS (1989). Waterkwaliteitsbeheer. Deel B : zuivering van afvalwater 1987. Milieustatistieken.

(3) Dutch Minister of The Environment, Physical Housing and Planning (Letter of october 10-th 1988). Preventie en hergebruik van afvalstoffen. Tweede Kamer, vergaderjaar 1988-1989, 20877 , no .2.

(4) Nagelhout,D.,Wieringa,K.,Joosten,J.M.(1989). Afval 2000. Een verkenning van de toekomstige afvalverwerkingsstructuur. RIVM 738605002.

(5) Webers,H.A.A.M., Rooymans,P.J.(1988). Slibstrategie studies: strategische plannen voor de verwijdering van zuiveringsslib. H2O, 23.

(6) Vesilind,P.A.,Hartman,G.C.,Skene,E.T.(1986). Sludge management and disposal for the practicing engineer. Lewis Publishers Inc.

(7) Francis,C.W.,White G.H.(1987). Leaching of toxic metals from incinerator ashes. JWPCF, 59, 11, p. 979-986.

(8) Witteveen+Bos (1989). Slibstrategieplan : toekomstige verwerking van zuiveringsslib in LImburg. Waterschap Zuiveringschap Limburg.

(9) Witteveen+Bos (1989)..Rookgasemissies bij slibverbranding. Werkdocument nr. 89.098x. Ministerie VROM en DBW/RIZA.

(10) Witteveen+Bos (1990). Addendum Rookgasemissies bij slibverbranding. Werkdocument nr. 90.024x. Ministerie VROM en DBW/RIZA.

MATHEMATICAL MODELLING OF THE SINGLE AND TWO STAGE SEWAGE SLUDGE TREATMENT

H.SIEGRIST, D.RENGGLI AND W.GUJER
Swiss Federal Institute for Water Resources and Water Pollution Control (EAWAG), CH-8600 Dübendorf, Switzerland

Summary

A mathematical model was develloped to describe the dynamical behaviour of the anaerobic mesophilic and thermophilic digestion. Special emphasis was given to the acetate degradation kinetics, as the step, that determines the stability of anaerobic digestion. Due to the experimentally verified high maximum growth rate (0.55 d^{-1}) and decay rate (0.18 d^{-1}) for mesophilic treatment, process stability increased not significantly for sludge detention times above 10 to 15 days. Two stage treatment leads to better sludge characteristics (thickening, viscosity, COD, TSS) of the treated sludge. However two stage treatment is more sensitive to load variation.

1. INTRODUCTION

In Switzerland about 45 % of the treated sewage sludge was applied to agricultural land in 1987. The sewage sludge treatment on most treatment plants consists of mesophilic digestion. Due to the increased demand to sewage sludge treatment (hygienisation, dewatering and storage), the mesophilic digestion was supplemented by more than sixty plants (2.2 million inhabitant equivalents) by a thermophilic pretreatment (mostly aerobic) with a short detention time of about one day.

The aim of this study was to develop a mathematical model for the single and two stage treatment, based on the reaction scheme (Fig.1) described in (1). Special emphasis is given to the acetate degradation kinetics, as the step, that determines the stability of anaerobic digestion.

2. MATHEMATICAL MODEL OF ANAEROBIC DIGESTION

a) Model processes

Seven distinct processes may be identified in an anaerobic digester (Fig. 1, Tabs. 1,2):

1) Stripping of the biogas (methane, carbon dioxide and hydrogen)
2) Hydrolysis of the biopolymers (proteins, carbohydrates, lipids) into the monomers (amino acids, sugars and long chain fatty acids)
3) Fermentation of amino acids and sugers
4) Anaerobic oxidation of long chain fatty acids and alcohols

5) Anaerobic oxidation of intermediary products such as volatile fatty acids (except acetate; in the mathematical model described in Tabs. 1,2 only propionate is included)
6) conversion of acetate to methane (acetotrophic methanogenesis)
7) conversion of hydrogen to methane (hydrogenotrophic methanogenesis).

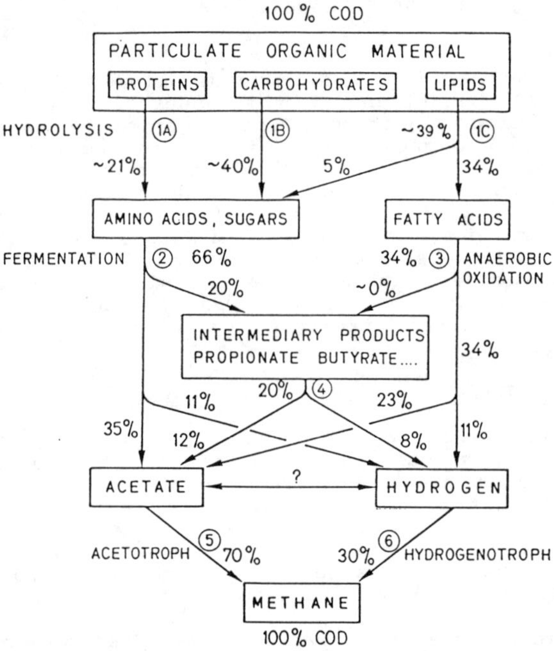

Fig. 1 Proposed reaction scheme for anaerobic digestion of domestic sludge (1). Percentage indicates substrate flow in the form of COD equivalents. In the model a sligthly different stoichiometry was chosen (Tab. 1).

The model includes an additional five processes (8 to 12) to describe the decay of the five distinct microbial groups (processes 3 to 7) above. Decay rates seem to be higher than estimated in the literature and are therefore important to describe digester stability and adaptation to load variation.

To calculate the pH of the sludge in the digester two additional processes have been added:

13) Protolysis of bicarbonate to carbon dioxide and water
14) Deprotolysis of carbon dioxide and water to bicarbonate.

Protolysis and deprotolysis of volatile fatty acids are not yet included in the model, it is therefore not possible to simulate extreme acidification (pH<6) of a digester.

Tab. 1 Stoichiometric coefficients of the 14 processes defined for an anaerobic mesophilic digester.

Dissolved components (S_i)											Particulate organic material (X_i) Biomass, that degrades							
1	2	3	4	5	6	7	8	9	10	11	12	13	14	15	16	17	18	
H_2	CH_4	CO_2	HCO_3	H	NH_4	Ace-tate	Propio nate	Am.ac sugar	Fatty acid	Inert	de-grad.	Am.ac sugar	Fatty acid	Propio nate	Ace-tate	H_2	inert	Pro-cess
S_{H2}	S_{CH4}	S_{CO2}	S_{HCO3}	S_H	S_{NH4}	S_{ac}	S_{pro}	S_{as}	S_{fa}	S_{in}	X_S	X_{as}	X_{fa}	X_{pro}	X_{ac}	X_{H2}	X_{in}	
COD g/m^3	COD g/m^3	Mol/m^3	Mol/m^3	Mol/m^3	N g/m^3	COD g/m^3	COD g/m^3	COD g/m^3	COD g/m^3	COD g/m^3	COD g/m^3	COD g/m^3	COD g/m^3	COD g/m^3	COD g/m^3	COD g/m^3	COD g/m^3	
-0.084																		1_{H2}
	-0.014																	1_{CH4}
		-0.04																1_{CO2}
								0.50	0.45	0.05	-1.00							2
0.96		0.082	-0.051		0.187	3.29	1.42	-6.67				1.00						3
6.70		0.199	-0.202		-0.08	14.3			-22.0				1.00					4
8.20		0.163	0.004		-0.08	10.8	-20.0							1.00				5
	39.0	-0.007	0.619		-0.08	-40.0									1.00			6
-22.0	21.0	-0.291	-0.006		-0.08											1.00		7
			0.004		0.056						0.80	-1.00					0.20	8
			0.004		0.056						0.80		-1.00				0.20	9
			0.004		0.056						0.80			-1.00			0.20	10
			0.004		0.056						0.80				-1.00		0.20	11
			0.004		0.056						0.80					-1.00	0.20	12
		1.00	-1.00	-1.00														13
		-1.00	1.00	1.00														14

Tab. 2 Kinetics and parameters of the 14 processes of an anaerobic mesophilic digester

Process / kinetics	Parameters
1. Stripping of biogas (g COD/$m^3 \cdot d$ for H_2, CH_4 and Mol/$m^3 \cdot d$ for CO_2) $\rho_{1i} = Q_g \cdot S_i/(V_R \cdot R \cdot T)$	see section 2b)
2. Hydrolysis of the biopolymers (g COD/$m^3 \cdot d$) $\rho_2 = k_R \cdot X_S$	$k_R = 0.25\ d^{-1}$
3. Fermentation of amino acids and sugars (g BM-COD/$m^3 \cdot d$) $\rho_3 = \mu_{max,3} \cdot \frac{S_{as}}{(K_{S,as}+S_{as})} \cdot X_{as}$	$\mu_{max,3} = 5.0\ d^{-1}$, $K_{S,as} = 50$ g COD/m^3
4. Anaerobic oxidation of fatty acids (g BM-COD/$m^3 \cdot d$) $\rho_4 = \mu_{max,4} \cdot \frac{S_{fa}}{(K_{S,fa}+S_{fa})} \frac{K_{I,H}}{(K_{I,H}+S_H)} \frac{K_{I,ac}}{(K_{I,ac}+S_{ac})} \frac{K_{I,H2}}{(K_{I,H2}+S_{H2})} \cdot X_{fa}$	$\mu_{max,4} = 0.80\ d^{-1}$, $K_{S,fa} = 200$ g COD/m^3, $K_{I,H} = 5 \cdot 10^{-4}$ Mol/m^3, $K_{I,ac} = 2000$ g COD/m^3, $K_{I,H2} = 0.02$ Mol/m^3
5. Anaerobic oxidation of propionate (g BM-COD/$m^3 \cdot d$) $\rho_5 = \mu_{max,5} \cdot \frac{S_{H2}}{(K_{S,H2}+S_{H2})} \frac{S_H}{(K_{i,H}+S_H)} \frac{K_{I,H}}{(K_{I,H}+S_H)} \frac{K_{I,ac}}{(K_{I,ac}+S_{ac})} \frac{K_{I,H2}}{(K_{I,H2}+S_{H2})} \cdot X_{pro}$	$\mu_{max,5} = 4.2\ d^{-1}$, $K_{S,pro} = 10$ g COD/m^3, $K_{S,H} = 4 \cdot 10^{-5}$ Mol/m^3, $K_{I,H} = 1.6 \cdot 10^{-4}$ Mol/m^3, $K_{I,ac} = 1500$ g COD/m^3, $K_{I,H2} = 0.005$ Mol/m^3
6. Acetate conversion to methane (g BM-COD/$m^3 \cdot d$) $\rho_6 = \mu_{max,6} \cdot \frac{S_{ac}}{(K_{S,ac}+S_{ac})} \frac{S_H}{(K_{S,H}+S_H)} \frac{K_{I,H}}{(K_{I,H}+S_H)} \cdot X_{Ac}$	$\mu_{max,6} = 1.25\ d^{-1}$, $K_{S,ac} = 50$ g COD/m^3, $K_{S,H} = 4 \cdot 10^{-5}$ Mol/m^3, $K_{I,H} = 1.6 \cdot 10^{-4}$ Mol/m^3
7. Conversion of hydrogen to methane (g BM-COD/$m^3 \cdot d$) $\rho_7 = \mu_{max,7} \cdot \frac{S_{H2}}{(K_{S,H2}+S_{H2})} \frac{S_H}{(K_{S,H}+S_H)} \frac{K_{I,H}}{(K_{I,H}+S_H)} \cdot X_{H2}$	$\mu_{max,7} = 4.2\ d^{-1}$, $K_{S,H2} = 5 \cdot 10^{-4}$ gCOD/m^3, $K_{S,H} = 4 \cdot 10^{-5}$ Mol/m^3, $K_{I,H} = 1.6 \cdot 10^{-4}$ Mol/m^3
8.-12. Decay of biomass (g BM-COD/$m^3 \cdot d$) $\rho_i = k_{d,i} \cdot X_i$	$k_{d,8} = 1.0\ d^{-1}$, $k_{d,9-11} = 0.18\ d^{-1}$, $k_{d,12} = 0.4\ d^{-1}$
13.-14. Protolysis HCO_3 and deprotolysis of CO_2 (Mol/$m^3 \cdot d$) $\rho_{13} = k_{13} \cdot S_{HCO3} \cdot S_H$, $\rho_{14} = k_{14} \cdot S_{CO2}$	$k_{13} = 10 \cdot d^{-1}\ m^3$/Mol, $k_{13} = 0.0045\ d^{-1}$

b) Process kinetics and conversion rates

Process rates (ρ_j), kinetic expressions and parameters (maximum growth rate ($\mu_{max,j}$), Monod saturation ($K_{S,i}$) and inhibition constants ($K_{I,i}$)) are given in Tab. 2. Hydrolysis of particulate organics, biomass decay and deprotolysis of carbonic acid are first order and protolysis of bicarbonate second order kinetic.

Hydrolysis is a very complex process and only partly to be modelled by first order kinetics. A more suitable procedure would be to devide hydrolysis into two processes; a slow decay of complex particulate substrate into biopolymers and a fast hydrolysis of the biopolymers to the monomers as suggested by the IAWPRC task group on mathematical modelling (2).

The process kinetics of biomass (BM) growth is more complex. Acetate conversion to methane (process 6) is assumed to depend on Monod kinetic ($S/(K_S+S)$) for acetate and on a combination of Monod and inhibition kinetics ($K_I/(S+K_I)$) for the proton (10^{-pH}) with maximum growth at pH = 7.1:

$$\rho_6 = \mu_{max,6} \cdot \frac{S_{ac}}{(K_{S,ac}+S_{ac})} \frac{S_H}{(K_{S,H}+S_H)} \frac{K_{I,H}}{(K_{I,H}+S_H)} \cdot X_{Ac} \quad \left(\frac{\text{g BM-COD}}{m^3 \cdot d}\right)$$

The pH dependence is to some extent taken from literature data (3), but is not verified for the experiments described below.

The degradation kinetics of fatty acids and propionate have two additional inhibition terms for hydrogen and acetate, since these two processes (4 and 5) become inhibited if hydrogen or acetate concentrations are increased. These two inhibition terms are estimated from literature data for hydrogen (4) and by our own experiments for acetate, respectivelly.

The maximum growth rates selected in the model are higher than found in literature, this has two reasons:

- due to the inhibition terms the effective maximum growth rates are lower than the values in Tab. 2 (e.g. $\mu_{max,6}$ = 1.25 d^{-1} is reduced by the pH inhibition terms and is effectively equal 0.55 d^{-1} at pH = 7.1) and
- biomass decay is higher than given in the literature, therefore, assuming similar net growth rates, maximum growth rates have to be higher.

Stripping of methane, carbon dioxide and hydrogen is assumed to be linearly related to the total gas production and the concentration of the considered component. The concentration of the gas in the rising gas bubble is supposed to be thermodynamically in equilibrium with the dissolved gas concentration in the sludge. The stripping rate for component i is therefore:

$$r_{i,1} = -Q_g \cdot p_i/(V_R \cdot R \cdot T) = -H_i \cdot Q_g \cdot S_i/(V_R \cdot R \cdot T) = -H_i \cdot \rho_{1i}$$
($Mol/m^3 \cdot d$ for CO_2 or g $CSB/m^3 \cdot d$ for H_2, CH_4)

with: Q_g = total Gasflow (m^3/d) V_R = Reaktor volume (m^3)
T = Temperature in K
R = Gas constant = $8.2 \cdot 10^{-5}$ $atm \cdot m^3/Mol \cdot K$
S_i = concentration of component i in the reactor sludge
p_i = gas pressure of comp. i in equillibrium with S_i
H_i = Henry constant of component i
($atm \cdot m^3/Mol$ for CO_2 and $atm \cdot m^3/g$ CSB for H_2, CH_4)

The conversion rates for each component are calculated by summing up over all 14 processes the product of the process rate (ρ_j) and the stoichiometric coefficient ($\nu_{i,j}$):

$$r_i = \sum_{j=1..14} r_{i,j} = \sum_{j=1..14} \rho_j \, \nu_{i,j} \qquad (\text{g COD or Mol})/m^3 \cdot d)$$

c) Stoichiometry

The modell distinguishes 11 dissolved components (hydrogen, methane, carbon dioxide, bicarbonate, proton, ammonium, acetate, propionate, amino acid and sugar, long chain fatty acids, inert organic material) and 7 particulate organic substances (particulate degradable and inert material, 5 groups of microorganisms (processes 3 to 7)).

The stoichiometric coefficients $\nu_{i,j}$ (i = substance, j = process) are based on Fig. 1, the biomass yield and substance characterization (Tab. 3) by considering the continuity of COD, nitrogen, carbon and electrical charge.

Since the kinetics of acetate conversion to methane (process 6) will be discussed below, the stoichiometric coefficients of this process will be derived as an example. The biomass (BM) yield of this process was estimated with Y_{ac} = 0.025 g BM-COD/g acetate-COD converted (1). Therefore, the stoichiometric coefficient for acetate (ac) of process 6

$$\nu_{ac,6} = -1/Y_{ac} = -40 \text{ g COD/g BM-COD}, \quad \nu_{BM,6} = 1,$$

and from COD continuity ($\nu_{CH4,6} + \nu_{ac,6} + \nu_{BM,6} = 0$)

$$\nu_{CH4,6} = 40 - 1 = 39 \text{ g COD/g BM-COD}.$$

It is assumed that ammonium serves as nitrogen pool for biomass growth:

$$\nu_{NH4,6} = -\nu_{BM,6} \cdot i_{N,BM} = -0.08 \text{ g N/g BM-COD (Tab. 3)}.$$

The bicarbonate consumption is determined by the electrical charge balance:

$$\nu_{HCO3,6} = -\nu_{ac,6} \cdot i_{L,ac} - \nu_{NH4,6} \cdot i_{L,NH4} = $$
$$= 0.619 \text{ Mol } HCO_3/\text{g BM-COD},$$

and carbon dioxide production with the carbon continuity:

$$\nu_{CO2,6} = -\nu_{ac,6} \cdot i_{C,ac} - \nu_{BM,6} \cdot i_{C,BM} - \nu_{CH4,6} \cdot i_{C,CH4} - \nu_{HCO3,6} = -0.007 \text{ Mol } CO_2/\text{g BM-COD}.$$

The biomass decay of microbal acetate conversion (process 11) is assumed to produce 80% particulate substrate ($\nu_{Xs,11}$ = 0.80 g COD/g BM-COD) and inert particulate material (20%).

For thermophilic digestion the stoichiometric coefficients are assumed to be similar, except the ones for stripping ($\nu_{H2,1} = -H_{H2} = -0.086$ atm·m^3/g COD, $\nu_{CH4,1} = -0.017$ atm·m^3/g COD, $\nu_{CO2,1} = -0.058$ atm·m^3/Mol).

Tab. 3 Substance characterization. The nitrogen content of the particulate degradable substrate and the amino acids is low, because deamination takes partly place during transport and storage of the raw sludge.

Substance	specific content of		
	carbon $i_{C,i}$ Mol C/g COD	nitrogen $i_{N,i}$ g N/g COD	electrical charge $i_{L,i}$ Mol equiv./gCOD (or N)
Methane (S_{CH4})	0.0156		
Ammonium (S_{NH4})			-0.0714
Acetate (S_{ac})	0.0313		0.0156
Propionate (S_{pro})	0.0268		0.00893
Am. acid/sugar (S_{as})	0.0299	0.040	
Fatty acid (S_{fa})	0.0216		0.0012
Inert (S_{in})	0.030		
Part. degrad. (X_s)	0.030	0.020	
Biomass (X_{BM})	0.0313	0.080	
Part. inert (X_{in})	0.030	0.040	

d) Computer program

The simulation program, which is written in PASCAL, is based on a simple, first order forward integration with variable time step. Since the response time of hydrogen is in the order of 10^5 to 10^6 times shorter than that of most other components, the integration of the hydrogen concentration is separated out and is based on a short timestep, whereas the integration of the other components relies on a larger timestep. Besides the integration routine a relaxation routine is available, which allows for further differentiation of the required timesteps, by distinguishing between hydrogen, soluble and particulate species. Relaxation is designed to rapidly approach steady state, without providing information on time dependent intermediary results.

A simulation run usually starts with finding an approximate steady state situation with the aid of relaxation. Steady state results are then refined with forward integration. Finally, simulation with time dependent variation of temperature and feed concentrations is obtained with the integration routine. Results of a simulation run are stored in a form which may be used for further processing in spreadsheets.

In order to further accelerate the computations, the hydrogen concentration was artificially increased by a factor of 1000. Since the resulting concentration is still smaller than the concentration of all other dissolved components, this increase does not affect the results.

Besides the change of the state of the reactor, the program predicts gas flow and gas composition in the headspace of the digester.

Two stage treatment was simulated sequentially, by storing intermediary results in a form which could be used as input for the simulation of the second step.

3. LABORATORY SYSTEM AND CHARACTERIZATION OF SEWAGE SLUDGE

The laboratory treatment plant consists of three different lanes:

- single mesophilic digestion (Volume V = 28 l, T = 35°C),
- anaerobic thermophilic pretreatment (V = 16 l, T = 55°C) and anaerobic mesophilic main treatment (V= 28 l, T= 35°C),
- aerobic thermophilic pretreatment (V = 3.2 l, T = 60°C) and anaerobic mesophilic main treatment (V = 28 l, T = 35°C).

The sewage sludge used for the laboratory plant was a mixture of primary, secondary and tertiary sludge, taken from the treatment plant Uster (Zürich), stored in a 1.1 m^3 stirred vessel and kept cold at 0.4 °C. The raw sludge was added in 30 to 60 min intervalls. One batch of sludge was used up within 2 to 3 months. During this time the volatile fatty acids and the ammonium concentration did increase about 30 %, whereas the total COD concentration remained practically constant. The average composition of the raw sludge was 42 g total COD/l, 6 g dissolved COD/l, 2.2 g Acetate/l, 0.7 g propionate/l, 0.5 g NH_4-N/l, 2.0 g N_{tot}/l, 0.85 g P_{tot}/l, and pH 6.8. The bicarbonate was practically used up by the deprotolysis of the fatty acids. From the degradation data the particulate COD was estimated to be 60% degradable substrate and 40% inert organic material.

4. RESULTS AND DISCUSSION

a) Calibration of acetate conversion to methane

The hydrolysis of the particulate organic substances is the rate limiting step of anaerobic digestion and determines therefore the detention time of the sewage sludge in the digester (1). The microbial groups that degrade fatty acids, acetate and propionate are slow growing bacteria and since 70% (Fig. 1) of the methane is produced by acetate degradation, this process determines digester stability. If acetate production is greater than acetate degradation (during load variation or inhibition of acetate conversion), the acetate accumulates and lowers the pH, because of decreasing puffer capacity. This again could inhibit hydrogen, propionate and acetate degradation and the sludge becomes acidified.

To compare maximum acetate conversion of mesophilic digestion with the steady state conversion rate the experiment illustrated in Fig. 2 was carried out at three different hydraulic detention times. During normal feeding 2.8 to 3.8 g acetate/l sludge were added and gas production, gas composition, pH, alkalinity and volatile fatty acids were analyzed.

From increased methane production and the acetate profile the safety factor (SF) of the acetate conversion process was calculated (Fig. 3):

$$SF = \frac{\text{maximum acetate conversion}}{\text{average acetate conversion}} = \frac{\mu_{max,ac}}{\mu_{ac}} = \frac{\mu_{max,ac}}{\mu_{n,ac} + k_{d,ac}}$$

with: $\mu_{max,ac}$, μ_{ac} = maximum and average growth rate and
$k_{d,ac}$ = decay rate of acetate conversion (d^{-1})
$\mu_{n,ac}$ = $\mu_{ac} - k_{d,ac}$ = net growth rate (d^{-1})
θ = hydraulic detention time (d)

For steady state conditions the net growth rate is equal to the dilution rate $D_x = 1/\theta$ and:

$$SF = \mu_{max,ac}/(1/\theta + k_{d,ac}).$$

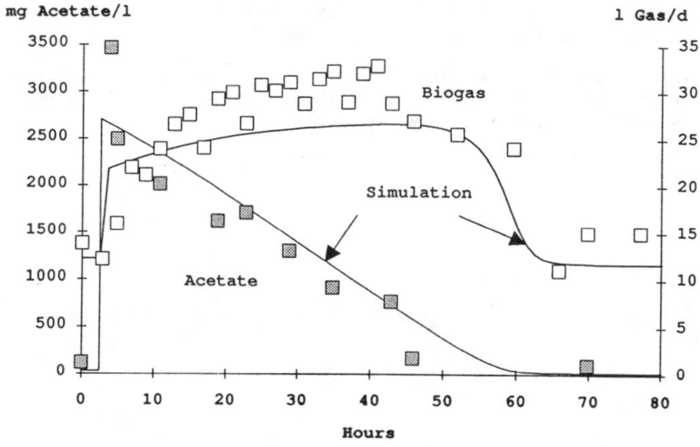

Fig. 2 Gasproduction and acetate concentration after addition of sodium acetate (θ = 22 days, T = 35°C).

For a very small decay rate the SF will increase linearly with the hydraulic detention time ($SF \cong \mu_{max,ac} \cdot \theta$). However, the experimental data suggest a substantial decay rate, since SF does not increase significantly for a detention time above 12 days. An experiment with a single mesophilic digester (Fig. 5), in which θ was lowered in two steps from 18 to 5.3 days within four weeks, would suggest that no other microbial population had time to adapt at the lower θ.

From the data presented in Fig. 3 a decay rate $k_{d,ac} \cong 0.18\ d^{-1}$ and a maximum growth rate $\mu_{max,ac} \cong 0.55\ d^{-1}$ was calculated for mesophilic digestion. This yields a maximum net growth rate of 0.37 d^{-1} as found in the literature. This is also confirmed by the results from Kapp (5) with an anaerobic mesophilic digester with 5 days detention time.

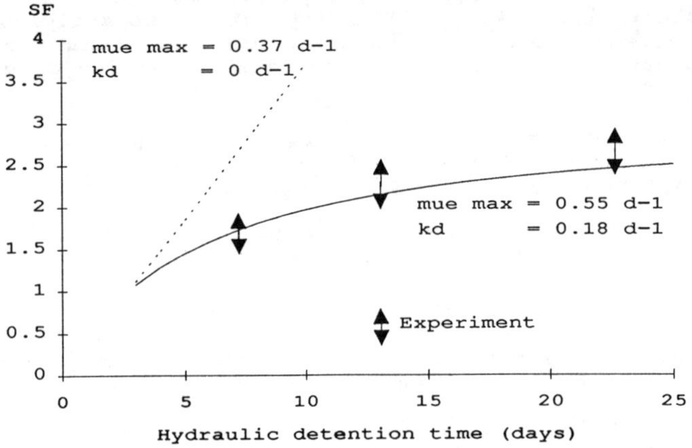

Fig. 3 Safety factor SF for acetate degradation in function of the hydraulic detention time θ.

Recent experiments by which the acetate degradation capacity was measured 0, 3 and 6 days after the influent was stopped, support this high decay rate.

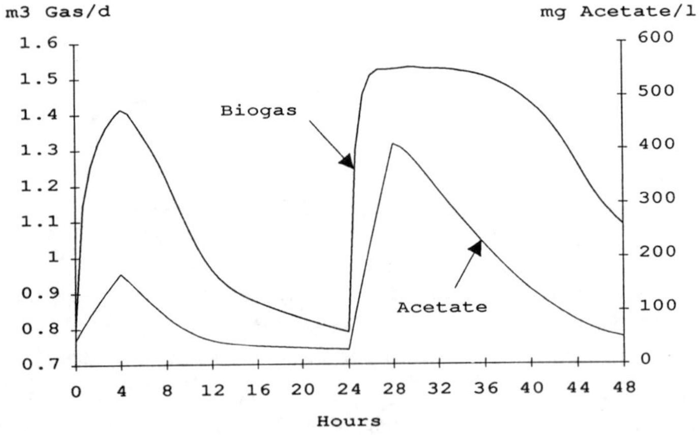

Fig. 4 Gasproduction and acetate concentration for a $1m^3$ reactor, that is fed once per day during four hours, with 100 and 250% of the average load at the first and second day, respectively. (θ = 12 days, T = 35°C, influent COD_{tot} = 42 g/l).

The propionate concentration was increasing after acetate addition. This confirms the inhibition of propionate degradation during high acetate concentration. Additional experiments to describe growth and decay of propionate degradation as well as short term pH and temperature effects of acetate and propionate degradation will be carried out in the future.

For anaerobic thermophilic digestion the kinetic parameters are different and not yet verified.

The mathmatical model is a powerful tool to simulate dynamic conditions. In Fig. 4 the behaviour of a mesophilic digester with a realistic load variation is simulated. The raw sludge was only added once per day during four hours, 100% and 250% of the average load at the first and second day, respectively. As seen from the gas production and acetate profile acetate degradation was saturated for about 16 hours. From the accumulated volatile fatty acid the bicarbonate puffer got decreased and pH was lowered from 7.2 to 7.1.

b) Verification for two stage treatment

In Figs. 5 to 7 the mathematical model is verified with experimental data from single step and two stage treatment. The three lanes, described in section 3, were simultaneously fed with the same raw sludge and after four weeks of constant sludge addition the hydraulic load was doubled for another four weeks.

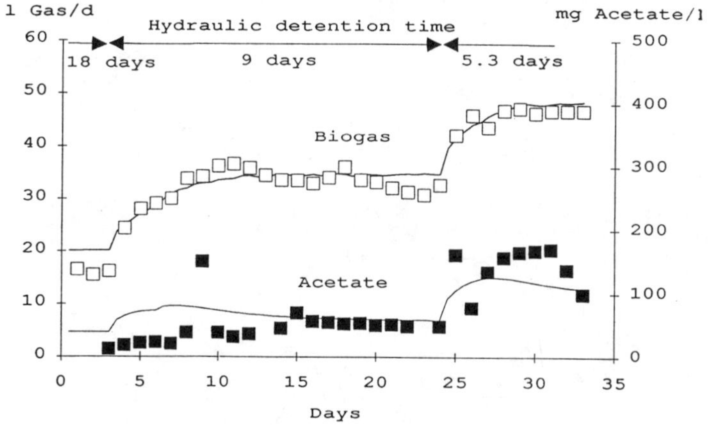

Fig. 5 Gasproduction and acetate concentration of an anaerobic mesophilic digester during a two step decrease of θ from 18 to 9 and 5.3 days (influent COD_{tot} = 42 g COD/l).

As seen from the acetate profile of Figs. 5 to 7 two stage treatment did react more sensitive to changes of hydraulic load than single step treatment, since changes of COD load (see also gas production) is significantly higher than for single step treatment. There is no explanation yet for the deviation of the acetate profile of the experiment and of the simulation in Fig. 5 after the first increase of the hydraulic load.

Effluent sludge of the two stage treatment shows significantly better stabilization data (thickening, viscosity, COD, TSS) than single treatment due to the thermophilic pretreatment (Tab. 4). Better thickening of suspended solids lowers the required storage volume, if during winter time agricultural use of sewage sludge is prohibited.

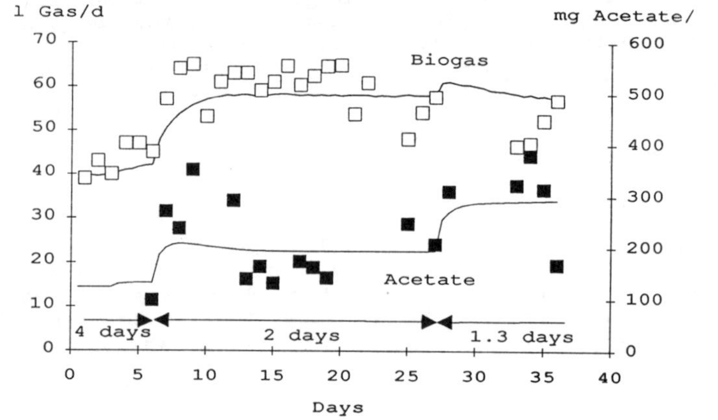

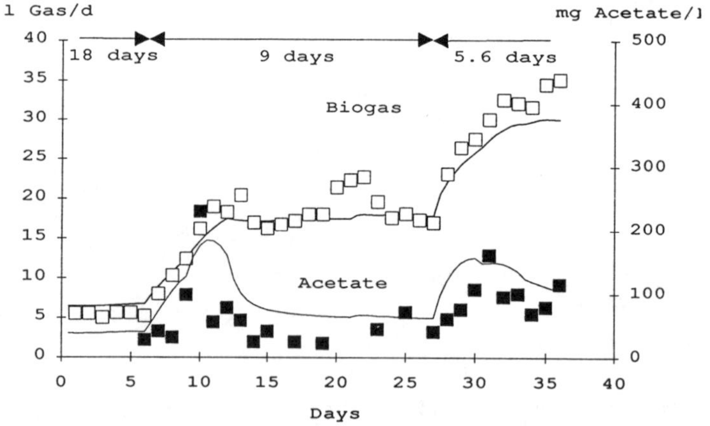

Fig. 6 Behaviour of the two stage anaerobic thermophilic (top)/ anaerobic mesophilic digestion (bottem) during a decrease of θ from 4, 2 to 1.3 and 18, 9 to 5.6 days, respectively. Influent COD for pretreatment was 42 g COD/l; influent COD for mesophilic digestion did increase from 25 to 32 g COD/l after change of θ.

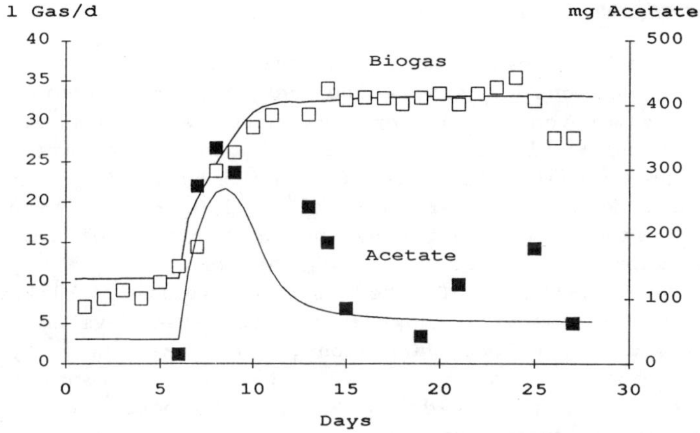

Fig. 7 Behaviour of an anaerobic mesophilic digester with an aerobic thermophilic pretreatment during a decrease of θ from 1 to 0.5 and 18 to 9 days, respectively. Parallel to the increase of θ the aeration capacity in the pretreatment was reduced. Influent COD for pretreatment was 42 g COD/l; influent COD for mesophilic digestion did increase from 33 to 40 g COD/l after change of θ.

Tab. 4 Sludge characteristics of treated sewage sludge for single and two stage treatment (settled volume after 7 days in a 1 liter cylinder; η_S, η_W: dynamic viscosity of sludge, thickened to half of the volume, and water at the change from laminar to turbulent flow).

System	θ of 1./ 2.step days	COD g/l	TSS g/l	Settled volume %	Viscosity η_S/η_W -
raw sludge		42.1	38.8		
anaerob mesophil	18 9	22.3 26.3	27.5 31.1	49 49	20 24
anaerob thermophil	4 2	25.0 32.0	30.5 35.4	42 46	19 17
anaerob thermophil anaerob mesophil	4/18 2/9	20.0 23.5	26.7 30.5	38 40	14 15
aerob thermophil anaerob mesophil	1/18 0.5/9	18.3 24.8	25.8 31.5	37 41	13 14

5. CONCLUSIONS

The mathematical model is a powerful tool to simulate dynamic conditions of single and two stage treatment. Since acetate conversion is the process that determines digester stability, the kinetics of this process were evaluated, within the hydraulic detention time of 5 to 20 days. Maximum growth and decay rate were 0.55 d^{-1} and 0.18 d^{-1}, respectively, corresponding to a maximum net growth rate of 0.37 d^{-1}. The thermophilic pretreatment of the sewage sludge yields better stabilization of the treated sludge (thickening, viscosity, COD removal). However the two stage system reacts more sensitive to load variations. Therefore their dynamic behaviour should be tested, if the hydraulic detention time of the second step is reduced due two the better stabilization of the two stage treatment.

6. REFERENCES

(1) Gujer, W. and Zehnder, A.J.B. (1983). Conversion processes in anaerobic digestion. Wat. Sci. Tech. 15, 127-167.

(2) Henze, M., Grady, C.P.L.Jr., Gujer, W., Marais, G.V.R. and Matsuo, T. (1987). A general model for single sludge wastewater treatment systems. Wat. Res. 18, 255-276.

(3) Tschui, M. (1989). Dynamisches Verhalten der mesophilen anaeroben Schlammstabilisierung. Diss. ETH-Zürich Nr. 8958.

(4) Kaspar, H.F. (1977). Untersuchungen zur Kopplung von Wasserstoff- und Methanbildung im Faulschlamm. Diss ETH-Zürich Nr. 5984.

(5) Kapp, H. (1987). Die anaerobe Schlammfaulung. Stuttgarter Berichte zur Siedlungswasserwirtschaft Bd. 97.

OPERATOR ASSISTED OPTIMIZATION OF SLUDGE DEWATERING

H. GRÜTTNER AND J. LA COUR JANSEN,

Water Quality Institute
Science Park Hørsholm
11, Agern Allè
DK-2970 Hørsholm
Denmark

SUMMARY

On a municipal wastewater treatment plant using af decanter-centrifuge for dewatering of anaerobic digested sludge an operator assisting system for sludge dewatering was developed. The system is based on a database used to collect data on sludge properties and operational conditions which is added by the operator. By graphical presentation and an advisory service these data are used to support the operator in his dewatering operations and to secure a running optimization of the sludge dewatering.

Evaluations show that this system is a useful tool for data collection and presentation and that the data collected seem to reflect the actual situation at the plant. In the future such systems are expected to be used as tools for education of operators, transfer of knowledge from one operator to another and for a continous optimization of dewatering operations.

INTRODUCTION

Optimization of sludge dewatering is a troublesome task. The main reasons are:

- The sludge dewatering mechanisms are not fully understood.
- Sludge properties and dewatering equipment vary much from plant to plant.
- Optimization goals are often conflicting and have in many cases no clear priority.
- Operators's skills are different and in many cases need improvement.

Based on these findings, a knowledge based system for sludge dewatering has been set up and implemented on two full-scale treatment plants. The system, which is completely based on in-put from the operator, consists of three main parts:

- Data on sludge properties, operational conditions, results from the dewatering and operator observations make a database, which is the backbone of the system.

- Graphical presentation of data and observations from the database and correlations of data on the dewatering give the operator immediately access to the past experience with the function of the sludge treatment system.

- An advisory service is directly installed in order to support in case of operational problems. The advise is based on general information and knowledge on dewatering combined with the information from the database.

This paper describes the development of the system and the data collected during the period of implementing of the system.

THE SYSTEM

The overall philosophy has been to make it as easy as possible for the operator to use the system. It means that the same data are only put into the system once, and during operation he only needs to put in new data if a parameter is changed. But at the same time it is tried to illustrate for him which parameters are relevant to register when adjusting the operation or taking a sample.

Databases

Indata part of the system is structured around a typical dewatering run, where there is a start-up in the morning, a running through the day with some adjustments and samples, and finally closing down the machines in the afternoon.

The data registered in each part of the operation are shown in table 1.

As stated above a separation is made between parameters which are constant during the time of operation, and those which may be changed during operation and hence need to be updated by each adjustment. A very important part of this database is the "reason for adjustment" which the operator is asked to fill in by each adjustment, either by choosing from a list of typical reasons, or by creating a new.

The system can handle the fact that many adjustments are made without taking a sample, and in the situation where a sample is taken, the result will be registered some hours later.

As a service to the operator those of his registrations which are not in standard units are immediately transferred, e.g. the speed of a feed-pump is transferred to the actual flow. In addition calculations of some key parameters are made:

- amount of polymer used
- concentration of polymer added
- polymer dosing (g/kg DS) - if dry solids is known
- amount of dry solids feed to the decanter
- %-solids removal

During start-up	By adjustments	Sample-data
MACHINE - water level in bowl POLYMER - type of polymer - polymerconcentration SLUDGE - origin of sludge - remarks on sludge OPERATION - start-up time - initials of operator - hourcounter - flowcounter Plus the data in ADJUSTMENTS	REASON FOR ADJUSTMENT - standard reasons or explanation SAMPLE - yes/no (yes generates number) ADJUSTMENTS - Feed flow - Polymerflow - oilpressure in the automatic scroll-speed system READOUTS - level in storage tank - amounts of water used for polymerdilution - differention-speed	INLET SLUDGE - temperature - dry solids contents - CST CONDITIONED SLUDGE - CST REJECT WATER - pH - suspended solids DEWATERED SLUDGE - dry solids content

Table 1: Parameters used in the system.

Data-presentation

The data put into the system are presented as graphs and as reports. Some of the important parameters are presented on a time scale to illustrate the fluctuations. Other graphs can be considered as part of the advisory service since they present the relation between selected parameters.

Reports include monthly report on amounts, key parameters for dewatering and analytical data plus a daily report including nearly all data put into the system on that particular day.

The general principle has been that all information put into the system must be presented or used in the system since that is the most direct way to encourage the operator to supply data into the system.

Advisory service

Two sorts of advises are supplied; a rule-based strategy for what to do in a particular situation and a presentation based on the actual data collected in the databases.
The situations supported are:

- low solids in dewatered sludge
- high solids in dewatered sludge
- high suspended solids in reject
- too high capacity
- too low capacity
- polymer consumption too high (optimization)

The same motives are used in the indata-part as "reason for adjustment" and in that way the actual changes made can be used. In addition when making an adjustment the operator is asked to evaluate whether the changes made are considered a succes or not, and hence only data considered a succes are later used in the advisory service described above.

The rule-based advises are based on a general strategy commonly used:

- Keep the amount of dry solids to the decanter at a constant level.

- Dry solids content in dewatered sludge should be controlled by the differential scrollbowl-speed.

- Reject quality should be controlled by the polymer addition.

From this very simple strategy an explanation on what to do in the above mentioned situations is derived.

In addition to this feature suggestion for start-up adjustments is offered. If the operator gives information to the system on the sludge origin, the dry solids content and the temperature, the system will seek the databases for a similar situation. In this search the operator is asked to give priority to the different operational strategies:

- High dry solids in dewatered sludge
- Low suspended solids in reject
- Low polymer-use
- High capacity.

RESULTS

The system described above was developed and implemented on a small wastewater treatment plant (15.000 PE) north of Copenhagen during the first eight months of 1990. Dewatering of the anaerobic digested sludge is done by one decanter from DDS equipped with a hydraulic system for automatic backdrive torque control, called Viscotherm Rotodiff.

The development has involved three phases:

1) Passive collection of data (first period).
2) Collection of data when operator is supported by graphs and reports (second period).
3) Collection of data when advisory service is supplied (third period).

Data from the first and second period will be presented by:

- Evaluation of the use of the system.
- Presentation of the data collected.

The use of the system

Table 2 shows the number of data-set collected. In table 3 the number of samples taken are shown together with the number of determinations of dry solids. Please note, that sample in this relation means that one or several samples are taken on one of the four positions indicated in table 1.

	Number of data-set	Start-ups	Adjustments	Adjustments per day
First period	152	51	101	2,0
Second period	118	29	89	3,1
Total	270	80	190	2,4

Table 2: Number of data-set and adjustments.

The system has been in use on all "dewatering days" and on average 2-3 adjustments have been made each day. There is clearly an increase in the number of adjustments made each day and in the number of samples taken per dewatering run from the first to the second period.

		Dry solids determined in:		
	Number of samples	Inlet	Dewatered sludge	Reject
First period	34	25	24	4
Second period	33	28	31	24
Total	67	53	55	28

Table 3: Number of samples collected and determination of dry solids.

In table 4 the adjustment made is distributed on the different reasons for adjustments offered in the two first periods. It may be noted, that in the first period the major concern was on the solids content in the dewatered cake while later on some concern on reject quality has come up. On the background of the many adjustments with other explanations or no explanation the list of possible reasons were expanded to the list mentioned above (Advisory service).

Reasons for adjustments	First period	Second period	Total
Start-up	51	29	80
Low solids in dewatered sludge	45	50	95
High solids in reject	1	6	7
Other reasons	21	17	38
No explanation	34	16	50

Table 4: Reasons for adjustments.

A way to check what use the operator has had from the system is to test what use is made of the analytical results. As an example reactions on dry solids in inlet can be desribed. Each time a value for dry solids contents in the inlet is put into the system the specific polymer dosing is shown on the screen. During the first period this information provoked only one adjustment, but during the second period seven adjustments were made. This indicates that the operators are getting more conscious on the feed-back from the systems.

Another way to evaluate the use of the system and the operators behavior, is to see which adjustments the operators made in the different situations. As an example table 5 shows the adjustments made to cure "low solids in dewatered sludge".

	UP	DOWN	TOTAL
Feed flow	15 (8)	8 (4)	23 (12)
Polymer addition	5 (3)	7 (5)	12 (8)
Differential speed	1 (0)	2 (2)	3 (2)

Table 5: Adjustments made to get a higher solids concentration in the dewatered sludge. Value in parenthesis is second period.

As it can be seen from the table the adjustments made were not very systematic. The most reasonable things to do - to lower the feed-flow or lower the differential speed - were only done in 10 out of 38 adjustments. It is expected that this situation will change when the advisory-service has been in use for a period.

Data registered by the system:

In figure 1 and 2 some of the important parameters during the first six months of operation are presented.

As it can be seen the oilpressure in the Viscotherm, the feed-flow and the polymer addition have been kept at a quite constant level for the whole period. At the same time the differential speed vary to compensate for the variations in the amounts of dry solids fed to the decanter. This leads to a high variation in the specific polymer dosing and as a result the solids removal is not as good as it could be.

On the surface the system seems very quiet and steady, but when investegated deeper it shows big fluctuations. This again has had two consequences. First the polymer use has been higher than necessary and periodically the reject quality has been much poorer than necessary. Both consequences have a significant economic effect on running cost of the plant.

It seems that the general strategy suggested above may be very useful at least at this plant. However, this strategy creates a need for quick determination of dry solids. During the first months dry solids has been determined with a weight combined with a drying lamp, but this procedure takes around

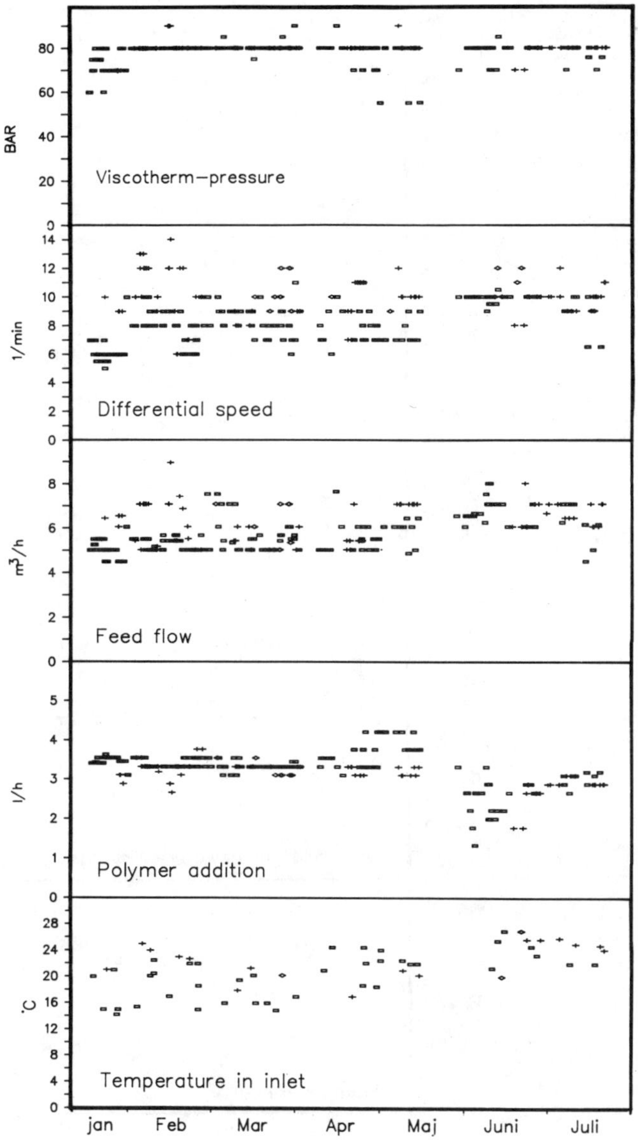

Figure 1: **Selected parameters registered by adjustments. Data from the first six months of data collection at Sjælsø treatment plant.**

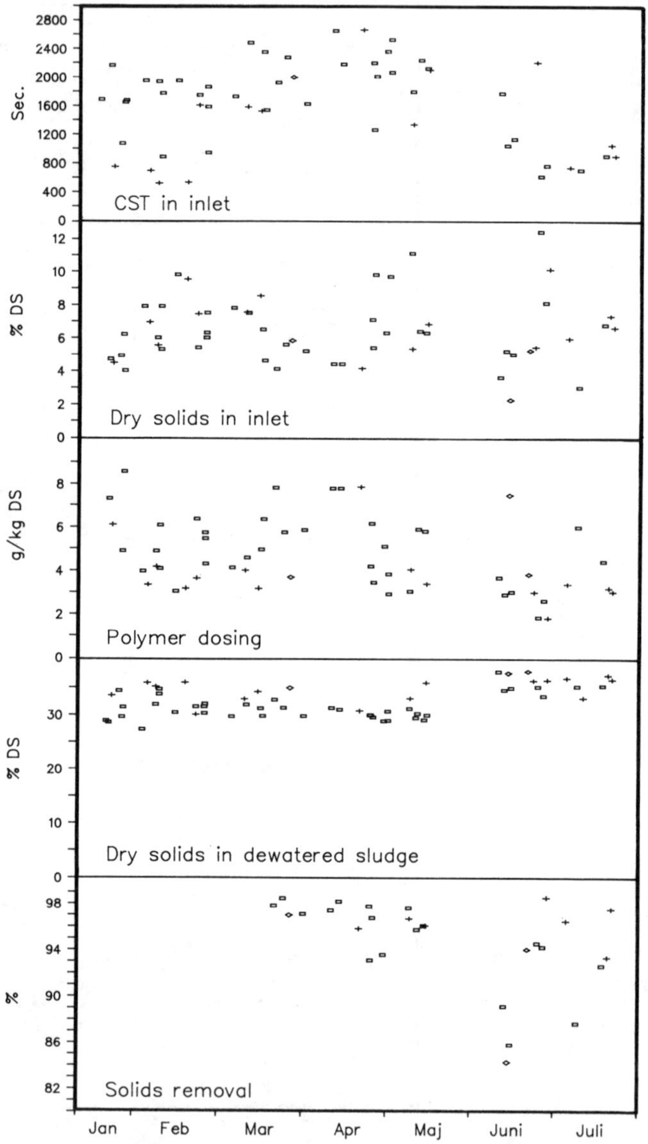

Figure 2: **Selected analytical results and derived figures from the first six months of data collection at Sjælsø treatment plant.**

one hour. An alternative has been introduced using a laboratory-centrifuge spinning 3000 Rpm for 3 minutes and reading the percent of sedimented material. It seems that this method correlates very well with dry solids and should make it possible to make much faster adjustments of the system.

General relations in the data

Another way to test the material is to see whether correlations between variables and dewatering results are in agreement with general theory. The result of such a test for linear correlations (SAS-procedure: CORR) is shown in table 6.

	Dry solids in dewatered sludge	Dry solids in reject	Percent solids removal
Temperature	0.40	0.65	0.48
CST	0.57	-0.74	0.53
Feed-flow	0.46	0.59	-0.61
Polymer addition	-0.74	-0.52	0.65
Specific polymer-dosing	-	-0.81	0.56
Differential speed	0.57	0.77	-0.76

- : No significant correlation (95% level).

Table 6: Correlation coefficients between sludge characteristics/adjusting parameters and dewatering results.

Dry solids content in the dewatered sludge is obviously a function of CST, polymer addition and differential speed while the specific polymer dosing shows no correlation. This may be explained by the fact that the bigger particle will settle and be compacted even if no polymer is added.

Dry solids in reject is clearly correlated to specific polymer dosing, while several other parameters seems to have some influence. This is in good agreement with the general understanding that the polymer addition is supposed to bind the small particle together and when too little polymer is added the reject quality becomes poorer.

Percent solids removal seems to be a function of differential speed and is influenced by sludge characteristics as described by CST and temperature.

These findings are in good agreement with general accepted theory for the performance of decanter centrifuges, and this indicates that the data collected reflects reality fairly good.

Conclusions on evaluation of data:
The first six months full operation has shown:

- The data collection system can be used properly by operators of the sludge dewatering system.

- Graphical presentation of measurements and observations are useful for the daily operation and has increased the use of the operator supporting the system.

- There are some difficulties in keeping a constant polymer addition due to variations in dry solids in inlet.

- The data collected during the two first periods of operation shows a correlation pattern in good agreement with general theory on decanter performance.

DISCUSSION OF THE APPROACH

The general situation for many small wastewater treatment plants in Denmark, is that the people are present at the plants, but they need support and education. It means that introduction of an automatic control system, described by several authors (e.g.: 1, 2, 3), may optimize the dewatering on a short range, but in the long run it is better to give the operators an understanding of the mechanisms controlling the processes they are working with.

In the future there will be a need for systems like the one described above, used as a tool for supporting the unskilled operator, guiding the skilled operator through further optimizations and used also as a way to transfer knowledge from one operator to another. In addition such systems may be a usefull way to give information to people developing equipment on the real-life performance of their equipment.

The systems for automatic control based on several on-line measurements may be used on bigger plants to control the individual machines, and may to our oppinion take over the control only within specific limits or be used as a consulting system giving information and advises to the operators.

With the price level on computers it is very likely that in the future each plant will have its own model for dewatering like the one described in (4). Such an empirical model might be used as a tool for giving advises to operators and the data collected will provide a continous calibration.

ACKNOWLEDGEMENT

This project is supported by the National Agency for Environmental Protection and The Board of Technology.

REFERENCES

(1) WIKSELL, H. and de BESCHE, S. (1978). Automatic control of a centrifuge process. Vatten 1, 10-14.

(2) CAMPBELL, H.W. and CRESCUOLO, P.J. (1988). Control of polymer addition for sludge conditioning: A demonstration study in: What's new in wastewater technology? Eds.: Environment Canada and Canadian Water and Wastewater Association. 25-40.

(3) WINTHER, F. (1990). Optimierte Flokkungsmitteldosierung am Beispiel der Schlammeindickung/Schlammentwässerung. Korrespondenz Abwasser 4, 296-392.

(4) HERATH, B., GELADI, P. and ALBANO, C. (1989). Developing an empirical model for dewatering... Filtration & Separation Jan/Feb, 53-56.

TECHNICAL AND ECONOMIC ASPECTS OF MANURE PROCESSING

P.J.W. TEN HAVE

Institute of Agricultural Engineering (IMAG)
P.O. Box 43, 6700 AA Wageningen, the Netherlands

Summary

Since 1976 treatment of veal calf manure in central processing plants has been practised in the Netherlands. In 1989 180,000 m^3 with 1.6% DM was treated in three plants. The capacity is to be doubled to 360,000 m^3 by 1991. The process applied is biological treatment by means of activated sludge, mainly to remove COD, N and P. Dissolved salts (such as K and Cl) are discharged with the final effluent.

Despite the fact that the Government requires that the annual processing capacity is to be at least 6,000,000 t in 1995, the start of pig and poultry manure processing is far from being swift. About ten companies are planning to build and run processing plants. Nearly all the processes are based on the conversion of slurry into a dry product, although there are considerable differences between the individual processes. The effluent is practically free of salts. Evaporation and drying are important stages in the processes. Much attention is paid to minimizing the energy consumption and to preventing the emission of volatile components such as fatty acids and ammonia.

The costs of veal calf manure treatment amount to approx. Dfl 14/m^3 at a capacity of 180,000 m^3/year. The drying of pig slurry costs approx. Dfl 15-35 at a capacity of 1,000,000 m^3/year.

1. INTRODUCTION

The last few decades the Netherlands has seen a strong increase in the production of pigs, poultry and veal calves. Moreover, this production is concentrated in a few regions. The amount of manure applied to the land tends to be high, which is a consequence of this. As a result, the nitrate concentrations in the groundwater are too high in many places and the soil is becoming over-saturated with phosphate. Since a few years the Government is putting limits to the amounts of phosphate allowed to be spread on arable land and grassland as manure. Table 1 shows the schedule for the gradual reduction of maximum permissible quantities for the years to come.

Table 1: Maximum amounts of phosphate in animal waste to be applied (kg P_2O_5/(ha year)

Application	1987	1991	1995	ca. 2000
Arable land	125	125	125	?
Grassland	250	200	ca. 175	?
Mais	350	250	ca. 175	?

After the introduction of the standard for 1991, the manure-producing farms are expected to have a total surplus of approx. 18,000,000 t slurry containing approx. 86,000 t P_2O_5 (1). How much of this can actually be applied to other farms, will depend on the willingness of other farmers to use animal waste to cover their need for minerals and organic matter. The

degree of acceptance of animal waste for land-spreading is not known. It is estimated, however, at 11,000,000-16,000,000 t. For the remainder, estimated at 2,000,000-7,000,000 t, another solution has to be found. It is intended to apply large-scale processing, with the following objectives:
- export of the final product,
- replacement of mineral fertilizer on the home market.

Recently, the Dutch Government has announced that measures will be taken to reduce the livestock population if the farming sector will not have succeeded in realizing a processing capacity for at least 6,000,000 t manure per year in 1994.

2. WHAT IS PROCESSING?

Generally, processing implies the separation of water and dry matter. The water is discharged as effluent into the surface water. The product consists mainly of the dry matter contained in the waste. Its application can be compared with that of untreated manure.

More and more stringent requirements are imposed on emissions into air and water. Emission into the air mainly concerns odour and ammonia. The BOD (biochemical oxygen demand) as well as the N and P concentrations in the effluent have to be reduced very much. Furthermore, in many regions of the country it is required that most of the salts dissolved be removed. In contrast to the condition of most wastewater sludges, the water component of slurry contains much dissolved matter, an important part of which is salts. Table 2 shows a few indicative values of effluent discharges into the rivers Rhine and Maas. The average values in pig slurry are also stated. The standards for effluent discharge are based on experiments on the removal of dissolved matter by means of reverse osmosis.

Tabel 2: Average concentration of some elements in pig manure and proposed maximum values for effluent discharge to the rivers Rhine and Maas in mg/l

Element	Pig manure	Effluent
P	2,000	2
N	7,000	20
K	6,000	400
Cl	1,500	150

The advantages of dry manure over liquid manure are:
- more economical in transport and storage,
- easier handling, clean product,
- free of odour,
- low NH_3 emission.

Furthermore, the following objectives are to be aimed at:
- sterile,
- free of weeds,
- guaranteed composition and effectiveness,
- good spreading properties,
- low heavy metal contents.

In most processes the mineral content is enriched, as in some cases the process requires so. In other cases the N:P:K ratio is altered as the market demands so. In a number of processes a part of the organic matter is anaerobically decomposed or incinerated, mainly with a view to minimizing the energy requirements of the processing plant. Occasionally, organic matter is added. Generally, it is tried to convert manure into a type of 'organic fertilizer'. Such a fertilizer makes up for the organic losses in the soil. In addition, the minerals N and P shall be released quickly for plant growth, i.e. for the most part during the first year.

It is expected the dry manure, when exported, will primarily be used in wine, fruit and vegetable growing in France and the Federal Republic of Germany. The quantities estimated are between 235,000 and 550,000 t/year. To produce these quantities, about ten times as much pig slurry is needed. As a matter of fact, attempts will be made to find more outlets than the two countries just mentioned. For example, recently a shipment of 1000 t of dried pig slurry was sent to Spain.

Attempts are made to make more out of slurry than just dry manure. A few different options are:
- NH_4 salt,
- carbon black,
- protein,
- amino acids,
- fish and livestock feed,
- algae.

3. PRESENT SITUATION IN THE NETHERLANDS

3.1. *VEAL CALF MANURE*

Since 1976 treatment of veal calf manure in central processing plants has been practised. The first processing plant had an initial capacity of 18,000 m^3/year. The capacity of this plant has meanwhile been extended to approx. 27,000 m^3/year. The second plant was commissioned in 1985, and a third in 1986. In 1989 a total of 183,000 m^3 veal calf manure was treated which had an average DM content of 1.6%. Next to these three plants, the construction of a fourth has been started recently, with an annual processing capacity for 180,000 m^3. The principle applied is the activated sludge process, in which the emphasis is laid on achieving the most complete possible nitrification and de-nitrification. The load is approx. 0.05 kg BOD/(kg sludge d) and 0.02 kg NH_4-N/(kg sludge d). The COD and N concentrations are high; therefore, the retention time is long (15-25 days) and special attention has to be paid to the temperature control. Surface aerators would excessively cool down the contents of the aeration tanks in the winter, so that bubble aeration must be applied.

Except for in the oldest plant, which is to be closed down in due time, phosphate is removed by adding $Ca(OH)_2$ to the aeration tanks. Plans exist to raise the total veal calf manure processing capacity to approx. 650,000 m^3/year.

Desalting has not been planned for these plants, which are all located in the same region (Veluwe in the Province Gelderland). Research into desalting, however, has been performed by Van Tongeren (2), which deals with the post-treatment of effluent by reverse osmosis. In extended time tests it appeared that the effluent could be concentrated at most 6.5 times. Cleaning had to be done not more than once every two weeks. An uncertain factor, however, is the life of the membranes. Table 3 shows the average composition of effluent and permeate. A recent publication by Ten Have et al. provides more information (3).

Table 3: Application of reverse osmosis on effluent biological treatment [1) 2)]

Parameter	Unit	Effluent	Permeate R.O.
COD	mg/l	600	15
BOD	mg/l	100	< 5
N kj	mg/l	80	5
NH_4-N	mg/l	50	5
P total	mg/l	40	0.2
K	mg/l	3,000	100
Cl	mg/l	2,500	100
pH		8 -> 6 [2)]	6
Volume [1)]	%	100	82

[1)] Concentration factor 5.5
[2)] Acidification with H_2SO_4

3.2. *PIG MANURE*

Agricultural cooperatives established two enterprises a few years ago, which have become pilot plants. In the south of the country this is the Promest pilot plant, with a capacity of 100,000 m^3/year, and north of the great rivers MeMon had run a pilot plant for two years, with an hourly capacity of 400 kg slurry.

Recently, *Promest* decided to build a plant for 500,000 m^3/year in Helmond, at the same location as the pilot plant. It is to be commissioned by the end of 1992. In 1994 its capacity shall be 1,000,000 m^3/year. Moreover, it is intended to have one more plant operating of the same capacity, at a different location.

Figure 1 shows the Promest process diagram. It is based on research which authorities and industry together have been performing at Sterksel and Nistelrode since 1980. The pig slurry (approx. 10% DM) is first digested, primarily to remove volatile organic matter which otherwise would be emitted in the drying process. A digestion by-product is biogas. The value of biogas alone does not justify this process step. Anaerobic treatment has the following additional advantages:

- Differences in concentration are levelled so that following process steps are improved.
- The oxygen demand of the nitrification stage is considerably lower.
- The separation of slurries is favourably affected.

Next, a centrifuge separates most of the suspended solids from the slurry. The remaining moisture in the cake (approx. 25% DM) is removed in a drier. The drier is of the indirect type, with a division between the heating medium and the substance to be dried. An advantage of this arrangement is the low volume of gas flow to be treated.

The liquid effluent from the centrifuge (approx. 75% by volume and 3% DM) is then passed on to a biological nitrification step where the ammonia is converted into nitrate to prevent it escaping when heated. CaO is added to compensate for the accompanying reduction in pH value. Not only the effluent from the centrifuge, but also the ammonia-rich condensate from the drier is sent to the aeration tanks.

The effluent after nitrification is concentrated to approx. 25% DM in a multi-stage vacuum evaporator. The condensate is discharged into the

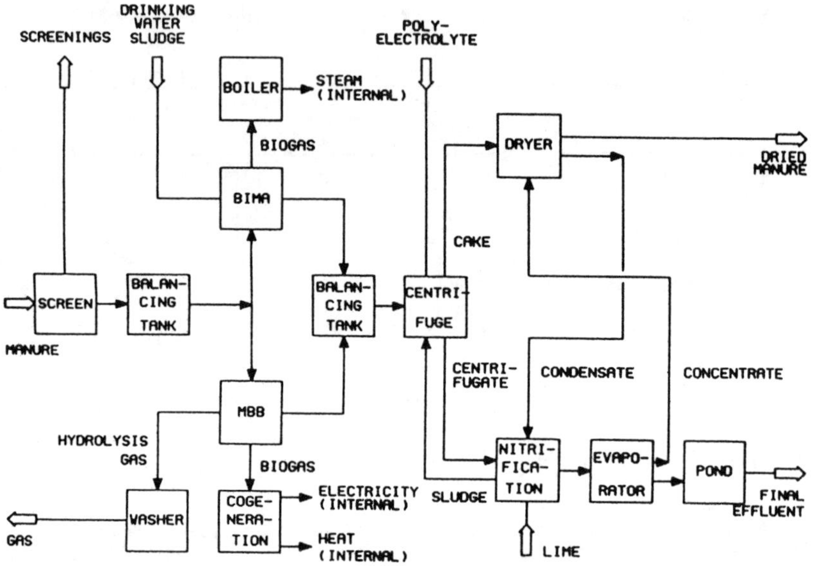

Fig. 1: Flowdiagram Promest

surface water. Thanks to the anaerobic and the aerobic biological decomposition as well as the ammonia conversion into nitrate, the condensate from the evaporator hardly contains any more pollutants when discharged.
All the tanks are of the closed type except the nitrification tanks. Air is extracted from storage tanks and buildings, and acid washers combined with compost filters are used to remove odour and ammonia from the exhaust air.

In May 1990, *MeMon* closed a two-year period during which research was carried out with a Carver Greenfield evaporation plant at the premises of the TNO research institute of Apeldoorn. It is expected that within a few months a start can be made with the erection of a plant with a capacity of 500,000 m^3/year.

The basic process diagram of MeMon is shown in Figure 2. NH_3 emission is strongly reduced by previous acidification of the slurry. For this purpose, MeMon uses sulphuric, nitric or phosphoric acid. The slurry is mixed with an ample amount of carrier oil and dried in a low-energy process in a two-stage vacuum evaporator. The presence of a carrier oil with high boiling point allows the slurry to be completely dried in an evaporator whilst the mixture remains completely liquid, which is a condition for the proper performance of an evaporator. After the water has been evaporated from the slurry, most of the carrier oil is separated by centrifugal action. The remaining oil in the solid cake is evaporated in a drier of the indirect type. The separated oil circulates through the system as a heat transfer and slurry transport medium. Most of the water is separated in the first evaporator stage, which works on the principle of mechanical vapour recompression.

About 10% of the dry matter dissolves in the carrier oil but can be drained off from the plant after the carrier oil has been distilled. Consequently, this process has two final products: powdered dry manure and 'pig oil'.

The water from the slurry is released as a condensate. It contains practically all the volatile fatty acids (VFA) from the original slurry (approx. 15-20 kg/m^3) and part of the ammonia. With more acid being added, the ammonia concentration is lower. At a pH of 4-4.5, the condensate contains about 15-20 kg VFA and 0.5 kg NH_4-N per m^3. Research has shown that post-purification can be performed by converting most of the VFA into biogas in a UASB (Upflow Anaerobic Sludge Bed) reactor. In a downstream activated-sludge unit, the ammonia can be removed by means of nitrification and denitrification. A nutrient mixture must be added, though this causes the salt content to be slightly raised.

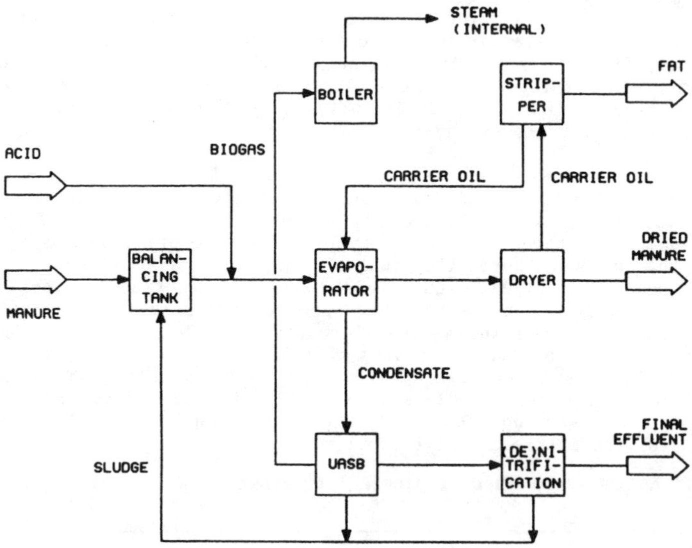

Fig. 2: Flowdiagram MeMon

At least seven more pig slurry processing initiatives can be named. Two of these are planned to be realized soon in plants for 500,000 m^3: STIMT and Ecotechniek. The process to be applied by *STIMT* resembles that of Promest very much, with two major differences. In the first place, there is no biogas plant. The nitrified liquid passes a reverse osmosis unit prior to being further concentrated by an evaporator unit. *Ecotechniek* has developed a process claiming to be able to make fish and livestock feedstuffs, algae and phosphate pellets as final products.

There are a few more processes, which have not yet been dealt with here; they appear in the plans of potential slurry processors not named in this survey. The processes are:

- Wet oxidation. A substantial part of the organic matter is burned.
- Electrophoresis. Proteins and heavy metals are separated from the slurry.

- Pyrolysis. Slurry is converted into carbon black, oil, gas and water.
- NH_3 stripping. Separation of ammonia from the slurry by means of stripping with air or vapour, after which the ammonia is caught in acid.
- Composting. *Ferm-O-Feed* ferments a mixture of broiler manure and organic residuals which is claimed to be suitable as a basic material for cattle feed.

3.3. OTHER MANURES

Poultry droppings can be dried up to 50% DM on the farm by means of exhaust air. This process is practised already, and some of the product is being exported, especially to France. The costs of post-drying in central plants are fairly low, expressed per unit weight of final product. Various companies that intend to process pig slurry, are also planning to dry poultry droppings. Generally, driers of the indirect type are preferred, as these are able to keep the gas flow low and to recover heat, e.g. by mechanical vapour recompression. A large-scale poultry manure drying plant, however, does so with a drier of the direct type with gas recirculation and after-burning; the relevant company claims to be very satisfied with it.

About 70% of animal wastes in the Netherlands are produced by cattle, but hardly any surpluses of cattle slurry occur because cattle farming is closely linked to land use. Consequently, no plants are planned for large-scale cattle slurry processing.

4. COSTS

Much is uncertain, and there are differences in opinion about costs and yields. In general, the initial estimates of net costs (supply and production minus proceeds) are rather low. When the plans are being worked out, it appears that the cost estimates have to be raised.

The costs of veal calf manure processing, as described in Section 3.1, are fairly accurately known, because there has been quite some experience with it. Table 4 gives the estimated costs of a new plant to be built in Ede. The plant is to have a capacity of 180,000 m^3/year. In this plant the P concentration is to be lowered from approx. 700 mg/l to some 30 mg/l by adding lime to the aeration tanks. The effluent sludge from the activated sludge unit is centrifuged to be concentrated to approx. 10% DM after being conditioned with poly-electrolyte.

Table 4: Estimated costs of biological treatment

Capacity	m^3/year	180,000
Investment	Dfl * 10^6	9
Processing costs		Dfl/m^3 manure
- Interest	8%	2.00
- Depreciation	15 yr	3.30
- Maintenance	2%	1.00
- Operators	3 @ kf 70	1.20
- Electricity	12 kWh @ f 0.13	1.60
- $Ca(OH)_2$	2.5 kg @ f 0.20	0.50
- Poly-electrolyte	0.07 kg @ f 10	0.70
- Sludge disposal	0.1 m^3 @ f 25	2.50
- Tax on effluent discharge		0.70
- Various		0.50
- Total, excl. transport and overhead		14.00

The capital outlays required for this plant are estimated at Dfl 9,000,000. From 1 m^3 slurry supplied with 1.5% DM, about 0.1 m^3 sludge remains. To the costs of sludge to be marketed a fairly conservative estimate applies of Dfl 25/m^3 sludge. At this price the sludge can be sold to arable farmers far away or be sent to a slurry processing plant for drying, together with pig slurry. Through a pressure pipe of about 1 km the effluent is sent to a sewage treatment plant. For the remaining COD and Nkj a fee of Dfl 0.70/m^3 has to be paid. The total costs are, rounded, Dfl 14/m^3 veal calf manure, disregarding the costs of manure transport to the plant and overhead costs.

In several studies into the processing of pig slurry it has been shown that the lowest costs of supply and processing together are achieved at an annual capacity of between 500,000 and 1,500,000 m^3, for the Dutch livestock density conditions. If the capacity is lower, the costs of capital and personnel are too high. At a capacity of more than 1,500,000 m^3, the increase in supply costs would not be compensated for by a decrease in processing costs.

The investments in a 500,000-m^3 plant for the processes as applied by Promest and MeMon, are estimated at Dfl 75,000,000. For a capacity of 1,000,000 m^3/year the estimate would be Dfl 125,000,000. The capital costs make out a considerable portion of the total processing costs. The period of depreciation will play an important role. For the first generation of production lines, it will be relatively short. Based on a 15-year depreciation period and an annual capacity of 1,000,000 m^3, the gross processing costs will be about Dfl 40/m^3 pig slurry with 8% DM. From 1 m^3 of this slurry about 100 kg of dry manure can be made. The proceeds from 1 t organic fertilizer are estimated at a minimum of Dfl 150 ex works. The net processing costs are then Dfl 40 - 15 = 25/m^3 manure, ranging between Dfl 15 and Dfl 35. Such a plant, which is run fully continuously (8000 h per year), employs at least 30 people. The energy consumption is strongly reduced as the evaporated water is condensed and the released heat is made use of.

The costs of transport to a plant for 1,000,000 m^3/year amounts approx. Dfl 6.00, provided this plant is centrally located in a surplus area.

Per pig produced, the costs of slurry will decrease as the dry matter content increases, for less water has to be evaporated. Furthermore, the transport costs per pig become lower. It has appeared that the DM content at many farms has increased from 6-8% to 10-14% in the last few years. Related to the amount of slurry, the processing costs increase by Dfl 0.50-1.00 for every 1% the DM content rises. But in that case the proceeds from the sold final product also increase by at least Dfl 1.50, so that with the increase in DM content the net costs per m^3 of supplied slurry will fall.

Besides the increase in DM content, a few more changes can be expected which will affect the costs. These are:

- stricter environmental requirements, resulting in higher costs,
- optimization of the treatment process, resulting in lower costs,
- product quality improvement, resulting in higher proceeds,
- development of new products with higher proceeds, e.g. amino acids.

5. FINAL REMARKS

In the introduction it was stated that the Dutch Government demands from the farming sector that they will have a minimum processing capacity installed for 6,000,000 m^3/year by 1994. To realize this, great efforts have to be made. The main difficulties are:

- uncertainty about the processing costs,
- uncertainty about the supply of slurry by farmers,
- uncertainty about the possibilities of the market,
- time-consuming licensing procedures.

There is no agreement as to the final capacity required (by the year 2000?). A figure is mentioned of 20,000,000 m^3/year. The need will also depend on other solutions, such as:

- less minerals in animal feedstuffs,
- improved digestibility of animal feedstuffs,
- stimulated application of animal waste in arable farming.

Besides, a reduction in the livestock population will have to be taken into account, not only as a result of potential measures by the Government, but also considering the costs livestock farmers are due to make to reduce the NH_3 emissions.

REFERENCES

(1) LUESINK, H.H., and VAN DER VEEN, M.Q. (1989). Twee modellen voor de economische evaluatie van de mestproblematiek. Landbouw-Economisch Instituut LEI (Agricultural Economics Research Institute) Report 47.

(2) VAN TONGEREN, W.G.J.M. (1990). Omgekeerde osmose van effluent van een biologische kalvergierzuivering (duurproeven). MT-TNO Research Report 90-098.

(3) TEN HAVE, P.J.W., WILLERS, H.C. and VAN TONGEREN, W.G.J.M. (1990). Purification of veal calf manure. Proceedings International Symposium on New Trends in Veal Calf Production, Wageningen.

SESSION REPORT

J.H.VOORBURG, IMAG, Wageningen

1. SLUDGE PROCESSING

There was agreement on the importance of the contribute of COST 681 to the progress made in the processing of sewage sludge. It cannot be said that certain innovations are the results of activities. The main effect was that we learned quickly from each other and so contributed to a faster improvement.

Especially, progress in aerobic and anaerobic digestion should be mentioned.

The same holds for the use of poly-electrolytes.

A part of the discussion was devoted to the increasing interest in incineration.

This is due to the fact that in an increasing number of situations no land is available for utilisation in agriculture or in which the sludge is not suitable for land application.

Moreover, in the Netherlands the severe standards for the treatment of exhaust air cause an increasing interest in sludge drying.

These standards have been made more severe because of the fear for the emission of dioxine.

At this session, however, there were no papers handling the risks of dioxine in the exhaust air of the incineration plant.

From the point of view of energy cost, drying is less attractive. The meeting concluded, however, that the calculation of economic aspects of sludge treatment should not only be based on costs and profits. An important aspect should be the benefits of systems especially from the point of view of the protection of ecosystems.

The calculations of technical experts however should be based on technical information.

It is the responsibility of local and national governments to impose regulations for environmental protection which should be taken into account.

2. MODELS

There is an increasing interest in the use of models in research on and management of sewage treatment.

In research, models prove to be an important mean for an exact description of processes and for a better understanding of the relation between the relevant factors.

Very different are the models which are applied as expert systems for the management of a sewage treatment plant. Such models prove to be very useful to assist the operator of the plant. As far as the model is based on data collected at the plant, it can only be used for optimal management of this plant. As soon as it is possible to develop more sophisticated expert systems, a more general application of the models will be possible.

In the discussion on the data to be collected on sludge properties it was realised that from its start COST 681 has tried to develop one single and simple parameter for the characterization of the properties of sewage sludge.

These properties, however, are so variable that there are many doubts about the chance that such a parameter will ever be found.

3. PROCESSING OF ANIMAL MANURE

Though animal slurries and sewage sludge have very much in common, the discussion has shown that experts in these two fields tend to live in different worlds so that there is very much misunderstanding between them.

Also combined processing of both materials was not considered promising. This has not only to do with differences in technical properties but also with the lack of agreement in the appreciation of the end-products.

Anyway, as far as the processing of animal manure is concerned priority is giving to drying systems. This is based on the supposition that there is a market for dried manure. In general, this product is considered to have a composition which makes it more suitable for utilisation as a fertilizer than sewage sludge.

In spite of these differences in approach, experts of the two fields proved to have had a fruitful cooperation in the COST 681 programme.

A very relevant difference between animal manure and sewage sludge is the water content of the raw material. The water content of manure can be influenced strongly by the production systems.

Different from sewage sludge, the liquid fraction of animal slurries contains high concentrations of soluble nutrients, such as N and K, and of soluable organic materials, such as volatile fatty acids. In processing slurries this urges to a different processing of the separated liquid.

Session II

LONG–TERM ENVIRONMENTAL EFFECTS IN CONNECTION WITH THE USE OF SEWAGE SLUDGE AND LIQUID AGRICULTURAL WASTES

Chairman : H. HAENI (CH)

Rapporteur : J.E. HALL (UK)

ENVIRONMENTAL EFFECTS OF LONGTERM SLURRY AND SEWAGE SLUDGE APPLICATION: REASONS AND MEASURES

A.F. SIEGENTHALER and W. STAUFFER

Swiss Federal Research Station for Agricultural Chemistry and Hygiene of Environment, 3097 Liebefeld - Berne

Summary

Due to industrial animal husbandry on the basis of purchased feed and a large production of sewage sludge in urban areas very large amounts of slurry and sewage sludge are often applied to a limted surface. This can lead to nutrient as well as pollutant accumulation in the soil. Enrichment of specific elements could lead to an unbalanced nutrient supply to plants and soil microorganisms and even to toxic concentrations.

In long-term experiments with excessive slurry and sewage sludge application to various crops (rotation: silage maize, cereals, grass-clover mixture and permanent grassland) effects of physical, chemical and biological characteristics on the soil, on yield and quality of the plants were studied. Up to the present, plant yields revealed that high application rates of sewage sludge and pig slurry are tolerated by silage maize and grass, but not by clover. With high application rates of pig slurry and sewage sludge, the P-balance is positive, which leads to a high P-accumulation mostly in the uppermost soil horizon. With high loads of sewage sludge and pig slurry the zinc (Zn) and copper (Cu) content of the soil increased. The mercury (Hg) content is influenced only by excessive sewage sludge application. Possible reconstruction measures were conidered.

1. INTRODUCTION

The quantity of sewage sludge and pig slurry has increased in Switzerland in the past twenty years. Because of the large amounts of plant nutrients and organic matter, recycling through fertilization must be considered. The recycling of these large quantities of nutrients in agriculture presents increasing difficulties on the regional level because the surface area necessary for this is often unavailable and application is limited to certain times. This is the case because not enough suitable surface area is available for best utilization with respect to crop production (as in the case of pig slurry) or because unlimited, large quantities cannot be applied (as in the case of sewage sludge).

Individual farms with large feed purchases are often confronted with the problem of over-fertilization of their fields, because intensive pig breeding is mostly based on commercial feed.

Community wastes are found mostly in urban areas and because of the absence of agricultural land, this can lead to a waste disposal problem in these areas.

Over-fertilization with both of the above-mentioned substances can lead to a loading of surface waters with phosphate (Stadelmann, 1984) as well as to significant changes in soil characteristics, especially enrichment of P, heavy metals and organic pollutants, as well as to leaching of nitrate into ground water.

During the past few years the solution to these problems has been dealt with through the adoption of various measures (establishment of maximum animal populations by law, enlargement of storage capacity, research and consulting in the areas of storage and utilization of farmyard manure and community wastes, the improvement of sewage sludge quality). In spite of these efforts, the problem of over-fertilization of agricultural land in various parts of Switzerland (and other european countries) could not be sufficiently reduced. The question of the consequences for man and the environment has lost nothing in it's urgency.

In order to shed some light on the effects of too much fertilization with sewage sludge and pig slurry, a soil stress experiment was set up in 1976 at the Federal Research Station in Liebefeld. Through extensive experimentation, the effects of large applications of pig slurry and sewage sludge on soil, plants, and water would be investigated.

2. MATERIALS AND METHODS

2.1 Fertilization plan and application

Relatively large quantities of organic matter are applied to fields through fertilization with sewage sludge and pig slurry. Therefore, an important question would be how the resulting large quantities of organic matter from over-fertilization affect soil and soil organisms? For this reason the doses of sewage sludge and pig slurry were based on quantities of organic matter:

2 t organic matter per ha and year as a rather light over-fertilization and 5 t per ha and year as a hardly and extreme over-fertilization. In comparison, a control without any fertilization as well as a treatment with standard mineral fertilization were established.

These considerations lead to the following six treatments:

1. 0 = Control
2. Min = Standard mineral fertilization (reference treatment)
3. SS_2 = 2 t org. mat. per ha and year as liquid sewage sludge (approx. 70 m^3)
4. SS_5 = 5 t org. mat. per ha and year as liquid sewage sludge (approx. 180 m^3)
5. PS_2 = 2 t org. mat. per ha and year as pig slurry (approx. 70 m^3)
6. PS_5 = 5 t org. mat. per ha and year as pig slurry (approx. 180 m^3)

The composition of sewage sludge and pig slurry is not constant and significant fluctuations appear from sample to sample (Table 1). If the level of normal fertilization was not achieved with the applied quantities then supplements in the form of mineral fertilization followed. This was frequently necessary for potassium whereas the content of P in the normal application of 50 kg/ha usually was acceptable. In the case of nitrogen, only the short-term effective nitrogen was taken into account, and, when necessary supplemented.

Table 1: Annual nutrient application (Average)

Fertilization Plan		0	Min	SS_2	SS_5	PS_2	PS_5
Applications	org. matter planed	0	0	2000	5000	2000	5000
(kg/ha a)	applicated	0	0	1984	5004	2173	5033
	Kjeldahl-N	0	220	232	418	369	714
	org. N	0	0	110	280	66	182
	NH_4-N	0	0	58	138	207	490
	P	0	49	163	423	62	138
	K	0	250	253	257	252	289
	Ca	0	212	435	1012	89	182

SS_2 = Sewage sludge 2t org. mat. per hectare and year, SS_5 = Sewage sludge 5t org. mat. per hectare and year, PS = Pig slurry.

2.3 Soil Conditions

Liebefeld, average soil with 16% clay, pH 6.4, 2.2% C_{org}. In order to invetigate the effect of soil type, the experiment was set up at two other location on light (with 5% clay) and heavy soil (with 40% clay). Results of this paper are in relation only with the experiment Liebefeld.

2.4 Weather Conditions

Weather data for the experimental location is shown in Table 2.

2.5 Crops

The experimental plots were divided into 4 sub-plots. The sub-plots A, B, and C were planted with a 3-year crop rotation (silage maize, cereal and grass-clover). Sub-plot D was reserved for permanent grassland (perennial grass-clover mixture with white clover, orchard grass, English and Italian ryegrass).

After harvest of the cereal (winter wheat or dinkel), an annual grass-clover mixture was planted every August (red clover, white clover and Italian ryegrass), which was cut in autumn.

3. RESULTS AND DISCUSSION

3.1 Yields

Silage maize:

The smaller applications of sewage sludge and pig slurry lead to the yield level achieved with standard mineral fertilization. The larger applications of sewage sludge and pig slurry (by a factor of 2.5) resulted in a slight yield increase of about one ton per hectare (see Figure 1).

Grass-Clover:

The effect of fertilization on botanical composition is noteworthy.

Table 2: The average weather data for the experimental location Liebefeld.

Precipitation (mm/a)	1141
Air temperature (Annual average, ^{o}C)	8.0
Sunshine (h/a)	1446

The percentage of clover in the mixture was largest when fertilization (mostly N) was at a minimum. The control plot produced a considerable dry matter yield because of a large percentage of clover and the P and K reserves in the soil at the beginning of the experiment. An increase in the applications of sewage sludge and pig slurry resulted in a decrease in the percentage of clover, but heavily stimulated the growth of grasses. The total yield was always highest in the case of the highest fertilization.

Permanent grassland:

The botanical composition of the pasture changed quickly under the influence of fertilization. The large percentage of clover (30 - 60%) in the control plots was remarkable. In fertilized plots, however, this was mostly very small. The percentage of herbs in fertilized plots was between 10-20%, whereby no clear effect of fertilization was observed. In the case of satisfactory botanical composition, the dry matter yield was highest in the treatment with high pig slurry application.

Figure 1: Dry matter yield for silage maize (total plant), grass-clover and permanent grassland (1988).

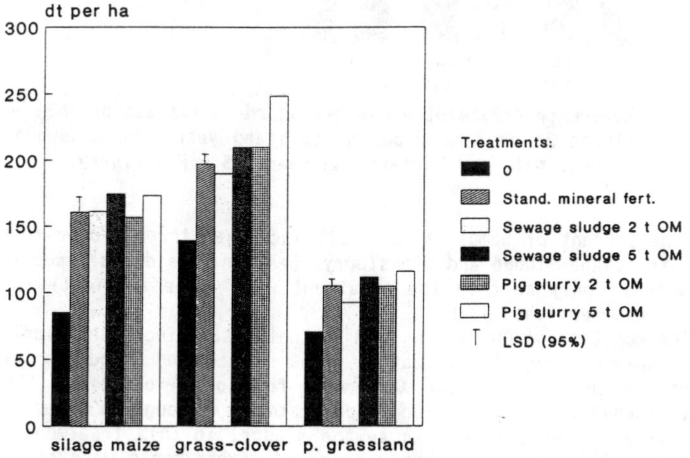

Reference treatment = mineral standard fertilization, SS_2 = Sewage sludge 2t org. mat. per hectare and year, SS_5 = Sewage sludge 5t org. mat. per hectare and year, PS = Pig slurry.

Wheat:

Grain yields in the control plots were surprisingly high. The absence of fertilization, however, had a clear effect on straw yield.

The weight of 1000 grains was highest in the control treatment. No difference between the other treatments was observed.

3.2 Soil Analysis

As shown in Table 3, 11 years of high or excessive applications of sewage sludge lead to an increase in pH, pig slurry applications on the other hand, to a decrease in pH. The increase or respective decrease depended on the quantities of sewage sludge or pig slurry. It must be taken into account that the pH value of the soil decreased in the case of no fertilization as well (see control treatment).

Figure 2: Yield parameters for wheat (1987)

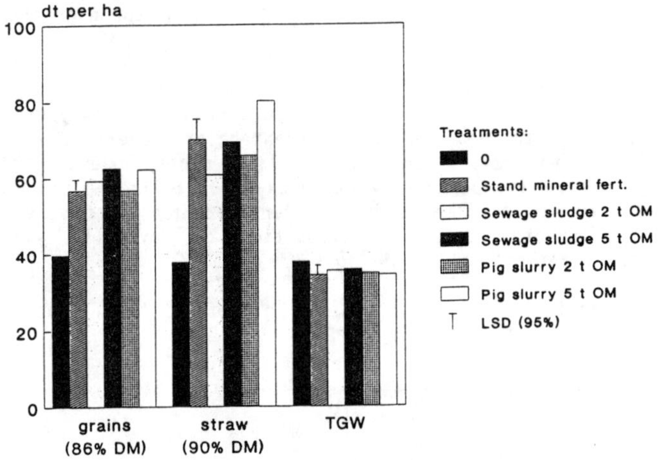

Reference treatment = min. standard fertilization, SS_2 = Sewage sludge 2t org. mat. per hectare and year, SS_5 = Sewage sludge 5t org. mat. per hectare and year, PS = Pig slurry.

The content of humus in the soil increased through over-fertilization with sewage sludge and pig slurry. Despite considerable supplementation with P, only soil in the treatment with large applications of pig slurry became enriched in P.

The contents of Cu, Zn, Hg and Pb in the soil increased significantly with sewage sludge fertilization. The recommended standards for total element contents set in the Ordinance for Soil Protection (1986) was almost reached with sewage sludge application. Although Zn input was much less with pig slurry than with sewage sludge, in this treatment, standards for soluble Zn were clearly exceeded. Experiments with celery root at the Federal Research Station in Liebefeld showed that there is no relationship between total heavy metal content in the soil and the heavy metal content in the edible part of the root. On the other hand, the zinc content in the plant correlated well with the content of soluble zinc (Stadelmann, 1987).

The altered heavy metal solubility in soils can essentially be explained by the cation exchange capacity (CEC), which can be varied by pH and the quantity of organic matter.

Table 3: Effects of long-term (1976-1986) increased sewage sludge and pig slurry application on soil characteristics (0-20 cm) and heavy metals.

Fertilization plan	0	Min	SS_5	PS_5	guidevalues[3]
$pH(H_2O)$	5.2^{**}	5.7	6.4^{***}	5.2	
org C (%)	1.4	1.6	2.0^{***}	1.8^{*}	
P-Test value[1]	6.9^{***}	23.5	20.7	94.0^{***}	
K-Test value[2]	1.2^{***}	5.0	4.1	2.9^{***}	
Cu (mg/kg d.m.)	18.2	18.5	36.9^{***}	24.5^{***}	50
Cd (mg/kg d.m.)	0.25	0.25	0.7^{***}	0.25	0.8
Pb (mg/kg d.m.)	21.3	21.3	48.2^{***}	23.5	50
Zn (mg/kg d.m.)	47.7	48.4	150.9^{***}	68.3^{***}	200
Zn soluble (mg/kg d.m.)	0.7^{***}	0.3	0.3	2.0^{***}	0.5

* = 0.05, ** = 0.01, *** = 0.001, Reference treatment = mineral standard fertilization, SS_5 = Sewage sludge 5t org. mat. per hectare and year, PS = Pig slurry.

1 P-Test value: 8.0-16.0 sufficient, 16.1-32.0 Storage, >32.0 enriched

2 K-Test value: 2.0-4.0 sufficient, 4.1-8.0 Storage, >8.0 enriched

3 Guidevalues in Swiss ordinance on soil polution

3.3 Soil Organisms

Fertilization of the soil with organic matter leads to an inreased quantity of aerobic bacteria and actinomycetes. The differences are not only very significant with pig slurry, but also in the case of sewage sludge (P = 0.01). This increase in the titer of heterotrophic, mostly saprophytic soil microorganisms (bacteria, actinomycetes, filamentous fungi) may be primarily due to the input of organic matter contained in sewage sludge and pig slurry.

The number of autotrophic soil algae was significantly decreased by sewage sludge application (P = 0.05). This decrease is possibly due to an altered micro-climate in the surface layer of the soil. Occasional coverage of the soil surface with dark sewage sludge can cause an increase in temperature and hinder the penetration of sunlight into the surface layer. It is also conceivable that this phenomenon has something to do with the negligible content of inorganic nitrogen in sewage sludge.

Enzymatic activities (catalase, alk. phosphatase) were higher than in mineral fertilized or unfertilized soils not only for sewage sludge but also for pig slurry. Also, an increased C and N mineralization rate was determined in organically fertilized soils. These observations are largely in agreement with those of comparable investigations.

Table 4: Influence of fertilization with sewage sludge and pig slurry on selected microbiological characteristics in soil of permanent grassland (Means from eight sampling dates and four replications).

Fertilization plan	0	Min	SS_5	PS_5
Heterotrophic MO				
Aerobic bacteria (10^6/g soil d.m.)	93	118	232**	171**
Actinomycetes (10^3/g soil d.m.)	1187	1285	2601**	2014**
Filamentous fungi (10^3/g soil d.m.)	190	174	225*	177
Autotrophic MO				
Algae (10^3/g soil d.m.)	1832	2923	1467*	2596
Enzymatic activities				
Catalase Units	12.5	12.4	17.8*	21.1**
Mineralization				
C-Mineralization (10^{-6}g CO_2-C/g soil-dm/24h)	36.8	34.5	70.4*	45.7
N-Mineralization (10^{-6}g NH_4-N/g soil-dm)	102	96	112*	120**

* = 0.05, ** = 0.01 * = 0.05, Min = Reference treatment = mineral standard fertilization, SS_5 = Sewage sludge 5t org. mat. per hectare and year, PS = Pig slurry.

4. CONCLUSIONS

In order to insure their income, many farmers in Switzerland have been induced to increase their numbers of livestock during the past 10 to 30 years. This increase - mostly of pigs - has lead to an increased concentration of livestock on individual farms and also in some entire regions.

In 1988, about 90% of all Swiss residents were connected to sewage treatment plants (STP). The total sewage sludge quantity from 900 facilities came to about 4 million m^3. Approximately one half of that quantity is utilized in agriculture. For financial reasons and because of limited reserves of raw materials, this is worthwhile. This way, about 2000 tons of phosphorus (P) as well as large amounts of nitrogen, organic matter and lime are returned to agriculture through fertilization.

Industrial animal husbandry on the basis of purchased feed, together with a large production of sewage sludge in urban areas, has resulted in the application of very large amounts of slurry and sewage sludge to surface areas of limited size. This can lead to nutrient as well as pollutant accumulation in the soil. Enrichment of specific elements could lead to an unbalanced nutrient supply to plants and soil microorganisms and even to toxic concentrations. Higher concentrations in soils increase the risk of contamination of ground and surface waters. Over-fertilization is, however, not harmless and short- or long-term negative effects on soil, plants and waters cannot be ruled out.

In a longterm experiment with high slurry and sewage sludge application the effects on physical, chemical and biological characteristics of the soil on yield and quality of the plants as well as nutrient leaching were studied.

Up to the present, plant yields revealed that high rates of sewage sludge and pig slurry application are tolerated by silage maize and grasses, but not by clover. Clover disappears almost completely, whereas grass and silage maize show increases in yield.

With high rates of application of pig slurry and sewage sludge, the P-balance is positive, which leads to a high P-accumulation mostly in the uppermost soil horizon. On steep slopes, erosion can lead to the runoff of enriched material which, in turn, can cause an undesirable contamination of surface water. Also, with high loads of sewage sludge and pig slurry, zinc (Zn) and copper (Cu) contents increase as well as mercury (Hg) contents in the case of over-application of sewage sludge.

Microorganisms are obviously capable of dealing with even excessive applications of organic fertilizers. Bacteria and actinomycetes are strongly stimulated by over-fertilization of soils with organic wastes, while mites and, to some extent algae, are strongly suppressed. The problem of over-fertilization is not so much a disruption of soil microbiological activity, but rather the consequence of an over-activity of soil microorganisms (e.g. nitrate leaching, gaseous N-losses).

The following possible reconstruction measures can be considered:

- A limitation in animal densities down to three large animal units (LU), as this is laid out in a revision of the water protection law.
- Reduction of the number of animals to an acceptable level with respect to plant production.
- A reduction in element contents in animal feed for pigs and cattle to that which is really necessary.
- Single farm balance sheets for individual elements with the goal of a correct utilization of farmyard manure, sewage sludge and commercial fertilizers.
- Manure utilization distributed among different farms.

5. REFERENCES

FURRER, O.J., STADELMANN, F.X. und LEHMANN, V.: Die Wirkung steigender Gaben von Klärschlamm und Schweinegülle in Feldversuchen. I. Versuchsprogramm und Auswirkungen auf den Pflanzenertrag. Schweiz. Landw. Fo. 21 (3/4), 1982.

STADELMANN, F.X.: Die Wirkung steigender Gaben von Klärschlamm und Schweinegülle in Feldversuchen. II. Auswirkungen auf Population und Aktivität von Bodenmikroorganismen. Schweiz. Landw. Fo. 21 (3/4), 1982.

STADELMANN, F.X., FROSSARD, R. FURRER, O.J., LEHMANN, V. und MOERI, P.B.: Wirkung und Nachwirkung langjähriger hoher Klärschlamm- und Schweinegüllegaben auf die Qualität von Knollensellerie. VDLUFA-Schriftenreihe, 23, Kongressband 1987, 857-882, 1988.

ECOLOGICAL EFFECTS OF AMMONIA DEPOSITION

N. van Breemen
Department of Soil Science and Geology,
Agricultural University,
POB 37, 6700 AA Wageningen, The Netherlands

ABSTRACT

Atmospheric deposition of inorganic N, mainly ammonium volatilized from manure produced in intensive stock-breeding, on sensitive terrestrial and aquatic ecosystems in The Netherlands is in the order of 40 to 80 kg/haJyr. This deposition causes (i) eutrophication with N, leading to floristic changes (ii) acidification of base-poor sandy soils and of moorland pools, leading to higher concentrations of dissolved, potentially toxic metals such as Al3 +, and (iii) increased levels of nitrate in groundwater below woodlands. In acid forest soils, but not in soils under heathland, nitrification and leaching of nitrate is common. However, in very poor sandy forest soils and at very high ammonium inputs, nitrification may be too slow to prevent the development of high concentrations of ammonium. Both excessive acidification and excessive levels of ammonium probably play an important role in the general forest decline, which is most severe in the southern and central parts of the country, where ammonium inputs are highest.

INTRODUCTION

This paper is based largely on an article by Van Breemen and Van Dijk (1988) which has been updated for the purpose of this publication.

The estimated atmospheric inputs of NHx and NOx to terrestrial and aquatic ecosystems reported in The Netherlands are amongst the highest in the world. There is direct evidence for enhanced N deposition as a major cause of certain changes observed in those ecosystems, notably acidification, increased availability of N, and increased predominance of nitrophilous plants at the expense of plants tolerant of low N supply. Most probably, enhanced N deposition is also implied in the general decline of the Dutch forests. This review describes the levels of atmospheric deposition of N in The Netherlands and its distribution, as well as the abiotic changes in soils, groundwater and soft water pools resulting from enhanced N inputs. Next the biotic changes possibly related to excess N deposition and evidence for causal links between these biotic changes and excess N deposition will be discussed.

DEPOSITION OF N IN THE NETHERLANDS

In 1989 mean wet plus dry deposition of N was about 50 kg/haJyr, of which about 35 % was NOx (standing for NO3-, HNO3 plus all nitrogen oxide gases) and the rest was NHx (standing for gaseous NH3 plus NH4+) (Van Aalst and Erisman, 1990). Dry deposition on exceeded wet deposition and made up of the total deposition (Table 1).

According to Asman et al. (1987), the total deposition of NHx in The Netherlands was about 27 % of the present-day value in 1870, 37 % in 1923, and 50 % in 1950. These changes illustrate the rapid increase in NH4+ deposition over the past decades, concurrent with the increase in intensive stock- breeding, which is by far the dominant source of atmospheric NHx. The increase in deposition of NHx is probably fairly similar to that of SOx and of NOx over the same period (Asman et al., 1987).

The deposition of NHx on forests is higher than the average deposition in the country due to (i) increased dry deposition of NHx on the canopies of trees and heather, (ii) the proximity of important sources of NH_3 (intensive animal production units) to most woodlands. Moreover, those sources of NH_3 are situated mainly in areas with poor sandy soils, with a low acid buffering capacity and, originally, low levels of available N (Figure 2). The total (dry + wet) deposition on ecosystems in areas with a relatively high atmospheric pollution is probably reflected reasonably well by the canopy throughfall. Table 2 shows that the annual throughfall fluxes of NH_4^+ and SO_4^{2-} in woodlands are considerably higher than the bulk deposition in open collectors adjacent to the woodlands, and are also higher than the country-wide total deposition of NHx and SOx in Table 1. Bulk deposition in the open field as well as throughfall deposition of NO_3^- are lower than the estimated total deposition of NOx according to Table 1. Throughfall fluxes of NH_4^+ and SO_4^{2-} exceed bulk deposition in the open field by a factor of almost 2 in heathland, by 2 to 3 times in deciduous woods, and by 3 to 5 times in coniferous woods (Table 2). For NO_3^- those factors are closer to 2 in forests, and close to 1 in heathland.

The proximity of poor sandy soil areas with wood- and heathland to intensive stock-breeding farms is related to the history of the land use (Figure 2). Particularly in such areas, smallholders used to eke out a living by extracting soil fertility (sods and manure from sheep and cattle) from a large areas of heathland and woodland before the advent of fertilizers. In the past decades they have managed to secure a reasonable income by turning to intensive stock-breeding. In addition to wood- and heathland, other acid-sensitive ecosystems found in those areas are isolated moorland pools.

ABIOTIC CHANGES IN ECOSYSTEMS AS A RESULT OF EXCESS N DEPOSITION

- SOILS

Enhanced N deposition may lead to an increase in the pool of organic N in the soil system, and in the availability of N for plants and soil biota. Data on N accumulation in forest ecosystems in The Netherlands under the influence of atmospheric deposition are fragmentary. Winkels (1985) estimated accumulation rates of organic N in the forest floor alone, to be about 17 to 28 kg/haJyr (between 1960 and 1985) in the oak-birch woodlands near Warnsveld, where the present-day throughfall inputs of N are 60 kg/haJyr (see Table 2). Soil solutions and aqueous extracts of acidic forest soils in The Netherlands generally indicate very high levels of dissolved nitrate, from 1.2 to 4.2 mmol/l in the surface soil to 0.2 to 6.5 mmol/l below the root zone (mean annual values for sixteen sites, Van Breemen and Verstraten, 1990). Reference levels from the times that atmospheric inputs were lower are lacking. From comparison with literature data on similar soils in less polluted areas, as well as from chemical composition profiles in groundwater under woodlands, however, it is obvious that these levels have increased dramatically over the past decades. High nitrate concentrations are associated with high concentrations of Al^{3+} (10-20 mg/l). In many cases, Al^{3+} concentrations are more strongly correlated with concentrations of NO_3^- than of SO_4^{2-}. Laboratory studies, as well as field studies involving ^{15}N, showed that nitrification rates in acid forest soils (pH 3.5) were sufficiently high to account for transformation of essentially all atmospherically deposited NH_4^+ to HNO_3 (Marnette, 1987; Stams and van Breemen, unpublished). Autotrophic nitrification is widespread in acidic (pH3-4) forest and heathland soils (De Boer et al., 1989). Uptake of NH_4^+ and or nitrification of NH_4^+ to NO_3^- (Van Breemen et al., 1987; Van Dobben and Mulder, 1988) are now the dominant sources of H^+ in Dutch forest soils. As a result of a combination of former litter removal and of atmospheric deposition of S and N over the past century, most of these soils are so strongly acidified that the base saturation is generally below 10 %.

Roelofs and co-workers observed NH_4^+/NO_3^- molar ratios in aqueous extracts of most surface soils of forests generally exceeding 1, suggesting little nitrification. The fact that Roelofs et al. find a dominance of NH_4^+ while the others observe mostly NO_3^-, is probably due in part to differences in methodology and partly to true differences between the sites studied by the various groups. Aqueous extracts have been shown to give higher NH_4^+ concentrations due to preferential desorption of NH_4^+ during dilution and to release of NH_4^+ from organic N. (Van Breemen and Verstraten, 1990). There is some evidence that nitrification is inhibited under conditions of particularly high NH_x deposition and very poor soil conditions. Inhibition of nitrification at high levels of NH_4^+ in acid soil was observed in laboratory incubation studies (Van der Maas, 1985).

- GROUND WATER

Aquifers underlying the Pleistocene sandy areas in The Netherlands are traditionally important for collection of high- quality drinking water. Ground water in these areas has become increasingly polluted with nitrate, not only from arable land and pastures but also from forest areas. Nitrate concentrations in the molar range , found below the root zone of forest stands, are now frequently observed also in the upper meters of the underlying aquifers (Appelo, 1985 ; Oosterom and Van Schijndel, 1979; Verheyen, 1985 ; Krajenbrink, 1987). Much lower values are generally reported under heathland. Occasionally, high concentrations of NH_4^+-N (up to 20 mg/l) have been observed below the root zone of Scots pine stands in areas heavily impacted by atmospheric inputs of NH_x (Smeets, 1988), indicating that locally breakthrough of ammonium to the groundwater may be imminent.

- SOFT WATER POOLS

In the middle, southern and eastern part of the country and in the coastal region many small (< 0.1-5 ha) isolated pools with poorly buffered water are found. They are situated on highly quartzitic, sandy, Pleistocene deposits. Originally, the pool-waters were only slightly acid (pH 5-7) with small amounts of bicarbonate ($\leq$ 1 mmol (±)/l). Most of them receive water mainly from precipitation falling directly on their surface, while water stagnates on an impermeable or slowly permeable layer.

The majority of these pools are now acidic (Leuven et al., 1986c): 65% have pH values below 5.0, and 35% below 4.0. Van Dam (1987), and Schuurkes & Leuven (1986) showed that most pools have acidified in the past half century. The pH decline generally lies between 0 and 2 units. Sulphuric acid, partly from oxidation during very dry years of (originally atmospherically derived) S stored as sulphide and organic S in bottom sediments, partly derived from dry and wet deposited ammonium sulfate after assimilation of ammonium, is the major cause of water acidification (Van Dam, 1987; Schuurkes, 1986). The present-day contribution of NH_4^+ to the potential acidifying effect of wet deposition is more than 80% in the southern part of the country and approximately 60% in the north (Schuurkes, 1986).

BIOTIC CHANGES POSSIBLY RELATED TO EXCESS N DEPOSITION

The dominant ecosystems that are potentially sensitive to adverse effects of atmospheric N and S deposition are situated mainly in the areas with base-poor, sandy soils in the central, eastern and southern parts of the country. They include forests, heathlands and isolated soft water pools.

- FORESTS

The condition of the Dutch forests is alarming. In 1987 only 43% of the stands were vital and 21% were hardly vital or not vital (Anonymous, 1987). Air pollution is generally accepted as a major cause. The observed geographical pattern of damage does not fit in very well with the occurrence of well-known pollutants like O3, SO2 (Den Boer, 1986) and NOx. The greatest problems occur in areas with intensive stock-breeding, indicating that deposition of NH3 is an important factor.

Symptoms shown by the trees include considerable loss of needles or leaves and discolourations. Pine trees are often infected by the parasitic fungi *Brunchorstia sp.* and *Sphaeropsis sapinea* (Anonymous, 1987). *Brunchorstia* causes loss of mature, green needles, *Sphaeropsis* prevents the maturation of young shoots. *Sphaeropsis* can enter the branches and stems, causing the trees to die (Smits, 1986). In the southern part of the country complete stands of Corsican and Austrian pine have been killed by this parasite (Anonymous, 1985).

Numbers of both species and fruiting bodies of mycorrhizal fungi have considerably decreased in Dutch forest during the last decades (Arnolds, 1985, Arnolds & Jansen, 1987). The decline is most pronounced in coniferous stands and oak forest on dry, acidic, sandy soils poor in humus and N. Terrestrial saprophytic fungi, too, are disappearing from these habitats. Saprophytic fungi living on humus and nitrogen rich soils have increased, however, as have saprophytic and parasitic fungi growing on wood, especially in the southern part of the country where N and S deposition are highest.

The herb layers show a general shift towards a species composition typical for N-enriched sites (Ellenberg Sr., 1979). Species like *Dryopteris dilatata, Corydalis claviculata, Galeopsis tetrahit* and *Rubus spp.*, are becoming increasingly abundant. Plants like *Solanum nigrum, Senecio vulgaris* and *Stellaria* media, formerly confined to arable land, are now frequently observed in the forests. Also the grasses *Molinia caerulea* and *Deschampsia flexuosa* are expanding strongly. Similar changes have been reported from other European countries (Ellenberg Jr., 1985, Tyler, 1987). Several species of epiphytic lichens, that had become rare in the past decades, are spreading again. These species are considered to be nitrophilous (Van Dobben, 1987).

- HEATHLANDS

Grasses have expanded in heathlands at the expense of the heather species *Calluna vulgaris* and *Erica tetralix*. Most important in this respect is again *Molinia caerulea*, followed by *Deschampsia flexuosa* and *Festuca ovina*. In 1980, 20% of the Dutch heathlands were covered exclusively by grasses and another 25% had equal coverage of grasses and heather. Only 55% consisted of pure heather vegetation (Schneider & Bresser, 1987). Expansion of grasses continued during the past 7 years (Table 3).

Many other plant species that are characteristic for dry and wet heathlands are becoming increasingly rare too. In the past such decline could be ascribed to reclamation and hydrological measures. Nowadays most plants are confined to nature reserves and their decline is probably due mainly to air pollution (Van Dam et al., 1986; Roelofs, 1986).

- SOFT WATER POOLS

Dramatic vegetational changes have been observed in small soft water pools during the last decades. Plant communities of low productivity, involving isoetid species such as *Littorella uniflora, Lobelia dortmanna, Isoetes lacustris* and *I. echinospora*, and many other soft water plants, that used to dominate vegetation in these pools, have almost completely vanished (Roelofs, 1983, Arts, 1987). Arts (1987) reported a striking similarity between patterns of disappearance of plants and increase in acid deposition. In many pools the species that have disappeared have been replaced by nitrophilous and acid- tolerant species like *Sphagnum spp., Drepanocladus fluitans, Juncus bulbosus. Agrostis canina*, has spread into the water from the banks (Roelofs, 1983, Arts, 1987). Submerged macrophytes have disappeared completely from some pools.

Microflora diversity has strongly decreased and a general shift towards acidobiotic species has been observed (Van Dam & Beljaars, 1984; Geelen & Leuven, 1986). Microfauna is affected in a similar way (Geelen & Leuven, 1986). Some groups of invertebrate macrofauna have declined strongly in acidified pools, notably *Ephemeroptera, Nematocera, Trichoptera* (Leuven et al., 1986d, Leuven et al., 1987a), *Chironomidea* (Buskens et al., 1986; Leuven et al., 1987b), and *Tricladida* (Schuurkes & Leuven, 1986). Others, like *Gastropoda*, freshwater *Bivalvia* and some groups of *Crustacea* disappear completely when the pH drops below 5 (Schuurkes & Leuven, 1986). Fish and amphibians have been affected too (Leuven et al., 1986a). On the other hand numbers of species or individuals of *Heteroptera, Coleoptera, Collembola*, and larvae of *Megaloptera, Diptera* and *Odonata* (Leuven et al., 1986d, Schuurkes & Leuven, 1986, Verbeek et al., 1986) apparently have become more abundant as a result of acidification. Particularly the rarer species disappear, while commoner ones remain. An even distribution of many species is replaced by dominance of a few acid-tolerant and nitrophilous organisms. As a result dissimilarity between different pools is greatly reduced.

CAUSAL RELATIONS BETWEEN BIOTIC CHANGES AND EXCESSIVE N-DEPOSITION

- FORESTS

Excessive N deposition may interfere with tree vitality in several ways. Soil pH and Ca/Al ratios in soil solutions are often in the range described as dangerous or harmful by (Rost-Siebert, 1985; Murach, 1984). Smit et al. (1987a,b) observed a significant negative correlation between the Ca/Al ratio at the adsorption complex and root development of naturally propagated Douglas fir seedlings in six Dutch stands. Greenhouse work confirmed the hypothesis that high concentrations of dissolved Al depresses growth of 1-year old Douglas fir seedlings. Stronger growth reduction in plants fed with NH4+, compared to NO3--fed plants, could be attributed to extra soil acidification (and related Al mobilization) in the rhizosphere of NH4+-fed plants (Fig. 1). On the other hand, vitality of 30- to 50 year old Douglas stands was not correlated at all with concentrations of Al3+ and NO3- in the soil solution, while some relatively healthy stands showed very high levels of those solutes (De Vries and Zuidema, Soil Survey Inst., Wageningen, pers. commun.). Also Van Lier & Smeets (1985) did not find clear correlations between Al and Ca/Al ratios and the vitality of Scots pine. A problem with interpreting such findings is that, while high levels of Al may tend to hamper tree growth and vitality, well-growing trees can be expected to transpire more water, and so cause higher evaporative concentration of the soil solution, with concomitant higher solute concentrations.

Roelofs et al. (1985), Roelofs (1986) and Boxman et al. (1987b) observed negative correlations between the condition of Corsican pine, Douglas fir and native oak and the ratios of NH4+ to K+, Mg2+ and Ca2+ in aqueous soil extracts. This correlation suggests that disturbed nutrient balances may be responsible for the decrease in tree vitality. The affected trees exhibit discolourations and premature shedding of the older needles, low leaf or needle densities and bare branches. Needles and leaves contain high N levels and low amounts of Ca, Mg and particularly K and Mg.

Hydroculture experiments with two-year old Corsican pine trees demonstrated that high NH4 to base cations ratios impair uptake of base cations (Boxman & Roelofs, 1986). When the NH4+ to K+ ratios exceed 50 even a loss of Ca and Mg from the tree roots is observed. High NH4+ concentrations exert also a negative effect on the NO-3 uptake, but enhance uptake of phosphate.

Indirect impacts of $NH4+$ on plant nutrition include enlarged shoot to root ratios and effects via mycorrhizas. Container-grown seedlings of pines, Douglas fir and spruce showed a negative correlation between amounts of $NH4+$ applied and mycorrhiza formation (Van Dijk, unpublished results, Meyer, 1986, Van der Eerden & Wit, 1987). Mycorrhizal fruiting bodies have indeed declined recently in the southern part of the country (Arnolds, 1985, Arnolds & Jansen, 1987).

Nutrient deficiencies do not only arise from impaired uptake from the soil. Experiments with excised pine needles demonstrate that $NH4+$ uptake by the needles involves increased losses of other cations, like $K+$, $Mg2+$ and $Ca2+$, thus furthering the development of nutrient deficiencies (Roelofs et al., 1985). Electron microscope studies reveal that $(NH4)2SO4$, like gaseous $NH3$, affects the cuticular wax layer of Douglas fir needles (Van der Eerden & Wit, 1987). Reddish-brown discolouration of conifer needles near strong sources of atmospheric $NH3$ in winter time (Janssen, 1982) has been attributed to direct damage by $NH3$, when the $NH3$-detoxifying capacity of the trees is lowered by low temperatures (Van der Eerden, 1982). Scots pine trees in the so- called 'Peel'-area in the southern part of the country, showing yellowing of the young needles, have high levels of total N in the needles and a considerable part of this is stored in free amino acids, predominantly arginine (Van Dijk & Roelofs, 1987). Krauss et al. (1986) found that accumulation of soluble N compounds is connected with a strong growth decline. Infection by the parasitic fungus Sphaeropsis sapinea may be enhanced by imbalances in nutrition. Needles of infected Corsican pine trees had higher N levels than those of uninfected trees (Roelofs et al., 1985). Because N contents of soils under infected stands are relatively high too, the high N content in the needles is presumably the cause, and not the result of the fungal infection (Van Dijk, unpublished results). The increased N availability also stimulates nitrophilous grasses and herbaceous plants, particularly in forest stands with decreased vitality, where canopies have opened up. The renewed spreading of some epiphytic lichens is ascribed to the neutralizing influence of $NH3$, counteracting the acidification of tree bark by $SO2$ (Van Dobben, 1987).

- HEATHLANDS

The change of heathlands into grasslands is undoubtedly caused by an increased availability of nutrients in the soil (De Smidt, 1983; Roelofs et al., 1984a), which in turn is the result of an increased nutrient input from atmospheric deposition and decreased output due to decreased utilization of heather. Fertilization experiments confirm that high N levels favour grasses at the expense of heather species (Heil & Diemont, 1983, Berendse & Aerts, 1984 and Heil & Bruggink, 1987). Mass development of grasses occurs at N inputs exceeding 20 kg/haJyr (Roelofs, 1986). *Calluna* and *Erica* may survive at less fertile places, where the competitive ability of the grasses is small (Berendse & Aerts, 1984; Berdowski et al., 1987).

The heather beetle (*Lochmaea suturalis*) further contributes to the observed vegetational changes (Brunsting & Heil, 1985). This insect severely damages *Calluna* stands. Mineralization of dead plant material, beetles and faeces supplies considerable amounts of N during the first years after an infestation (e.g. 24.3 kg/ha.a; Heil & Diemont, 1983). Grasses in the understorey of closed heather stands may take over dominance when, due to a heather beetle infestation, the canopy opens up, taking advantage of the rise in nutrient availability (Brunsting & Heil, 1985 and Berdowski & Zeilinga, 1987). Berdowski & Zeilinga (1987) report an increased abundance of *Deschampsia flexuosa* at the expense of *Calluna vulgaris* after an infestation.

By increasing the biomass, the litter production and N content of the heather leaves, atmospheric N inputs stimulate growth and reproduction of the beetles (Brunsting & Heil, 1985, Schildwacht & De Smidt, 1985). Vegetative regeneration is impaired after an attack of the heather beetle, because the increased supply of N depletes carbohydrate reserves (Berdowski & Siepel, 1987). Little is known about possible $NH4+$ toxicity or effects of nutrient imbalances on heather plants or on mycorrhizas associated with heather.

Many rare plant species, of heathland ecosystems are adapted to slightly acidic soils of low N content. Due to high NH4+ deposition, followed by nitrification and acidification, these plants tend to disappear from the heathland ecosystem.

- SOFT WATER POOLS

An application of (NH4)2SO4 exceeding 20 kg/haJyr of N to artificial moorland pools brought about vegetational changes similar to those observed in the field (Schuurkes et al., 1986a). Ammonium treated *Littorella uniflora* produced a high biomass (but few flowers), was infected by fungi' and retreated to the banks. *Luronium natans* did not flower at all and its leaves turned yellow. *Pilularia globulifera* disappeared completely. *Juncus bulbosus*, *Sphagnum spp.* and *Agrostis canina*, however, strongly increased in biomass production both above and below the water surface.

Laboratory experiments studies indicate, that changes in inorganic carbon and nitrogen availabilities are responsible for these changes (Roelofs et al., 1984b, Schuurkes et al., 1986b).

A cleaning-up operation of the largest moorland pool in The Netherlands (100ha), the so-called 'Beuven', in 1985 - '86 provides indirect evidence for the effects of eutrophication (Buskens & Zingstra, 1988). The Beuven was strongly eutrophied by nutrient-rich water from a small stream, while some shallow parts were acidified. The pool was drained and a thick layer of organic mud was removed. Buffered water was let in to counteract acidification by atmospheric inputs. Isoetid plants and many other species typical for soft water pools strongly expand in the restored pool, and *Isoetes echinospora*, that had disappeared for many years, has returned.

Experimental evidence of effects of excessive N deposition on microflora, microfauna and macro-invertebrates is scanty. In addition to various direct physiological effects of acid water on fish (Wendelaar Bonga & Dederen, 1986, Leuven et al., 1987c), and amphibians (Leuven et al., 1986a) changes in vegetation may play a role too. *Sphagnum* produces phenolic substances, that may affect the reproduction of amphibians. Some species of amphibians are found more abundantly in moderately acidified pools, possibly due to reduced predation and competition by fish (Leuven et al., 1986a).

DISCUSSION AND SUMMARY

The deposition of N from the atmosphere in The Netherlands is presumably one of the highest in the world. Moreover, the sensitivity of soft water pools and most of the soils under forest and heathland, to the acidifying and eutrophying effects of the present-day atmospheric deposition is high. This is due to (i) the low levels of acid-neutralizing substances in the predominantly sandy, highly quarzitic soils in those areas, and (ii) the originally low levels of N in these ecosystems. The situation is further aggravated by the proximity of the dominant sources of ammonia (intensive stock-breeding) to most forest areas.

Marked changes in ecosystems, possibly caused by air pollution in general and by high deposition of N in particular include (i) changes in the chemistry of soils, ground waters and soft water pools, (ii) changes in the floristic composition of the herb layers of forests, of heathlands, and of moorland pools, (iii) changes in microfauna and microflora of moorland pools, and (iv) a general decline of the vitality of both coniferous and deciduous forests. Of these changes, (i), (ii) and (iii) are clearly caused mainly by excessive atmospheric deposition of N. The links between forest decline and air pollution are more complicated, and causal relationships have not yet been proven. It is very likely, however, that high levels of N, and intensive soil acidification associated with inputs of (NH4)2SO4 (often aggravated by nitrification) are major factors involved in the forest decline.

The leaching of N from some forest ecosystems is as high as, or higher, than the atmospheric input. Such N-saturated systems may develop as a combined effect of high inputs and poor vitality, leading to decreased retention of N by the ecosystem. Excessive leaching of NO_3^- implies accelerated soil acidification due to nitrification. This may result in a destabilizing, positive feedback loop, involving increased decline, and increased mobilization of N from the pool of organic N in the forest floor. There are some indications that the pool of soil N is locally acting already as a dominant source of further soil acidification (Hauhs, 1985 ; Paces, 1985; Van Breemen et al., 1987).

To stop the harmful effects of high N deposition, measures are urgently needed. Critical deposition levels for N deposition have been summarized by Schneider and Bresser (1987). To prevent groundwater NO_3^--N levels from exceeding the norm of 11.3 mg/L, N deposition should be below 2.2 (coniferous forests) to 50 (heathland) kg/haJyr. In order to return to presumably unharmful NH_4/K and Ca/Al ratios in soil solutions of the rooting zone of trees, deposition of NH_x and total acid (NH_x + NO_x + $2SO_x$) should decrease below 1.5 (coniferous trees) to 2.4 (deciduous trees) kmol/haJyr, respectively. For the return of vegetations typical of the ecosystems low in N, that abounded decades ago, N deposition would have to decrease to about 6 kg/haJyr.

Present-day N deposition on forests is in the order of 40 to 80 kg/haJyr, and a decrease to less than half of that value is not likely in the near future. To prevent further harm to trees, adding nutrients such as K, Ca, Mg and P, that may now be limiting tree growth, and liming to prevent excessive soil acidity, is being considered (Van Diest, 1987). If indeed excess N supply and extreme soil acidification are the main factors hampering tree growth, fertilization and liming will enhance biomass accumulation, and may help to tie up atmospheric N in the biomass and decrease groundwater pollution. If, however, other factors such as direct effects of air pollutants on leaves and poor water availability, are more important, fertilization may be useless or even harmful. In forests, heathlands, and moorland pools, other management procedures, such as removal of biomass by grazing, harvesting and litter removal may be considered as well. For the relatively simple heathland ecosystems (and for at least one moorland pool) litter removal has lead to remarkable results. Unfortunately, we know too little about the biogeochemistry of forest ecosystems and about the reaction of forest biota to changed conditions to predict the effect of such actions on forests with much confidence.

REFERENCES

Anonymous (1985). Verslag van het landelijk vitaliteitsonderzoek 1985. De vitaliteit van het Nederlandse bos 3. Staatsbosbeheer (State Forest Service) 1985-16. Afdeling Bosontwikkeling. pp.

Anonymous (1987). Verslag van de landelijke inventarisatie 1987. De vitaliteit van het Nederlandse bos 5. Staatsbosbeheer (State Forest Service) 1987-27, 22pp.

Appelo, C.A.J. (1985) Silikaatverwering als buffer tegen verzuring van bodemvocht en grondwater. Internal rept.,Dept. Earth Sciences, Free University, Amsterdam, The Netherlands.

Arnolds, E.J.M. (ed.) (1985). Veranderingen in de paddestoelen- flora (mycoflora). Wetenschappelijke Mededeling K.N.N.V. nr. 167.

Arnolds, E.J.M. & Jansen, A.E. (1987). Qualitative and quantitative research on the relation between ectomycorrhiza of Pseudotsuga menziesii. Vitality of the host and acid rain. Dutch priority programme on acidification, report 25-01. Wijster. Biological Station. Communication nr. 334.

Arts, G.H.P. (1987). Geschiedenis van de verzuring van zwak gebufferde wateren in Nederland onder invloed van atmosferische depositie. Dutch priority programme on acidification, report 28- 01. Nijmegen. Laboratorium voor Aquatische Oecologie. Katholieke Universiteit.

Asman, W.A.H., Drukker, B., & Janssen, A.J.(1987). Estimated historic concentrations and depositions of ammonia and ammonium in Europe and their origin (1870-1980). Inst. for Meteorology and Oceanography (IMOU), Utrecht, Rept R-87-2. 88 pp.

Berdowski, J.J.M. & Siepel, H. (1987). Vegetative regeneration of Calluna vulgaris at different ages and levels of fertilization. In: The catastrophic death of Calluna vulgaris in Dutch heathlands. J.J.M. Berdowski, PhD. Thesis, 79-87. University of Utrecht.

Berdowski, J.J.M. & Zeilinga, R. (1987). Transition from heathland to grassland: damaging effects of the heather beetle. J. Ecol., 75, 159-175.

Berdowski, J.J.M. & Hengeveld, R. & Sosef, M. (1987). Changes in aggregation of a temporal transition of heathland into grassland. In: The catastrophic death of Calluna vulgaris in Dutch heathlands. J.J.M. Berdowski, PhD. Thesis, 53-62. University of Utrecht.

Berendse, F. & Aerts, R. (1984). Competition between Erica tetralix L. and Molinia caerulea L. Moench as affected by the availability of nutrients. Acta Oecologica/Oecol. Plant., 5, 3- 14.

Boxman, A.W. & Roelofs, J.M. (1986). Some physiological and effects of NH4+ and Al3+ on pine forest ecosystems. In: Texte Wiss. Symp. 'Neue Ursachenhypothesen', ed. by H.-D. Gregor, M. Hübner, E. Niegengerd, L. Ries, M. Schneider and W. Schröder, 407-414. Berlin, U.B.A.

Boxman, A.W., Dijk, H.F.G. van & Roelofs, J.G.M. (1987b). Some effects of ammonium sulphate deposition on pine and deciduous forests in The Netherlands. In: Acid rain: scientific and technical advances, ed. by R. Perry, R.M. Harrison, J.N.B. Bell and J.N. Lester, 680-687. London, Publications Division, Selper Ltd. ISBN 0 948411 03 1.

Brunsting, A.M.H. & Heil, G.W. (1985). The role of nutrients in the interactions between a herbivorous beetle and some competing plant species in heathlands. Oikos, 44, 23-26.

Buijsman, E., Maas, H.,& Asman, W.H.M. (1984). Een gedetailleerde ammoniakemissiekaart van Nederland. Inst. for Meteorology and Oceanography (IMOU), Utrecht, Rept V-84-20 124 pp.

Buskens, R.F.M. & Zingstra, H.L. (1988). Beuven: verwording en herstel. De Levende Natuur, accepted for publication.

Buskens, R.F.M., Leuven, R.S.E.W., Velden, J.A. van der & Velde, G. van der (1986). The spatial and temporal distribution of chironomid larvae in four lentic soft waters differing in acidity. In: Proceedings of the 3rd European congress of entomology, 103-106. Amsterdam, Nederlandse Entomologische Vereniging. ISBN 90-71912-0109.

De Boer, W., Klein Gunnewick, P.J.A. & Troelstra S.R. (1989). Two types of chemolithotrophic nitrification in acid heathland soils. Plant and Soil, 119:229-235.

Den Boer, W.M.J. (1986). Ammonia, not only a nutrient but also a cause of forest damages. In: Texte Wiss. Symp. 'Neue Ursachenhypothesen', ed. by H.-D. Gregor, M. Hübner, E. Niegengerd, L. Ries, M. Schneider and W. Schröder, 405-406. Berlin, U.B.A.

De Smidt, J.T. (1983). Heathland management in The Netherlands. Scientific and social aims. Acta Bot. Neerl., 32, 252.

Ellenberg Jr., H. (1985). Veränderungen der Flora Mitteleuropas unter dem Einfluss von Düngung und Immissionen. Schweiz. Z. Forstw., 136. 19-39.

Ellenberg Sr., H. (1979). Zeigerwerte der Gefässpflanzen Mitteleuropas. Scripta Geobotanica 9, 2nd. ed., Göttingen.

Erisman, J.W., 1990.

Geelen, J.F.M. & Leuven, R.S.E.W. (1986). Impact of acidification on phytoplankton and zooplankton communities. Experientia, 42, 486-494.

Hauhs, M. (1985). Wasser- und Stoffhaushalt im Einzugsgebiet der Langen Branke (Harz). Bev. Forschungszentrums Waldöko-systeme/Waldsterben, Univ. Göttingen, FRG, Bd 17, 206 pp.

Heil, G.W. & Bruggink, M. (1987). Competition for nutrients between Calluna vulgaris (L.) Hull and Molinia caerulea (L.) Moench. Oecologia, 73, 105-107.

Heil, G.W. & Diemont, W.H. (1983). Raised nutrient levels change heathlands into grasslands. Vegetatio, 53, 113-120.

Janssen, Th.W. (1982). Intensieve veehouderij in relatie tot ruimte en milieu. Dutch National Forest Service, Utrecht.

Konsten, C.J.M., Tiktak, A. & Bouten, W. (1988) Soil Chemistry and soil physics of two Douglas fir stands affected by acid atmospheric deposition on the Veluwe, The Netherlands. Rept. 03-01, Dutch priority programme on acidification. RIVM, Bilthoven, The Netherlands.

Krajenbrink, G.J.W.(1987) Stofbelasting van de onverzadigde zone en het bovenste grondwater onder bouwland en bos in de beschermingszones van waterwingebied Putten. RIVM, Bilthoven, The Netherlands, rept. nr 728472004, 90 pp.

Krauss, H.H., Heinsdorf, D., Hippeli, P. & Tölle, H. (1986). Untersuchungen zu Ernährung und Wachstum wirstschaftlich wichtiger Nädelbaumarten im Tiefland der DDR. Beitr. f. d. Forstwirtsch., 20, 156-164.

Leuven, R.S.E.W., Hartog, C. den, Christiaans, M.M.C. & Heijligers, W.H.C. (1986a). Effects of water acidification on the distribution pattern and the reproductive success of amphibians. Experienta, 42, 495-503.

Leuven, R.S.E.W., Schuurkes, J.A.A.R., Arts, G.H.P & Roelofs, J.G.M. (1986c). Oorzaken, omvang en effecten van waterverzuring in Nederland. In: Waterverzuring in Nederland en België, ed. by R.S.E.W. Leuven, G.H.P. Arts and J.A.A.R. Schuurkes, 87-102. Nijmegen, Laboratorium voor Aquatische Oecologie, Faculteit der Wiskunde en Natuurwetenschappen, Katholieke Universiteit. ISBN 99-9001195-1.

Leuven, R.S.E.W., Velde, G. van der, Vanhemelrijk, J.A.M. & Eeken, R.L.E. (1986d). Impact of acidification on the distribution of aquatic insects in lentic soft waters. In: Proceedings of the 3rd European congress of entomology, 103-106. Amsterdam, Nederlandse Entomologische Vereniging. ISBN 90-71912- 01-9.

Leuven, R.S.E.W., Vanhemelrijk, J.A.M. & Velde, G. van der (1987a). The distribution of Trichiptera in Dutch soft waters differing in pH. In: Proceedings of the 5th international symposium on Trichoptera, ed. by M. Bournaud and H. Yachet, 359- 365. Dordrecht, Dr. W. Junk Publishers. ISBN 90-6193-620-9.

Leuven, R.S.E.W., Velden, J.A. van der, Vanhemelrijk, J.A.M. & Velde, G. van der (1987b). Impact of acidification on chironomid communities in poorly buffered waters in The Netherlands. Ent. scand. Suppl., 29, 269-280.

Leuven, R.S.E.W., Wendelaar Bonga, S.E., Oyen, F.G.F. & Hagemeijer, W. (1987c). Effects of acid stress on the distribution and reproductive success of freshwater fish in Dutch soft waters. In: Ecophysiology of acid stress in aquatic organisms, ed. by H. Witters and O. Vanderborght. Annls. Soc. r. zool. Belg., 117, suppl. 1, 231-242.

Marnette, E.C.L.(1987). Nitrificatie in zure en kalkrijke bosgronden. Unpubl. MSc Thesis, Dept. Microbiology, Agric. University, Wageningen, The Netherlands. 55 pp.

Meyer, F.H. (1986). Root and mycorrhiza development in declining forests. In: Indirect effects of air pollution on forest trees. Root-rhizosphere interactions. Proceedings workshop Comm. Eur. Comm. and Kernf. Anlage Jülich. 139-151.

Murach, D. (1984). Die Reaktion der Feinwurzeln von Fichten (Picea abies, Karst.) auf zunehmende Bodenversaurung. Göttingen Bodenkdl. Ber. 141, 1-126

Oosterom, H.P. & Van Schijndel, J.H.W.M. (1979). De chemische samenstelling van het bovenste grondwater bij natuurlijke begroeiingen op kalkarme grond. Nota 1075, ICW, Wageningen, The Netherlands.

Paces, T. (1985) Sources of acidification in Central Europe estimated from elemental budgets in small basins. Nature, 315,31-36.

Roelofs, J.G.M. (1983). Impact of acidification and eutrophication on macrophyte communities in soft waters in The Netherlands. I. Field observations. Aquat. Bot., 17, 139-155.

Roelofs, J.G.M. (1986). The effect of airborne sulphur and nitrogen deposition on aquatic and terrestrial heathland vegetation. Experientia, 42, 372-377.

Roelofs, J.G.M., Clasquin, L.G.M., Driessen, J.M.C. & Kempers, A.J. (1984a). De gevolgen van zwavel- en stikstofhoudende neerslag op de vegetatie in heide- en heidevenmilieus. In: Zure regen. Oorzaken, effecten en beleid, ed. by E.H. Adema and J. van Ham, 134-140. Pudoc, Wageningen.

Roelofs, J.G.M., Schuurkes, J.A.A.R. & Smits, A.J.M. (1984b). Impact of acidification and eutrophization on macrophyte communities in soft waters. II. Experimental studies. Aquat. Bot., 18, 389-411.

Roelofs, J.G.M., Kempers, A.J., Houdijk, A.L.F.M. & Jansen, J. (1985). The effect of airborne ammoniun sulphate on Pinus nigra var. maritima in The Netherlands. Plant Soil, 84, 45-56.

Rost-Siebert, K. (1985). Untersuchungen zur H+- und Al- Ionentoxizität an Keimpflanzen von Fichte (Picea abies Kartst.) und Buche (Fagus sylvatica L.) in Lösingskultur. Ber.d. Forschungszentrums Waldökosysteme/Waldsterben, Univ. Göttingen,FRG, Bd 12,1-219

Schildwacht, P. & Smidt, J.T. de (1985). De invloed van nutriëntenaanbod en -beheer op de soortensamenstelling van een droge heide. The Utrecht Pl. Ecol. News Report, 1, 47-52.

Schneider, T. & Bresser, A.H.M. (1987). Verzuringsonderzoek eerste fase. Tussentijdse evaluatie augustus 1987, Dutch priority programme on acidification, report nr. 00-04, R.I.V.M., Bilthoven, The Netherlands, 75 pp.

Schuurkes, J.A.A.R. & Leuven, R.S.E.W. (1986). Verzuring van oppervlaktewateren. Publikatiereeks Lucht 53. Ministerie van Volkshuisvesting, Ruimtelijke Ordening en Milieubeheer. Den Haag. ISBN 90 346 0966 9.

Schuurkes, J.A.A.R., Elbers, M.A., Gudden, J.J.F. & Roelofs, J.G.M. (1986a). Effects of simulated ammonium sulphate and sulphuric acid on acidification, water quality and flora of small-scale water systems. Aquat. Bot., 28, 199-226.

Schuurkes, J.A.A.R., Kok, C.J. & Hartog, C. den (1986b). Ammonium and nitrate uptake by aquatic plants from poorly buffered and acidified waters. Aquat. Bot., 24, 131-146.

Smeets, W. (1988). Vergelijkingsonderzoek m.b.t. verschil-lende bodemvochtanalysen. Unpubl. MSc Thesis, Dept Aquatic Ecology, Catholic Univ. Nijmegen, The Netherlands.

Smit, H., Van Breemen, N. & Keltjens, W.G. (1987a). Effects of soil acidity on Douglas fir seedlings. I. Rooting characteristics of natural regeneration of Douglas fir in strongly acid forest soils. Neth.J.Agr.Sci., 35, 533-536.

Smit, H.,Keltjens, W.G. & Van Breemen, N. (1987b). Effects of soil acidity on Douglas fir seedlings. II. The role of pH, aluminium concentration and nitrogen nutrition (pot experiment). Neth.J.Agr.Sci., 35, 537-540.

Smits, T.F.C. (1986). Sphaeropsis sapinea. Wageningen, Rijksinstituut voor onderzoek in de bos- en landschapsbouw. "De Dorschkamp".

Tyler, G. (1987). Probable effects of soil acidification and nitrogen deposition on the floristic composition of oak (Quercus robur L.) forest. Flora, 179, 165-170.

Van Aalst, R.M. & Erisman, J.W. (eds.). Atmospheric impacts. Thematic report of the Dutch acidification programme (draft), May 1990.

Van Breemen, N., Mulder, J.& Van Grinsven, J.J.M.(1987). Impacts of acid atmospheric deposition on woodland soils in The Netherlands: II N-transformations. Soil Sci.Soc.Amer.J. 51: 1634-1640.

Van Breemen, N., Visser, W.J.F. & Pape, Th. (eds.) (1988). Biogeochemistry of an oak woodland affected by acid atmospheric deposition. Agr.Res.Repts 930, PUDOC, Wageningen, The Netherlands.

Van Breemen, N. & Dijk H.F.G. van (1988). Ecosystem effects of atmospheric deposition of nitrogen in The Netherlands. Environm. Pollution 54: 247-274.

Van Breemen, N. & Verstraten, J.M. (1990). Soil acidification and nitrogen cycling: Summary of research in the Dutch priority programme on acidification. Draft, 60 pp.

Van Dam, H. (1987). Verzuring van vennen: een tijdverschijnsel. PhD. Thesis, Wageningen, Agricultural University.

Van Dam, H. & Beljaars, K. (1984). Historische veranderingen in microflora en chemie van vennen in relatie tot zure neerslag. In: Zure regen. Oorzaken, effecten en beleid, ed. by E.H. Adema and J. van Ham, 156-161. Pudoc, Wageningen.

Van Dam, D., Dobbén, H.F. van Braak, C.F.J. ter & Wit, T. de (1986). Air pollution as a possible cause for the decline of some phanerogamic species in The Netherlands. Vegetatio. 65, 47- 52.

Van der Eerden, L.J.M. (1982). Toxicity of ammonia to plants. Agric. Environm., 7, 223-235.

Van der Eerden, L.J. & Wit, A.K.H. (1987). Effecten van NH3 en NH4+ op planten en vegetaties; relevantie van effect- grenswaarden. In: Studiedag effecten van NH3 op organismen, ed. by A.W. Boxman and J.F.M. Geelen, 45-51. Nijmegen, Laboratorium voor Aquatische Oecologie, Faculteit der Wiskunde en Natuurwetenschappen, Katholieke Universiteit. SBN 90-9001588-4.

Van der Maas, M.P. (1985) De invloed van de vegetatie en de pH op de nitrificatie in zure bosgronden. Unpubl. MSc Thesis, Dept. Soil Science & Plant Nutrition, Agric. University, Wageningen, The Netherlands.

Van Diest, A. (1987). Bosbemestingsonderzoek in Harderwijk, bodemvruchtbaarheidsaspecten. Presentation at the symposium "Forests and fertilization", November 20, 1987, Reehorst Ede, The Netherlands.

Van Dijk, H.F.G. & Roelofs, J.G.M. (1987). Effects of airborne ammonium on the nutritional status and condition of pine needles. In: Direct effects of dry and wet deposition on forest ecosystems - in particular canopy interactions. Air pollution research report 4, commission of the European Communities, Eur 11264, Brussels.

Van Dobben, H.F. (1987). Effecten van ammoniak op epifytische korstmossen. In: Studiedag effecten van NH3 op organismen. ed. by A.W. Boxman and J.F.M. Geelen, 52-61. Nijmegen, Laboratorium voor Aquatische Oecologie, Faculteit der Wiskunde en Natuurwetenschappen, Katholieke Universiteit. ISBN 90-9001588- 4.

Van Dobben, H.F., & Mulder, J. (1988) Impacts of acid atmospheric deposition on the biogeochemistry of moorland pools and the surrounding terrestrial environment. Report in prep., Institute of Nature Mangement, Leersum, The Netherlands.

Van Grinsven, J.J.M. & Van Breemen, N. (1986). Effecten van zure depositie op bodem en grondwater. Handboek voor Milieu Bescherming, bodembescherming, D2100; 1-18.

Van Lier, J. & Smeets, W. (1985) Effecten van verzurende depositie op Pinus sylvestris in Nederland. Unpubl. MSc Thesis, Dept. Soil Science and Biogeology, Catholic University, Nijmegen, The Netherlands, 86 pp.

Verbeek, P.J.M., Velde, G. van der, Krekels, R.F.M. & Leuven, R.S.E.W. (1986). Occurrence and spatial distribution of odonate larvae in four lentic soft waters of varying pH in The Netherlands. In: Proceedings of the 3rd European congress of entomology, 103-106. Amsterdam, Nederlandse Entomologische Vereniging. ISBN 90-71912-01-9.

Verheyen, L.A.H.M. (1985) Ontwikkeling van de nitraat- concentratie van het grondwater in de omgeving van het pomp- station Vierlingsbeek (WOB). KIWA rept. SWE-84.015. KIWA, Rijswijk, The Netherlands.

Wendelaar Bonga, S.E. & Dederen, L.H.T. (1986). Effecten van verzuurd water op vissen. In: Waterverzuring in Nederland en België, ed. by S.E.W. Leuven, G.H.P. Arts and J.A.A.R. Schuurkes, 87-102. Nijmegen, Laboratorium voor Aquatische Oecologie, Falculteit der Wiskunde en Natuurwetenschappen, Katholieke Universiteit, ISBN 99-9--1195-1.

Winkels, H.J.(1985). Kwantitatieve aspecten van de koolstof- en stikstofhuishouding van de strooisellaag in de gronden van het Hackfortproject. Unpubl. MSc Thesis, Dept. Soil Science and Geology, Agric. Univ., Wageningen, The Netherlands.

	kmol/haJyr		
	SOx	NHx	NOx
wet	0.23	0.66	0.30
dry	0.46	1.65	0.85
total	0.69	2.31	1.15

Table 1. Wet plus dry deposition of N and S in The Netherlands (1989) (Van Aalst and Erisman, 1990). Values are country-wide means, from measured wet deposition plus dry deposition estimated from measured gas- and aerosol concentrations.

	H	NH_4	SO_4	NO_3	period	ref
Oak-birch						
Warnsveld(C), bulk	0.2	1.2	1.3	0.5	1980-83	1
throughfall	0.1	3.0	3.2	1.0		
Mixed deciduous-spruce						
Winterswijk(E) bulk	0.2	1.1	1.2	0.4	1979-85	2
throughfall	0.2	2.4	3.6	0.6		
Scots pine						
Boxtel(S), bulk	0.2	0.8	0.9	0.4	1982-84	3
throughfall	0.1	3.7	3.5	0.9		
Oisterwijk(S), bulk	0.2	0.8	1.0	0.4	1983-84	3
throughfall	0.1	4.2	3.8	0.7		
Kootwijk(C), bulk	0.2	0.8	0.9	0.4	1982-84	3
throughfall	0.2	4.3	4.0	1.2		
Kliplo(N), bulk	0.3	0.7	0.8	0.4	1982-84	3
throughfall	0.2	3.7	3.3	1.1		
Douglas fir						
Speuld(C), bulk	0.2	0.8	0.8	0.4	1987	4
throughfall	0.1	3.5	3.5	1.1		
Calluna						
Valkenswaard(S)bulk	0.2	0.8	0.9	0.4	1983-84	3
throughfall	0.1	1.4	1.7	0.3		

Table 2. Annual values of bulk deposition in the open field and in canopy throughfall (in kmol(+/-)/haJyr) of free H+, sulphate and inorganic nitrogen for a number of ecosystems in the southern (S), central (C), eastern (E) and northern (N) part of The Netherlands.

References : 1 ; Van Breemen et al., 1988
2 ; Van Breemen and Verstraten, 1990
3 ; Van Dobben and Mulder, 1988
4 ; Konsten et al., 1988

	year	pure heather	heather > grass	heather = grass	heather < grass	pure grass
Location at "Het Gooi"	1980 1986	40 --	50 30	-- 30	-- --	10 20
Location at "De Veluwe"	1980 1986	40 --	-- --	10 --	-- 40	50 60

Table 3. Areal proportions (in %) of heather and grasses in the vegetation cover in two Dutch nature reserves in 1980 and 1986 (Schneider & Bresser, 1987).

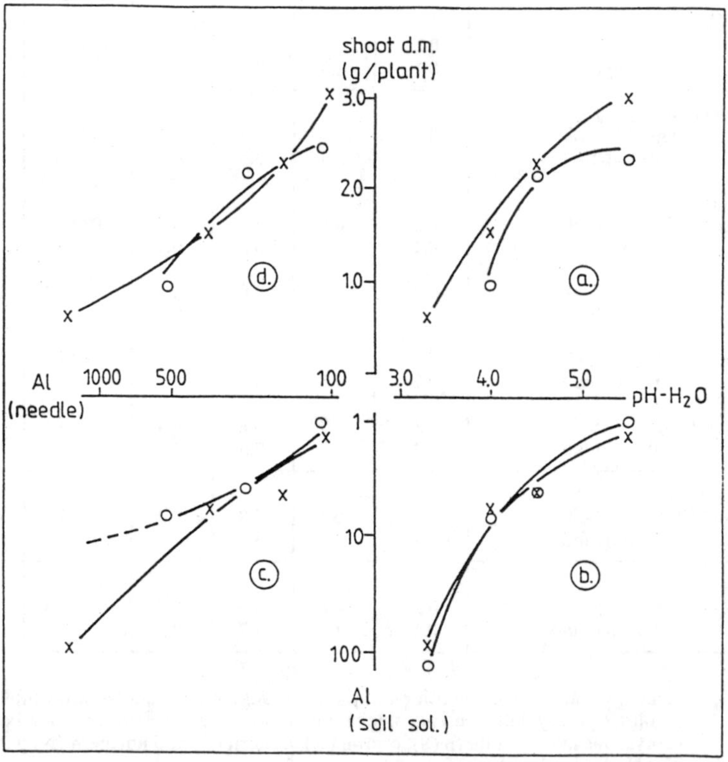

Figure 1. Relationships between soil pH, shoot growth, Al content in needles (mg/kg) and Al concentration in the soil solution (mg/l) for one-year old Douglas seedling grown on acidified (pH \ 4) or limed (pH > 5) sandy soil, with 100 kg/ha of N supplied as NO3 (x) or as NH4 (o) plus nitrification inhibitor (from Smit et al., 1987b).

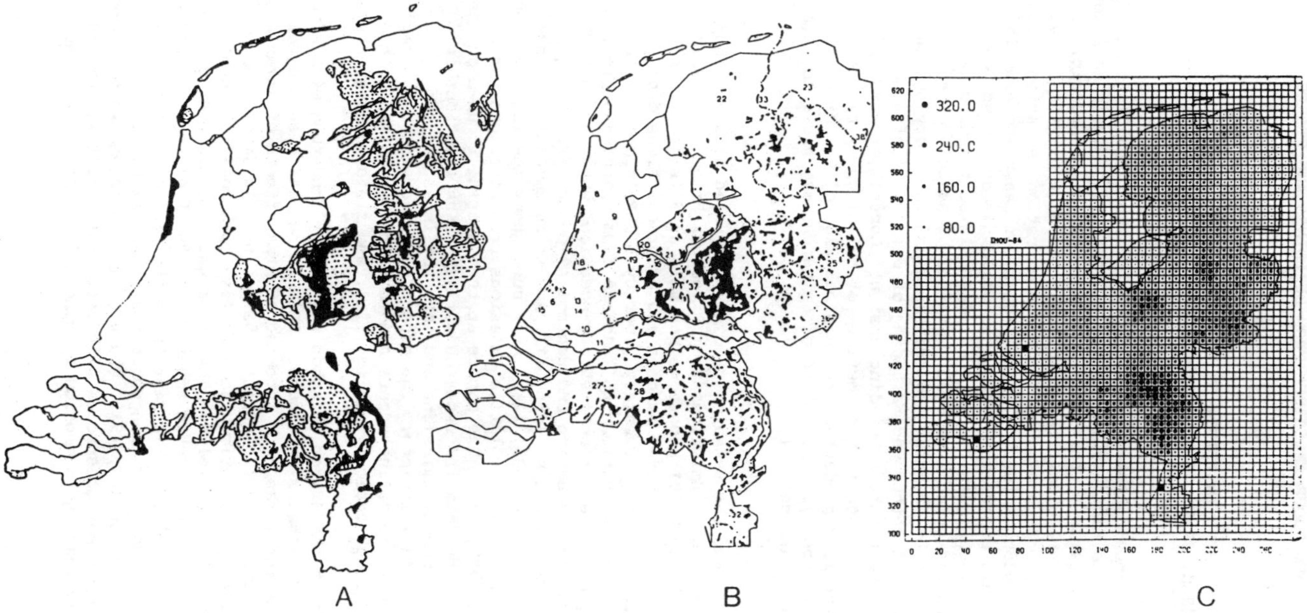

Figure 2. Maps of The Netherlands showing (A) soils sensitive to acid deposition (black: very sensitive, hatched: moderately sensitive) (B) Forested areas and (C) Emission of NH3; values are metric tons of NH3 per 5 x 5 km2 per year (sources: Van Grinsven and Van Breemen, 1986; Buijsman et al., 1984).

AMMONIA AND ODOUR EMISSIONS FROM BUILDINGS AND STORES

V C NIELSEN

ADAS UNIT, INSTITUTE OF ENGINEERING RESEARCH
WREST PARK, SILSOE, BEDFORDSHIRE, MK45 4HS

Summary

There is considerable evidence that livestock buildings and manure stores are responsible for a third of total ammonia emissions from livestock farms. These same source points are responsible for at least 50 percent of all the complaints of odour nuisance. Methods have been developed to measure odour concentration, emission rates and determine odour intensity. Techniques for measuring ammonia emissions within controlled ventilated buildings have also been established. Techniques for naturally ventilated buildings and open manure stores are being developed. Control of ammonia emissions from within buildings by flushing and treatment and rapid drying are under investigation. A better understanding of dispersion of ammonia by modelling may help future building design. Control of ammonia emissions from the ventilation outlets of buildings by bioscrablers and biofilters is dependent on the manipulation of the nitrification process but is costly. Prevention of emissions may prove to be a cheaper option than control.

1 INTRODUCTION

This paper reviews the research findings presented at a workshop entitled Ammonia and Odour Emissions from Livestock Production, which was organised within the COST 681 programme and held at Silsoe, UK, in March 1990. The papers reviewed cover ammonia and odour emissions from buildings and stores. A complementary paper by B F Pain deals with emissions from slurry applications to land and from grazed pastures.

The expert group have met on previous occasions when they directed their attention to the prevention and control of odours from organic sludge and livestock production 1985, (1), and to recommendations on olfactometric measurements 1989, (2). At a subsequent meeting the group widened its study to include volatile emissions from livestock farming and sewage operations 1987, (3). The papers at that meeting focused on the need to carry out more detailed research work into the relationship between odours and ammonia emissions and into the impact on the environment of ammonia emissions. The expert group agreed that other parameters which help to define odours needed to be harmonised such as odour intensity through category scaling, (4), and field measurements with odour panels.

The emphasis at the Silsoe workshop was mainly concerned with ammonia emissions with less emphasis on odours. This change has been the result of intense international research into the adverse effects of ammonia on the environment. There are two effects that of ammonia deposition on natural and cultivated vegetation close to livestock buildings causing leaf burn and loss of crop yields. The other which occurs over greater distances is caused by the chronic adverse effects of nitrogen enrichment and acidification of soils. This results in the binding of essential minerals leading to long term changes in the botanical composition of dunes, heathlands and forests, (5). The economic consequences of these changes are a replacement of species with deep matted root systems in sand dunes with grasses leading to erosion by wind and the sea. In forests the effects are to weaken the resistance of trees to attacks by invasive organisms and stresses caused by frost, droughts and ozone leading to death. It has been estimated that ammonia emissions in the Netherlands need to be reduced by 70 percent to avoid these effects. The influence of housing and livestock on ammonia release from buildings was reviewed (6). It can be broken down into a number of components which together account for approximately 30 percent of emissions from livestock. This is a significant source in that the emissions are concentrated in a small area and are continuous throughout the year. The effects on vegetation are therefore more severe that that from land spreading which is transitory covering a period ranging from a few hours to not more than 2 days.

The effect of the species of livestock on the volume of ammonia emitted has been calculated for the Netherlands based on the total volume of neat livestock manure produced each year. This shows that cattle produce 53 percent of the total ammonia emissions, pigs 24 percent and poultry 23 percent. On this basis cattle are the most important source of ammonia and it could be concluded the least efficient at converting feed nitrogen into livestock products. However, when the ammonia emission rate is calculated on the basis of ammonia loss per kilogramme of liveweight, poultry produce significantly more ammonia than pigs and cattle and are the least efficient nitrogen converters.

Measures to control and decrease ammonia emissions from buildings need to consider the following points:-

Restricting feed protein by constructing better balanced amino acid contents to match the species requirements.

General hygiene in buildings, clean dry pens, controlled ventilation and humidity, control of livestock numbers to a given liveweight loading per metre of floor area.

Ammonia emissions cam be effected by manipulating the pH of stored manure using additives, encouraging fast air drying in poultry houses, or by rapid removal from buildings by flushing.

2 METHODS

There are few exact methods of measurement of ammonia emissions for buildings, this is due to the fact that experimentation is very recent and well tested techniques which provide reliable results are still being developed. Odour measurements from livestock buildings and small scale stores have been monitored for a longer period, but reliable methods for monitoring naturally ventilated buildings and large open stores are still being developed.

Automatic ammonia sampling and analysis

This equipment is used to determine the ammonia concentration in the atmosphere at set intervals of time. It is possible to monitor a number of closely associated experimental chambers sequentially. The ammonia concentration together with continuous monitoring of the air flow through the outlet ventilators of chambers will provide data on the concentration of ammonia in grammes per cubic metre of air, over any desired period of time.

Adsorption flasks or traps for sampling ammonia

This method of monitoring ammonia by adsorption and trapping with dilute acids is used frequently in field experimentation, (7). It has the disadvantage that the ammonia collected is cumulative and to monitor daily cycles requires frequent changing of the collection devices. Its advantage is that samples can be collected from a range of situations and at farm scale. Air flow measurements from the outlet ventilators of the building together with the ammonia concentration provide the measurement of ammonia in grammes per cubic metre of air.

Improved analytical techniques using high performance liquid chromatography allow very many samples to be collected and tested daily, thus reducing costs and speeding up measurements, (8).

Lindvall boxes

These have been adopted for collecting odours and ammonia from floor surfaces in naturally ventilated buildings, (9), and for monitoring emissions from the surfaces of large slurry stores, (10).

Odour Concentration

Odorous air is collected in teflon bags and subjected to measurements based on organoleptic methods using olfactometers. The expert group have published recommendations on the methodology and interpretation of results (2).

Odour Intensity

This is a measure of the persistency of an odour to be detected by panellists and is not directly linked to concentration. It is measured by a group of panellists determining its value by category scaling. The methodology used and interpretation of results has been the subject of intense debate by the expert group, the technique recommended has been published (4).

Field measurement of odorous air pollution

Field measurement of odorous air pollution with panels has been reviewed by an ad hoc working group (11), to harmonise methods and make recommendations on techniques and interpretation of results. This method of assessing odour nuisance is very practical and can be used without resorting to expensive laboratory equipment. It has uses in situations where source points of odours, such as livestock buildings which are close to communities, can be monitored. Comparisons can be made both before and after efforts have been made to abate an odour nuisance to determine how effectively the control system operates.

3 RESULTS

Ammonia emissions from livestock housing

Modelling the dispersion of ammonia in livestock buildings

This was based on fluid mechanics and has been used for demonstration purposes (11). Storage beneath slatted floors has not yet been modelled. Air flow with and without obstacles and for various positions of the air inlet and outlet can be modelled. The study so far has found that there is no homogenous distribution of ammonia within animal houses. The fluid flow pattern must be calculated for different geometrical and suction conditions. To make use of the flow fields the results have to be condensed into simplified zone models. Compartment models may be useful to develop a control technique. Output of ammonia from animal houses into the environment can be minimised by control of suction velocity by means of inlet cells.

The effect of housing on ammonia emissions

Work has been carried out to compare the effects of different housing systems on ammonia emissions (12), 200 sites were monitored. Measurements were based on a livestock unit (L.U. = 500 kg live mass) and on time. Ammonia emissions are recorded as gramme of ammonia per hour and livestock unit (L.U.) ($g\ NH_3\ h^{-1}\ LU^{-1}$). Ammonia was measured as spot records using Drager-Capillary tubes. Factors affecting emission rate were the nitrogen content of the feed, ventilation, animal activity and day/night variations. The latter had a very significant effect on ammonia emissions over 24 hours in battery laying hen houses.

The effect of the ambient air temperature was to increase ammonia emissions for all species of stock as the temperature increased. The effect of animal housing on emission rate depended on the surface area of manure or manure covered floor area and the presence of manure. Semi-slatted pens had lower ammonia losses than completed slatted floors. Twice daily flushing reduced ammonia emissions.

Broiler housing systems based on in/out manure management showed that ammonia emission increased by 30:1 over a period of 35 days as the live mass increased and manure built up. After 15 days ammonia emissions had risen to 2 to 3g NH_3 h^{-1} LU^{-1} at the end of the crop life day 35 this had risen to 30 g NH_3 h^{-1} LU^{-1}. Odour emissions followed a similar pattern 25 odour units (OUs) $^{-1}$ LU^{-1} at day 15 and at 35 days 225 OUs^{-1} LU^{-1}.

Ammonia emissions from a dairy cow house

Ammonia emissions from a dairy cow house and from three methods of flushing pig houses were monitored in the Netherlands (9). The study in the cattle house found large variations in ammonia emission rates during the day. Peak emissions were found to occur just before morning milking when cattle moved from laying in cubicles to a collection yard. This was accompanied by a lot of dunging and urination. A method of measuring emissions in naturally ventilated buildings was used to compare floor surfaces. Measurements of ammonia emissions found that there was little difference in volumes emitted between soiled slatted floors and the surface of concrete covered in manure, scraping had very little difference on total emission values. Flushing floors did reduce emission values.

Three systems of slushing pig fattening pens were evaluated. There was no significant difference between the systems. Flushing systems which have a system for retaining a shallow layer of flush water in the collection channel and prevent drying can achieve a 70 percent reduction in ammonia emissions over conventional slurry collection systems. Such systems require a complex treatment system in order to recycle treated flushing water containing nitrates which can be used to sustain aerobic conditions.

The effect of the method of storing poultry manure

Tests were carried out at two poultry farms for laying hens with two different operational systems, for the partial drying of poultry manure (13). At one farm composting was used to achieve this objective and on the other a ventilated deep pit. Comparisons were made of the characteristics of each system with the aim of controlling ammonia emissions. The results showed that composting produce an ammonia loss of 0.95 to 1.63 g NH_3 hen^{-1} d^{-1} while the ventilated pit was very much lower at 0.17g NH_3 hen^{-1} day^{-1}.

The effects of storage and various methods of covering slurry stores

The effects of storage and various methods of covering slurry stores containing pig and cattle slurry were monitored over a year for ammonia and odour emissions (10). The results of the programme of research found the odour emission concentration was low in the experimental tanks but higher in a farm sized tank. Odour emissions could be effectively reduced by covering the tanks, the greatest reductions were in summer when odour concentrations were higher. Pig slurry was more odorous than cattle slurry throughout the year. Ammonia emissions from stored slurry tanks is closely related to season and temperature. The summer emissions were two to three times higher than winter. Ammonia emissions were twice as concentrated as cattle slurry. Covering tanks could result in reductions of ammonia emitted by 70 to 90 percent. A lot depended on how well the cover fitted the store; where air could circulate emission losses were greater. Crust formation on cattle slurry stores helped reduce ammonia losses by 60-70 percent. Covering tanks caused an increase in the concentration of ammonia in the slurry in the store.

Controlling ammonia and odour emissions from buildings and stores

Research has concentrated on the use of bioscrubbers and biofilters to remove ammonia and odour emissions from livestock buildings. Costs are a major factor and simple cost effective systems need to be developed.

Experience in the Netherlands (14), based on many years of applying air scrubbing techniques to the exhaust air from livestock buildings has identified the most important causes of failure. The biochemical reactions which occurs in a bioscrubber due to the adsorbtion of ammonia and odorous chemical compounds are closely related to the process of nitrification. This in turn is dependent on a narrow range of pH the limits being between 6 and 10 with the optimum between 7 and 8. Research has shown that manipulation of the pH to maintain it at 7 to 8 can be achieved by the addition of dilute acids or alkalines. Marbel grains has been used successfully in farm systems. Other important factors are the addition of topping up water and the removal of effluent to maintain a balance of dissolved ammonia.

Work at Rosmalen Pig Research Station into the practical development of two biofilters, one made of peat and heather, the other of compost and bark have shown that both will reduce ammonia emissions by 85 percent and odours by 75 percent. Costs of construction and maintenance are high and have been estimated to be 25 guilders per pig reared.

Development of bioscrubbers to reduce odour and ammonia emissions have resulted in an improved design for new buildings (15). The research has shown that a large single bioscrubber which can accept airflows of 800 m^3h^{-1} are cheaper to build and easier to maintain. Monitoring has indicated that such scrubbers can achieve a 90 percent reduction in odour emissions. Work to achieve a similar reduction in ammonia emissions has concentrated on maintaining the pH between 7 and 7.5, and the development of nitrification within the plant.

Work to develop a successful bioscrubber for existing buildings has concentrated on providing small but effective scrubbing chambers for each ventilation outlet linked to a central bioscrubbing plant. So far the experimental plant has achieved 80 percent reduction of the odour emissions. However costs are higher than for the single bioscrubber, development work is continuing to achieve better performance and lower costs.

Research into the prevention of the development of odours in broiler houses (16) has indicated that controlling the moisture content of the litter can have a significant effect. The results of olfactometric techniques used to assess odour concentration, emission rate and odour intensity from commercial boiler houses under different management regimes showed that odour concentration and emission rate increased sharply after 30 days of age. Low moisture content litters produced significantly lower odour concentrations than high moisture 50 percent content litters.

Odour intensity measurements indicate that the exhaust air odour concentration must be reduced by >97 percent to reduce its intensity category to faint odour. This may be difficult to achieve with present air cleaning devices.

4 CONCLUSIONS

Between 30 and 40 percent of ammonia emissions arise from livestock buildings and manure stores. These are fixed source points and make a significant local effect on both natural and cultivated vegetation.

The collection of representative samples of ammonia and odour emissions from controlled ventilated buildings have been described and used successfully.

Collection of representative samples of odorous air and ammonia from naturally ventilated buildings and open manure stores are being developed.

A technique has been developed for assessing the nuisance level of odours from buildings and stores using panels which can be set up with using laboratory techniques.

Modelling of ammonia dispersion in buildings based on fluid mechanics has shown possibilities for the future design of livestock buildings.

Systems for controlling ammonia emissions by the rapid removal of manure from animal houses or by quick drying in situ show promise.

Biological scrubbers to remove ammonia and odours have achieved 90 percent reduction. The control of the process of nitrification within the scrubbers is necessary. High costs must be reduced if the systems are to be used on livestock farms.

Prevention of odour development by controlling the environment within the building indicates that this may be a cheaper method of control.

5 FUTURE WORK

Future work on the measurement, control and prevention of ammonia and odour emissions should include:-

Agreement on more precise recommendations on olfactometry with regard to collection, frequency and the accuracy of the mean of the sample and to the means of expressing the results.

Harmonisation of the methods of collection and evaluating the concentration of ammonia and the means of expressing the results.

Work on reducing odour and ammonia emissions must be directed to realistic and cost effective systems which can be used on farms. At present control systems are still at laboratory and pilot scale systems which are too expensive to be put into practical use.

ACKNOWLEDGEMENTS

The contributions of all the delegates who attended the workshop are gratefully acknowledged. I would personally like to thank my Senior Officer Mr P L Redman for allowing me the time to organise the workshop and the Director of the Institute of Engineering Research Professor J Matthews for all the facilities and assistance. Finally to Mr J H Voorburg for chairing the group and Mr P L'Hermite for his quiet and helpful assistance on behalf of the Commission.

REFERENCES

(1) "Odour Prevention and Control of Organic Sludge and Livestock Farming." NIELSEN, V.C., VOORBURG, J.H., and L'HERMITE, P (eds) 1986, Elesevier Applied Science Publishers, London.

(2) HANGARTNER, M.J., HARTUNG, J., PADUCH, M., PAIN. B.F. and VOORBURG, J.H. 1989 Improved recommendations on olfactometric measurements. Environmental Technology Letters 110, 231-236.

(3) "Volatile Emissions from Livestock Farming and Sewage Operations", NIELSEN, V.C., VOORBURG, J.H., and L'HERMITE, P., (eds) 1988, Elesevier Applied Science Publishers, London.

(4) HANGARTNER, M. 1989. Scaling of odour intensity. Measurement of odour emissions, Concerted Action SL/130/89 XII/ENV/S/89, Commission of the European Communities, Waste Recycling Programmes, 200 Rue de la Loi, Brussels.

(5) *ROELOFS, J.G.M. and HOUDIJK, A.L.F.M. Ecological effect of ammonia.

(6) *HARTUNG, J. Influence of housing and livestock on ammonia release from buildings.
(7) DENMEAD, O.T. 1983 Micrometeorological methods for measuring gaseous losses of nitrogen in the field. In: Gaseous losses of Nitrogen from Plant-Soil systems. FRENEY, J.R. AND SIMPSON (eds). MARTINUS NIJHOFF, The Hague.
(8) *SCHMIDT-VAN RIEL, C.J.M. Experiences with automatic ammonia analysis.
(9) *OOSTHOEK, J., KROODSMA, W., and HOEKSMA, P Ammonia emission from dairy and pig housing systems.
(10) *DE BODE, M.J.C. Odour and ammonia emissions from manure storage.
(11) *KRAUSE, K.H. and JANSEN, J. Modelling the dispension of ammonia within animal houses.
(12) *MANNEBECK, H and OLDENBURG, J. Comparison of the effects of different systems on ammonia emissions.
(13) *VALLI, L. PICCININI, S and BONAZZI, G. Ammonia emission from two poultry manure drying systems.
(14) *SCHOLTENS, R and DEMMERS, T.G.M. Biofilters and air scrubbers in the Netherlands.
(15) *SCHIRZ, S. Practical application of bioscrubbbing technique to reduce odour and ammonia.
(16) *CLARKSON, C.R. and MISSELBROOK, T.H. Odour emissions from broiler chickens.

* Papers in Ammonia and odour emissions from livestock production. NIELSEN, V.C., HARTUNG, J and PAIN, B.F. (eds) Proceedings of a COST 681 Workshop, Silsoe U.K. 1990

AMMONIA AND ODOUR EMISSIONS FROM SLURRY APPLICATIONS TO LAND AND FROM GRAZED PASTURES

Brian F. Pain
Institute of Grassland and Environmental Research, Hurley, Maidenhead, Berkshire SL6 5LR, UK

Summary

It is well established that livestock production is a major source of ammonia and odour. A range of methods were developed for quantifying rates of ammonia and odour to emission from grass and arable land. Emissions were influenced by a wide range of factors including the weather, soil structure and condition, cropping and slurry composition. Ammonia emission was reduced by alternative landspreading techniques (e.g. injection into soil), by rapid incorporation into arable land, by diluting slurry to increase the rate of infiltration into the soil and by using (e.g. acid to decrease pH) chemicals to modify slurry properties. Several of these methods were also effective in reducing odour although some (e.g. addition of acid) decreased ammonia but increased odour emission and vice versa.

1. INTRODUCTION

This paper reviews the research findings presented at a workshop entitled 'Ammonia and Odour Emissions from Livestock Production'. The meeting was organised within the framework of COST 681 and held at Silsoe, UK in March 1990. A complementary paper by V. C. Nielsen deals with emissions from livestock buildings and slurry or manure stores.

Previously, an expert group focused attention on harmonizing methods for the sampling and assessment of odours. This activity resulted in the publication of improved recommendations on olfactometric measurements in 1989 [1]. In addition, there were discussions and proposals for the harmonization of methods for measuring odour intensity through category scaling and for field measurement of odours with a panel.

As the consequences of ammonia volatilization from livestock production became increasingly well recognised, more attention was paid to this process. It is well established that livestock production, especially the aspects concerned with slurry and manure, is the major source of ammonia emission over much of Northern Europe. These emissions not only represent a significant loss of utilizable nitrogen [2] but more particularly a major source of environmental pollution. Ammonia is the most prevalent alkaline gas in the atmosphere and plays an important role in atmospheric chemistry and acid deposition. As ammonium aerosols, it can be transported over long distances so constituting a pollutant on an international scale [3]. It has been estimated that ammonia emissions in Europe have increased by about 50% since 1950 and doubled in some countries such as The Netherlands. Airborne ammonia can affect natural ecosystems both directly and indirectly [4]. The former is due to the poor detoxification of plants, e.g. coniferous trees, at low temperatures and is mainly restricted to cold climates. The indirect effects of ammonium deposition are more widespread and serious, resulting in acidification in weakly buffered soils and nitrogen enrichment. Such effects can result in change from natural heath -or peatlands into grassland. In forests, imbalance in the mineral composition of soils can follow leading to potassium and/or magnesium deficiencies and secondary symptoms related to stress factors.

2. METHODS

In order to calculate the total or rates of odour or ammonia emission, it was necessary to measure gas concentration and volumetric air flow or wind speed. In the field, therefore, it proved feasible to use similar methods for both types of emission during or following the application of livestock slurries to grassland or arable soils.

2.1 Ammonia concentration in air

This was measured mostly by drawing an air sample of known volume through absorption flasks or traps containing dilute acid. Ammonia concentration in the solution was analyzed by titration against acid [5], a colorimetric procedure [6] or by ion chromatography [7].

2.2 Odour concentration in air

Since no suitable chemical analyses exist for odour, measurements were usually based on organoleptic methods using olfactometers. Such methods were used to determine the odour threshold value, the number of odorous air: clean air dilution at which 50% of people can just perceive an odour, which is a dimensionless number. Although a weakness in the current recommendations on olfactometric procedures, this was frequently expressed as a concentration (Odour Units m^{-3} air) in order to calculate rates of emission. Samples of air were collected in bags (mostly of Teflon film) and measurements conducted in accordance with the recommendations in Hangartner *et al.* (1989).

2.3 Field experiments

Two main methods were employed in field experiments:

Micrometeorological methods have the advantage of allowing measurements to be made under ambient conditions, of not influencing the processes involved in the volatilization of ammonia or odorous compounds and of allowing experiments to be conducted on relatively large land areas using farm scale machinery. Experiments with **wind tunnels** are simpler to conduct, allow certain variables such as wind speed to be controlled and are particularly useful for comparative studies.

2.3.1 Micrometeorological methods. Mass balance methods were used in the measurement of emissions from relatively large (up to 0.2 ha) land areas to which slurry was applied [6,7,8] or on which livestock grazed [9]. For ammonia, wind speed and ammonia concentration were measured at 5 to 6 heights (e.g. between 0.2 and 3.5 m) at the upwind and downwind boundaries of experimental areas and emission calculated from the vertically integrated product of these two parameters divided by the fetch [10]. The fetch was the distance (m) between the upwind and downwind boundaries and, where feasible, circular areas were used to avoid variation in length with wind direction. A simplified method allowed gaseous emissions from the land surface to be inferred from measurement of wind speed and gas concentration at one pre-determined height termed ZINST [11]. Since, unlike chemical analyses for ammonia, the number of olfactometric measurements possible per day was always limited, the latter method was used in the measurement of odour emission from land following the application of slurries.

Preliminary experiments were conducted with an instrument (OPSIS) based on optical absorbance spectroscopy for measuring ammonia concentration in the air above slurry-treated land. The results obtained were comparable to those from the micrometeorological mass balance method.

A new type of passive sampler for ammonia was also developed [12]. This has the advantages over the micrometeorological mass balance method of not requiring wind speed and wind direction measurements, pumps or air flow meters and of being easy to operate with low labour requirements. An array of samplers, each consisting of two pairs of glass tubes connected in series and with the insides coated with oxalic acid, were assembled around the periphery of an experimental area. The samplers were designed to collect ammonia in an amount that was proportional to the product of ammonia concentration and wind speed. Comparable results between results from this method and the micrometeorological method were obtained for emissions from experimental plots with known ammonia emission rates.

An additional technique was developed for measuring emissions during slurry application with farm machinery [6]. The spreading machine was driven directly upwind and was followed by a modified Land Rover equipped with masts supporting an array of sampling points for odour and ammonia. Samplers, pumps etc were mounted in the back of the vehicle.

2.3.2 Wind tunnel methods. Small wind tunnels of varying design [5,7,13,14,15, 16,17] were used mainly for comparative studies on small plots. The tunnels were designed to have minimal effects on natural conditions and to allow the air flow rate across plots to be controlled. Emission was calculated from the product of volumetric air flow and ammonia or odour concentration in air leaving the tunnel.

2.4 Laboratory experiments

Experiments were conducted under controlled environmental conditions using a closed dynamic system [8,16]. This comprised a scrubber containing water to moisten the air, an acid trap for collecting ammonia and small volatilization chambers containing soil and slurry.

3 RESULTS

3.1 Odour and ammonia emission during spreading

Relatively few measurements were reported for emissions during spreading. In one series of experiments [6] in which pig slurry was applied with 5 different types of machine, ammonia emissions ranged from a mean of 0.09 g NH_3 m^{-3} slurry applied with a deep injector (150 mm deep), to 1.7 g NH_3 m^{-3} slurry applied with a conventional vacuum tanker and up to 22.5 g NH_3 m^{-3} slurry applied with a cable driven irrigator fitted with shrouding curtains. This high rate of emission was considered to be caused by the high speed discharge of slurry through small nozzles encouraging loss of ammonia through the formation of fine droplets of slurry.

Similarly, odour emission ranged from a mean of 6500 Odour Units m^{-3} slurry applied with the same irrigator to < 100 for low trajectory spreaders or injectors. For both odour and ammonia, the total emission during spreading represented a very small proportion of the total occurring during the two to three days after spreading. Nevertheless, rates of odour emission were high during spreading and using an injector or low trajectory spreader may still reduce complaints from the public because of the visual or psychological advantage. With these types of machine, it is far less obvious that slurry spreading is in progress.

3.2 Ammonia emissions following slurry application

After application to the land surface, high initial rates of emission during the first few hours decreased exponentially with time to much lower rates which were characterised by small diurnal fluctuations with changes in wind speed and temperature [7]. In general, over 50% of the total loss occurred within one day and the rate of loss was negligible within 5 days of application. Losses of ammonia equivalent to between 8 and 100% of the NH_4^+-N applied were recorded with an average of 45% for experiments conducted in different countries under wide ranging conditions.

3.3 Factors influencing ammonia loss

The amount of ammonia through volatilization following application to the land surface was influenced by a wide range of factors. These included environmental factors (rainfall, wind speed, temperature) soil conditions (water content, porosity, pH), the type of cropping, slurry type and composition (class of livestock, total solids content, pH, NH_4^+-N content) and rate of application.

A mathematical model was developed [18] to study the interaction of these chemical, physical and environmental factors influencing volatilization. Many of the important factors depend upon weather conditions, which cannot be predicted. Similarly, soil pH, a key factor in determining the equilibrium between NH_4^+ and NH°_3, is difficult to predict. Hence the model could not be used to predict amounts of ammonia lost but more to aid understanding of the processes involved.

Despite the complexity of the processes involved, a number of factors were identified as having a dominant influence over other factors in both field and laboratory experiments.

Temperature had a major effect on the total amount of ammonia lost after slurry application and could also influence changes in the rate of loss with time. In controlled laboratory experiments[8], increasing temperature from 0°C to 25°C increased total losses from cattle slurry from 28% to 45% of the NH_4^+-N applied and from 4% to 25% for pig slurry. This had important implications for the time [7,18] and season of application [5,16]. Highest total losses occurred from spreading in the summer during warm, dry weather conditions and the lowest from winter applications to cold (< 5°C) or partially frozen soil. However, very warm, drying conditions during the day resulted in rapid crust formation, especially on cattle slurry, which inhibited gaseous exchange and so reduced total losses [7].

The total solids content of the slurry frequently had an overriding influence on volatilization losses [8,14,16]. Under the same environmental conditions, losses were generally greater for cattle than for pig slurry and this difference ascribed to the lower TS content and viscosity of the latter resulting in more rapid infiltration into the soil and a sorption of NH_4^+ ions on to the soil colloids. Field experiments with wind tunnels showed that the six day cumulative losses from cattle slurry increased with increasing slurry TS content (r^2 = 81) [14]. Other studies demonstrated the interaction between slurry TS content and soil porosity which was especially important for more dilute pig slurry. In the latter instance, losses were much greater following application on compacted soils with impeded drainage than on well structured soils. For similar reasons, losses were usually greater from grassland than from bare arable soils. In addition, the presence of grass or crop residues tended to enhance losses by increasing the surface area of the applied slurry. Rainfall after spreading tended to reduce losses by washing slurry off the leaves and into the soil.

3.4 Methods for reducing ammonia emissions

Four main methods showed particular promise for reducing ammonia emission from slurry applications. These were landspreading techniques, techniques for incorporating slurry into arable soils, dilution of slurry prior to spreading and the use of chemicals to modify slurry properties. There was evidence that reducing losses of N as ammonia enhanced losses through other pathways e.g. as nitrous oxide through denitrification.

3.4.1 Landspreading techniques. As to be expected, deep (150 mm) injection of slurry into the soil was the most effective method [2,6,7,8]. For example, immediately after application to grassland, deep injection resulted in a rate of ammonia emission which was only 3% of that from more conventional spreading with a vacuum tanker. Similar results were obtained for a shallow injector and, since this is likely to be less costly to operate, warrants further investigation. The benefits of injection were generally greatest when rates of emission from surface application were high, for example, during warm weather conditions. Application by means of a spreader fitted with a series of hoses to place slurry directly on the soil surface resulted in a 30% reduction in emission compared with a conventional baffle plate spreader.

3.4.2 Incorporation. On arable land, incorporation of slurry [7,8,16] into the soil was effective in reducing emissions by up to 90%. However, since rates of emission were highest during the first few hours, the operation needed to be carried out soon after application. Burying slurry (e.g. by ploughing) or mixing slurry and soil (e.g. by rotovating) to ensure close contact between NH_4^+ ions and soil colloids, were the most effective methods if undertaken within 3 to 6 hours of spreading.

3.4.3 Improving flow properties of slurries. Under suitable soil conditions, diluting slurry [7,8] or mechanically separating [8] to produce a free-flowing liquid prior to application was effective in reducing ammonia emissions. This effect was due to the increased rate of infiltration into soil for more dilute, less viscous slurries. In The Netherlands, a dilution of slurry to water of 1 : 3 was proposed [7]. This was particularly effective for cattle slurry (TS% *ca.* 10%) and reduced ammonia loss from 65% of the NH_4^+-N applied to 19%. It was recognised that this procedure would increase the volume of slurry requiring transportation. An alternative was to irrigate from an adjacent ditch or borehole immediately after spreading.

3.4.4 Chemicals to modify slurry properties. Volatilization of ammonia from slurry can be represented by:

$$NH_4^+ + HCO_3^- \qquad NH^o_3 + CO_2 + H_2O$$

Adding strong acid [13,15] to the slurry neutralizes the alkalinity from bicarbonate derived from the hydrolysis of urea and so reduces ammonia volatilization. Experiments showed that reducing slurry pH, which was normally > 7, to ca. 5 before application to land resulted in negligible ammonia losses. Other experiments demonstrated that adding $CaCl_2$, $MgCl_2$ or formaldehyde also reduced ammonia volatilization but that large amounts were needed to obtain a worthwhile reduction. A mixture ('Kapto') containing carefully chosen proportions of these three materials gave much better results. The total ammonia loss was reduced by 70, 85 and 44% for cattle, pig and poultry slurry, respectively [5].

3.5 Ammonia emissions from grazed pasture

The major source of ammonia in grazed systems is from the nitrogen excreted in dung and urine, particularly that derived from enzymic hydrolysis of the urea contained in the urine. On average, 10% of the nitrogen deposited in urine was lost as ammonia [17]. Other sources of less significance include release from plants *via* their stomata, from senescing foliage or from cut grass under humid conditions and from applied fertilizer. The mechanisms controlling the extent of net volatilization are complex and poorly understood but field experiments have shown that grazed grassland is a source of ammonia and that the magnitude of the annual losses increase with increasing nitrogen input from fertilizer or biological fixation [9]. However, recent calculations suggest that the combined losses from buildings, slurry stores and slurry spreading can be from 3.4 to 11.4 times greater per animal than those during grazing.

3.6 Odour emission from landspreading

Although far fewer experiments were reported, it would appear that the factors influencing odour emission from land are similar to those for ammonia volatilization. Much greater variation in mean values for odour emission made it more difficult to draw clear conclusions. Nevertheless, it proved possible to establish for untreated slurries good, positive linear relationships between odour and ammonia emission rates both during ($r = 0.85$) and following ($r = 0.92$) slurry application [13].

After spreading, changes in the rate of odour emission with time were as reported for ammonia. High initial rates for a few hours were followed by much lower rates with small diurnal fluctuations. Generally, odour was detectable for a shorter period (48 - 72 h) than ammonia, but this may have been due to less sensitive measurement methods for the former. Total emission was influenced by temperature, soil conditions and the type and composition of the slurry.

3.7 Methods for reducing odour emission

Several of the methods used for reducing ammonia emission were also effective for odour. These including injection and incorporation into arable land [7]. For the former, immediately after application to grassland, injection resulted in an odour emission rate which was 27% of that for a low trajectory spreader, and 15% of that for a conventional vacuum tanker [6]. Alternatively, some methods for reducing odour increase ammonia loss and vice versa. Adding 5.0 to 7.5 kg CaO m^{-3} slurry decreased the odour concentration following application to land from 27552 to 640 (Odour Units m^{-3} air), but the resulting increase in slurry pH increased ammonia concentration from 1172 to 134483 μg NH_3-N m^{-3} [15]. Conversely, adding nitric acid to reduce slurry pH markedly reduced ammonia concentration but increased odour. Aerobic treatment of slurry was very effective in reducing odour but, under certain conditions, the process could result in increased ammonia loss from land through increasing slurry pH [13].

4. CONCLUSIONS

4.1 A range of methods were developed for use in the measurement of both ammonia and odour emission from grassland and arable land. Odour concentration was determined by dynamic dilution olfactometry.

4.2 Although rates of emission were high, total emission during slurry application represented a very small proportion of that following application and was dependent on the type of spreader used.
4.3 Following slurry application to land, rates of emission were high for a few hours then decayed exponentially with time with small diurnal fluctuations.
4.4 Rates of emission following slurry application were influenced by a wide range of factors including temperature, rainfall, soil structure and conditions, the type and composition of the slurry.
4.5 Both odour and ammonia emission were reduced by using alternative landspreading techniques such as injection into soil, by rapid incorporation into arable soils or by dilution of slurry to increase the rate of infiltration into soil. Other methods for reducing ammonia emission (e.g. acidification) increased odour emission and vice versa.
4.6 Grazed grassland was shown to be a source of ammonia emission, the magnitude of the annual losses increasing with increasing nitrogen input from fertilizer or biological fixation. On average, 10% of the nitrogen deposited in urine was lost as ammonia. Losses from slurry were much greater than those from grazing.

5. FUTURE WORK

Future work on emissions of volatile compounds from livestock production should include:
5.1 Further improvements to the recommendations on olfactometry, especially to increase the precision of measurements and to find more valid means of expressing results.
5.2 Harmonization of methods for measurement of volumes of polluted air, sampling techniques and analyses. This is urgently needed for ammonia measurements.
5.3 To more fully develop methods for reducing ammonia emission for use on farms.
5.4 To investigate the effects of reducing ammonia volatilization from slurry or manure applications on other losses of nitrogen e.g. of nitrous oxide and nitrates.
5.5 To more fully understand factors involved in the formation and release of odorous compounds from slurries and manures.

ACKNOWLEDGEMENTS

All the contributions from the delegates to the Silsoe workshop are gratefully acknowledged, especially the chairmanship of the expert group by Ir J. H. Voorburg, organisation of the workshop by Mr V. C. Nielsen and administration by Mr P. L'Hermite.

REFERENCES

1. HANGARTNER, M. J., HARTUNG, J., PADUCH, M., PAIN, B. F. and VOORBURG, J. H. 1989. Improved recommendations on olfactometric measurements. Environmental Technology Letters **10**, 231-236.

2. *CARTON, O. T. and TUNNEY, H. Policy for ammonia emissions in Ireland.

3. *APSIMON, H. M. and KRUSE-PLASS, M. The role of ammonia as an atmospheric pollutant.

4. *ROELOFS, J. G. M. and HOUDIJK, A. L. F. M. Ecological effect of ammonia.

5. *VLASSAK, K., BOMANS, H. and VAN DEN ABEEL, R. Ammonia emission and control after landspreading livestock waste.

6. *PHILLIPS, V. R., PAIN, B. F. and KLARENBEEK, J. V. Factors influencing the odour and ammonia emissions during and after the landspreading of animal slurries.

7. *KLARENBEEK, J. V. and BRUINS, M. A. Ammonia emissions after landspreading of animal slurries.

8. *DOHLER, H. Laboratory and field experiments for estimating ammonia losses from pig and cattle slurry following application.

9. *JARVIS, S. C. Grazed pastures as a source of ammonia.

10. DENMEAD, O. T. 1983. Micrometeorological methods formeasuring gaseous losses of nitrogen in the field. In: Gaseous Losses of Nitrogen from Plant-soil Systems. Freney, J. R. and Simpson, J. R. (eds). Martinus Nijhoff, The Hague.

11. PAIN, B. F., CLARKSON, C. R., PHILLIPS, V. R., KLARENBEEK, J. V., MISSELBROOK, T. H. and BRUINS, M. A. 1990. Odour emission arising from application of livestock slurries on land: Measurements following spreading using a micrometeorological technique and olfactometry. Journal of Agricultural Engineering Research (in press).

12. *FERM, M. Full investigation of methods to measure ammonia volatilization.

13. *PAIN, B. F. and MISSELBROOK, T. H. Relationships between odour and ammonia emission during and following the application of slurries to land.

14. *SOMMER, S. G. and CHRISTENSEN, B. T. Effect of dry matter content on ammonia loss from surface applied cattle slurry.

15. *KLASINK, A., STEFFANS, G. and KOWALENWSKY, H. H. Odour and ammonia emissions from grassland and arable land.

16. *AMBERGER, A. Ammonia emissions during and after land spreading of slurry.

17. *VERTREGT, N. and RUTGERS, B. Ammonia emissions from grazing.

18. *CHARDON, W. J., VAN DER MOLEN, J. and VAN FASSEN, H. G. Modelling ammonia emissions from arable land.

*Paper in 'Ammonia and odour emissions from livestock production'. NIELSEN, V. C., HARTUNG, J. and PAIN, B. F. (eds). Proceedings of COST 681 workshop, Silsoe, UK., 1990.

LONG-TERM HEAVY-METAL LEACHING FROM SOILS, SEWAGE SLUDGE AND SOIL/SEWAGE SLUDGE MIXTURES

P.J. van Erp and P. van Lune
Institute for Soil Fertility Research,
P.O. Box 30003, 9750 RA Haren (Gr.), the Netherlands

Summary

The effect of applying sewage sludge to soils on leaching of Cd, Cr, Cu, Ni, Pb and Zn was assessed in a long-term pot experiment. The treatments were a sandy and a heavy clay soil, replaced for 0, 10 or 100% (by volume) by sewage sludge. The largest amounts of heavy metals were leached from the 100% sewage sludge treatment. Compared with untreated soils, leaching from the soil/sewage sludge mixtures was reduced in the sandy soil mixture for Cd, Cr, Ni, Pb and Zn, and from the clay soil mixture only for Cr. Leaching of the other metals was greater for the soil/-sludge mixtures. The cumulative amounts of the heavy metals leached during 14 years could be fitted very well to a quadratic equation. Leaching of Cd, Cr, Ni and Zn in the treatments, especially in the 100% sewage sludge treatment, leveled off in the course of time, while that of Cu and Pb increased. Groundwater quality standards were frequently exceeded, especially the background level and the B standard. Most striking was the frequency of Cu exceeding the guidelines, this in contrast with Pb.

1. INTRODUCTION

Sewage sludge contains several heavy metals in relatively high concentrations. Therefore, the agricultural use of sewage sludge as a fertilizer or soil conditioner may result in leaching of heavy metals from the soil into the groundwater. The mobility, retention and release of heavy metals is governed by precipitation, complexation, adsorption and oxido-reduction (1).

Little is known about the long-term effects of leaching of heavy metals from agricultural soils into the groundwater after applying sewage sludge or substituting it for soil, and the soil-chemical processes involved. More is known about short-term effects (3, 4).

In 1972 a long-term pot experiment was started in the open air to assess the effect of applying sewage sludge to soils on leaching of Cd, Cr, Cu, Ni, Pb and Zn. This paper gives an overview of the experimental results of the period 1972-1986.

2. MATERIALS AND METHODS

In 1972 150-l pots with a diameter of 50 cm and a height of 75 cm were filled with a light sandy soil or a heavy clay soil, replaced for 0, 10 or 100% (by volume) by air-dried sewage sludge. The 0 and 10% sewage sludge treatment of the sandy soil and the clay soil are referred to as S0 and S10, and C0 and C10, respectively, and the 100% sewage sludge treatment as SS. The experiment was executed in triplicate. Characteristics of soils and sewage sludge are given in table 1. The bottom of the pots was covered with a 5-cm deep gravel layer, which was covered with nylon sheet. Just above the bottom of the pots an outlet pipe was attached discharging the drainage water into 2-litre polyethylene storage bottles. At irregular intervals the bottles with drainage water were collected, the amount of drainage water was recorded and the samples were stored at 4°C. During the 14-year period 21 samples of drainage water were analyzed according to standard methods (5).

During the experiment various crops were grown in the pots; these were fertilized at commonly used rates to ensure good crop growing conditions. Water was added to the pots only during extremely dry periods. Soil pH was not adjusted until 1984.

Table 1. pH and composition (on a dry-matter basis) of soils and sewage sludge in 1972.

		Sandy soil	Clay soil	Sewage sludge
pH-KCl		5.1	5.2	6.5
Organic matter	%	3.4	3.2	40.6
CaO	%	0.07	0.48	2.00
<16 μm*	%	5.2	56.6	42.4
Cd	mg/kg	0.3	1.1	2.9
Cr	"	19	69	193
Cu	"	2	77	933
Ni	"	8	42	46
Pb	"	17	56	538
Zn	"	14	151	2394

* Analyzed in samples taken in 1987

To assess the effect of applying sewage sludge to soils on leaching of heavy metals, the average heavy metal concentrations in the drainage water of the treatments were compared in the same sampling period; the differences between the treatments were recorded as an increase or a decrease. Differences between the treatments over the full 14-year period were statistically tested ($P < 0.05$) by using the sign test.

The average heavy metal concentration in the drainage water samples multiplied by the amount of drainage water gives the amount of heavy metal leached from the pots in the period concerned. Summation of these amounts gives the total amount of heavy metals leached in the course of the experimental period. The cumulated amounts of heavy metals as a function of time were fitted to a quadratic equation.

The heavy metal concentrations found in the drainage water during the 14 years of the experiment were evaluated on the basis of Dutch standards for characterizing groundwater quality (table 2). The percentage of observations exceeding these standards was calculated.

Table 2. Percentage of observations exceeding the A, B and C standards of the Dutch guidelines (2) indicating the degree of groundwater contamination.

Heavy metal	Standard*	Treatment**				
		S0	S10	C0	C10	SS
Cadmium	A (1)	14	14	14	24	33
	B (2.5)	0	0	5	14	14
	C (10)	0	0	0	0	5
Chromium	A (20)	32	37	26	16	42
	B (50)	5	5	5	0	11
	C (200)	0	0	0	0	5
Copper	A (20)	67	100	57	100	100
	B (50)	5	100	5	86	95
	C (200)	0	0	0	0	43
Nickel	A (20)	10	5	0	14	81
	B (50)	5	5	0	10	29
	C (200)	5	0	0	5	5
Lead	A (20)	5	0	0	0	0
	B (50)	0	0	0	0	0
	C (200)	0	0	0	0	0
Zinc	A (50)	62	67	5	29	48
	B (200)	0	5	0	5	5
	C (800)	0	0	0	0	0

* () The standards in μg/l ()
A = natural background concentration
B = possible contamination
C = contamination, decontamination is necessary

** S0 and S10 = sandy soil + 0 or 10% (by volume) sewage sludge, respectively
C0 and C10 = clay soil + 0 or 10% (by volume) sewage sludge, respectively
SS = 100% (by volume) sewage sludge

3. RESULTS AND DISCUSSION

The concentration range of most heavy metals in the drainage water was wide. The highest concentrations of Cd, Cr, Cu, Ni, Pb and Zn were found in the beginning of the experiment. Later the concentration decreased sharply to a "steady" level, depending on the heavy metal, treatment and year of the experiment. In the experiment the average annual amount of water leached was about 300 mm.

Statistical analysis of the effects of soil type and soil replacement by sewage sludge on the concentration of heavy metals in the drainage water showed significant differences:

- Zn concentration in the drainage water in the clay soil was lower than that in the sandy soil;

- replacement of the sandy soil by 10% sewage sludge resulted in higher Cu concentrations; on the clay soil replacement resulted in higher Cu, Ni and Zn concentrations;
- compared with the sandy soil, treatment SS (100% sewage sludge) resulted in higher Cu and Ni concentrations; compared with the clay soil, treatment SS resulted in higher Cr, Cu, Ni and Zn concentrations;
- treatment SS resulted in lower Zn and higher Cu and Ni concentrations than the S10 treatment, and in higher Cr, Cu, Ni and Pb concentrations than the C10 treatment.

The average annual leaching of heavy metals was greatest in the SS treatment; an exception was Zn in the sandy soil (table 3). The average annual leaching of heavy metals, except copper, in treatment S0 was always higher than in treatment S10. The opposite occurred on the clay soil, with chromium as an exception. A possible explanation is the pH of the sandy soil.

Table 3. Average and median concentrations and concentration range (µg/l) of heavy metals in drainage water and the calculated average annual leaching (g/ha) of heavy metals during the 14-year period.

Heavy metal	Treatment*				
	S0	S10	C0	C10	SS
Cd average	0.8	0.8	0.6	0.9	1.0
median	1.0	1.0	1.0	1.0	1.0
range	0.2-3.0	0.2-3.5	0.1-4.4	0.1-8.5	0.1-11
leaching	2.6	2.0	2.8	3.5	5.1
Cr average	14	18	15	11	24
median	13	20	14	10	20
range	2-100	4-82	2-110	1-79	3-217
leaching	47	39	55	36	93
Cu average	23	79	24	63	198
median	26	80	21	60	150
range	5-103	57-109	7-85	40-103	49-805
leaching	78	179	87	199	774
Ni average	8	9	7	13	39
median	7	10	7	11	33
range	1-76	2-52	1-38	1-192	5-305
leaching	47	27	29	75	172
Pb average	4.3	4.9	4.0	4.1	5.4
median	5.0	5.0	5.0	5.0	5.0
range	0.8-21.8	1.1-22.7	1.0-16.0	1.1-15.6	1.4-21.0
leaching	13.7	12.7	13.9	14.9	19.1
Zn average	60	70	17	39	51
median	80	69	20	37	50
range	10-353	17-288	1-212	5-297	9-288
leaching	207	173	88	154	197

*Treatments as in table 2

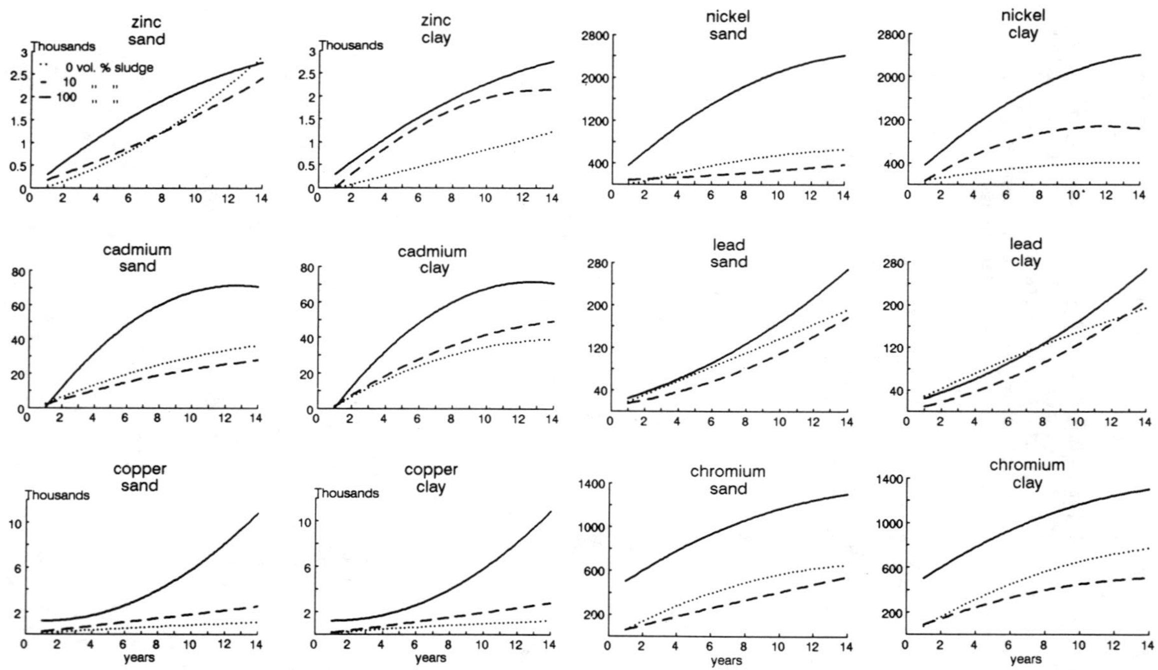

Figure 1. Total cumulative amounts of Cd, Cr, Cu, Ni, Pb and Zn (g/ha) leached from a sandy and a clay soil, replaced for 0 or 10 volume % by an air-dried sewage sludge, and in pure sewage sludge (100 volume %) under agricultural use.

.... 0 volume % sewage sludge
-- -- -- -- 10 " " "
———— 100 " " "

During the experiment no soil chemical or soil physical factors or parameters were measured. Differences in leaching behaviour of the heavy metals and soil/sewage sludge mixtures can therefore not be explained.

The fitting of the cumulative amounts of heavy metals to a quadratic equation resulted in a highly significant (F-test, 1% level) description of the leached amount of heavy metals during the 14-year period. The results are presented in figure 1. Some treatments showed minor differences between the calculated and actual heavy metal leaching in the first two years. This unexpected behaviour may have been caused by the treatment of the drainage water, the method of heavy metal analysis, the aftereffects of pot-filling and the incorrect decision to describe leaching as a quadratic function in the first two years.

The treatments, especially treatment SS, showed that the long-term leaching behaviour of the heavy metals differed (figure 1). Total cumulative amounts of leached Cd, Cr, Ni and Zn leveled off, which means that the concentrations in the drainage water decreased in the course of time, whereas the Cu and Pb concentrations increased. Because Cu and Pb are strongly bound to organic matter, decay of organic matter and the mobility of soluble organic matter may have caused this.

Assuming that under natural conditions drainage water will gradually become groundwater, and that the leaching pattern and heavy metal concentrations measured in this experiment will also occur under natural conditions, it is possible to calculate the percentage observations expected to exceed the A, B and C standards of the Dutch guidelines. The results of these calculations are presented in table 2.

All heavy metals, except Pb, showed a high percentage of analytical results exceeding the A standards (background level). The percentage of analyses exceeding the B standards was lower, but especially in treatments C10 and SS a considerable percentage of the observations exceeded the B standards. Exceptions were Pb and Zn. The high percentage of drainage water samples exceeding the B standard for Cu is striking, especially in the soil/sludge and 100% sewage sludge treatments. The C standard was exceeded only in the case of Cu in treatment SS.

4. LITERATURE

(1) Ahtiainen, M. (1986) Groundwater quality in relation to land application of sewage sludge. In: Factors influencing sludge utilisation practices in Europe. Proceedings seminar CEC held in Liebefeld. Eds.: R.D. Davis, H. Haeni and P. L'Hermite. Elsevier Appl. Sci. Publ., pp. 51-64.

(2) Anonymous (1983). Leidraad bodemsanering, afl. 1

(3) Gerritse R.G, R. Vriesema, J.W. Dalenberg & H.P. de Roos (1982) Effect of sewage sludge on trace element mobility in soils. J. Environ. Qual. 11: 359-364.

(4) Huylebroeck, J. (1981) Review of research projects on groundwater pollution from agricultural use of sewage sludge. Treatment and use of sewage sludge. Commision of the European Communities. Cost project 68 Bis. Final report. III Technical annexes, pp. 491-521.

(5) Vierveijzer, H.C., A. Lepelaar en J. Dijkstra (1979) Analysemethoden voor grond, rioolslib, gewas en vloeistof. Instituut voor Bodemvruchtbaarheid, Rapport

SESSION REPORT

J.E. HALL, WRC, Medmenham, Marlow (UK)

Questions to A Siegenthaler

F Huysman In Belgium we have carried out similar microbiological tests with pig manure and observed decreases in the copper resistance of micro-organisms and in the nitrogen fixation of blue green algae. Don't you agree that you need to look at activity rather than numbers of micro-organisms present?

A Siegenthaler For soils in Switzerland, copper is generally not a problem. I agree in principle but whilst total microbial activity does tell you much, diversity and inbalance in populations tells more.

K Aichberger How much leaching of phosphate did you find? With the addition of pig slurry, did other elements increase in the soil other than zinc which you showed?

A Siegenthaler We only looked for accumulation in top soil. There was a tendency for increases deeper but we don't know whether this was due to leaching or not. We made monthly measurements and the many sampling holes may have contributed to phosphate movement. As regards other elements in the soil, these did increase, particularly copper.

R Leschber Your application rates of 2 and 5 tonnes of organic per hectare and year are six times higher than the German standard for sewage sludge. We know the consequences of too high applications of sludge, and pig slurry will have similar effects but there is no directive controlling it.

A Siegenthaler Agreed. When these studies were planned in 1973, the objectives were rather different from today - how much land area for application was required for slurry production, and consequently rates of addition were high.

S Sakellariadis Was there increase in the mobility of the heavy metals from the fertiliser?

A Siegenthaler Ammonium acetate extractable zinc increased as well as total zinc.

P van Lune What were the heavy metal levels in the crops?

A Siegenthaler They were not very high with no dramatic increases except in vegetables, particularly celery.

Questions to N van Bremen

Scheidegger Did you apply ammonium sulphate to the leaves or the soil?

N van Bremen Both.

H Siegrist Nitrification normally stops at pH 6 but your results suggest pH 4.

N van Bremen Most nitrification takes place in the forest floor and tends to have a slightly higher pH. Autotrophic nitrification has been shown to occur down to pH 3 only recently.

M de Bertoldi Which method did they use to separate autotrophic and heterotrophic nitrification?

N van Bremen I think they used nitrification inhibitors.

H Siegrist What method was used for measuring aluminium concentration?

N van Bremen Soil solution was extracted with suction cups.

Questions to V C Nielsen

R Unwin What are your thoughts on the relative emissions from solid bed systems for animals and their future in livestock systems?

V C Nielsen No techniques have been developed for measuring ammonia emissions from straw-bedded systems. Most research has been done on slurry systems. Some work has been done on poultry manure composting in Italy which had high ammonia losses. By implication, bedding systems may also have high losses.

N van Bremen Has any thought been given to establishing ammonia resistant trees around livestock units?

V C Nielsen An interesting thought that is worth investigating.

J H Voorburg Some modelling work has been done with hedgerows but only as a means of physical dispersal rather than treatment.

J A Hansen What extent area biofilters and scrubbers used in Europe?

V C Nielsen The greatest concentration of biofilters is in Bavaria. There are a few in the Netherlands mostly on experimental farms, a famous one in Sweden, one or two in the UK. They are not common due to high cost and used principally for odour control but they are more efficient at removing ammonia.

N van Bremen What is the relationship between odour and ammonia? Is it possible to measure odour analytically?

V C Nielsen The relationship between odour and ammonia is specific to a particular source. You cannot transfer co-relationships between sites since different mixtures occur on different farms. Many chemicals make up odour and intensive research has been carried out to find marker chemicals but unfortunately it seems to be site specific.

J Hannan If odour filters cost 25 Guilders per pig reared in the Netherlands, what is the profit margin per pig? This cost is very high.

V C Nielsen That is why odour filters are only used on experimental farms!

Questions to B F Pain

J A Hansen Loss of ammonia from urine can be high. Can the composition of the feed be altered to reduce losses?

B F Pain With pigs, this can be done by fine tuning the feed composition with artificial amino acids. It is not really possible with ruminants, particularly cows.

S Sakellariadis When you mentioned low and high porosity of soil, do you refer to pore size or pore space?

B F Pain Rate of infiltration would be a better term.

Questions to P van Lune

G A Fleming Copper uptake continued to increase with time unlike the other metals.

P van Lune Copper is bound to organic matter, thus delaying release.

R Chaney You showed figures for chromium. Was this as chromate or chromic since it is only chromate which is the problem? Chromic is not toxic.

P van Lune This was chromic.

R Chaney It is normally expected that the greatest risks due to heavy metals are when the organic matter levels decrease but your results showed the short term to be the problem.

P van Lune Agreed. Most metals were released in a short period and then declined, except copper.

N van Bremen I was surprised that there was no difference between clay and sandy soils. Clay would usually be better structured, possibly giving rise to short circuit flow whereas sand provides more even infiltration.

P van Lune It is important to recognise that the data for leachate is as total amounts, ie, concentration times volume.

H Siegrist What is the percent leached of the total soil content?

P van Lune We did not look at that.

J A Hansen I wonder if your experiment is really relevant to normal applications of sludge?

P van Lune In 1972 the problems were seen to be different. It was an approach to simulating long term effects. Lower application rates are required now.

Unknown How much did you water your pot experiment? What soil moisture did you maintain?

P van Lune The pots were watered only when required.

Session III

ANALYSIS AND RISK ASSESSMENT OF ORGANIC MICROPOLLUTANTS

Chairman : L.E. DUVOORT–VAN ENGERS (NL)

Rapporteur : N. HARKNESS (UK)

ORGANIC SUBSTANCES IN SEWAGE SLUDGES STATE OF THE ART

R. LESCHBER
Institute for Water, Soil and Air Hygiene, Federal Health Office, Berlin

SUMMARY

In the last years a number of investigations on the occurrence, fate and disposal of organic contaminants in sludges were undertaken. The reason was to improve the knowledge about these substances, some of them were thought to be potential soil pollutants.

These investigations led to the results that organic contaminants in sludges will cause no problems to soil and growing plants if sludges are used in agriculture in accordance to legal regulations. Systemic transfer of organics from soil to plant could not be observed at currently found concentrations. Among organic contaminants which may be present in sludges in elevated concentrations, due to the effects of industrial effluents, highly persistent organohalogen compounds play an important role. Therefore the revised German Sewage Sludge Regulations will comprise limitations for these substances. Mean concentrations presently found in sludges indicate that there may be no harmful effects for agricultural land.

Indeed, the problem of organic substances needs further observation especially if sludge is applied to grassland and forage crops. On the one hand future investigations under the conditions of good agricultural practice may improve the knowledge about the occurence and fate of organics. On the other hand all measures must be undertaken to minimize the input of harmful substances into waste waters thus leading to sewage sludges which may be applied to land according to a careful soil protection.

1. INTRODUCTION

The ECE Directive 86/278 of 1986 (3), covering the agricultural use of sludge followed similar national regulations which had come into force quite earlier and were set up to contribute to soil pollution prevention. All these regulations during the early 80s limited the input of heavy metals via sludge into soil, at this time known as major pollutants. There was only less knowledge about the role of organic pollutants. During the following years the things changed and research work concentratedno longer only on the occurence, behaviour and fate of heavy metals but on the problems of organic priority pollutants as well.

At present, in some countries the organic pollutants seem to play a dominant role in discussions on the agricultural use of sewage sludge. The reasons for that are on one hand that questions of heavy metals have been answered by investigations during the last years and on the other hand that there is a wide variety of organic compounds used in households and industry which find their way into sewage sludges via waste water treatment. A large number of them is now detectable and may be identified according to progress in analytical chemistry made in recent years.

However, actual research work has made it clear that the problem may reduced to only a few groups of contaminants which are of main interest due to larger quantities which are brought into the environment or due to the persistency and toxicological behaviour of these substances.

When going into detail,it seemed that the organochlorine compounds were playing a major role, among them especially polychlorinated biphenyls (PCB) and polychlorinated dibenzodioxins/dibenzofuranes (PCDD/PCDF). Moreover, some other groups of organics like surface active agents, polycyclic aromatic hydrocarbons (PAH) and phthalates are of interest as well. During the last two years new results of research investigations on all groups of pollutants named before were reported at special scientific conferences like the Joint Workshop COST 641/681 held in Brussels, 26 - 27 October 1988 (21), the Sludge Section of the 6th International Recycling Congress, Berlin, 28 - 30 November 1989 (19), the Joint Seminar COST 681, Working Parties 4 and 5 held in Braunschweig, 6 - 8 June 1990 (13) and the 102. VDLUFA-Congress Berlin, 17 - 22 September 1990 (23). From all reports it can be concluded that earlier findings on organic pollutants in sewage sludges were confirmed. Under normal conditions e.g. those laid down in the ECE Directive or in national regulations on the agricultural use of sewage sludges, and if extremly polluted sludges with a dominant industrial influence will be excluded, no harmful effects to soil and plants caused by organic contaminants may occur. In the following the most important groups of organics will be discussed with special regard to their potential for the food chain.

2. POLYCHLORINATED DIBENZODIOXINS AND FURANES (PCDD, PCDF)

Discussions about the importance of PCDD/PCDF for the agricultural use of sludge in Germany came up when H. Hagenmaier and co-workers presented results of investigations from 26 sewage treatment works in a report to the German Federal Environmental Agency (12). The findings in 17 treatment works in various areas of the Federal Republic and 9 in the upper Rhine area (districts of Waldshut and Lörrach) revealed in some cases extremely high values especially for the Waldshut-Lörrach region. In general, the values ranging between 47 and 108 ng TE/kg dm (equivalents of toxicity per kilograms of dry matter) did not cause immediate concern. These results have been discussed in detail quite earlier

(16) with special emphasis on the high persistency of PCDD/PCDF. In the meantime a great number of investigations has been performed by the German Länder and by several communities to make clear whether agricultural use of sludge may be critical or not due to high PCDD levels. The results of investigations at more than 250 sewage treatment plants showed values far below 100 ng TE/kg. Thus the mean values of investigations for plants in Bavaria, Hesse and Lower Saxony were in the range 30 - 50 ng TE/kg. Taking into account common agricultural practice, the amount of sludge applied to land due to the restrictions of the Regulations (5 tons dm/3 yrs · ha), and the half-life of PCDD in soil which may be calculated as 10 to 20 years, it may be assumed that there will be no harmful effects to agricultural land. Two points may be taken into account:

- Because some sludges may have higher contents of PCDD, the levels of these substances in representative sludge samples have to be controlled,
- the application of sludge to grassland should be avoided or subjected to stringent regulations due to the fact that grazing cattle may directly ingest contaminated grass or soil.

3. POLYCHLORINATED BIPHENYLS (PCB)

Kampe (15) and Witte et al. (24) have published results of investigations as well as Drescher-Kaden et al. (10) and Schaaf (22) where the mean values for single PCB congeners 28, 52, 101, 138, 153 and 180 in sludges were 0,1 mg/kg and in soils ranged from 1,4 µg/kg (normal application) to 100 µg/kg (higher amounts of sludge applied). When sludge was applied in an amount 200 times higher than allowed by the German Regulations 339 µg/kg PCB were observed in soil. But there were no differences in PCB contents in plants from the sludge areas and the controls so that PCB transfer from soil to plant may be excluded. Although PCB are very persistent in the environment it has been reported that PCB concentrations in sludges have decreased from 4 to 1 mg/kg during the last two decades as a result of restrictions for PCB use in the Federal Republic (2).

4. ORGANOCHLORINE PESTICIDES

The authors named before (10, 15, 24) have reported data for chlorinated pesticides with mean values 0,02 - 0,5 mg/kg in sludge dry matter which characterize the present situation and may allow calculations of the hygienic risk from this group of substances. These results did not indicate harmful effects on agricultural land, where the concentrations in soil for DDT, aldrin, dieldrin and lindane were quite low, only hexachlorobenzene showed an elevated concentration, with a mean of 7 µg/kg when sludges with average concentrations of 0,1 mg/kg of the chlorinated substances had been applied to agricultural land.

5. AOX, EOX, POX

In the last years, cumulative parameters for halogenated substances have got more and more importance for the investigations of waste water and sewage sludge. Corresponding methods for the determination of AOX and EOX in waste water have been standarized in the Federal Republic of Germany in 1984 and 1985 (6, 7) and for sludge and sediments in 1989 (8, 9).

In a co-operative research project (18) 170 representative sludge samples from sewage treatment plants from all parts of the Federal Republic were investigated. There generally 0 to 10 mg/kg for purgeable organohalogen compounds (POX) and only in three cases higher values were found. Substances extractable by hexane (EOX) were present in 95 % of all cases in the 0 to 70 mg/kg range where 72 % of all values lay below 30 mg/kg. The highest value was 285 mg/kg. Organically bound halogens either absorbed or incorporated in the sludge matrix, measured as AOX, were found to be between 60 and 400 mg/kg with a maximum of results in the 200 to 300 mg/kg range. AOX values showed a pattern like a Gaussian distribution up to 900 mg/kg. Three values were much higher (1600, 1800 and 3000 mg/kg) resulting from the influence of specific industrial waste waters.

Evaluation of these results seemed to be difficult because there was no information about the nature of substances which had been determined and sufficient data about single organohalogen substances in sludges were not available to establish correlations.

Nevertheless, it had been stated already that

- smaller sewage treatment plants do not give the lowest values,

- small and medium size treatment plants with population equivalents < 10.000 show "normal" loads of 200 to 400 mg/kg for AOX and 10 to 30 mg/kg for EOX,

- there seems to be a similar tendency for AOX and EOX values in these normal ranges,

- there is no correspondence between very high values of AOX and EOX pointing to different sources of these two types of pollutants.

The experience gained from these investigations has led to the conclusion that AOX as well as EOX may be suitable tools to characterize the usefulness of sewage sludges for agricultural purposes. It is the advantage of the determination of AOX that this quite simple method covers nearly all organically bound halogens and is thus useful for routine analysis because it has been proven in a collaborative study with a great number of participating laboratories. The EOX method, which is a bit more complicated to perform, covers only a part of organohalogen compounds, e.g. nonpolar and weakly polar substances. In contrast to the AOX method, however, there is a possibility of analyzing the hexane extract by gaschromatography/mass spectrometry to obtain more detailed information about single substances.

6. SURFACE ACTIVE AGENTS

This group of substances covers compounds of quite different origin from households and industry found in waste waters and sludges in quite high amounts as compared to other anthropogenic organics (11). Contrary to organohalogen compounds, surfactants are aerobically degradable in waste water as well as in soil. Half-life is reported to be some days. There were investigation programmes dealing with degradation studies of alkylbenzene sulphonates (ABS) (1, 14) where final concentrations of some µg/kg in soil have been reported after repeated application of sewage sludge with initial concentrations of ABS in the g/kg range and there were Swiss experiments on grassland where relatively high amounts of sludge had been applied resulting in final ABS concentrations in the mg/kg range although nonionic surfactants (nonylphenol derivatives) were degraded more completely (20). This may be interpreted in such a way that "normal" application of sewage sludge in ariculture in accordance with current Regulations (5) may result in a more complete biodegradation and that surplus application to land must be avoided in view of the need for a sufficient degradation of surfactants.

7. POLYCYCLIC AROMATIC HYDROCARBONS (PAH)

PAH have been known as xenobiotics for quite a long time. Investigations during the last years have contributed to a better knowledge of this group of substances especially with a view of the agricultural use of sludge (15, 25). These results are quite in agreement with own findings from investigations with sludges from sewage treatment plants in northern and western Germany. The mean concentrations were in the range 0,1 to 1,5 mg/kg with the exception of fluoranthene (0,3 to 3,5 mg/kg). Sewage sludges are not the only source of PAH in soil, where average concentrations of 5 to 40 µg/kg are normally present. Application of sewage sludge may raise PAH levels in soils to 30 - 300 µg/kg. These concentrations were reported not to cause harmful effects because there was no observation of plant uptake, which may start at concentrations above 3 to 5 mg/kg in soil. It can be concluded that the normal load of PAH in sewage sludge is of no importance for agricultural use.

8. PHTHALATES

Finally another group of substances of anthropogenic origin shall be discussed: phthalates, a number of substances which are used mainly as plasticizers in plastics' production. Present knowledge is quite limited. From these substances bis-ethylhexyl-phthalate (DEHP) plays a dominant role because its concentration in sludges is much higher as of other phthalates. In our investigations we found DEHP concentrations of 100 to 500 mg/kg in sludges (normal concentrations in soil are 1 to 25 µg/kg and in sludge amended soils 30 to 650 µg/kg) (15), although DEHP was found in concentrations of 5 mg/kg in soil when an amount of sludge exceeding the permissible one under the Regulations (5) was applied. As phytotoxicity will become noticeable only at concentrations above 200 mg/kg, phthalates will not mean problems if sludge is applied to land following the respective legislation.

9. SUMMARY

During the last years information about existence, degradation and final fate of anthropogenic organic substances in sludges has been improved by several investigation programmes. These investigations have confirmed that substances which may be present in sludges used in agriculture in accordance with legal regulations will cause no problems to soil and growing plants. Systemic transfer from soil to plant could not be observed at currently found concentrations of organics in sludges.

Certainly there may be elevated concentrations of one or the other substance or group of substances in sludge due to influence on sludge quality from industrial sources. Here especially organohalogen compounds of high persistency play an important role. As it is impossible to analyze a large number of single substances, the determination of cumulative parameters in sludges for purposes of monitoring may be a suitable tool to guarantee adequate sludge quality. Among the group of organohalogen compounds, polychlorinated dibenzodioxins and furanes need special observation because of their toxicity and persistency. Concentrations presently found in sewage sludge indicate that there may be no harmful effects for agricultural land. Moreover special attention must be given to the fact that the situation is different if sludge is applied to grassland where grazing cattle may directly ingest contaminated vegetal and soil material.

Generally, the problem of organic substances needs further observation. Future investigation programmes should study the presence and fate of organics in sludges and soils under the conditions of good agricultural practice.

Moreover, all measures must be taken to minimize the input of potentially harmful substances from households and industry to improve the quality of waste water resulting in sewage sludges which can be applied to land without having to fear a contamination of agricultural soils.

10. REFERENCES

(1) Berna, J.L., Ferrer, J., Moreno, A., Prats, D. and Ruiz Bevia, F.: The fate of LAS in the environment. Tenside Surfactants Detergents 26 (1989), No. 2, 101 - 107

(2) Capel, P.D., Lichtensteiger, T. and Brunner, P.H.: The use of sludge-only landfills as historical records of persistent organic chemicals and heavy metals in sewage sludge. Water Research 23 (1989), No. 4, 525 - 527

(3) Commission of the European Communities: Council Directive of 12 June 1986 on the protection of the environment, and in particular of the soil, when sewage sludge is used in agriculture (86/278/EEC). Off. J.of the Europ. Communities 1986, No. L 181/6 - L 181/12 (4.7.1986) incl. annexes

(4) Crathorne, B., Donaldson, K., James, H.A. and Rogers, H.R.: The determination of organic contaminants in UK sewage sludges. In: D.Quaghebeur, I.Temmerman and G. Angeletti (Editors): Organic contaminants in waste water, sludge and sediment: Occurrence, fate and disposal, 45 - 65, Elsevier Applied Science Publ., London - New York, 1989

(5) Der Bundesminister des Innern: Klärschlammverordnung - AbfKlärV vom 25. Juni 1982, Bundesgesetzblatt 1982, Teil I, 734 - 739 (incl. annex)

(6) DIN - Deutsches Institut für Normung e.V. and Fachgruppe Wasserchemie in der Gesellschaft Deutscher Chemiker: Bestimmung der extrahierbaren organisch gebundenen Halogene (EOX) DIN 38 409 Teil 8 / Deutsche Einheitsverfahren zur Wasser-, Abwasser- und Schlamm-untersuchung (DEV H 8), Beuth Verlag, Berlin and VCH-Verlagsges. mbH, Weinheim - New York

(7) ibid.: Bestimmung der absorbierbaren organisch ge-bundenen Halogene (AOX) DIN 38 409 Teil 14 (DEV H 14)

(8) ibid.: Bestimmung von ausblasbaren und extrahierbaren organisch gebundenen Halogenen, DIN 38 414 Teil 17 (DEV S 17)

(9) ibid.: Bestimmung von absorbierten organisch gebundenen Halogenen (AOX) (DEV S 18)

(10) Drescher-Kaden, U., Matthies, M. and Brüggemann: Orga-nische Schadstoffe in Klärschlämmen. Das Gas- und Wasserfach - GWF Wasser - Abwasser 130 (1989), No. 12, 613 - 620

(11) Giger, W., Brunner, P.H., Ahel, M., McEvoy, J., Marcomini, A. and Schaffner, Ch.: Organische Wasch-mittelinhaltsstoffe und deren Abbauprodukte in Abwasser und Klärschlamm. Gas - Wasser - Abwasser 67 (1987), No. 3, 111 - 122

(12) Hagenmaier, H., (Project leader): Untersuchungen der Gehalte an polychlorierten Dibenzodioxinen, polychlorierten Dibenzofuranen und ausgewählten Chlorkohlenwasserstoffen in Klärschlämmen. Bericht zum Forschungsvorhaben 103 03 305 des Umweltbundesamtes im Umweltforschungsplan des Bundesministers des Innern, 58 p. und 9 p. annex, February 1988

(13) Hall, J.E. and Sauerbeck, D. (Editors): Effect of Organic contaminants in sewage sludge on soil fertility, plants and animals. Proc. of the Joint Seminar COST 681 held in Braunschweig, 6 - 8 June 1990 (in press)

(14) Holt, M.: The levels and fate of linear alkylbenzene sulphonate in sludge amended soils. In: D. Quaghebeur, I. Temmerman and G. Angeletti (Editors): Organic contaminants in water, sludge and sediment: Occurence, fate and disposal, pp. 161 - 182, Elsevier Applied Science Publ., London - New York, 1989

(15) Kampe, W.: Organische Stoffe in Böden und Pflanzen nach langjährigen intensiven Klärschlammanwendungen. Korrespondenz Abwasser 34 (1987), No.8, 820 - 827

(16) Leschber, R.: Organische Stoffe im Klärschlamm. Beihefte zu Müll und Abfall H. 28 (1989), 12 - 15, Erich Schmidt Verlag, Berlin

(17) Leschber, R.: Organic substances in sewage sludges. Recycling International Vol. 2, 785 - 792, Proc. of the 6th Int. Recycling Congr., Berlin, 28 - 30 Nov. 1989, EF-Verlag f. Energie- und Umwelttechnik GmbH, Berlin, 1989

(18) Leschber, R., Mergler-Völkl, R. and Zimmermann, U.: Organische Halogenverbindungen im Klärschlamm - Bedeutung für die landwirtschaftliche Schlammverwertung. Gewässerschutz - Wasser - Abwasser (Aachen) 112 (1990), 551 - 579

(19) Loll, U. (Editor): Recycling von Klärschlamm 2. XVIII and 406 p., EF-Verlag für Energie- und Umwelttechnik GmbH, Berlin, 1989

(20) Marcomini, A., Capel, P.D., Lichtensteiger, T., Brunner, P.H. and Giger, W.: Fate of organic pollutants in sludge amended soil and sludge-only landfills: Linear alkylbenzene-sulphonates, nonylphenols and polychlorinated biphenyls. In: D. Quaghebeur, I. Temmerman and G. Angeletti (Editors): Organic contaminants in waste water, sludge and sediment: Occurence, fate and disposal, 105 - 123, Elsevier Applied Science Publ., London - New York, 1989

(21) Quaghebeur, D., Temmerman, I. and Angeletti, G. (Editors): Organic contaminants in waste water, sludge and sediment: Occurrence, fate and disposal. Proc. of a Workshop held in Brussels, 26 - 27 October 1988, IX. and 213 p., Elsevier Applied Science Publ., London - New York, 1989

(22) Schaaf, H.: Datensammlung und statistische Auswertung über Gehalte an chlorierten Kohlenwasserstoffen und polychlorierten Biphenylen in Klärschlämmen, Böden und Pflanzen und tierischen Produkten. VDLUFA-Schriftenreihe 31/1990, 243 p., VDLUFA-Verlag, Darmstadt, 1990

(23) Verband Deutscher Landwirtschaftlicher Untersuchungs- und Forschungsanstalten (VDLUFA): Landwirtschaft im Spannungsfeld von Belastungsfaktoren und gesellschaftlichen Ansprüchen. VDLUFA-Schriftenreihe 32/1990, ISBN-Nr. 3-922712-41-X, VDLUFA-Verlag, Darmstadt (in press)

(24) Witte, H., Langenohl, T. and Offenbächer, G.: Untersuchungen zum Eintrag von organischen Schadstoffen in Boden und Pflanze durch die landwirtschaftliche Klärschlammverwertung, Teil A: Belastung des Klärschlamms mit organischen Schadstoffen. Korrespondenz Abwasser 35 (1988), No. 5, 440 - 448
Teil B: Einfluß der Klärschlammverwertung auf die Gehalte organischer Schadstoffe in Böden und Pflanzen. ibid. 35 (1988), No. 6, 570 - 581

(25) Witte, H., Poletschny, H., Langenohl, T. and Offenbächer, G.: Untersuchungen zum Eintrag von organischen Schadstoffen in Boden und Pflanze durch landwirtschaftliche Klärschlammverwertung. ATV-Dokumentation und Schriftenreihe aus Wissenschaft und Praxis No. 22, 208 p., Gesellschaft zur Förderung der Abwassertechnik e.V., D-5205 St. Augustin, 1989

RISK ASSESSMENT FOR ORGANIC MICROPOLLUTANTS: U.S. POINT OF VIEW.

R.L. CHANEY[1], J.A. RYAN[2], and G.A. O'CONNOR[3]

[1]Soil-Microbial Systems Laboratory, USDA-Agricultural Research Service, Beltsville, MD, USA; [2]Risk Reduction Engineering Laboratory, US-Environmental Protection Agency, Cincinnati, OH, USA; and [3]Department of Soil Science, University of Florida, Gainesville, FL, USA.

Summary

Basic research and monitoring of sludge utilization programs have identified specific Pathways by which potentially toxic constituents of sewage sludge can reach and cause toxicity to livestock, humans, plants, soil biota, wildlife, etc. In the process of preparing a new regulation for land application of sewage sludge in the US, a Pathway approach to risk assessment was undertaken. Two Pathways were found to comprise the greatest risk from persistent lipophilic organic compounds such as PCBs: 1) direct ingestion of sludge by children; and 2) adherence of sludge to forage/pasture crops from surface application of fluid sludge, followed by grazing and ingestion of sludge by livestock used as human food. Each Pathway considers risk to Most Exposed Individuals (MEIs) who have high exposure to sludge. Because 1990 sewage sludges contain very low levels of PCBs, the estimated risk level to MEIs was $<<10^{-4}$; low sludge PCBs and low probability of simultaneously meeting all the constraints of the MEI indicate that MEIs are at $<10^{-7}$ lifetime risk. We conclude that quantitative risk assessment for potentially toxic constituents in sewage sludge can be meaningfully conducted because research has provided transfer coefficients from sludges and sludge-amended soils to plants and animals needed for many organic compounds.

1. INTRODUCTION

U.S. approaches to sludge regulations have changed over the years as scientific information about the fate and effects of potentially toxic constituents has become more complete. Until the 1970s sludge was regulated only as a source of infection or as a public nuisance due to odors. Then, a program of constructing sewage treatment works which would increase sludge production led to consideration of how sludge could be handled safely. Research began to evaluate risks from toxic constituents of sludge in different "ultimate disposal" environments (agricultural land, incinerators, landfills, ocean). In 1979, the United States Environmental Protection Agency (34) published a draft regulation on land application and landfill disposal of sludge for public comment.

The 1970s were a period of intense research on land application of sewage sludge. One of the most important areas of risk from sludge utilization/disposal identified was from sludges sold or given to individual citizens (Distributed or Marketed, D&M) without controls. Many Publicly Owned Treatment Works (POTWs) dried sludge on sand beds during part of the year, and dried sludge was often given to anyone who would take it. As in Europe, research found examples of highly polluted sludges (Cd, PCBs, etc.) being given away for use on lawns, gardens, and farms (5, 10, 24, 25). This information dramatized the need for regulation of sludge use in situations other than farmland; either the practice should be controlled or prohibited (14). In 1980, US-EPA started efforts to regulate D&M and to finalize the 1979 proposed regulations, but this effort was never completed.

While reauthorizing the Clean Water Act in the mid-1980's, Congress decided that further sludge regulations were needed, partly because the first rule dealt with only a few pollutants, and partly because sludge D&M had not been regulated and still continued in many cities. At the same time, other Congressional and US-EPA actions have improved industrial pretreatment regulations for many industries which discharged unwanted pollutants to the sewers. Protection of the quality of sludge was judged to be a valid reason to require pretreatment; if public acceptance of sludge application, or meeting state regulations on sludge utilization required better industrial pretreatment, states and cities now had the power to protect sludge quality. Based on the recent National Sewage Sludge Survey (39), sludge utilization in agriculture has increased since 1980 (Table 1), and concentrations of contaminants in most sludges are much lower than found in previous surveys (Table 2). This survey (39) found many "not detected", partly because solids content in sludge samples was not considered in taking wet sludge sample weights. For those samples with many "non detected" sludges, the Maximum Likelihood Estimation procedure was used to provide a reasonable estimation of the geometric distribution of a constituent in dry sludges.

Table 1. Methods of utilization and disposal of sewage sludge in the United States based on the National Sewage Sludge Survey (39) of 479 POTWs in 1988-1989, and 208 POTWs in Analytical Survey.

Use/Disposal Method	Information Survey US-EPA, 1989 (39) # POTWs	Fraction	Analytical Survey US-EPA, 1989 (39) # POTWs	Fraction	Estimated US Distribution # POTWs	Fraction
		%		%		%
Land Application	161	33.61	78	37.50	3542	31.05
Distribution&Marketing	27	5.64	10	4.81	308	2.70
Municipal Landfills	76	15.87	51	14.90	1851	16.23
Monofills	28	5.85	6	2.88	203	1.78
Surface (e.g. lagoons)	59	12.32	13	6.25	3147	27.59
Incineration	63	13.15	23	11.06	294	2.58
Ocean	21	4.38	6	2.88	115	1.01
Other	24	5.01	25	12.02	1526	13.38
	459	95.83				

During the 1980s, it became clear that addition of equal amounts of a pollutant in sludge or as a pure chemical caused different effects. Sludge was found to add specific adsorption capacity for metals and organics to the sludge-soil mixture (15). In the case of sludge metals, the plateau response of plant metal concentrations to increasing application rates of a sludge (13, 15) showed that the concentration of a pollutant in sludge did affect potential for risk. Previously it had been believed that all sludges had equal effect at equal cumulative applied amounts of a constituent, and that limiting the cumulative amount of metals or organics was the proper form of regulations. With this new information that the concentration of a constituent in a sludge could strongly affect the potential transfer to plants or animals, there was a scientific basis for regulating sludge constituent concentrations. The final effect of these

Table 2. Composition of U.S. Sewage Sludge based on 200 POTWs sampled during 1988-1990 (8, 39).

Constituent	Geometric Mean	Median	90th	95th	98th	Maximum	Minimum
	------------------ mg/kg dry solids --------------------						

Pollutant	Comparison with the 1990 Nat. Sewage Sludge Survey				
	Normal Statistics			Maximum Liklihood[5]	
	Median	95th	98th	Median	98th
	--------------mg/kg dry sludge---------------				
As	6	43	62	5	33
Cd	7	21	25	4	19
Cr	40	635	1960	39	409
Cu	463	1940	2490	456	2180
Hg	4	17	43	2	19
Mo	11	42	56	5	32
Ni	29	223	438	18	159
Pb	106	296	444	76	373
Se	5	28	51	3	16
Zn	725	4100	4760	755	3270
PCB-1248	0.21	0.67	1.5	0.02	0.21

[1] Adjusted downward for pretreatment considerations.
[2] No adverse effects reported for any Cr^{3+} level in municipal sludge.
[3] Valid for all sludge uses except mushroom production.
[4] Mo limit raised because Mo slowly leaches from alkaline soil.
[5] Maximum Liklihood Estimation Procedure, assuming multicensored lognormality.

changes is the development of the NOAEL (the No Observed Adverse Effect Level) sludge approach in which sludge constituents are limited to concentrations which comprise low risk to Most Exposed Individuals (MEIs) under worst-case scenarios (27, 22, 8). Application of at least 1000 Mg of a NOAEL sludge/ha has been found to comprise no risk to agriculture, livestock, humans, or wildlife/ecosystems based on data from field studies (27,8). The comprehensive risk analysis effort which allowed development of the NOAEL sludge approach may be helpful to other nations which are considering sludge regulations. Further, the detailed analyses identified specific areas of needed research, research not yet conducted in the agricultural or environmental research communities. Thus, this paper summarizes the new US approach to sludge regulation and the general conclusions reached. We conclude that if sludges can be beneficially used in sustainable agriculture with so low risk to agriculture or environment, that utilization of farmland should be the preferred method of "Ultimate Disposal". Pretreatment of industrial and non-industrial sources of some pollutants will often be required to achieve the NOAEL sludge quality. Technology is presently available to achieve the needed pretreatment.

The details of the arguments and the data used for the risk analyses have been described in several papers (30, 27, 8) and official documents

(36). Details of the Cd risk analysis were recently described in Chaney (8). We have also reviewed plant uptake of toxic organics, and discussed at length the potential errors in research methods and methods of analysis which have often caused over-estimation of the risk of toxic organics added in sludges (26). After working on this reconsideration of the scientific basis for sludge regulations for the US, we believe that a detailed example of risk analysis for PCB transfer illustrates the efforts currently underway in the United States. The paper shows the principles of the Pathway Method of risk analysis. Data on PCBs are nearly as complete as needed. Further, as with many of the persistent potentially toxic organics in sludge, sludge concentrations have decreased with the prohibition of PCB use or disposal in 1979. Based on these evaluations, we conclude that PCB concentrations will not limit utilization of nearly any sludge in the US.

PATHWAY APPROACH FOR ANALYSIS OF THE RISKS TO HUMANS AND THE ENVIRONMENT FROM TOXIC ORGANICS IN LAND-APPLIED MUNICIPAL SEWAGE SLUDGES.

In the ongoing effort to develop regulations to protect the environment during land application of sewage sludge, EPA developed a "Methodology" to assess the potential risk of toxic organic transfer to humans or the environment from all identified pathways for this transfer (36). The EPA Pathways (Table 3) were selected after consultation with the research community (35). Under the Pathway Approach, risk assessment is conducted in such a way as to protect the Most Exposed Individual (MEI) who ingests sludge or sludge-grown foods as a high portion of his diet. The MEI can be a human child eating sludge or soil-sludge mixture, a human adult consuming foods grown on sludge-amended land for 70 years, livestock, earthworms, soil bacteria and fungi, birds consuming large amounts of earthworms, etc.

Previous reviews have considered the possible transfer of toxic organics from sludge to humans (e.g. 6, 9, 16, 19, 26, 28). These reviews have focused on the direct ingestion of Distributed or Marketed (Pathway 2-D&M) sludge products by children, and on human consumption of a large fraction of ingested meats from livestock grazing sludge sprayed pastures (Pathway 4-Surface) (19). It has become generally accepted that all other Pathways transfer organics to humans at much lower levels than do Pathways 2-D&M and 4-Surface. However, a quantitative estimation of the allowed applications of sludge PCBs under all the pathways should clearly demonstrate why these two Pathways are so important. If these are the critical pathways for all persistent TOs, then research on other TOs should focus on the critical transfer coefficients involved in these Pathways.

Assessing Risk From Soil/Sludge Ingestion (Pathway 2-D&M; Sludge→Human).

Direct sludge ingestion by humans is the simplest Pathway for pollutants in sludge to reach humans. Research on Pb risk from ingested soil and dust (2, 32, 12) has shown the importance of hand-to-mouth play and of *pica* (intentional ingestion of non-food items) on transfer of soil-Pb and other strongly soil-adsorbed metals and TOs to children. Clearly, soil Pb comprises much greater risk to children with *pica* than any other way soil Pb can reach humans (e.g. plant uptake by garden food crops [12]). Because of the concern about soil ingestion by children, information needed to assess Pathway 2-D&M risks has been accumulated.

In the EPA Methodology, the algorithm for calculating Pathway 2 starts by calculating the allowed daily ingestion of PCB so as to protect against development of cancer in the exposed population. Cancer from excessive

Table 3. Pathways for risk assessment of potential transfer of sludge-applied trace contaminants to humans, livestock, or the environment, and the Most Exposed Individual to be protected by regulation to be based on the Pathway Analysis (36, 27).

	PATHWAY	MOST EXPOSED INDIVIDUAL
1	Sludge→Soil→Plant→Human	General food chain; 2.5% of all food for lifetime.
1F	Sludge→Soil→Plant→Human	Home garden 5 yr after last sludge application; 50% of garden foods for lifetime.
1D&M	Sludge→Soil→Plant→Human	Home garden with annual sludge application; 50% of garden foods for lifetime.
2F	Sludge→Soil→Human Child	Residential soil, 5 yr after last sludge application; 200 mg soil/d.
2-D&M	Sludge→Human Child	Sludge product; 200 mg sludge/d.
3	Sludge→Soil→Plant→Animal→Human	Rural farm families; 40% of meat produced on sludge amended soil, for lifetime.
4-Surface	Sludge→Animal→Human	Rural farm families; 40% of meat produced on sludge sprayed pastures, for lifetime.
4-Mixed	Sludge→Soil→Animal→Human	Rural farm families; 40% of meat produced on sludge amended soils, for lifetime.
5	Sludge→Soil→Plant→Animal	Livestock fed feed forages and grains, 100% of which are grown on sludged land.
6-Surface	Sludge→Animal	Grazing livestock on sludge sprayed pastures; 1.5% sludge in diet.
6-Mixed	Sludge→Soil→Animal	Grazing Livestock; 2.5% sludge-soil mixture in diet.
7	Sludge→Soil→Plant	"Crops"; vegetables in strongly acidic sludged soil.
8	Sludge→Soil→Soil Biota	Earthworms, slugs, bacteria, fungi in sludged soil.
9	Sludge→Soil→Soil Biota→Predator	Birds; 33% of bird diet earthworms affected by sludge.
10	Sludge→Soil→Airborne Dust→Human	Tractor operator.
11	Sludge→Soil→Surface Water→Human	Water Quality Criteria; fish bioaccumulation, lifetime.
12	Sludge→Soil→Air→Human	Farm households.
12W	Sludge→Soil→Groundwater→Human	Farm wells supply 100% of water used for lifetime.

D&M refers to sludge products distributed or marketed.

ingested PCBs is the most sensitive risk endpoint for humans from chronic ingestion of PCBs. The q_1* is the cancer potency slope from the Cancer Assessment Group of EPA. In this assessment of the cancer potency of a compound based on lifetime feeding experiments with rats and other animals at the maximum tolerated dose, the upper 95th percent confidence limit of the slope is used as the limit to add further protection. This method of estimating potential cancer incidence after lifetime exposure of humans is generally agreed to lead to very conservative regulatory controls (1). The q_1* is used with other assumed parameters (RL, BW, RE) to calculate the Adjusted Reference Intake (RIA). The RIA is the average daily intake of PCBs that may not be exceeded if Risk Level is to be $\leq 10^{-4}$.

$$RIA = \left(\frac{RL \cdot BW}{q_1^* \cdot RE}\right) \cdot 10^3 = \left(\frac{10^{-4} \cdot 10\ kg}{7.7 \cdot 1}\right) \cdot 10^3 = \frac{1.00}{q_1^*} = 0.130\ \mu g/day$$

where RIA = Adjusted Reference Intake (μg/day).
q_1* = human cancer potency slope ([mg/kg BW/day]$^{-1}$).
RL = risk level = 0.0001 of MEIs for EPA Proposed 503 Rule.
BW = body weight of 1-5 year old child = 10 kg.
RE = relative effectiveness, or bioavailability of pollutant in ingested sludge compared to pure chemical form (used in cancer potency slope assessment) added to diet.

The Reference Soil Concentration (RLC) which cannot be exceeded at the selected risk level is then calculated from the RIA. In this case the Pathway 2-D&M RLC is the maximum allowed sludge PCB concentration for a sludge D&M product. Pathway 2F (ingestion of pollutants 5 years after cessation of mixing sludge with soil) would provide appreciably less exposure than Pathway 2-D&M.

The average soil ingestion by children has been estimated by several researchers. The Calabrese et al. (7) paper shows the most reliable published estimation at this time. The 95th percentile (log normal) of soil/dust ingestion by 2 year old children was about 0.5 g/day, while the geometric mean soil ingestion was < 50 mg soil/day. We believe that the peak soil ingestion period by children is about 2 years long, and that use of the 0.5 g sludge/day provides a conservative estimate of risk (27). As it happens, use of either 0.2 g/day (the number used in the EPA Superfund program [37]) for 5 years, or the 0.5 g/day for 2 years are the same amount of sludge ingestion exposure.

The Duration Adjustment (DA) must be made when cancer risk slopes for 70 year lifetime exposure are used to estimate allowed exposures for soil/sludge ingestion which lasts only a part of the lifetime. Thus, the calculation can use 0.2 g/day for 5 years (DA = 0.0714) or 0.5 g/day for 2 years (DA = 0.0286). Values of RIA, I_s, and DA are used to calculate the RLC.

$$RLC = \frac{RIA}{I_s \cdot DA} = \frac{0.130\ \mu g/day}{[(0.5 \cdot 0.0286) = 0.0143] \text{ or } [(0.2 \cdot 0.0714) = 0.0143]} = 9.09\ \mu g \text{ PCB/g dry sludge.}$$

RLC = Reference Soil/Sludge Concentration of the pollutant (μg/g DW).
DA = Duration Adjustment for < 70 year: 5/70=0.0714; 2/70=0.0286.
I_s = Soil Ingestion Rate (g DW per day).

Thus, for PCBs, the RLC = 9.09 μg/g DW. The Pathway 2-D&M risk assessment, in which children are assumed to ingest sludge directly for 2-5 years, indicates that the required sludge PCB concentration limit is much higher than present levels of PCBs in sludges (Table 2). Based on data discussed below for bioconcentration of PCB from sludge ingested by cattle compared to bioconcentration from pure PCB added to cattle diets, the relative effectiveness of sludge PCB may be about 0.50, which would increase the RLC estimate accordingly.

Assessing Risk From Sludge Ingestion by Grazing Livestock Used as Food by Humans (Pathway 4; Sludge→Soil→Animal→Human and Sludge→Animal→Human)

Fat in the human diet which comes from livestock which graze in sludge-amended pasture fields is the source of the human exposure calculated in Pathway 4. Pathway 4 has two quite different kinds of exposure, and we believe they should be considered separately. The first involves direct ingestion of sludge by livestock, where sludge has been surface applied to pasture crops. Livestock can ingest sludge adhering to the crops, or sludge lying on the soil surface. Each year the grazing livestock are presumed to be exposed to freshly applied sludge with no time for dissipation of the organic chemicals (this is *Pathway 4-Surface Application*). Alternatively, sludge can be injected into the soil or mixed with the plow layer soil, and the grazing livestock ingest the soil-sludge mixture. Injected sludge is not at the soil surface, so injection minimizes exposure until the plow layer soil is mixed. The highest exposure under the *Pathway 4-Mixed* scenario would be livestock ingesting soil while grazing a crop seeded immediately after mixing the applied sludge into the plow layer soil.

Sludge ingestion by grazing livestock was expected after Chaney and Lloyd (11) observed that sludge adhered to forages for a prolonged period if the fluid sludge was not washed off the plants immediately after application (23). If sludge is applied to standing forages, the rate of application and solids content of the sludge affect sludge adherence. N-fertilizer rate applications of typical fluid sludges containing 5% solids to standing forage crops caused sludge to reach about 15% of the forage dry matter. This evidence led to prohibition of sludge application on standing forages (34), and thus reduced the potential for sludge ingestion by livestock during sludge utilization on pastures.

Chaney et al. (9) reviewed several studies in which cattle grazed pastures to which sludge was surface applied using good practices. In the Decker et al. (17) study, fluid sludge was sprayed on the pastures after mowing, and the crop grew for 21 days before cattle were allowed to graze; the study used four "rotation paddocks" so that each 7 days the cattle were moved to the next pasture which had grown for 21 days after sludge application. By analysis of the cattle feces after 7 days grazing on the treated pastures, they found that the animals ingested about 2.5% sludge in their diets. Metals served as a label for sludge; concentration of metals such as Fe, Pb, and Cu were much higher in treated forages and feces than in control forages and feces of control treatment cattle. Similar results were observed by Bertrand et al. (4). However, when sludge compost was applied the previous grazing season rather than during each "rotation", the cattle ingested only about 1.0% sludge (17).

Another possible consideration is the effect of having only part of a farm treated with sludge each year. The EPA (36) Methodology considered that for the rural farm family who consumed the greatest fraction of "home-grown" livestock (the MEI in Pathway 4), all grazing fields were treated with sludge each year for a 70 year lifetime. However, if sludge were applied intermittently, the fraction of sludge in the chronic grazing animal diet is lower because adhering sludge exceeded sludge ingested from the soil surface in season-long grazing studies with sludge (see 9). Based upon our experience and discussion with regulatory officials in several states, we estimate that in any one year, the normal maximum fraction of a farm treated with sludge may be 33% rather than 100%. If one presumes the cattle are rotated among several pasture fields, the actual fraction of the diet which is sludge will be lower than the 2.5% measured with continual exposure (this lifetime approach is necessary because the cancer endpoint protected against requires that exposure appropriate for a chronic lifetime model be estimated). To take this into account, the allowed RFC for sludge-treated forage should be adjusted for the fraction of livestock diet affected by sludge application. Using 1% sludge ingestion during non-application periods (2/3 of time), and 2.5% ingestion during sludge application periods (1/3 of time), the long term average would be 1.5% sludge in diet.

To estimate sludge PCB transfer from grazing livestock to humans, the lifetime average amounts of fat of meat and dairy products from grazing livestock ingested per day are needed. The US-EPA (36) "Methodology" estimated the amounts of fat in the human diet from each class of livestock, for a range of age and sex groups of the US population. The original data were described by Pennington (31). One can calculate a lifetime diet composition by assuming the number of years each age range of consumption covers. Table 4 shows the estimated average lifetime daily consumption of fat from meats from different livestock. Other foods considered in this paper are also included in Table 4.

The MEI for Pathway 4 is believed to be members of a rural farm family ingesting much of their lifetime meat consumption from home grown livestock (which consume sludge on forages or soil-sludge mixture) for their 70 year lifetime exposure. The calculation starts with estimation of the maximum allowed daily PCB intake to limit 70 year lifetime cancer risk:

$$RIA = \left(\frac{RL \cdot BW}{q_1^* \cdot RE} \right) \cdot 10^3 = \left(\frac{10^{-4} \cdot 70\ kg}{7.7 \cdot 1} \right) \cdot 10^3 = 0.909\ \mu g\ PCB/day$$

RIA = adjusted reference intake (μg/day)
q_1^* = human cancer potency ($[mg/kg/day]^{-1}$)
RL = risk level = 0.0001 for TSD
BW = body weight = 70 kg
RL = Relative Effectiveness or bioavailability.

After calculating the RIA, one calculates the maximum allowed feed concentration of PCBs (RFC) for Pathway 4 (see Table 5):

$$RFC = \frac{RIA}{\Sigma\ (UA_i \cdot DA_i \cdot FA_i)}$$

RFC = reference feed concentration of the pollutant (μg/g DW)
RIA = adjusted reference intake (μg/day)
UA_i = uptake response slope of pollutant in the animal tissue food group i for organics, on a fat basis, = 2 (μg PCB/g fat)$\cdot$(μg PCB/g feed DW)$^{-1}$ for *sludge-borne* PCBs added to test diets.
DA_i = Daily dietary consumption of animal food/fat group i, g dry wt.
FA_i = fraction of the food group i assumed to be derived from amended soil (in this case, from fat in tissues of or milk from cattle consuming 1.5% sludge averaged over their lives). Assumes that a high fraction of the diet came from cattle raised on the sludge-treated pastures (44% for meat fat, and 40% for dairy fat for a 70 year lifetime).

Table 4. Human consumption of selected foods (fat from meats derived from several classes of livestock; garden crops) (g dry weight/day) for different age groups, and estimated lifetime average food intakes for 70 kg US adult citizens. The child age group (not reported by Pennington [31]) was assumed to consume the average of that consumed by toddlers and teen-agers. Meats in mixed foods assigned to source by US-EPA (36).

Food Group	Age Grouping						
	Baby	Toddler	Child	Teen-Agers	Adult	Older Adult	Estimated Lifetime
Age:	0-1	1-6	6-14	14-20	20-45	45-70	0-70
<u>Grazing Livestock:</u>							
Beef Fat	2.45	6.48	11.34	16.22	20.40	14.07	15.50
Beef Liver Fat	0.05	0.07	0.08	0.10	0.29	0.33	0.25
Lamb Fat	0.14	0.08	0.07	0.06	0.31	0.22	0.21
Dairy Fat	38.99	16.48	20.46	24.43	18.97	14.51	18.13
<u>Non-Grazing Livestock:</u>							
PorkFat	2.01	8.19	10.47	12.75	14.48	13.04	12.73
PoultFat	1.10	0.83	1.12	1.41	1.54	1.31	1.34
EggFat							0.96
<u>Garden Food Crops:</u>							
Potatoes	5.67	10.03	14.72	19.40	17.28	14.79	15.60
Leafy Veg.	0.84	0.49	0.85	1.22	2.16	2.65	1.97
Legume Veg.	3.81	4.56	6.51	8.45	9.81	9.50	8.75
Root Veg.	3.04	0.67	1.20	1.73	1.77	1.64	1.60
Garden Fruit	0.66	1.67	2.57	3.47	4.75	4.86	4.15

For Teen, Adult, and Older Adult categories, FDA (31) intakes of females and males were averaged; for other groups, no separation of data by sex was reported. Child intake was set equal to the average of Toddler and Teen.

The best value for UA for PCB transfer from feed to fat has been discussed by Fries (19). Based on his own and other research, Fries concluded that the UA was 4 for beef fat and 4.8 for dairy products. However, this was for pure PCB added to cattle diets or provided in corn oil. Baxter et al. (3) tested the bioaccumulation ratio for PCB in a dried anaerobically digested sludge containing 24 mg PCBs/kg DW. Because sludge adsorbs PCBs strongly, the UA for sludge-borne PCB was only 1.9. This value was the average for beef cows and steers fed 10% sludge in their complete diet for 270 days. It should not be unexpected that sludge PCB adsorption properties and oils could reduce the bioavailability of sludge PCBs. Rozman et al. (33) found adding mineral oils to feeds increased the excretion of hexachlorobenzene from cattle. Fairbanks and O'Connor (16) found soil organic matter strongly adsorbed PCBs, and the adsorbed PCBs showed strong hysterisis (limited release) of PCBs in sludge.

$$RFC = \frac{RIA}{\Sigma(UA_i \cdot DA_i \cdot FA_i)} \quad = \quad \text{or } RFC = \frac{0.909}{27.13} = 0.0335\ \mu g/g\ DW$$

The fraction of sludge in the diet enters into the calculation so the maximum amount of PCB allowed in the sludge can be estimated for *Pathway 4-Surface Application*. Reference Sludge Concentration (RSC) can be calculated as specified in the Methodology:

Table 5. Calculation of $\Sigma(UA_i \cdot DA_i \cdot FA_i)$ needed to estimate human exposure to PCB in fat of meat from grazing livestock. UA value from Baxter et al. (3), based on feeding cattle 10% sludge for 270 days at which time equilibrium of PCB in fat with the diet would have been reached according to Fries (19). DA is the lifetime daily average intake of fat from grazing livestock. FA is the fraction of total lifetime ingestion of these meats by the MEIs (members of a rural farm family.

	Sludge PCB in Diet of Grazing Livestock			
Food Group	DA g/day	UA	FA	$(UA_i \cdot DA_i \cdot FA_i)$ Estimated Lifetime
Beef fat	15.50	1.9	0.44	12.96
Beef liver fat	0.25	1.9	0.44	0.21
Lamb fat	0.21	1.9	0.44	0.18
Dairy fat	18.13	1.9	0.40	13.78
ΣFat from grazing livestock	34.09			27.13

RSC = RFC/FS

RSC = Reference Sludge Concentration
RFC = Reference Feed Concentration
FS = Fraction of livestock diet comprised of sludge (chronic lifetime exposure = 0.015 according to discussion above.)

RSC = RFC/FS = 0.0335/0.015 = 2.23 µg PCB/g DW in applied sludge.

This same number is used for estimating allowed sludge PCB applications under *Pathway 4-Mixed With Soil*, only now it is 2.23 µg PCB/g dry soil-sludge mixture which is chronically ingested by the grazing livestock at 1.5% of diet (annual average basis; see also [9, 19, 21]). In the case of sludge mixed into soils, however, the regulatory approach is quite different because PCBs can volatilize or be biodegraded over time after application. The 2.23 µg PCB/g dry soil-sludge mixture which must not be exceeded is considered to be the equilibrium reached when PCB dissipation during one year equals the annual PCB application. Persistent higher chlorinated PCBs have about a 6-10 year half-life; for these calculations, the $T_{0.5}$ is presumed to be 10 yr ($k = 0.693/T_{0.5} = 0.0693$/yr).

An equilibrium PCB concentration is reached after about 5.6/k = 81 years (36), and the formula used to calculate the annual PCB application mixed into the soil (RP_a) needed to reach this equilibrium RSC after 81 continuous annual applications is:

$$RP_a = RLC \cdot (2000 \text{ Mg soil/ha}) \cdot 10^{-3} \cdot [1e^{0 \cdot 0.0693} + 1e^{-1 \cdot 0.0693} + \ldots + 1e^{-81 \cdot 0.0693}]^{-1}$$
$$= 2.23 \cdot 2 \cdot [14.9]^{-1} = 0.299 \text{ kg PCB/ha/yr.}$$

This could be applied by 10 Mg/ha/yr of a sludge containing 29.9 mg PCB/kg:

$$\frac{0.299 \text{ kg PCB}}{\text{ha} \cdot \text{yr}} \cdot \frac{1 \text{ ha} \cdot \text{yr}}{10 \text{ Mg sludge DW}} = \frac{0.0299 \text{ kg PCB}}{\text{Mg sludge}} = 29.9 \text{ mg PCB/kg sludge DW.}$$

Thus, Pathway 4-Surface Application comprises significantly greater potential for PCB transfer to humans than does Pathway 4-Mixed. Pathway 4 risk assessment remains a very conservative estimation of allowable toxic organic concentrations in sewage sludge. The exposure scenario, a rural farm family ingesting, for a 70 year lifetime, a high proportion of home grown meat and dairy products from livestock which often graze sludge-amended pastures, is conservative. In *Pathway 4-Surface Application*, the sludge is assumed to remain on the surface of pastures continuously for 70 years in contrast with known agronomic practices which require that the surface be intermittently tilled with the plow layer soil to incorporate pH-modifying agents, fertilizers, and organic matter into the surface soil. The MEI is conservatively assumed to consume 44% of meat of grazing livestock and 40% of dairy products from livestock raised on sludge-using farms, but few farms raise both lamb and beef, and dairy cattle are fed feed supplements to improve production efficiency (21). Further, it is likely that current data on food consumption by rural farm families would show that a lower fraction of locally grown livestock is consumed by the highest consuming families.

The alternatives to surface application of sludge on pastures include use of sludge injection, and mixing sludge with the soil before forage crops are established. These would all lead to much lower estimations of risk than found with Surface Application. *Pathway 4-Mixed With Soil* calculations indicate that much greater protection of humans from sludge-applied PCBs can be achieved by avoiding surface application of sludge on pastures. The use of sludge injection minimizes sludge ingestion by grazing livestock, conserves sludge nutrients, and prevents malodors and unsightliness of surface applied sludge. In any case, the MEI is well protected against chronic health effects of sludge PCBs by limitations appropriate for the method of application.

Assessing Risk From Uptake of Soil PCB Residues by Forage and Feed Crops For Livestock Used as Food by Humans (Pathway 3; Sludge→Soil→Plant→Animal→Human)

For Pathway 3, sludge is presumed to be uniformly mixed with the plow layer soil, and forage and feed crops are grown on the sludge-treated soil. As in Pathway 4, the MEIs are rural farm families assumed to consume a high proportion of their lifetime meat and dairy products from livestock which eat only crops grown on sludge-treated soils. This Pathway is similar in approach to Pathway 4-Mixed, but all meats and dairy products are considered. Fat from non-grazing livestock classes (poultry, swine) are included in Pathway 3 because feed grains can be harvested from sludge-amended soils in addition to hay and silage harvested for ruminant livestock. Transfer of PCBs from sludge-amended soil to edible parts of crops is quite low (28, 40) and annual sludge applications are limited by the N fertilizer supplied by the sludge, resulting in the steady state concentration of PCB in soil being quite low compared sludge lying on the soil surface in Pathway 4. Pathway 3 therefore, would always be expected to provide lower transfer of persistent PCBs to humans via animal fat.

For Pathway 3 (uptake of PCBs by forage/feed plants, or contamination with PCB independent of sludge or soil ingestion), the EPA (36) Methodology presumed that the PCB uptake slope for forage crops was the appropriate transfer slope to feeds for all livestock. However, in calculating the estimate of toxic organic compound transfer to humans by this Pathway, we believe it is important to also discriminate between grain crops fed to some classes of livestock and forage crops fed to ruminant livestock. Further, milk cows are supplied grains and other "concentrates" to provide nutrients and energy for high production rates, and it is inappropriate to presume 100% forage diet for dairy products (21). A modified approach to calculating this limitation is shown below which allows correction of this error in the risk assessment Pathways.

For 70 year lifetime exposure, the EPA (36) algorithm to calculate RIA for Pathway 3 is the same as in Pathway 4). After calculating the RIA, one calculates the RFC for the feed as under Pathway 4, but adds the fat from other foods as shown in Table 6:

$$RFC = \frac{RIA}{\Sigma\ (UA_i \cdot DA_i \cdot FA_i)}$$

RFC = reference feed concentration of the pollutant (μg/g DW)
RIA = adjusted reference intake (μg/day)
UA_i = uptake response slope of pollutant in the animal tissue food group i for organics, on a fat basis, = 4 for PCB mixtures added to test diets; does not use the data from Baxter et al. used for Pathway 4 because sludge is not present to adsorb PCB during transit through the gut.
DA_i = Daily dietary consumption of the animal tissue food group i
FA_i = fraction of the food group i assumed to be derived from amended soil (in this case, from fat in tissues of or milk from cattle consuming feeds which were all grown on soils which received sludge annually.

The RLC is calculated from the RFC by dividing by the uptake slope for the compound for the crop used to estimate uptake (UC):

$$RLC = (RFC \div UC) + BS$$

where:
RLC = Reference Soil Concentration (μg/g DW)
UC = linear response slope of forage crop [μg/g crop DW (μg/g soil DW)$^{-1}$]
BS = background soil concentration of pollutant (μg/g DW) (assumed to be 0 since the cancer risk to be estimated is the incremental risk, not the absolute risk.

Our method of calculation sums a new product of UC_i and DF_i (Diet Fraction) times the old product (Table 6) to directly calculate RLC from RIA:

$$RLC = \frac{RIA}{\Sigma((UC_i \cdot DF_i \cdot UA_i \cdot DA_i \cdot FA_i)} = \frac{0.909}{0.04954} = 18.3 \text{ mg/kg soil-sludge mixture DW}$$

The Methodology calculation then estimates an annual sludge PCB application rate taking into account the rate of dissipation (volatilization or degradation) of PCBs ($T_{0.5}$ = 10 yr) so that the RLC is reached at steady state between application and dissipation.

The formula was introduced above.

$$RP_a = RLC \cdot (2000/1000) \cdot (14.9)^{-1} = 2.46 \text{ kg PCB/ha/yr.}$$

Table 6. New approach to estimate PCB limitation for Pathway 3; $\Sigma(UC_i \cdot UA_i \cdot DA_i \cdot FA_i)$ is calculated to allow discrimination between PCB transfer to livestock which consume grain and those consuming forage.

Animal Tissue		Diet	Uptake(UC_i)				
Group	Feed Type	Fraction	Slope	UA	DA	FA	Σ
Beef Fat	Forage	1.00	0.001	4.0	15.50	0.44	0.02728
Beef Liver Fat	Forage	1.00	0.001	4.0	0.25	0.44	0.00044
Lamb Fat	Forage	1.00	0.001	4.0	0.21	0.44	0.00036
Pork Fat	Grain	1.00	0.0001	4.0	12.73	0.44	0.00224
Poultry Fat	Grain	1.00	0.0001	4.0	1.34	0.34	0.00018
Dairy Fat	Forage	0.50	0.001	4.8	18.13	0.40	0.01740
Dairy Fat	Grain	0.50	0.0001	4.8	18.13	0.40	0.00146
Eggs Fat	Grain	1.00	0.0001	4.0	0.96	0.48	0.00018
Total				-	-	-	0.04954

Uptake slopes (UC_i) from [28].

At an annual application rate of 10 Mg dry sludge/ha, the sludge could contain as much as 246 mg PCBs/kg.

$$\frac{2.46 \text{ kg PCBs}}{\text{ha}\cdot\text{yr}} \cdot \frac{1 \text{ ha}\cdot\text{yr}}{10 \text{ Mg sludge DW}} = \frac{0.246 \text{ kg PCBs}}{\text{Mg sludge DW}} = \frac{246 \text{ mg PCBs}}{\text{kg sludge DW}}.$$

This high allowed annual application of sludge PCBs will not practically limit sludge application because modern sludges contain low levels of PCBs, with median levels < 0.1 mg/kg DW.

Another way to view this high annual allowed PCB application and equilibrium soil PCB concentration under Pathway 3 is that Pathway 4 would estimate unacceptable risk levels for the same soils. During periods of sludge use on pastures, the limitations of Pathway 4 must limit sludge PCB application. These considerations confirm the summary of relative risk from Pathways 3 and 4 noted in the introduction.

Calculations for Pathway 3 also remain very protective for the presumed MEIs. Sources of protection include: 1) the assumption that the rural farm family consumes a large portion of "home-grown meats" for their 70 year lifetime exposure; and 2) that sludge is applied annually to 100% of the farmland used to produce forage and grain crops used as animal feed.

Assessing Risk From Food Crops (Pathway 1F; Sludge→Soil→Plant→Human):

The production of food crops for humans on sludge-amended soils will transfer some sludge constituents into human diets. The MEI for this Pathway is the home gardener who grows a large fraction of his diet on sludge amended soil, for 70 years. Sludge is presumed to be applied and mixed into the soil annually until the PCB concentration reaches an equilibrium with PCB loss (81 years as noted above). The exposure is assumed to be at the equilibrium level for the whole 70 year consumption of garden crops grown on the treated soil. Pathway 1F-D&M could be estimated by assuming a D&M product is applied annually, so no waiting period is allowed. In the Agricultural Use Pathway 1F, a five year waiting period is assumed before crops are grown on the treated soil, presuming a waiting period followed by conversion of the site to residential use. Because

Pathway 1-D&M estimates the highest possible exposure under Pathway 1, it is used for an example. Exposure from Pathway 1F-D&M is much lower than from Pathway 2-D&M or Pathway 4-Surface. Pathway 1F-D&M will not practically limit utilization of sludges with low PCB concentrations presently available.

The method of calculation of food chain transfer of toxic organics according to the Methodology (36) (home garden scenario) uses several steps to complete the calculation. The first step is the calculation of daily allowed TO intake (RIA) based on the cancer potency value. One then sums the amount of each food group consumed times the slope of plant uptake for the compound times the fraction of food group grown on sludge amended soil. This value is divided into the RIA to produce the RLC, the soil PCB concentration which sludge application may not be allowed to exceed.

From calculations above, we know the RIA is 0.909 µg PCB/day. The next step sums the amount of PCB ingested if FC (fraction of the diet) were grown on sludge amended soil, if DC (g DW of foods) were consumed daily on average for 70 years, and if UC were the transfer coefficient from soil to each food group (µg PCB/g dry crop per µg PCB/g dry soil) grown in the garden.

Calculation of the $\Sigma(UC_i \cdot DC_i \cdot FC_i)$ for PCBs relies on the values in Table 7. The DC for garden vegetables lifetime ingestion estimates were shown in Table 4. Table 7 combines the new food intakes, corrected crop PCB uptake slopes, and fraction of diet from the US-EPA Proposed 503 Rule (US-EPA, 1990a) to calculate the $\Sigma(UC_i \cdot DC_i \cdot FC_i)$ for PCBs. Carrot accounts for 53% of the dry matter of the root vegetable grouping according to Chaney et al. (9). Thus, the overall root vegetable group is shown as equal to 53% of the uptake slope of carrot. Although most people peel carrots before consuming them, there is no assurance that carrots will be peeled. Therefore, the uptake slope for unpeeled carrots was used. O'Connor et al. (29) found little PCB entry to carrot deeper than the normal peel depth.

Table 7. Using corrected UC slopes, food intakes (DC), etc., for the calculations for Pathway 1F home garden analysis for PCBs. The absolute slope used for **unpeeled carrot** was (0.075 µg/g DW carrot)·[µg PCB/g soil DW]$^{-1}$; from O'Connor et al. (29), after review of PCB uptake by carrot and lettuce from sludge-PCB treated soil. Other data are best judgement values because slopes are not generally available from properly conducted field experiments with sludge-applied PCBs.

Food Group	Food Intakes DC	Uptake Slope UC	Fraction home-grown. FC	$(UC_i \cdot DC_i \cdot FC_i)$ µg/day / µg/g soil DW	
	g DW/day	µg/g(µg/g)$^{-1}$			%Total
Potatoes	15.60	0.025Carrot=0.00188	0.45	0.0132	24.9
Leafy Veg.	1.97	≤0.0010	0.60	0.00118	2.2
Legume Veg.	8.752	0.001Carrot=0.000075	0.29*	0.00019	0.4
Root Veg.	1.60	0.53Carrot =0.03975	0.60	0.03816	72.1
Garden Fruit	4.15	0.001Carrot=0.000075	0.60	0.00019	0.4
			Σ =	0.05292	100.0

*0.17 for dried legume vegetables (8.412 g DW/day) and 0.60 for fresh legume vegetables (3.340 g DW/day) gives 0.29 for weighted total legume vegetables.

Using the corrected ($UC_i \cdot DC_i \cdot FC_i$), the estimated RLC is:

$$RLC = \frac{0.909\ \mu g\ PCB/day}{0.05292\ \mu g\ PCB/day\ (\mu g\ PCB/g\ soil\ DW)^{-1}} = 17.2\ \mu g\ PCB/g\ dry\ soil$$

This is the maximum allowed concentration of PCB in soil at any time that garden vegetables are grown on the soil. This can be converted to an annual sludge-PCB application as done above in Pathway 4-Mixed With Soil, and Pathway 3. The RP_a would be $17.2 \cdot 2 \div 14.9 = 2.31$ kg PCB/ha/yr. A sludge applied at 10 Mg/ha/yr could contain 231 mg PCB/kg DW.

Table 8 compares the original calculated limitations for the Proposed 503 Regulation (38), and the estimated limitations if the appropriate algorithms and transfer coefficients shown in this paper were used (28, 8).

Table 8. Comparison of PCB application limits for each pathway for data from the 503 Proposed Rule and the corrected versions reported in this document. The last column lists concentrations of PCBs which could exist in sludges applied as N fertilizer annually at 10 Mg/ha.

mg/kg Pathway	Original EPA Estimate Limit Units	Limit value	Corrected Approach Limit Units	Limit value	Sludge Limit, @ 10 Mg/ha/yr
1	kg/ha•yr	4.1			
1D&M	kg/ha•yr	0.264	mg/kg Soil Max.	17.2	
1D&M	kg/ha/yr	2.31		231.	
2D&M	kg/ha•yr	7.3	mg/kg sludge DW	9.09	9.09
3	kg/ha•yr	0.0056	mg/kg Soil Max.	18.3	
			kg/ha/yr	2.46	246.
4-Surface Application	kg/ha•yr	0.019	mg/kg sludge DW	2.23	2.23
4-Mixed With Soil	kg/ha/yr	0.019	mg/kg Soil Max.	2.23	
4-Mixed With Soil	kg/ha/yr	0.019	kg/ha/yr	0.299	299.

FUTURE

The effort to complete development of the Clean Water Act 503 Regulations for sewage sludge is continuing. Final regulations are expected to be issued in 1992, and all 28 contaminants being considered in the present regulation will undergo risk assessment for all Pathways.

LITERATURE CITED

1. Ames, B.N. and L.S. Gold. 1990. Too many rodent carcinogens: Mitogenesis increases mutagenesis. Science 249:970-972.

2. Baker, E.L., Jr., D.S. Folland, T.A. Taylor, M. Frank, W. Peterson, G.Lovejoy, D. Cox, J. Housworth, and P.J. Landrigan. 1977. Lead poisoning in children of lead workers. Home contamination with industrial dust. New England J. Med. 296:260-261.

3. Baxter, J.C., D. Johnson, E. Kienholz, W.D. Burge, and W.N. Cramer. 1983. "Effects on Cattle

From Exposure to Sewage Sludge. PB83-170589." (EPA-600/2-83-012)

4. Bertrand, J.E., M.C. Lutrick, G.T. Edds, and R.L. West. 1981. Metal residues in tissues, animal performance, and carcass quality with beef steers grazing Pensacola bahiagrass pastures treated with liquid digested sludge. J. Anim. Sci. 53:146-153.

5. Bergh, A.K., and R.S. Peoples. 1977. Distribution of polychlorinated biphenyls in a municipal wastewater treatment plant and environs. Sci. Total Environ. 8:197-204.

6. Beyer, W.N., and C.D. Gish. 1980. Persistence in earthworms and potential hazards to birds of soil applied DDT, dieldrin, and heptachlor. J. Appl. Ecol. 17:295-307.

7. Calabrese, E.J., R. Barnes, E.J. Stanek, III, H. Pastides, C.E. Gilbert, P. Veneman, X. Wang, A. Lasztity, and P.T. Kostecki. 1989. How much soil do young children ingest: An epidemiologic study. Regulat. Toxicol. Pharmacol. 10:123-137.

8. Chaney, R.L. 1990. Research provides basis for quantitative risk assessment to make appropriate regulations for land application of sewage sludge: 1) Twenty years of land application research. BioCycle 31(9):54-59. 2) Public health and sludge utilization. BioCycle 31(10):68-73.

9. Chaney, R.L., R.J.F. Bruins, D.E. Baker, R.F. Korcak, J.E. Smith, Jr., and D.W. Cole. 1987. Transfer of sludge-applied trace elements to the food-chain. pp. 67-99. In A.L. Page, T.J. Logan and J.A. Ryan (eds.) Land Application of Sludge – Food Chain Implications. Lewis Publishers Inc., Chelsea, MI.

10. Chaney, R.L., S.B. Hornick, and P.W. Simon. 1977. Heavy metal relationships during land utilization of sewage sludge in the Northeast. pp. 283-314. In R.C. Loehr (ed.). Land as a Waste Management Alternative. Ann Arbor Science Publishers, Inc., Ann Arbor, MI.

11. Chaney, R.L., and C.A. Lloyd. 1979. Adherence of spray-applied liquid sewage sludge to tall fescue. J. Environ. Qual. 8:407-411.

12. Chaney, R.L. and H.W. Mielke. 1986. Standards for soil lead limitations in the United States. Trace Subst. Environ. Health. 20:355-377.

13. Chaney, R.L., S.B. Sterrett, M.C. Morella, and C.A. Lloyd. 1982. Effect of sludge quality and rate, soil pH, and time on heavy metal residues in leafy vegetables. pp 444-458. In Proc. Fifth Annu. Madison Conf. Appl. Res. Pract. Municipal and Industrial Waste. Univ. Wisconsin-Extension, Madison, Wisconsin.

14. Comptroller General. 1978. Sewage sludge: How do we cope with it? 38pp. USGAO Report CED-78-152.

15. Corey, R.B., L.D. King, C. Lue-Hing, D.S Fanning, J.J. Street, and J.M. Walker. 1987. Effects of sludge properties on accumulation of trace elements by crops. pp. 25-51. In A.L. Page T.J. Logan and J.A. Ryan (eds.) Land Application of Sludge. Lewis Publishers Inc., Ann Arbor, MI.

16. Davis RD, Howell K, Oake RJ, Wilcox P (1984) Significance of organic contaminants in sewage sludges used on agricultural land. In: Proceedings of the International Conference on Environmental Contamination. CEP Consultants Ltd, Edinburgh, Scottland, pp 73-79.

17. Decker, A.M., R.L. Chaney, J.P. Davidson, T.S. Rumsey, S.B Mohanty, and R.C. Hammond. 1980. Animal performance on pastures topdressed with liquid sewage sludge and sludge compost. p. 37-41. In Proc. Natl. Conf. Municipal and Industrial Sludge Utilization and Disposal. Information Transfer, Inc., Silver Spring, MD.

18. Fairbanks, B.C. and G.A. O'Connor. 1984. Effect of sewage sludge on the adsorption of polychlorinated biphenyls by three New Mexico soils. J. Environ. Qual. 13:297-300.

19. Fries, G.F. 1982. Potential polychlorinated biphenyl residues in animal products from application of contaminated sewage sludge to land. J. Environ. Qual. 11:14-20.

20. Fries, G.F., G.S. Marrow, and P.A. Snow. 1982. Soil ingestion by dairy cattle. J. Dairy Sci. 65:611-618.

21. Fries, G.F. and D.J. Paustenbach. 1990. Evaluation of potential transmission of 2,3,7,8-tetrachlorodibenzo-*p*-dioxin-contaminated incinerator emissions to humans via foods. J. Toxicol. Environ. Health 29:1-43.

22. Jacobs, L.W., A.C. Chang, R.L. Chaney, C. Frink, R. Horvath, J.A. Ryan, L. Tabatabai, J.B. Weber, and J. Werner. 1989. Distribution and Marketing. pp. 103-122. In A.L. Page, T.J. Logan, and J.A. Ryan (eds.) W-170 Peer Review Committee analysis of the Proposed 503 Rule on sewage sludge. CSRS Technical Committee W-170, Univ. California-Riverside.

23. Jones, S.G., K.W. Brown, L.E. Deuel, and K.C. Donnelly. 1979. Influence of rainfall on the retention of sludge heavy metals by the leaves of forage crops. J. Environ. Qual. 8:69-72.

24. Landrigan, P.J., C.W. Heath, Jr., J.A. Liddle, and D.D. Bayse. 1978. Exposure to polychlorinated biphenyls in Bloomington, Indiana. Report EPI-77-35-2. Public Health Service, CDC, Atlanta, GA.

25. Logan, T.J. and D.E. Cassler. 1988. Case study for correction of a widespread cadmium soil contamination problem using local funds. Trace Subst. Environ. Health 22:337-345.

26. O'Connor, GA (1989) Degradation, crop uptake, and risk of micropollutants in sewage sludge. Proc Sewage Sludge Conf: Qual Aspects and Risk in Connection with Land Application. Swedish Water and Wastewater Assoc (April 11-12, 1989).

27. O'Connor, G.A., R.L. Chaney, J.A. Ryan, D.E. Baker, K. Barbarick, A.C. Chang, R.B. Corey, R.H. Dowdy, P.R. Fitzgerald, T.D. Hinesly. 1989. Land Application - Agricultural Land. pp. 27-83. In A.L. Page, T.J. Logan, and J.A. Ryan (eds.) W-170 Peer Review Committee analysis of the Proposed 503 Rule on sewage sludge. CSRS Technical Committee W-170, Univ. California-Riverside.

28. O'Connor, G.A., R.L. Chaney, and J.A. Ryan. 1991. Bioavailability to plants of sludge-borne toxic organics. Rev. Environ. Contam. Toxicol. In press.

29. O'Connor, G.A., D. Kiehl, G.A. Eiceman, and J.A. Ryan. 1990. Plant uptake of sludge-borne PCBs. J. Environ. Qual. 19:113-118.

30. Page, A.L., T.J. Logan, and J.A. Ryan (eds.) W-170 Peer Review Committee analysis of the Proposed 503 Rule on sewage sludge. CSRS Technical Committee W-170, Univ. California-Riverside.

31. Pennington, J.A.T. 1983. Revision of the total diet study food lists and diets. J. Am. Diet. Assoc. 82:166-173.

32. Roels, H.A., J.P. Buchet, R.R. Lauwreys, P. Bruaux, F. Claeys-Thoreau, A. Lafontaine, and G. Verduyn. 1980. Exposure to lead by the oral and the pulmonary routes of children living in the vicinity of a primary lead smelter. Environ. Res. 22:81-94.

33. Rozman, K., T. Rozman, H. Greim, I.J. Nieman, and G.S. Smith. 1982. Use of aliphatic

hydrocarbons in feed to decrease body burdens of lipophilic toxicants in livestock. J. Agr. Food Chem. 30:98-100.

34. U.S. Environmental Protection Agency. 1979. Criteria for classification of solid waste disposal facilities and practices. Federal Register 44(179):53438-53464.

35. U.S. Environmental Protection Agency. 1985. Summary of environmental profiles and hazard indices for constituents of municipal sludge: Methods and results. US-EPA, Off. Water Reg. Stand., Washington, DC.

36. U.S. Environmental Protection Agency. 1989a. Development of risk assessment methodology for land application and distribution and marketing of municipal sludge. EPA/600/6-89/001.

37. U.S. Environmental Protection Agency. 1989b. Interim Final Guidance for Soil Ingestion Rates. OSWER Directive 9850.4; Jan. 27, 1989.

38. U.S. Environmental Protection Agency. 1989. Standards for the disposal of sewage sludge; proposed rule. Federal Register 54(23):5746-5902.

39. U.S. Environmental Protection Agency. 1990. National Sewage Sludge Survey; Availability of information and data, and anticipated impacts on proposed regulations; proposed rule. Federal Register 55(218):47210-47283.

40. Witte H, Langenohl T, Offenbacher G (1988) Investigation of the entry of organic pollutants into soils and plants through the use of sewage sludge in agriculture. Part A. Organic pollutant load in sewage sludge. Part B. Impact of the application of sewage sludge on organic matter contents in soils and plants. Korrespondenz Abwasser 13:118-125, 126-136.

Surfactants in Sewage Sludges

Prof. Dr. Malz, Franz / Dr. Korber, Heinz-Günter,
Emschergenossenschaft/Lippeverband
and Main Committee of Detergents,
Post-Office Box 101161, 4300 Essen 1, Germany

Summary

Considering the present discussion upon organic contaminants in municipal sewage sludges the topic "Surfactants in sewage sludges" has become quite acute. Anionic, nonionic and cationic surfactants get into waste water and subsequently into sewage sludges by use of washing agents in households and industries. While ionic surfactants are found at concentrations between 1 and 15 g/kg d.m., the respective concentration of nonionics is about 1 g/kg d.m.

A valuation of these concentrations has to be performed by the analytical methods being applie, mainly because non-specific parameters like MBAS (methylene blue active substance) and others cannot compete with results of specific methods, such as HPLC. Furthermore, the chemical behaviour of surfactants in sewage sludges and soils has to be included. In accordance with the at present available results, anionics and nonionics are aerobically biodegraded in sludges and soils. Presumably, this is also true for cationics, but further investigations are requested to confirm this.

Hazardous effects on ground water, soil and plants due to the presence of surfactants in sewage sludges are not likely to exist.

Synthetic detergents and soap are of prominent importance for water quality affairs with regard to their production amounts. Nowadays, an annual consumption of about 7 mio tons of surfactants and 8 mio tons of soap is assumed.

In West Germany about

- 132.000 tons of anionics
- 93.500 tons of nonionics
- 23.000 tons of cationics
- 5.500 tons of ampholytics

have been consumed in 1988. These chemicals go into public waste water treatment plants, where they are extensively retained and thus kept away from surface waters. This applies for nearly 95 % of Western German inhabitants.

The overall elimination process for surfactants in treatment plants consists of the following steps:

- biological degradation within canalization
- adsorption onto primary sludge
- biological aerobic degradation in treatment plants
- adsorption onto activated sludge

The overall elimination amounts to 95 %.

This paper deals with the amount of surfactants arising from adsorption processes onto the primary and excess sludge, which finally leads to surfactants being present in the digested sludge. This topic is important because of ecological consequences originating from the agricultural utilization of digested sludge.

Within the above mentioned groups of surfactants the following compounds are most important (Consumption amounts in brackets).

Anionics : LAS (74.000 tons/a)
Nonionics: AE (64.000 tons/a)
Cationics: DSDMAC (19.000 tons/a)
Soap : 101.000 tons/a

Soap as a widespread surfactant has to be included here owing to its slow aerobic biodegradation.

As a consequence, it is adsorbed onto the primary sludge fraction mainly as a calcium-salt. In contrast to this, anaerobic degradation has been shown to occur easily.

For Europe and the U.S. surfactants in digested sewage sludges cover nearly the same range of concentration. Anionics are detected in concentrations between 1 - 12 g/kg d.s., nonionics between 0,2 - 1,2 g/kg d.s., while cationics amount to 4,3 - 8,1 g/kg d.s. (1). Comparing these data with their concentrations in influents of waste water treatment plants the following table can be established.

Table 1: Influent and Sewage Sludge Concentrations of Surfactants in West Germany *

		Anionics	Nonionics	Cationics
Influent of Treatment Plant:	[mg/l]	10	8	4
	[%]	46	36	18
Digested Sewage Sludge	[g/kg d.s.]	6	0,7	6
	[%]	47	6	47

* All data are average values

While adsorption of nonionics onto the sludge is obviously of minor importance, it is the preferred mechanism of elimination for ionic surfactants. Especially the fate of LAS as the very anionic surfactant in German treatment plants has been subject to intensive investigations [2]. Painter has located up to 20 % of LAS in the primary clarification units of British treatment plants (3). Even in the early sixties Malz and Jendreyko pointed out the adsorption of anionics onto primary sludge to an extent of up to 30 %, being independent of a possible aerobic biodegradation (4). The same was reported in Britain, where the agricultural utilization of sewage sludge had reached a higher degree of application. Adverse effects of surfactants in digested sludge being applied to agriculturally utilized acres could not be established that time.

As far as we know today, also DSDMAC as the most important representative for cationics is not harmful to agriculturally utilized acres. A study performed by the German Main Committee of Detergents yielded an overall elimination of DSDMAC in a Lippeverband treatment plant being 96 %, while 61,7 % were due to adsorption processes (5). Presently, there are discussions in Germany on substituting DSDMAC by other products being better biodegradable.

Summarizing all these data and provided that one inhabitant produces 200 litres of waste water and 55 g of digested sludge, one may conclude that

- 3 - 33 % of anionics
- 0,6 - 4 % of nonionics
- 30 - 56 % of cationics

being present in waste water actually enter the digested sludge.

The fate of surfactants in soil material, expecially their aerobic biodegradability, plays a dominant role in discussing the agricultural utilization of sewage sludge. The following remarks pay attention to this subject.

Provided that 5 tons of sludge d.s. are applied to one hectar within a three years' period, a thinning-factor of 860 will result if the sludge is mixed up with a 30 cm's soil-layer. Let us now claim the sludge to contain an average concentration of 0,7 g/kg d.s. nonionics and 6 g/kg d.s. ionic surfactants, respectively. This will result in soil concentrations of 7 mg/kg d.s. in case of the ionic species and 0,8 mg/kg d.s. for nonionics. Just keep in mind that these values are estimated! If you want to form an opinion about this you have to consider that already a loss of the surface-active property of surfactants surely will have positive effects on their ecological behaviours. LAS was shown by Schöberl (6) to be strongly fixed to soil particles, which significantly reduces its availability for complete mineralization. In contrast to these findings other investigators undoubtfully have proven LAS to be aerobically biodegraded in soil within a hundred days' period, thus excluding an enrichment of LAS in soil material caused by the agricultural utilization of digested sewage sludge (7). The same fortunately applies to nonionics, and even cationics are aerobically biodegradable according to preliminary results of the Main Committee of Detergents (8). Presently, cationics in soil material are subject to stringent studies performed by the German Höchst AG.

The results I have referred to so far consequently allow the conclusion that surfactants in agriculturally utilized sludges probably do not impose any risk on farming or human and animal food chains.

Ecological risk assessments for surfactants in sludge and soil materials immediately lead over to their chemical analyses. Few years ago, the so called "sum parameters" such like MBAS (methylene blue active substances) and BiAS (bismuth active substances) for anionics and nonionics, respectively, were of prominent importance. Also cationics meanwhile are subject to the sum parameter "DSBAS" (disulfine blue active substances). While application of the quoted methods to water samples had long been established, new sample preparation methods had to be developed for sediments and sludges. Especially ionic surfactants are strongly adsorbed onto solid matter and therefore require sophisticated extraction methods. In the meantime, the latter ones have been designed by the Main Committee of Detergents.

Quantification methods originating from "High Pressure Liquid Chromatography" have proved to be superior to sum parameter analyses. One reason for that among others is constituted by the fact that widespread appearing compounds (such as biological amines in case of a MBAS-measurement) are also detected, which consequently falsifies the results. (The other above mentioned parameters BiAS and DSBAS share the same problem with respect to other falsifying components).

In order to provide an idea of modern chromatographic procedures the HPLC-analyses of LAS and Methylalkylquaternaris (such like DSDMAC) in sludge including the sample preparation methods are depicted in the following scheme 1 (9, 10).

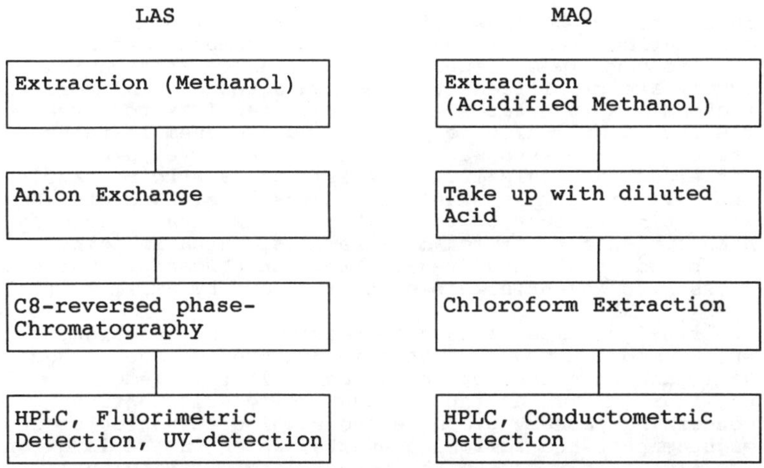

Scheme 1: HPLC of LAS and MAQ

In case of LAS, a methanol extraction is followed by an anion exchange step. The subsequent C_8-reversed phase chromatography can be interpreted in terms of a prepurification, which is in aduance of the final HPLC-analysis combined with a fluorimetric detection. The overall recovery for the C_{10}-C_{14} homologues is 84 %, the detection limit being 100 ppb.

Methylalkylquaternaries, for instance DSDMAC, are extracted using acidified methanol, and afterwards taken up in mineral acid. An extraction with chloroform preceeds the final HPLC, whilst quantification is guaranteed by conductometric detection. 97 % of MAQ compounds can be recovered; the detection limit amounts to 0,5 ppm.

In case of LAS, an interlaboratory test performed by the Main Committee of Detergents has shown the HPLC to yield significantly lower values of concentrations in sewage sludges than MBAS did. Furthermore, the statistical characteristics such as standard deviation and others are much better. It must not be forgotten to say that comfortable methods for analyzing nonionic surfactants still are lacking today. Recently, much work has been done on this.

In general it becomes evident that analytical data originating from the so-called sum parameter methods represent maximum values, whilst actual concentrations surely are significantly lower. On account of historical reasons and in order to ensure comparability to former results it will also be possible to use them in future, especially if waste water and sludge analyses are concerned. Ecological risk assessments are to be carried out performing more exact procedures, i.e. HPLC etc. Now it is time to mention that LAS and DSDMAC have recently been subjects to direct measurements, which in principal are based on UV- and UV-derivative spectroscopy. These works have essentially been carried out by Hellmann (11).

The final problem we have to tackle is the question, which importance surfactants in digested sewage sludge will gain with respect to the agricultural sludge utilization. Taking into account new developments concerning laundry products and washing technologies one might expect the following events, which in the long run are also partly due to the possibility of individual dosing of the washing agent ingredients.

- The consumption of LAS will decline, fatty alcohol sulphates will possibly gain growing importance.

- Substitution of anionics by nonionics is likely to occur owing to the increased usage of liquid detergent products. The overall consumption of surfactants won't be affected.

- The use of cationics is going to be constant, showing a long-term tendency to decrease.

- The prominent role of soap will be maintained.

It has clearly to be stated that surfactants in sewage sludge do not have a negative impact on any environmental system. Manufacture of surfactants being even more biodegradable than today's products will progress, which, on the other hand, will also favour an aerobic degradation in canalization systems and treatment plant influents. Furthermore, degradation processes are supported by the nowadays diminished BOD_5 loading in modern waste water treatment plants which offers a better mineralization both in the water and terrestrial sphere. The first step of mineralization is equivalent to a loss of the characteristic surface activity, which is a precondition for the riskless presence of surfactants in soils.

If we take all these new findings into consideration, it is a must to conclude, that surfactants in sewage sludge and soil put a rather low risk on our global environment.

References

1. **Bortlisz, J., Korber, H.-G., Malz, F.: Organische Komponenten und Schwermetalle in kommunalen** Klärschlämmen und landwirtschaftlich genutzten Böden, Abwassertechnik 4, 3 - 6 (1989)

2. International Status Seminar LAS-Environment: Tenside, Surfactants, Detergents 2 (1989).

3. Painter, H.A., Zabel, T.: cf. (2).

4. Malz, F., Jendreyko, H.: Beseitigung von Detergentien aus Abwässern und Gewässern, Forschungsberichte des Landes Nordrhein-Westfalen.

5. German Main Committee of Detergents, Division of Analytics, 1986

6. P. Schöberl, cf. (2).

7. Gilbert, P.A., Kleiser, H.H., Tenside, Surfactants, Detergents 25 (1988), 128.

8. German Main Committee of Detergents, Division of Cationics, 1988.

9. Matthijs, E., de Henau, H: Tenside, Surfactants, Detergents 24 (1987), 193.

10. Matthijs, E., de Henau, H.: Vom Wasser 69 (1987), 73.

11. **Hellmann, H., Z. Wasser-Abwasser-Forsch. 22, (1989), 131.**

THE PERSISTENCE OF POLYNUCLEAR AROMATIC HYDROCARBONS IN SEWAGE SLUDGE-AMENDED AGRICULTURAL SOILS

S.R.WILD[1], M.L.BERROW[2] and K.C.JONES[1].

[1]Institute of Environmental and Biological Sciences, Lancaster University, Lancaster, LA1 4YQ, U.K.

[2] Macaulay Land Use Research Institute, Craigiebuckler, Aberdeen, AB9 2QJ, U.K.

SUMMARY

In 1968 five metal enriched sewage sludges containing different concentrations of polynuclear aromatic hydrocarbons (PAHs) were applied to different plots on field soils at two experimental sites, Luddington and Lee Valley in the UK. This resulted in substantial increases in soil ΣPAH concentrations in all plots. Since application compound specific losses have occurred, with the high molecular weight PAHs being more persistent. Calculated half-lives range from under 2 years for naphthalene to over 9 years for benzo[*ghi*]perylene and coronene. The loss of PAH compounds in these field experiments can be related, in part, to certain physico-chemical properties, namely water solubility and the octanol:water partition coefficients.

1. INTRODUCTION

Sewage sludges are contaminated with a wide array of organic compounds [1]. The fate of these compounds when sewage sludge is applied to agricultural soil is, however, still largely unknown. Uncertainties remain over soil persistence and potential groundwater contamination, plant uptake and livestock ingestion. All these fates can potentially lead to human exposure. Currently, however, there are no legally enforceable restrictions aimed at limiting the organic input to agricultural soils from sewage sludges.

Polynuclear aromatic hydrocarbons (PAHs) are ubiquitous organic contaminants in sewage sludge. PAHs are present in rural and urban sludges in the range 1-10 mg/kg [2] while UK soil concentrations are in the range 100-54000 μg/kg, with a mean and median of 2320 μg/kg and 300 μg/kg respectively [3]. Therefore, applying sludge to land normally increases soil PAH concentrations. PAHs are on the USEPA and the EC priority pollutant lists since some are known or suspected carcinogens/mutagens. They have been highlighted as being one of the most important organic groups in sewage sludge with respect to human exposure [4, 5]. Data regarding soil persistence would aid assessment of the risks these compounds pose.

In 1968 field-based experiments were set up on two different soil types at Lee Valley and Luddington to investigate the effect of sewage sludge applications on crop uptake of heavy metals. This paper presents the results from the PAH analysis of archived soil samples from these experiments. These samples provide a unique opportunity to observe the loss of sludge applied PAHs in the field on different soil types. An attempt is also made to relate this loss to physical and chemical properties. Emphasis must be given to the use of long-term, field based experiments as realistic data concerning long-term persistence of organic chemicals can only be obtained from such experiments.

2. MATERIALS AND METHODS

2.1 Field Experimental Details

In 1968 the U.K. Agricultural Development Advisory Service (ADAS) and the Macaulay Institute for Soil Research (now the Macaulay Land Use Research Institute) set up experiments at Lee Valley and Luddington to investigate the uptake of selected metals by crops on sewage sludge amended soils. In 1968 sewage sludges contaminated with either Cr, Cu, Ni, or Zn or a rural sludge were applied to different soil plots at several different rates. Details of these experiments can be found in Berrow and Burridge [6]. For the purposes of this investigation six plots from each experiment were chosen, an untreated control plot and five others which had been treated with one of the metal rich sludges at a rate of 125 tonnes per hectare (dry weight). Soil plots will be referred to on the basis of which metal-rich sludge was applied. Following application the soil was dug or rotivated in to a depth of 15 cm. Since then soil samples have been taken periodically up to the late 1980's. After sludge treatment the plots from both sites exceeded EC limits for the single metals that were enriched in the sludge, ie. at Lee Valley the Zn, Cu, Ni and Cr plot soils contained 640 mg Zn/kg, 300 mg Cu/kg, 350 mg Ni/kg and 400 mg Cr/kg, respectively.

2.2 Sample Extraction and Analysis

Samples of soil (10 g) or sludge (2 g) were soxhlet extracted for three hours with dichloromethane (DCM). The extracts were reduced to dryness, taken up with *ca.* 5 ml of DCM then cleaned up by passage through a Sep-Pak Florisil cartridge and a Millex 0.5 μm filter unit. Samples were then reduced under nitrogen and made up to known volumes, 1 ml for soil extracts and 5 ml for sludge extracts.

Extracts were analysed by high performance liquid chromatography with fluorescence and diode array detection. Standards were obtained individually as solids from Aldrich Chemical Co. and as solutions from Supelchem (UK). A 2 μl injection of the extract was separated with an acetonitrile/water gradient running over 30 minutes, at a flow rate of 1.5 ml/minute. Further details of the extraction and analysis procedure are given in Wild *et al* [7]. The full range of compounds quantified was naphthalene, acenaphthene/fluorene, phenanthrene, anthracene, fluoranthene, pyrene, benzanthracene/chrysene, benzo[*b*]fluoranthene, benzo[*k*]-fluoranthene, benzo[*a*]pyrene, dibenz[*ah*]anthracene, benzo[*ghi*]perylene and coronene. ΣPAH refers to the sum of these compounds.

3. RESULTS AND DISCUSSION

3.1 PAH Concentrations in the Sewage Sludges

At Lee Valley the Cu-sludge contained 89 mg ΣPAH/kg and was the most PAH contaminated sludge applied while, as expected, the rural sludge contained the least (15 mg ΣPAH/kg). At Luddington the Zn-sludge was the most contaminated, containing about 57 mg ΣPAH/kg. In most of the sludges fluoranthene was the most abundant individual compound, although in both the Cr sludges benzo[*ghi*]perylene was found to have the highest concentrations. Generally the pattern of PAHs was similar in each sludge, with phenanthrene, fluoranthene, pyrene, benzanthracene / chrysene, benzo[*b*]fluoranthene, benzo -[*a*]pyrene and benzo[*ghi*]perylene being present at the highest concentrations, and with naphthalene, acenaphthene / fluorene, anthracene and dibenz[*ah*]anthracene present at the lowest.

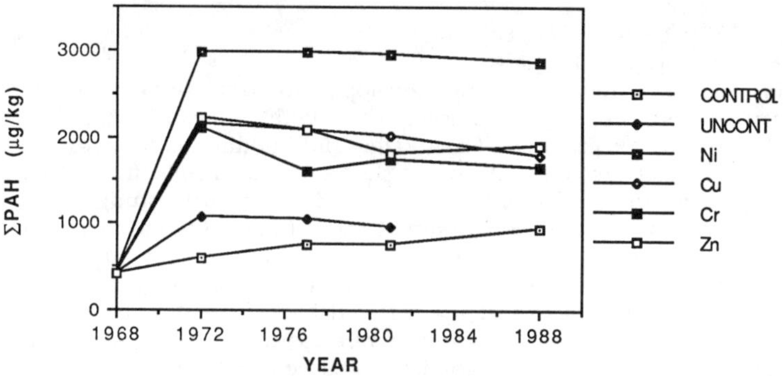

Figure 1. ΣPAH concentrations in Lee Valley soil plots through time (µg/kg).

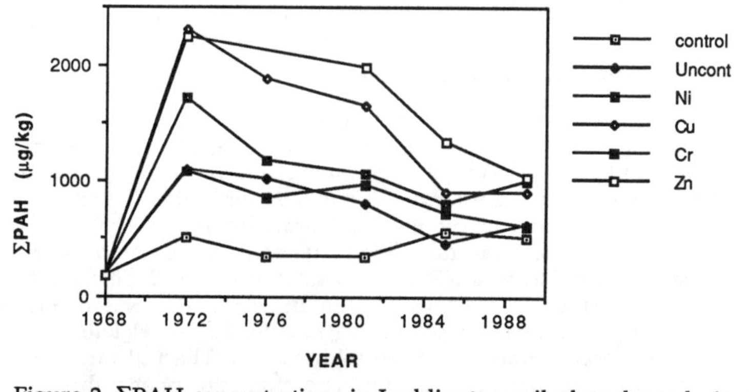

Figure 2. ΣPAH concentrations in Luddington soil plots through time (µg/kg).

3.2 PAH Concentrations in the Soils

Figures 1 and 2 illustrate the trends in ΣPAH observed in the experiments. At Lee Valley PAH concentrations in the control plot have steadily increased with time, from 415 to 934 µg ΣPAH/kg, a trend which has been observed previously in soils in a rural area [8]. PAH compounds in the control plot will be chiefly derived from atmospheric deposition. At Lee Valley between 1968 and 1988 there was a net gain of approximately 3.26 mg ΣPAH $m^{-2}yr^{-1}$. Concentrations of naphthalene, acenaphthene / fluorene and anthracene have all remained relatively constant through time. There is no evidence from the control plot to suggest contamination from the surrounding sludge-amended plots.

Application of sewage sludge caused significant increases in soil PAH concentrations in all plots of both experiments. By 1972 at Lee Valley, for example, the Ni plot had the highest ΣPAH concentrations (about 3000 μg/kg). In 1972 ΣPAH concentrations in the Lee Valley soil plots were Ni > Zn > Cu > Cr > rural. This order was still apparent in 1988. This broadly reflects the amount of PAHs applied to each of the plots, with the exception of the Cu plot which received the highest ΣPAH load.

At Luddington the ranking of soil plots in 1972 was different, namely Cu > Zn > Ni > rural > Cr. The Cu plot contained 2300 μg ΣPAH/kg. By 1989 this had changed to Zn > Ni > Cu > rural > Cr. Soil concentrations have been slightly lower at Luddington as the sludges were less PAH contaminated and the soil had a higher bulk density. Through time there is a clear decrease in PAH concentrations. The decrease at Lee Valley is less apparent, probably due to higher inputs of atmospheric PAHs. At both sites the higher molecular weight PAHs, such as benzo[*ghi*]perylene and coronene appear to be more persistent. This is discussed further below.

3.3 Possible Loss Mechanisms

The fate of PAHs in sludge-amended soils has been described in detail elsewhere [7]. In summary, PAH fates include leaching/ transport below the plough layer, plant offtake, abiotic degradation, volatilization, biological degradation and, potentially in these experiments, transboundary movement. Various observations suggest that plant uptake and migration to the sub-surface are inefficient processes for PAHs and unlikely to make an important contibution to the soil PAH mass balance. Soil lateral movement is also unimportant in these two experiments. Compound losses must therefore largely be due to abiotic processes (volatilization and chemical breakdown) and biotic degradation. Abiotic processes and volatilization were shown to be insignificant for compounds with three or more rings in a recent study [9]. Volatilization was only found to be significant for naphthalene. Strong retention to soil organic matter will reduce losses by volatilization [10].

It seem likely, therefore, that biodegradtion is the key process resulting in PAH losses from these soils [7, 9]. It's importance is well established for low molecular weight PAH compounds. Anthracene and phenanthrene, for example, can act as sole sources of carbon for micro-organisms [11]. Cometabolism of higher molecular weight species is also well established [11]. Persistence of PAHs in these soils varies markedly, and generally increases with increasing molecular weight [12]. Biodegradation is probably the most important process which can be invoked to explain this.

3.4 Calculation of PAH half -lives

Since the PAH concentrations in the sewage sludges and the sludge application rates (125 tonnes (D.W.) /ha) are known, the amount of PAHs applied to each plot in 1968 can be derived. In order to calculate the soil concentrations that would result from these applications an estimate of the soil mass is required, this necessitates soil bulk density (BD) values. For the Luddington sandy loam soil this can be calculated from an equation given by McGrath and Lane [13] which relates BD to soil percent carbon (%OC). However, this equation is not applicable for the silty loam soil such as that found at Lee Valley. Here BD estimates have been calculated from a relationship described by Jeffrey [14] between loss on ignition (LOI) and BD. With these estimates the soil concentrations that would be expected from sludge addition can be calculated and compared to the actual concentrations present in the soil to derive the percent remaining through time. These have calculated for all the sludge-amended plots. Several interesting features emerge from the data. Firstly, it is significant that by 1972 at Lee Valley, for example, most soil plots had lost about 60% of the PAHs applied in 1968. By 1988/89 percent losses had increased to about 80 -90%. At both sites the Ni plot experienced lower PAH losses; at Lee Valley 32% of the sludge-applied PAHs were still detectable in 1988, 35% at Luddington in 1989. The mass balance data also reveals some rapid loss of PAHs; in the Cu-plot at Lee Valley only about 19% of the PAHs applied in 1968 were detectable in 1972. The rural plot appeared to be back to control plot concentrations by 1985 at Luddington and by 1988 at Lee Valley.

There is evidence that persistence is related to compound structure. In all plots the higher molecular weight PAHs are less readily lost. Benzo[*ghi*]perylene and coronene are consistently the most persistent. From the data it is apparent that by 1972 the majority of some PAHs had been lost. Because of the long interval between sludge application and the first sampling, detecting differences between the losses of some individual compounds is problematic. However, estimates of half lives are possible for all the compounds. These can be calculated in two ways, although the full dataset is not shown here. The first method uses the mass balance data to simply calculate the time for 50% of the compound to be lost. The second uses a line of best fit from the soil time trend data to derive values for the half-life, assuming that the loss of organic chemicals from soil fit a first-order reaction [15]. Both methods give good agreement.

Table 1: PAH half-lives measured at Luddington and Lee Valley (years).

Compounds	Luddington	Lee Valley
Naphthalene	<2.1	<2.3
Ace/Fluorene	<3.2	<2.3
Phenanthrene	5.7	4.4
Anthracene	7.9	2.6
Fluoranthene	7.8	3.1
Pyrene	8.5	3.1
Benz/Chrysene	8.1	3.6
Benzo[b]fluoranthene	9.0	3.5
Benzo[k]fluoranthene	8.7	2.9
Benzo[a]pyrene	8.2	3.2
Dibenz[ah]anthracene	nd	6.8
Benzo[ghi]perylene	9.1	9.5
Coronene	16.5	9.3
ΣPAH	8.0	3.6

The mean half-life data is presented in Table 1. A structure-degradability relationship clearly exists for PAHs in the Lee Valley and Luddington soil, with half-lives ranging from about 2 years for naphthalene and acenaphthene/fluorene to over 6 years for dibenz[*ah*]anthracene and over 9 years for benzo[*ghi*]perylene and coronene. It should be emphasised that these half lives represent net losses of PAHs in the field due to the combination of all the mechanisms listed above. The mean half-life for ΣPAH was 3.6 ± 0.7 years at Lee Valley and 8 ± 4 years at Luddington. The loss rates also differ between the various metal-treated plots. The Ni plot has longer half-lives than the others; the ΣPAH half-life was between 3.5 and 5.3 years at Lee Valley and about 13 years at Luddington. The lowest half-lives were found on the Cu and Cr plots (3 - 6 years).

3.5 Comments on the persistence data

Several interesting points are evident, (i) the half-lives found at Luddington are significantly longer than the corresponding half-lives at Lee Valley. This is probably due to differences in soil type. The Luddington soil was a sandy loam of pH 5.8, with a low organic matter content (<2%), while the Lee Valley soil had a pH of 6.5, a higher organic matter content (>5%), higher cation exchange capacity and a lower bulk density. Because of this the Lee Valley soils may have had a more active microbial community than the Luddington soil. (ii) the PAH half-lives, particularly those from Luddington, tend to be significantly longer than those reported previously [see ref. 6]. The values reported in the literature are usually from laboratory studies where PAHs have been spiked into soils or sediments. Such experiments may yield shorter half-lives because of higher volatilizational/abiotic losses and/or increased bioavailability. Some experiments have also used acclimatised sites or soils inoculated with micro-organisms or surfactants. No half-lives have been reported previously for PAHs applied to soils in a sewage sludge matrix. At Luddington and Lee Valley the soils were both relatively pristine and the microbial community would not have been previously exposed to elevated PAH concentrations. (iii) There is consistent evidence from both sites to suggest that compound structure determines the long term persistence of PAHs in the soil. This is probably due to structure controls on loss mechanisms such as volatilization and biological degradation. (iv) At both Lee Valley and Luddington there appear to be lower PAH losses occurring on the plots treated with the Ni-rich sludge. The reasons for this are not clear, but this observation may be indicative of a metal suppression on microbial processes. It is not known whether this is due to the presence of high Ni, or some other contaminant(s). This is currently being investigated. However, it should be noted that the high concentrations of Ni measured in these soils exceed the EC statutory limits and will not be encountered in routine operational practice on sludge-amended soil.

3.6 Relating Physico-Chemical Properties to PAH Persistence

There is currently considerable interest in modelling the environmental behaviour of trace organic contaminants for predictive purposes [eg. 17, 18]. Various workers have shown, for example, that the adsorption of PAHs onto soils and sediments is a function of their octanol : water partition coefficients [eg. 19] whilst volatilization from the soil surface has been shown to be a function of the compounds Henry's Law constant (H_c) [18]. We have compared our field-derived estimates of soil persistence with various physico-chemical properties obtained from the literature.

The higher the H_c the greater the potential for volatilization from the soil surface. Compounds in the PAH group have a range of H_c figures, some such as naphthalene have relatively high H_c values (eg. 0.0562 Pa m^3/mol) and are expected to be rapidly lost from soil. If volatilization is the major loss mechanism affecting PAHs in these soils there should be a clear negative relationship between half-life and H_c. The field data for Luddington and Lee Valley does give a negative relationship with Hc, however, it is very weak with an R^2 of 0.16 for the combined data set. This implies that volatilization is not a major loss mechanism for all the compounds.

Water solubility is important when considering PAH persistence. Obviously, a higher water solubility infers greater mobility and renders the compound susceptible to greater leaching, plant uptake and biodegradation [20]. At Lee Valley and Luddington the log of water solubility and PAH half-lives are related giving R^2 values of 0.51 and 0.72 and correlastion coefficients of -0.72 and -0.85 respectively. All these are significant at the 95% confidence limit.

The octanol:water partition coefficient (Kow) describes the ratio of compound concentrations in n-octanol to that in water when at equilibrium, and provides a surrogate measure of a compound's tendency to partition between the aqueous phase and natural organic carbon in soils/sediment *etc.* or lipid phases in biota [17]. Regression of log Kow to half-lives from the two experiments described yields good relationships (see Figure 3). Correlation coefficients are 0.78 and 0.88 for Lee Valley and Luddington respectively, these are significant relationships at the 99% confidence limit. It is apparent that the higher the log Kow the longer the PAH's half-life, as would be expected. Compounds with a high log Kow are tightly bound to organic fraction in the soil which will render them less available for biodegradation, leaching, volatilization and plant uptake.

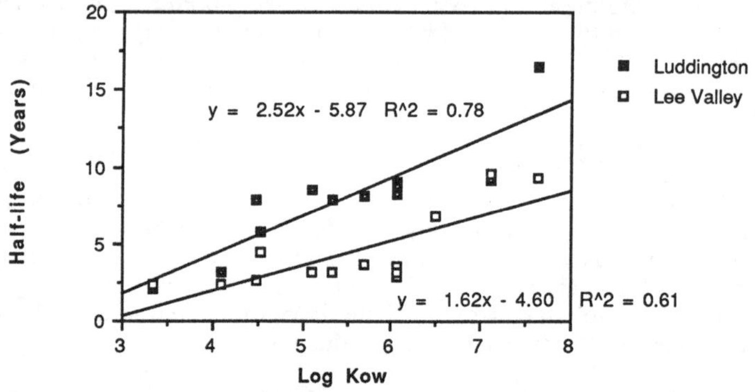

Figure 3. Relationship between Log Kow and half-lives found at Lee Valley and Luddington.

The high correlation coefficients between soil persistence and log Kow (and to a lesser extent water solubility) indicate that these parameters are important with respect to environmental fate. Low water solubility and a high log Kow appear to indicate long PAH persistence in these soils in the field. One would expect that other groups of organics such as polychlorinated biphenyls (PCBs) and polychlorinated dibenzo-p-dioxins and -furans (PCDD/Fs) which have similar properties to PAHs may behave in a similar manner in soils.

4. CONCLUSIONS

The addition of single sewage sludge applications to the Lee Valley and Luddington soil plots caused PAH concentrations to increase. The subsequent loss of these organic compounds was monitored and the soil half-lives calculated. Half-lives varied between compounds; the multi-ringed species were more persistent. In general, the calculated half-lives are longer than data from laboratory based studies reported in the literature. Our data also suggest that compound persistence is highly variable in the field. The persistence of PAHs in these field soils increases linearly with compound log Kow, and inversely with the compounds aqueous solubility. Relating compound behaviour in soil to simple physico-chemical parameters may be useful in helping to predict the persistence of other organic compounds in soil systems.

Acknowledgement

SRW is a research student funded jointly by the U.K. Natural Environment Research Council and the Water Research Centre. We are grateful to both organisations for their support.

5. REFERENCES

1. Jacobs, L. W., O'Connor, G. A., Overcash, M. A., Zabik, M. J. & Rygiewicz, P., (1987). In: Land Application of Sludge. A. L. Page, T. J. Logan and J. A. Ryan (eds). Lewis Publishers, Michigan.
2. Wild, S. R. & Jones, K. C. (1989). Chemosphere , **19**, 1765-1777.
3. Jones, K. C., Stratford, J. A., Waterhouse, K. S. & Vogt, N. B. (1989). Environ. Sci. Technol., **23**, 540-550.
4. Connor, M. S. (1984). Biocycle, **25**, 47-51.
5. Dean, R. B. & Suess, M. J. (1985) Waste Manage. Res., **3**, 251-278.
6. Berrow, M. L. & Burridge, J. C. (1990). Int. J. Environ. Anal. Chem., **39**, 173-177.
7. Wild, S. R., Waterhouse, K. S., McGrath, S. P. & Jones, K. C. (1990). Environ. Sci. Technol., (in press).
8. Jones, K. C., Stratford, J. A., Waterhouse, K. S., Furlong, E. T., Giger, W., Hites, R. A., Schaffner, C. & Johnston, A. E. (1989). Environ. Sci. Technol., **23**, 95-101.
9. Park, K. S., Sims, R. C., Dupont, R. R., Doucette, W. J. & Matthews, J. E. (1990). Environ. Toxicol. Chem., **9**, 187-195.
10. Fairbanks, B. C., O'Connor, G. A. & Smith, S. (1987). J. Environ. Qual, **16**, 18-25
11. Heitkamp, M. A. & Cerniglia, C. E. (1987). Environ. Toxicol. Chem., **6**, 535-546.
12. Bossert, I. D. & Bartha, R. (1986). Bull. Environ. Contam. Toxicol., **37**, 490-495.
13. McGrath, S. P. & Lane, P. W. (1989). Environ. Pollut., **60**, 235-256.
14. Jeffrey, D. W. (1987). Soil-Plant Relationships: an ecological approach. Croom Helm Ltd. England.
15. Jury, W. A., Focht, D. D. & Farmer, W. J. (1987a). J. Environ. Qual., **16**, 422-428.
16. Wild, S. R., Obbard, J. P., Munn, C. I., Berrow, M. L. & Jones, K. C. (in press). Sci. Total Environ.
17. Mackay, D. & Clark, K. E. (1991). In: Organic Contaminants in the Environment: Environmental Pathways and Effects. K. C. Jones (ed). Elsevier Appl. Sci. Pub.
18. Jury, W. A., Winer, A. M., Spencer, W. F. & Focht, D. D. (1987b). Rev. Environ. Contam. Toxicol., **99**, 119-164.
19. Means, J. C., Wood, G. S., Hassett, J. J. & Banwart, W. L. (1980). Environ. Sci. Technol., **14**, 1524-1528.
20. Stucki, G. & Alexander, M. (1987). Appl. Environ. Microbiol., **53**, 292-297.

SESSION REPORT

N. HARKNESS, Lichfield (UK)

The Chairman opened the session saying interest in organics had increased greatly in the last few years.

In presenting the first paper of the session, R. Leschber paid tribute to those who had contributed early on to COST 68 including in the second period when he became Chairman of Working Party 2, R. L. Chaney (USA) and the late H. M. J. Scheltinga (NL). Working Party 2 had initially concentrated on heavy metals but for the past 3 years had concentrated on organics. Much analytical data is available for organics but less toxicological information. Some exists for chlorinated substances also PAH, phthalates and detergents. Attention has been focussed on these substances since they are found in sewage sludges.

New legislation in the FRG proposes all sludge should be examined for dioxins and furans since high values had been found in some samples examined. Absorbed Organohalogen Compounds (AOX) is used as a screen and further examination for Extractable (EOX) and Purgeable (POX) may be required. Phthalates are always found but normally well below the standard.

It was concluded there are no problems if sewage sludge is used in accordance with the regulations. Uptake and bioaccumulation does not appear to be a problem with crops but ingestion of organics adhering to grazed grassland needs to be taken into account. Sewage sludge composition can mirror the general distribution of environmentally sensitive substances and as such can be used as an indicator.

In discussion points made were:

- that substances considered to be harmful from those found were listed but it could be environmentally helpful if those considered harmless were listed,
- that sewage sludges may accumulate substances not discharged into sewers, for example aerial depositions and surface water run-off, therefore education for general reduction of environmental contamination would be beneficial,
- that AOX may be a good general indicator but was not related to toxicity. Could some indication be given of particular compounds,
- could it be taken there was no danger in the European context ?

In reply it was said many of the substances which could be found analytically were being investigated and may or may not be harmful. Work was in progress on rainwater and gutters where indeed grass could grow. Substances are transformed and degraded in nature and can become of no significance to health. The nature of substances contributing to AOX are not yet fully defined but for example, removal of polychlorinated phenols used for leather impregnation reduced AOX by 50% in one instance. In relation to the public, there must be a response to their fears. These fears may have been fueled and it is not easy to reverse apprehensions. Proper research must be done, the media reasonable and an openness with the public. In the FRG 25% of sewage sludge is used in agriculture (and might increase back to 35%) although some people will only use organically grown vegetables without any sewage sludge. In East Germany it is proposed to keep separate primary and biological sludges: The latter may be used for agriculture while the former could be stored and possibly decontaminated. Money spent in the East will give greater benefit than in the West.

R. L. Chaney in discussing risk assessment made many interesting points with particular reference to draft revised US regulations. These included the greater risk of sludge adhering to plants from spraying and from ingestion of soil in some circumstances than from plant uptake. A recommendation could be injection rather than spraying. It was also believed tests with high metal content overestimated risk: There was little or no uptake of cadmium by lettuce from sludge low in the metal whereas but much uptake from high metal sludge. Especially important was the evidence of plateaux of uptake. There was also the possibility of adding metal binding substances to sludge for example, iron minimised cadmium uptake and zinc similarly causes speciation changes. In determining risk, pathway analysis is required and protection for the most exposed, NOAEL - No Observed Adverse Effect Level.

In discussion the following points were raised:

- can the plateau be relied upon, does it persist ? There is evidence that even a linear response will change to a plateau as organic matter is stabilised (miss out first year results). Iron from phosphate removal can help but what happens if sludge additions stop or there is a change to mineral fertiliser,

- on the sludge quality survey slide, 98 percentile values were similar to those in United Kingdom but the 50 percentile lower. How many plants were included was asked,

- the approach was liked but what is used as the toxicological end point may be an aspect of uncertainty as in the route to milk.

- there could be worry about the trend from the Environmental press and legislators against the type of risk analysis.

In reply it was said many studies of single sludge application had been made and the results were as expected even at low pH (<5.5). In the survey 250 plants were selected statistically and studied. Around 25% showed no pesticides but some were very low in solids (analysis on wet weight); if solids had been greater perhaps higher values than for samples shown to contain pesticides would have been obtained. Pathway 4 via milk was very significant especially with surface sludge application (cut grass before spreading). Cancer potency slopes (carcinogenicity) may be high because high exposure kills cells and regrowth cells are more sensitive. The proposed new regulations abandon the earlier numbers and are based on the better data given in the paper. There will always be some who worry; scientists must research well and protect on worst case. An open public approach is needed, ignorance not knowledge leads to fear and none of us want sludge utilisation to interfere with humans or wildlife.

On surfactants, H. G. Korber showed it was mainly anionic and cationic detergents which accumulated in primary sludge together with soaps as calcium salts (latter anaerobically biodegraded). These detergents generally appear to be biodegraded in soils and at the application rates in FRG do not pose any risk to farming and food chains. Analyses (as described) for specific substances in sludges are recommended.

In discussion matters raised were:

- the real confidence about sludges, treatment, aquatic environment and effect of nonionics in bioflocculation in activated sludge,

- data for biodegradation in sewers, toxicity of alkylnonyl phenols, status of use and metabolites.

In reply the author said nonionics may cause foaming but he had no experience on bioflocuIation. Calculation of inputs and quantities arriving at treatment works were made and degradation in sewer was indicated by low figures for treatment. Alkylnonyl phenols are banned in laundry products; some special industrial or imported products may be in use but are not considered a problem. Methods of analysis are still needed for some products and metabolites.

K. C. Jones described organics research on soils taken over many years from various sites including two which had received high metal sludge in 1968. Mass balance calculations showed substantial losses of PAH from 1968 in four years and continuing over subsequent years. Little or no difference has been found for plant material from sludged or control plots. In other surveys, dioxin increase from 1840 - 1980 in soil (not sludged) and crops was attributed to aerial deposition and only the higher chlorinated chlorobenzenes were persistent (though reduced); there was little evidence for direct plant uptake and sludge was a minor factor compared with aerial sources.

During discussion it was asked:

- how aerial deposition affected half life calculations,

- the effect of PAH on soil biota,

- how fast and what were the mechanisms of microbial degradation.

In response the author said control unsludged plots were used to allow for aerial deposition. He was not aware of any data showing adverse effects of PAH on soil biota. Data on rate of biodegradation was variable depending upon particular sites; experiments on highly enhanced sites (e.g. coal gas works) may give half lives of a few weeks.

The paper by J. Tarradellas was not presented and D. Scheidegger presented the paper on analysis for H. Giger. This reviewed analytical methods including extraction and clean-up procedures for environmental samples, metabolites and particular reference to alkyl polyethoxylate and alkylnonyl phenol detergents.

In discussion the usage of the different nonionic detergents was raised and also the possibility of halogenation in the GC injection system. In reply it was said nonyl phenol nonionics were still in industrial use in Switzerland.

The formal sessions closed with thanks to the authors and participants in the discussion.

Some of the poster papers were relevant to this session including "Micro pollutants in sewage sludge", J. J. Van Den Berg and L. E. Duvoort-Van Engers on dioxins and furans in wet oxidation of sewage sludge, "Mass flow through the Swiss Sewage System", V. Candinas, emphasising sludge as a monitor of environmental pollution and "Heavy Metals in urban sewage sludges", M. Angelidis and R. J. Gibbs considering distribution of metals in sludge with most of the metal in the particles being in the oxidisable phase.

In conclusion several themes came from more than one paper:

- the importance of aerial and other sources for some organics on agricultural crops as well as sewage sludges,

- the importance of ingestion of material on crop surfaces rather than uptake through the plants,

- a need for openness in public debates,

- the need for continuing development of analytical methods (including extraction procedures) for specific substances and metabolites,

- the need for toxicological data to interpret analytical findings.

Session IV

RISK ASSESSMENT OF THE OCCURENCE OF PATHOGENS IN SEWAGE SLUDGE AND EFFLUENTS FROM LIVESTOCK

Chairman : J. HANNAN (IRL)

Rapporteur : E.B. PIKE (UK)

RISK ASSESSMENT OF THE OCCURRENCE OF PATHOGENS IN SEWAGE SLUDGE AND EFFLUENTS FROM LIVESTOCK

PART 1 : PATHOGENS IN SEWAGE SLUDGE

EDMUND B. PIKE

WRc Medmenham
PO Box 16, Marlow, Bucks SL7 2HD, United Kingdom

Summary
Treatment and disposal of sludge is a major and essential part of sewage treatment. Because sludge can contain pathogenic micro-organisms and viruses derived from the community, treatment and disposal to land must be managed in a way to minimise the risks to humans and food animals, while, at the same time, maximising its fertilising value and obtaining good value for the total costs of disposal. Throughout the history of the COST 68/COST 681 collaborative action programmes, research information has been collected to provide assessment of risks and this has now been used in Community and national legislation and at an operational and local level to formulate procedures for risk management in use of sludge in agriculture. The paper deals generally with the principles of risk assessment and management in sludge treatment and disposal, but gives references to key review papers for more detail, which may be supplemented by Working Party 3's published seminars and contributions to the five international symposia held under the COST 68/COST 681 programmes.

1. INTRODUCTION

Sewage treatment, together with the provision of safe, piped drinking water, are the most important public health measures for controlling epidemic disease. Yet sewage will contain enteric pathogens excreted by cases and carriers in the community, many of which will appear by concentration in the sludge. Sludge disposal is an essential part of sewage treatment. Sludge however is, in its untreated form, a noxious and hazardous product, which must be treated and then disposed of in a manner involving the least risk. A major route is land disposal, which makes use of its properties as a fertiliser and soil improver. Sludge treatment and disposal is a very large element (about 40%) of the total cost of sewage treatment, hence there are strong challenges to control the costs of treating and disposing of sludge, while containing risks and maintaining a service to agriculture. This economic aspect of optimising costs must prevail, whether sewage treatment is viewed as a publically-operated service or as a business, because of the need to justify cost. In this consideration, the cost of harm, which may well be high in terms of financial loss or loss in goodwill, must also be included.

This line of thought is not new, but it does show that the principles of analysing, assessing and managing risk have always been inherent in the treatment and disposal of sewage and sludge. However, modern management techniques have developed and codified the procedures, which are now being widely applied in sewage treatment services. Throughout their history, the European Community COST programmes 68 and 681 have reviewed and evaluated research in treatment and use of sewage sludge and liquid agricultural wastes. The research has been valuable in identifying both the hazards and benefits from using sludge in agriculture and in describing the engineering and performance aspects of new processes and the economics of operating. It has therefore provided quantitative fundamental information for assessing and managing risk. Some of this research information was tendered and used during the drafting of the Directive on use of sludge in agriculture (1). The Directive and national legislation enforcing it identify constraints, which must be taken into account during management of risk and various national guidelines advise upon interpretation of this legislation. On the other hand, sewage and sludge disposal services must employ risk analysis and assessment in drawing up their local codes of practice and instructions for operators responsible for treating and disposing sludge.

2. DEFINITIONS OF HAZARD AND RISK

A hazard is a set of circumstances which could lead to harm. Harm in this paper means disease in man or food animals, although it could also mean revulsion to the odour of sludge and, in the wider sense, toxic effects of sludge to soil organisms and plant life and pollution of watercourses and groundwater.

Risk is the probability that the unfavourable event, in this case, disease, will occur as a result of exposure to the hazard.

This distinction between hazard and risk is important and was pointed out in the report of a WHO Working Group (2). A great variety of pathogens may be found in sludge, from time to time, particularly if sensitive and selective methods are used, but a risk exists only if it can be established that they cause disease under field conditions. This report also stated that the existence of risk would depend upon a great variety of local factors and individual circumstance and if pathogens were ingested in sufficient numbers to cause infection.

The terms 'risk assessment' and 'risk analysis' are widely used. In risk assessment, the likelihood of undesired events and the likelihood of harm or damage are evaluated and with them, value judgements are made concerning their significance. As part of this process, it is necessary to construct the sequence of events or faults which can lead up to the damage and then assess their probability. Risk analysis is carried out in different ways according to the investigator's involvement in the problem. Engineers and scientists will wish to quantify risks, based upon experimental evidence and ultimately to construct models of processes, such as infection or spread of disease. Such models can never be simple, because infection is a complex process, depending upon the invasiveness and pathogenicity of the disease agent, as well as on immunity of the subject, local conditions and state of community health. It has however, been attempted for Taenia saginata (3), with success. If such models are validated and are successful in predicting natural events, they can eventually be used in decision-making.

Economists may also use risk analysis, on the basis that safety is not a free commodity and that society will need to make decisions on allocation of scarce resources. Such analyses are dependent upon monetary values being placed upon the consequences of untoward events such as disease and loss of goodwill, such as willingness of farmers to accept sludge on their land.

3. RISK ASSESSMENT AND SLUDGE DISPOSAL

The finding of a particular pathogen is sludge does not mean that there is a risk of infection, for the reasons already stated. There are many difficulties in proving a causal relationship between disease and an event, such as use of sludge in agriculture. Nine criteria for assessing the strength of proof have been given by Bradford Hill (4), such as strength of statistical association, frequency of reporting, demonstration that the disease follows exposure after an expected incubation period, and not before exposure, that a positive dose-response relationship can be shown, that the association accords with the known biology of the disease and is not in conflict, that experiments support the causal relationship and that a fair analogy exists with a parallel context. Not all of these may be demonstrated, but the more that are, the more likely it is that a causal relationship exists.

Many of these criteria can be verified from national surveillance programmes of human and animal disease. These programmes have the advantages of enabling trends to be spotted rapidly and of permitting outbreaks to be investigated, with a view to establishing the causes. The statistical information they provide is, however, retrospective and there are often problems in obtaining accurate diagnoses and of identifying suitable control subjects with which to compare the histories of the diseased.

Prospective epidemiological studies, in which the health of subjects exposed to the hazardous agent and that of non-exposed, matched controls are compared after exposure, are necessarily very large and expensive, particularly if the risks are expected to be low. Of their nature, they can only investigate one or a few factors, defined at the outset of the trial. Another approach is a model infectivity study, designed to show, whether exposure under conditions resembling those experienced in the field can result in infection.

Examples of these various approaches were given in a COST 681 seminar (5). This seminar and a WHO Working Group (2) both considered the size of the risks associated with the various pathogens found in sludge and their significance for human and animal health. Although both were forward-looking in considering 'new' or 'emerging' pathogens as well as the established pathogens, *Salmonella* spp and the eggs of *Taenia saginata*, the prime importance of the latter were established.

4. SALMONELLOSIS

The relationships between use of sludge in European agriculture and salmonellosis in food animals were considered in detail (6). A major trend over the last 30 years or so has been for infections by particular host-adapted serotypes, such as *Salmonella dublin* in cattle and *S. cholerae-suis* in pigs, to have been replaced by those of *S. typhimurium* and others which affect a wide range of food animals and man indiscriminately. National surveillance programmes, such as those under the Zoonoses Order 1975 in Britain and large-scale studies in the Netherlands (7) have shown that complex routes for transmission of salmonellosis exist in agriculture and the environment. However, although much evidence exists to connect human salmonellosis with consumption of meat and dairy products, transmission from man to food animals is more fragmentary. In the review article (6), examples for transfer from human sources to farm animals included contamination of pasture by raw sewage and septic tank wastes, and from grazing and receiving sludge. In two of the sludge-related outbreaks, conditions were unusual. In a Swiss outbreak, cattle were infected with the rare serotype, S. tokoin after eating grass from pasture sludged 4 weeks earlier. This serotype was also recovered from the grass (8). In a Netherlands study of a dairy herd, grazed on land which received regular doses of sludge salmonella carriage in faeces was significantly in excess of the national average and S. panama was isolated from sludge and later from faeces(9). Sludge was applied at heavy rates at intervals throughout the spring and summer and cattle returned 8-21 days after sludging. In Britain, only one outbreak in cattle was identified with sludge application during 10 years of monitoring salmonellosis in food animals, and in that outbreak sludge containing a high proportion of poultry processing waste had been applied to grassland and cattle returned shortly afterwards(10). It was a conclusion of Panel A (Salmonella and other Bacteria) at the Seminar that if national guidelines or the provisions in the sludge in agriculture Directive (1) had been followed, none of the outbreaks of salmonellosis reviewed would have occurred (5, pp. 154-158). This panel also noted that bacteria other than Salmonella spp do not appear to be transmitted by sludge.

Model experiments in which cattle were fed strains of S. dublin of known infectivity in sludge or animal slurry suggested that they are not readily infected, unless there is prolonged exposure to heavy contamination or susceptibility of the animals is high(11,12).

5. BOVINE CYSTICERCOSIS

Bovine cysticercosis results from cattle (the secondary host) ingesting the eggs of the human beef tapeworm, Taenia saginata. Ingested eggs hatch out in the digestive juices, migrate through the gut wall and encyst, usually in muscle tissues. The cycle is completed by man eating raw or imperfectly cooked beef containing the cysts. There is thus no doubt that human excreta is the vehicle of infection, either voided directly on to pasture or contaminating it as sewage, sewage effluent or sludge.

The disease does not appear to be a major one in western Europe and the incidence has declined steadily over the past 30-40 years. In England and Wales, in 1978 the incidence was estimated at 0.041% of slaughterings(13) and in Denmark in 1986, 0.11%(14). The exact incidence in cattle is probably higher, because of the necessarily cursory nature of meat inspection. The incidence of human infections in Denmark, based upon sales of the drug, Niclosamide, was estimated at $3/10^5$(14) whereas, in England and Wales about 80-100 cases are referred annually to public health laboratories.

A Scottish study of sludge use on farms, where cysticercosis had been found, did not show that sludge was a major cause, being implicated in only 16 out of 370 farms (4.3%) and 26 out of 498 animals (5.2%)(15).

Experimental and theoretical considerations show that the settling velocities of T. saginata eggs in sewage are low (about 1 m/h) and that consequently they are poorly removed during primary sedimentation at sewage works (16,17). Light infections of carcasses are consistent with contact with streams carrying sewage effluent(14) or flooding of pasture with such water (17), whereas heavy infections ('cysticercosis storms') in Denmark, in most cases, originated with direct contact with septic tank waste or from grazing close to small sewage works(14).

6. OTHER PATHOGENS

Human ascariasis, although transmissible through land disposal of human excreta and eating of raw vegetables and salad crops, is now rare and has declined considerably since the last war in areas where it was endemic. Panel C (Viruses) of the Seminar (5, pp 161-163) could only identify two outbreaks of viral infection, both of Hepatitis A, one from cesspool infection of milk the other among sewage-sludge spreaders. Panel B (Parasites) (5, pp. 159-160) considered that further study was needed in the cases of the parasites, Giardia, Sarcocystis and Cryptosporidium.

7. RISK MANAGEMENT

In the context of sludge disposal in agriculture, risk management aims to reduce to a minimum at lowest cost, the disease, toxic and aesthetic hazards of sludge, while maximising its benefits.

In the first instance, this means that land disposal has been identified as the best practicable environmental option. Two principles are then used to reduce the risks to insignificant levels, namely identifying the ability of processes to reduce infectivity or viability of pathogens and matching these to the type of land and its agricultural use. Collectively these create barriers to the transmission of infection, by ensuring that the numbers which eventually reach the subject are well below infective levels. A major part of the work reviewed by Working Party 3 (Hygienic Aspects) during the life of COST 68/COST 681, has been concerned with obtaining fundamental information upon the decay of pathogens during sludge treatment and upon land treated with sludge. For example, Working Party 3 in its first report(17) formulated suggestions relating sludge treatment and land use and presented the opinions that

> "Economically and practically, a no-risk level cannot be attained, although it might be technologically possible", and "Definitions of levels of treatment of sludge that may provide reasonable safety are not easy to obtain, and they cannot be expressed in absolute terms, but would depend upon a number of economical, political and geographical factors".

It also remarked that if too strict rules were imposed, it might become impossible to place sludge on agricultural land.

These suggestions were made at the time when the Directive(1) was being drafted. The Directive imposes minimal restrictions, allowing for national legislation and local rules to be more severe, to prevent harmful effects, while encouraging correct use. The risks to man and food animals are prevented by permitting only 'treated' sludge to be used, unless injected or worked into the soil. 'Treatment' is defined as biological, chemical or heat treatment, long-term storage or any other appropriate process able to reduce significantly its fermentability and the health hazards resulting from its use.

This definition of treatment implies that pathogens are reduced and that the sludge is stabilised to reduce aesthetic problems in its use. No definition is made of the degree of pathogen reduction expected, because this is a decision which must be taken nationally and locally, having regard to particular needs. The US Environmental Protection Agency in its regulations has defined 'processes to significantly reduce pathogens' (PSRP) as those which reduce pathogens to a level comparable with a well-run anaerobic digester and which can achieve 99% reduction in faecal coliform bacteria or faecal streptococci and 90% reduction in total enteroviruses or salmonellae(18). 'Processes to further reduce pathogens' (PFRP) are also defined for use where the land application and the potential for human exposure cannot be adequately controlled; these processes must reduce pathogens below detectable limits.

The decision upon the severity of the process is thus one of risk management, bearing in mind the sensitivity of the land and its use, whether for grazing, growing cereals, vegetables or fruit or for recreation, if permitted and the likely pathogen content of the sludge. PSRP are usually effective in destroying salmonellae and eggs of Taenia saginata, but not the ova of Ascaris spp and other nematode parasites. In special cases, such as sludge from communities known to contain carriers of typhoid or paratyphoid fever or other serious infectious diseases, arrangements to treat sludges by PFRP may be considered prudent.

The scope of this paper does not permit detailed discussion of individual processes and their performances. Details are given in References 16 and 19-21. It must be emphasised that the process conditions, such as temperature, retention period, flow regime (plug-flow, batch storage or complete-mixing) and pH value will all influence the pathogen reduction obtained in practice. These must be specified in any definition of approved processes. Similarly, operation of any sludge treatment process must be properly controlled. The factors above are all physical and are readily monitored, so that they can be used to control processes either manually or automatically, and can be recorded to provide a history of operation, open to future inspection.

The disposal of sludge on agricultural land should receive equal attention from risk management. The emphasis should be based upon providing a complete service to the farmer or landowner, in terms of maximising the fertilising and soil-improving value of the sludge. A first step is inspection of the land with analysis of the soil. This will enable the optimal dose to be applied and the method of application to be selected. From the point of view of hygiene, this will involve decisions permitting the sludge to be applied cleanly, without creating aerosols or contaminating sensitive land and avoiding the possibility of polluting surface and ground-waters. Workmen applying the sludge should be provided with suitable clothing and facilities for personal hygiene, to avoid the possibility of acquiring infection.

Education and information is an important part of risk management. This includes instructing workmen in their duties, providing accurate information to farmers on the nature of sludge and the quantities applied and making sure that they fully understand the need for restrictions on land use after applying sludge. The goodwill of farmers is an important part of the economics of sludge disposal and it is one that is readily lost, as a result of accidents or bad practice.

8. CONCLUSION

Assessment of the risks in treating and using sludge in agriculture is an essential part of the wider concept of managing risks of disposal, since it provides the basic information upon which to base decisions upon capital investment in treatment and disposal and the conduct of the operation. Production of sludge and the need to dispose of it safely by the best practicable route are inherent and inescapable parts of sewage treatment. The minimum requirements for treatment of sewage, which form part of the proposed Directive on municipal wastewater treatment, COM(89) 518, will, if implemented, together with continuing improvements in the scope of sewage treatment, increase sludge production and thereby the necessity for managing its disposal in a modern and professional manner. Risk management is an obligatory part of this and should be conducted with the aim of satisfying the needs of the waste treatment authority or business and its customers and of landowners receiving sludge.

REFERENCES

(1) Council Directive of 12 June 1986 on the protection of the environment and in particular of the soil, when sewage sludge is used in agriculture (86/278/EEC). Official Journal of the European Communities, No. L 181, 4 July 1986, pp. 6-12.

(2) The Risk to Health of Microbes in Sewage Sludge Applied to Land. Report on a WHO Working Group. Stevenage, 6-9 January 1981. EURO Reports and Studies 54. World Health Organization, Regional Office for Europe, Copenhagen, 1981.

(3) GEMMELL, M.A. (1986). General epidemiology of Taenia saginata. In Reference (5), pp 60-71.

(4) BRADFORD-HILL, A. (1965). The environment and disease: association or causation? Proceedings of the Royal Society of Medicine **58**, 295-300.

(5) BLOCK, J.C., HAVELAAR, A.H. and l'HERMITE, P. eds. Epidemiological Studies of Risks Associated with the Agricultural Use of Sewage Sludge: Knowledge and Needs. Elsevier Applied Science Publishers, London.

(6) PIKE, E.B. (1981). The control of salmonellosis in the use of sewage sludge on agricultural land. In Characterisation, Treatment and Use of Sewage Sludge, ed. P. l'Hermite and H. Ott. D. Reidel Publishing Co., Dordrecht, pp 315-329.

(7) EDEL, W., SCHOTHORST, M. van, LEUSDEN F M van and KAMPELMACHER, E.H. (1978). Epidemiological studies on salmonella in a certain area ('Walcheren Project'). III The presence of salmonella in man, insects, seagulls and in foods, chopping block scrapings from butchers shops, effluent of sewage treatment plants and drains of butchers' shops. Zentralblatt für Bakteriologie und Hygiene. I Abteilung, Originale A, **242**, 468-470.

(8) HESS, E. and BREER, C. (1975). Sanitary effect of gamma eradication of sewage sludge. In Radiation for a Clean Environment, Munich, 17-21 March, 1975. International Atomic Energy Agency, Vienna, pp. 203-208.

(9) KAMPELMACHER, E.H. and NOORLE-JANSEN, L.M. (1974). Onderzoekingen over de bacteriële Kontaminatie van weilanden door bemesting met uitgegist slib. H_2O, 7, 418-422.

(10) BURNETT, R.C.S. and MACLEOD, A.F. (1980). An outbreak of salmonellosis in West Lothian. Communicable Diseases Scotland **14** (5), vi-vii.

(11) HALL, G.A. and JONES, P.W. (1978). A study of the susceptibility of cattle to oral infection by salmonellas contained in raw sludge. Journal of Hygiene, Cambridge, **80**, 409-414.

(12) JONES, P.W. (1986). Sewage sludge as a vector of salmonellosis. In Reference(s), pp. 21-33.

(13) BLAMIRE, R.V., GOODHAND, R.H. and TAYLOR, K.C. (1980). A review of some animal diseases encountered at meat inspections in England and Wales 1969-1978. The Veterinary Record, **106**, 195-199.

(14) ILSOE, B., KYVSGAARD, N., HENRICKSEN, S.A. and NANSEN, P. (1990) Taenia saginata and bovine cysticeriosis in Denmark - a review of the present epidemiological situation with special regard to the transmission by the agricultural dense of sewage sludge. In Proceedings of Symposium, Treatment and Use of Sewage Sludge and Liquid Agricultural Wastes, Athens, 1-4 October 1990. Commission of the European Communities, Brussels (preprint).

(15) REILLY, W.J., COLLIER, P.W., SHARP, J.C.M. and FORBES, G.I. Sewage sludge and bovine cysticercosis in Scotland. Communicable Disease Scotland 86.29, 19 July 1986, 5-7.

(16) PIKE, E.B. and CARRINGTON, E.G. (1986). Stabilisation of sludge by conventional and novel methods - a healthy future. In Symposium The Agricultural Use of Sewage Sludge - Is There a Future? Institute of Water Pollution Control, Maidstone. Paper D.

(17) WILKENS, S. (1981). Untersuchungen über die Ansteckungmöglichkeiten von Rindern mit Taenia saginata und Sarcocystis spp auf Abwasseregnungsflächern und über das Absetzverhalten von Helmintheiern in vitro. Inaugural Dissertation, Tierärtzliche Hochschule, Hannover.

(18) LUND, E. (1981). Activities of Working Party 3 "Biological Pollution of Sludge". Introductory remarks. In Characterisation, Treatment and Use of Sewage Sludge, ed. P. l'Hermite and H. Ott. D. Reidel Publishing Co., Dordrecht, pp. 300-303.

(19) US ENVIRONMENTAL PROTECTION AGENCY (1989) Environmental Regulations and Technology. Control of Pathogens in Municipal Wastewater Sludge. EPA/625/10-89/006. Center for Environmental Research Information, Cincinnati, Ohio 45268.

(20) STRAUCH, D. and BERTHOLDI, M. (1986). Microbiological specifications of disinfected sewage sludge. In Processing and Use of Organic Sludge and Liquid Agricultural Waste, ed. P. l'Hermite. D Reidel Publishing Co., Dordrecht, pp. 178-197.

(21) BRUCE, A.M., PIKE, E.B. and FISHER W.J. (1990). A review of treatment process options to meet the EC sludge Directive. Journal of the Institution of Water and Environmental Management, **4**, 1-13.

RISK ASSESSMENT OF THE OCCURRENCE OF PATHOGENS IN SEWAGE SLUDGE AND EFFLUENTS FROM LIVESTOCK

PART 2: EFFLUENTS FROM LIVESTOCK

D. STRAUCH

Institute for Animal Medicine and Animal Hygiene
University of Hohenheim
Stuttgart/Germany

Summary

The causative agents of infectious diseases are shed by the animals via faeces, urine and other excretions or secretions. Therefore they can be found in their manures and slurries. From that point of view all animal residues constitute a hazard for other animals and for the public. Despite this fact there are only very few documented records of disease transmission to animals or man through treated or untreated material. However, the always existing risk can be reduced to proportions that are acceptable if care is taken to treat and use slurry according to the guidelines which were issued by CEC-working parties.

Manures and slurries which are proven to be infectious can be disinfected. Manures by composting with a combination of biological and chemical influences, slurry by chemical disinfectants. The effect of chemical disinfectants on plants, soil and water is discussed as well as their influence on the biology of biogas production. Recommendations for manufacturers and users of disinfectants are given. The problem of slurry utilization in water protection areas is also shortly discussed.

Evidence was presented at the CEC-workshops to suggest that the European "Interim Minimum Guidelines" for slurry utilization are not sufficiently well known to farmers. Each country should ensure that steps are taken to make their contents widely known through the media - print, radio, television, agricultural extension service and meetings, shows and markets.

1. INTRODUCTION

It has been described that several hundred diseases are transmitted from animal to animal and more than 150 are transmitted between animal and man. The causative organisms of such diseases are excreted directly by the affected animal (e.g. nasal and pharyngeal secretions, faeces, urine, vaginal, placental and lochial secretions, milk, semen) or indirectly distributed via different agents (e.g. blood, carcasses, solid and liquid manure, milk and milk products, eggs and egg products, slaughterhouse offal). Most of these infectious direct excretions under agricultural conditions ultimately come into contact with the floor of the buildings. Therefore the floor of animal houses with its installations for the collection of the faecal residues constitutes the collecting basin for all pathogens of the animals which are kept in the respective enterprise. Thus the faecal matter in each case of infection within an animal population has to be considered as infectious and an inanimate vector for the respective pathogen.

Besides infectious diseases caused by obligatory infectious agents another group of diseases is advancing in larger animal production units, the so-called "transmissible factor diseases". They are characterised in that normally harmless infections lead frequently to clinically recognizable symptoms of disease released by non-microbial factors. The danger emerging from such factor diseases lies in the fact that in most cases the owner of such herds does only very late recognize clinical symptoms. But in the time before the infected animals do already excrete the microbial organisms which cause the disease. Therefore the owner usually does not take any action to prevent a spread of the infective agents in his surroundings. Hence he is not informed about the infective potential of the excretions of his animals and the manure or slurry to be utilized on his fields.

Another hygienic aspect of the utilization of slurry or manure is the possibility of leaching of nitrate into the groundwater. Thus a danger for young babies is emerging when their feed is prepared with water containing higher concentrations of nitrate, because an intoxication called methemoglobinaemia can occur and the children may die. Fortunately only babies up to an age of six month are susceptible to that disease. In areas with high nitrate concentrations in the drinking water therefore their mothers are supplied with water from other areas where the nitrate content is below the toxic level of 25 mg NO/l, which is recommended by the World Health Organization (1).

To better get to know these hygienic problems of the effluents from livestock CEC already in 1975 started a research programme.

2. CEC-ACTIVITIES

The CEC-Research Programme on Effluents from Intensive livestock started with a first seminar "Utilization of Manure by Land Spreading" (Modena/Italy 20-24 Sept. 1976). The current situation and the available knowledge in that field were assessed (2). According to a proposal made in Modena to establish more specialized groups a workshop was organized on "Animal and Human Health Hazards Associated with the Utilization of Animal Effluents" (Dublin/Ireland, 21-23 Nov. 1977). It was one of the results that there are still many gaps in the knowledge of the occurrence and the importance for animal and public health of pathogens especially in liquid manures. Nevertheless the overall success of the workshop is reflected by the specialist groups on microbiology, parasitology and toxicology being able to prepare a series of conclusions and recommendations which, after intensive discussion, evolved into a series of practical guidelines for the utilization of animal effluents from the respective point of view of each specialist group. These were condensed in "Interim Minimum Guidelines" and were subject to the reservations expressed in the individual group reports:

1. Slurry should be utilized on tillage crops (excluding crops for fresh consumption) wherever possible.
2. If slurry is spread on grassland then:
   2.1 Use on pasture for conservation, wherever possible
   2.2 If on grazing land
   - storage of all slurry for a minimum of 60 days before spreading
   - delay of 30 days before grazing
   - graze with adult or non-susceptible animals.
3. Utilization of slurry should be related to plant nutrient requirements (3).

The Dublin-meeting was followed by a workshop "Communicable Diseases Resulting from Storage, Handling, Transport and Landspreading of Manures" (Hannover/W.Germany, 4-6 Nov. 1980). There the "Interim Minimum Guidelines" of Dublin were approved, but it was stated that storage of slurry for 60 days in summer may be adequate, whereas in winter the storage time should be at least 90 days. It was further concluded that, despite of the wide variety of infective agents - bacteria, viruses, parasites, fungi - that can be present in slurry, there are only few published records of disease transmission to animals and man through slurry, treated or untreated. It was clear that a risk exists but it can be reduced to proportions that are acceptable if care is taken to treat and use slurry according to the guidelines mentioned above. Despite the relative lack of unequivocal evidence that slurry transmits disease to man or livestock it was recommended that the existing guidelines should be adapted to meet the considered needs of the individual countries. Further research is needed to

- determine the levels of contamination of pasture with pathogenic organisms that may result from treating it with slurry, and
- evaluate methods for the disinfection of slurry without decreasing its value as a fertilizer.

Spezial emphasis should be placed on Campylobacteriosis and Listeriosis as emerging diseases (4).

In a final workshop (Stuttgart/W.Germany, 11-13 Oct. 1982) the EC-Expert Group met with members of the FAO-European Network on Animal Waste Utilization-Subnetwork 1 "Influence of Animal Manure Handling and Utilization Systems on Animal and Human Health". The papers presented have demonstrated that we have by now a scientific base how animal manures should be handled in farms with specific infections. But the legislation on notifiable diseases of livestock have to be brought up-to-date in order to cover modern animal husbandry with large units of animals and new systems of handling animal effluents. It seems necessary to re-evaluate the efficacy of legally prescribed disinfectants and to elucidate pollutional aspects of these as well to develop new disinfectants which are biodegradable in the soil after spreading of slurry. To achieve comparable results by different investigators the use of "decimation rates" like e.g. T_{90}-values is strongly recommended. Such figures are generally applicable for advising public authorities or farmers (5).

After the incorporation of the CEC-Expert Group on Animal Effluents and of two representatives of Subnetwork 1 of the above mentioned FAO-European Network on Animal Waste Utilization into the COST 681 project "Treatment and Use of Organic Sludge" the liquid agricultural wastes fell somewhat behind the sewage sludge problem. But, nevertheless, the COST 681-Working Party No. 3 "Hygienic Aspects Related to Treatment and Use of Organic Sludge and Sanitary Aspects of Slurries and Manures" dealt with the aspects of animal excretions in a workshop held in Dublin/Ireland (21-23 Sept. 1987) and in the last workshop held in Lelystad/Netherlands (26-28 Sept. 1988). Besides national reports about the situation in the EC-member-states concerning animal manures, in the scientific reports emphasis was put on the problem of disinfection of large quantities of slurry and on problems possibly caused by spreading of slurry in water protection areas (6, 7).

To sum up the following can be said. Despite the fact that a lot of information is available about the occurrence of pathogenic agents in animal slurries and their survival in this substrate one did not succeed in proving their real threat to animal and human health. In the case of notifiable or reportable infectious diseases of animals their eradication is enforced by veterinary legislation. This is apparently so effective that no published reports are available about failures of the respective disinfection methods used. To prevent spread of disease by "normal" slurries from animal production sites with animals which are free of clinical signs of

disease the observance of the above mentioned "minimum guidelines" for slurry utilization apparently seems to be sufficient.

3. DISINFECTION OF MANURES

As already mentioned it must be expected in livestock farms that inapparent infections with bacteria, viruses or parasites may occur. In farms which still use bedding and store the solid manure in above-ground stores usually a composting process is beginning with a rise of temperature into the range of 50-70 °C. But it has been found that in modern animal husbandry the amount of bedding is reduced to 0.5 - 1.5 kg/cattle/day. This results in a too low oxygen supply and the interchange of gases is impossible because the pore volume is too low. Therefore the temperatures which are necessary for the inactivation of pathogens are insufficient. To overcome this disadvantage and to achieve a reliable disinfection the natural composting process has to be supported by external means. For that a method has been developed: the place where a compost pile/windrow is built should first be covered with a 25 cm thick layer of straw. This layer is covered with 10 kg/m^2 of lime hydrate ($Ca(OH)_2$) as a layer as closed as possible. The dung which is to be disinfected is loaded on a rear or side discharge spreader with not more than 2-3 m breadth of discharging. During loading granulated unslaked lime (CaO) is proportionately added to the manure in at least two layers. As reference amount an addition of 100 kg granulated unslaked lime per cubicmeter of manure is recommended. Subsequently the manure-lime mixture is slowly discharged from the spreader and thus a windrow of ca. 1,5 m height is piled. During discharging the emerging dust is hosed down by a powerful squirt. The pile is then covered with a strong foil of black plastic, which must be safely fastened around the basis of the pile. This windrow has to be stored for at least five weeks. Then it must immediately be ploughed into arable land. If this is not possible or if the manure is to be spread on pasture or cultivated forage land the windrow must be stored for at least 10 weeks.

In slurry no autothermic processes do occur. Therefore the disinfection of slurry must be achieved by other means. The most common procedure is the chemical disinfection. Many of the disinfectants which are internationally used may be toxic for crops, soil and/or ground water. In Table 1 some disinfectants are listed which can be used for slurry disinfection under the aspect of plant and environmental protection.

Disinfectant	Dosage* kg/m^3	Reaction time Days
Formalin	9-15	4
40 % lime milk	25-40	4
Sodium hydroxide (Na OH)	8-12	4
Peracetic acid 15 %**	40	4

* Lower value: effective against enveloped viruses
Upper value: effective against bacteria

** Only to be recommended for disinfection of small amounts.
Heavy formation of foam!

Table 1: For the disinfection of slurry suitable chemicals

The quantitative demands for chemical disinfectants in slurry could be considerably reduced when the solids of the slurry would be separated by technical means. The liquid phase is microbiologically and chemically relieved from the load of solids and so its disinfection is possible with a reduced use of disinfectants, in a shorter time and at lower cost. Under these circumstances one could think of the use of other and more efficient disinfectants which cannot be used for chemical or economical reasons as long as the solids are not separated (1).

It is important for all the recommended disinfectants that:

- the slurry is thoroughly homogenized before they are added
- the disinfectants are added at different positions of the mixing stream in the storage tank at the same time
- the disinfectant is added within six hours of mixing
- a minimum of 4 days reaction time is observed
- no fresh slurry is pumped into the tank during the disinfection procedure.

For a thorough mixing, storage tanks with a volume of 300 m^3 are considered to be optimal. Larger tanks, especially those above 500 m^3, are not very suitable, unless very capable mixing machinery is available. If this is not the case, for disinfection of very large amounts of slurry it should be tried to pump the infected slurry into smaller tanks for the duration of the reaction time (1).

In the context of chemical disinfection of slurry also the question is of interest whether the farms with biogas plants have to expect negative influences of the disinfected slurry on the biological processes during anaerobic fermentation. The available results of respective investigations can be summarized as follows:

- Disinfectants which are used for a routine disinfection of animal houses with liquid manure, as a rule, do not constitute a danger for the anaerobic process. An exception is one disinfectant which contains chloroform and carbon disulfide.
- A final disinfection of slurry can be carried out without longer lasting disturbances of the biogas production with NaOH and peracetic acid in concentretions up to 8 %, lime up to 6 % and formalin up to 4 % (8).

For reasons of environmental protection the following recommendations for manufacturers and users of disinfectants can be given:

- The manufacturers should produce disinfectants which are readily and without hazardous residues decomposed in biological systems like sewage treatment, composting of solid manure, aerobic-thermophilic stabilization of slurry or in soil.
- The customers should try to reduce the amounts of slurry by internal measures. If possible the solids should be separated. Disinfectants which are known not to do any harm to plants, crops, soil or water should be preferred. Physical or biological methods of disinfection should be favoured where energetical aspects do not conflict with that recommendation.

4. SLURRY AND GROUNDWATER PROTECTION

There is a tendency from medical authorities to prohibit the use of slurry in water protection areas. The main reason is to avoid contamination of groundwater with nitrate. But in case that slurry in reduced amounts in water protection areas should be allowed further by the legislation, the medical authorities demand a disinfection of slurry, but without chemicals.

A prerequisite for that is a definition of the term "disinfected slurry" or "hygienically safe slurry". This is a very difficult problem because the opinions of medical people and the agricultural organisations are widely divided on that matter. Still much work has to be done in this field. Table 2 gives an example of the restrictions for slurry utilization in the water protection areas of the Land Baden-Württemberg in the Federal Republic of Germany.

Protective Stipulations	Water Protection Zones Tapping Area Zone I	 Narrower Prot.Zone Zone II	 Broadened Prot.Zone Zone III
1. Prohibition to plow up grassland	in all protection zones		
2. Prohibition to utilize urine, slurry, soakage of silage, sewage, sewage sludge, night-soil and similar materials	permanently	permanently	- on grassland 15 October to 1 February - other land 1 October to 15 February
3. Prohibition to utilize solid/farm-yard manure on arable land	permanently	15 November to 1 February	15 November to 1 February
4. Prohibition to utilize nitrogen containing commercial fertilizer	permanently	- on grassland 15 October to 1 February - other land 1 October to 15 February	
5. Prohibition to utilize pesticides	permanently	as provided in a positive catalogue	

Maximal allowed pure-N content after harvest: 45 kg/ha
N-fertilization: 20 % risk deduction
Compensation for yield losses: 310,-- DM/ha

Table 2: Prohibitions for the application of organic and commercial fertilizers and pesticides in water protection areas (SchALVO Bad.-Württ.)

5. CONCLUSIONS

The results obtained by the CEC-working parties dealing with effluents from livestock give a good overall picture about the hygienic problems connected with manure and slurry. They show that the microbiological risk can be reduced to almost negligible dimensions. The risk emerging from manures and slurries which are proven to be infected can also be reduced to nearly zero by disinfection techniques based on biological, physical or chemical principles. Also the threat of chemical disinfectants to the environment can be minimized by selection of relevant chemical compounds. Yet the effect of established and newly developed disinfectants for slurry on the microflora in soils and the dimension of their biodegradability must be investigated further. These and some other important questions have to be dealt with even after the closure of the COST 681-project in the framework of one of the new EC-research programmes.

Nevertheless there is a demand for a microbiological definition of the term "hygienically safe slurry" as an essential prerequisite to obtain the approval of medical institutions, which are responsible for the quality of drinking water, for the utilization of "normal" slurry in water protection areas.

REFERENCES

(1) STRAUCH, D. (Ed., 1987). Animal production and environmental health. Series "World Animal Science", Vol. B6. Elsevier Science Publishers, B.V., Amsterdam

(2) VOORBURG, J.H. (Ed., 1977). Utilization of manure by land spreading. CEC-EUR 5672e, Luxembourg

(3) KELLY, W.R. (Ed., 1978). Animal and human health hazards associated with the utilization of animal effluents. CEC-EUR 6009 EN, Luxembourg

(4) WALTON, I.R. and WHITE, E.G. (Eds., 1981). Communicable diseases resulting from storage, handling, transport and landspreading of manure. CEC-EUR 7627 EN, Luxembourg

(5) STRAUCH, D. (Ed., 1983). Hygienic problems of animal manures. Inst. f. Tiermedizin u. Tierhygiene, Univ. Hohenheim, Postfach 70 05 62, D-7000 Stuttgart 70

(6) HANNAN, J.H. and L'HERMITE, P. (Eds., 1987). Hygienic aspects of the treatment and use of organic sludge and liquid agricultural wastes (I). CEC-SL/120/87-XII/ENV/14/87. CEC-DG XII, Rue de la Loi 200, B-1049 Brussels

(7) STRAUCH, D., OVER, H. and L'HERMITE, P. (Eds., 1989). Hygienic aspects of the treatment and use of organic sludge and liquid agricultural wastes (II). CEC-SL/129/89-XII/ENV/1/89. CEC-DG XII, Rue de la Loi 200, B-1049 Brussels

(8) STRAUCH, D. and WINTERHALDER, K. (1986). Effect of disinfectants, additives and antimicrobial drugs on anaerobic digestion. Israel Journal of Veterinary Medicine, 42 (3), 195-202.

PATHOGENS IN SEWAGE SLUDGE: AN EC PERSPECTIVE

ARIE H. HAVELAAR

Laboratory of Water and Food Microbiology
National Institute of Public Health and Environmental Protection
P.O. Box 1, 3720 BA Bilthoven, the Netherlands

Summmary

The use of sewage sludge on agricultural land creates potential hazards with respect to the dissemination of pathogenic micro-organisms. Proper pretreatment of sludges and management of treated land should be applied to reduce the risks to acceptable levels. The actions to be taken should take into account qualitative information from epidemiological studies and quantitative information from studies on the fate of relevant pathogens in sludge and on land. Local conditions are of great importance in determining the proper action to be taken.

1. INTRODUCTION

Sewage sludge contains a great variety of human pathogenic organisms which originate directly from the excreta of man or animals, or which have multiplied in the wastewater during transport to the treatment plant. Use of sludges in agriculture may therefore create risks to the health of man or animals. These health risks may be rather direct as through the consumption of contaminated crops, or indirect as through the contamination of food animals which may become healthy carriers of pathogenic agents. The routes of contamination of animals in their turn may be direct through contaminated feeds, or indirect through a general increase in the contamination of the environment with disease causing agents. Proper knowledge of the organisms concerned, of their life cycles and epidemiologal features is necessary to understand and manage the risks involved. The EEC Concerted Action 68(1) has provided scientists throughout Europe a framework to meet and discuss these problems, but more importantly to exchange information within a multidisciplinary group. This paper will review the benefits of this approach, its impact on European legislation and the personal benefits to the participants and their individual countries.

2. REDUCING HEALTH HAZARDS FROM LAND APPLICATION OF SEWAGE SLUDGE

Risk management of pathogenic organisms in sewage sludge involves basically three different options, which are:

-Restrictions on sludge use
-Restrictions on land use after sludge application
-Decontamination of sludge before application

It is important to realize that all pathogenic organisms can be effectively eliminated from sludges by appropriate disinfection methodologies but at relatively high costs. Hence, the central question in relation to pathogens in sludge is to find the most effective combination of the above mentioned general principles which leads to an acceptable risk at lowest costs. It is implied that in order to find these solutions, microbiologists and epidemiologists should work closely together with engineers and sludge producers, and this undoubtedly is one area where the COST 68(1) project has stimulated better communication. Within the framework of the action, scientists from different disciplines met at the International Symposia, and Working Parties 1 and 3 have organized a joint workshop in Zürich in 1982. Furthermore, the network created by the COST action led to the invitation of individual members of the Working Parties to national symposia, which would not have taken place otherwise.

Discussions on the most effective ways to reduce hazards in a European context have identified important differences between countries or regions within the Community, which will continue to exist and will lead to the adoption of different solutions in different countries. Key aspects are:

- The quantities of sludge produced in relation to the available agricultural land, to soil protection strategies and to the use of other fertilizers such as liquid manures
- The sludge treatment and disposal techniques that are generally used
- The availability of epidemiological data to identify hazardous pathogens, their impact on public health as well as the role of sludge in the overall epidemiology of infectious diseases

The importance of the first point can be illustrated by the situation in the Netherlands, where it is foreseen that in the near future land application of sewage sludge will virtually cease to exist because a new regulation on organic fertilizers that will be enforced in January 1991, will set strict limits on heavy metals and phosphates. It is also evident that the great surplus of liquid animal manure in the Netherlands is another driving force behind the rapidly diminishing agricultural use of sewage sludge.

The discussions in Working Party 3 have concentrated on the latter two aspects of risk management, which will be treated in some more detail below.

3. REDUCTION OF PATHOGENIC MICRO-ORGANISMS IN SLUDGE TREATMENT

Before sludge is applied to land, some form of stabilization is usually applied. There is a great variety in types of stabilization within the EC. Bruce and Fisher (1) distinguished between biological and chemical stabilization and identified only one type of stabilization that is well established and widely used in all countries: heated, anaerobic, mesophilic single-stage digestion. Seven other methods of stabilization were identified which were well established but much more common in some countries than in others. These included unheated anaerobic or aerobic digestion, stacking of sludge cakes, composting in windrows, drums or aerated vessels and lime addition. Eleven other processes were identified which were in limited use or only in the developmental stage. Taking into account this variety of methods and greatly differing local conditions, particularly climatic aspects, one must conclude that there is no way of indicating the general effect of stabilization on pathogenic micro-organisms. Working Party 3 has already indicated in the final report of the COST 68 bis phase that in general stabilisation processes can be grouped into three categories in relation to their sanitizing effect and has suggested applications appropriate to the achieved degree of pathogen reduction (Table 1). This approach, which has also been adopted in USA guidelines on land applications of sewage sludge appears to be realistic and cost-effective. Unfortunately, little of this knowledge can be found in the EC guidelines which simply require sludge to be "stabilized" before use in agriculture.

Table 1 Classification of sewage sludge stabilization processes in relation to pathogen reduction and land application
(Final report COST project 68 Bis)

Type of treatment / **Pathogen reduction** / Application
Stabilized liquid, dewatered or naturally dried; composting I
Especially vegetative bacteria; more resistant forms such as cysts, eggs and spores depending on time-temperature effects
Arable land during vegetation rest periods, vineyards and orchards; pasture and forage land after harvest and until the end of the year; grassland with no grazing period (6 months safe under all conditions); not on crops used uncooked for human consumption
Lime conditioning; low dose irradiation
Reduction except for parasites and spores, vegetative bacteria destroyed
Less restrictions on period of application than above
Pasteurization; (quick) lime treatment; composting II; high dose irradiation; aerobic thermophilic stabilization; high temperature treatment
Sanitized
No restrictions

Note: refer to original table for further details such as time-temperature combinations or radiation doses

Examination of the EC guidelines will also show that several key elements of the above table have been incorporated in the final text as such or in a modified form. The no-grazing period in the EC guidelines is 6 weeks rather than 6 months and it is left to the individual countries to extend the period, e.g. if the local climate requires this. Hence, although micro-organisms are not mentioned explicitly in the guideline, some articles have been prepared particularly to reduce risks of infection.

The above table also includes some processes which do not stabilize the sludge but have specifically been designed for disinfection of sewage sludge, such as pasteurization and irradiation. Their obvious disadvantage is that they do not stabilize the sludge, although their application may have favourable effects on the suitability of sludges for other processes like dewatering. The advantage of using these processes is that they can accurately be designed and controlled to obtain a desired degree of pathogen reduction, and they can be monitored by continuous physical methods, leading to a high degree of reliability. An enquiry into the use of "sanitizing" processes in EC member states again demonstrated the extreme variabilty encountered in this field (2).

The characterisation of the microbicidal effects of different sludge treatment processes has been an area of permanent attention in WP3, and much information has been exchanged. However, there are still several unresolved questions in this domain. One problem is related to the inherent difficulties of directly working with the pathogens of concern. Indigenous levels in sludges are often too low to be quantified by available methods and safety considerations preclude seeding experiments on large scales. It is often difficult to assess the viability and infectivity of organisms encountered in treated sludges, especially of parasites. Isolation and characterisation often require specialized techniques which are not routinely used by water authorities or even research institutes because they are labour-intensive and costly. Widespread application of pathogen assays are also hampered by the low priority that is assigned to them by policy makers and managers as compared to man-made chemical pollution of sludges. This problem certainly is not restricted to the field of sewage sludge hygiene, but is a general imbalance in many areas of environmental protection.

To overcome the problems with direct assays for pathogens, investigators have attempted to use model organisms instead. Principles and methods from drinking water and food microbiology were adopted, depending on local traditions and preferences. Although WP3 has attempted several times to harmonise the techniques used, no consensus has arisen. One drawback was that intercomparative studies were not the primary aims of national studies whereas research funds available within the EC were not available because the COST project only aimed at stengthening the bonds between already existing national programs. It has often been felt frustrating by members of the Working Party that agreed research priorities could not be effected because they were too big for one country and by definition excluded from EEC sponsored research.

4. EPIDEMIOLOGY OF SLUDGE-TRANSMITTED PATHOGENS

The priority microbial pollutants of sewage sludge have been defined as follows: Salmonella spp., Taenia saginata, Ascaris spp., and for certain applications human viruses. More information is needed on protozoan parasites such as Cryptosporidium and Sarcocystis. The agents involved are either transmitted directly between humans (viruses or Ascaris), or may involve food animals (zoönoses, such as Salmonella spp., Taenia saginata, Cryptosporidium and Sarcocystis). All agents have complex infectious cycles which may involve many routes of transmission of which sewage sludge is only one. The effect of sludge application on health can therefore not be assesssed accurately and will strongly depend on local conditions. One major drawback is that the incidence of disease caused by the above mentioned organisms is only poorly known. Salmonella may serve as an example in this respect. The official system of notification to the Medical Health Inspectorate leads to an incidence of ca. 200 cases annually in the Netherlands. Food Inspection Services register an annual number of 10-100 cases related to the consumption of contaminated food. Some major outbreaks like the 1984 "Eurotop" explosion in Maastricht may include up to 1000 cases. The real incidence of salmonellosis in the Netherlands is considerably higher as has been learned from an active surveillance study (A.M.M. Hoogenboom, unpublished results). General practitioners in an urban and a rural area, who served a population of 100,000 registered all cases of acute gastroenteritis on a continuous basis from 1987 onwards. An average incidence of 15 cases/1000 persons/year was found. This figure corresponds well to those obtained in other sentinel studies. All patients were asked to complete a questionaire and to submit faecal specimens for laboratory examination. Between 4 and 7% of samples yielded Salmonella, leading to a total estimate of 12,000 cases of salmonellosis which are seen by general practitioners in the Netherlands annually. As it is estimated that only 5% of all who suffer from acute gastro-enteritis consult their physician, the total number of cases of salmonellosis in the Netherlands may well be 240,000 per year instead of the officially reported few hundreds! The study also demonstrated that the majority of cases of salmonellosis is sporadic rather than outbreak-associated, and vehicles of infection arising from outbreak investigations may not be the most significant. Detailed case-control studies on non-outbreak related salmonellosis will be necessary to establish the major sources of infection, although it may safely be assumed that animal food products, in particular poultry and associated products will turn out to be the major source of infection. Even if this knowledge is available, it must be realized that land application of sewage sludge is only an indirect determinant in the epidemiology of human salmonellosis, leading to increased polution of the environment in general and possibly to direct infection of animals grazing on treated land. Working Party 3 has concluded that sufficient knowledge is available to indicate the general concern on salmonellas in sewage sludge, and that further epidemiological studies were not necessary. The fight against human salmonellosis requires a multifactorial approach and those producing or applying sewage sludge should be prepared to contribute their share by reducing Salmonella-levels in sludges as far as possible and applying the sludge in an approved manner. This approach should take into account local factors on land availability, climate, density of animal-rearing units etc. In this view the EEC Directive is a minimum approach, applicable to all countries under all circumstances and should not be seen as sufficient in most situations. Each country or region should adopt supplementary measures tailored to the local situation.

For the other priority agents the insight in their general epidemiology is less. It seems therefore appropriate to cite some conclusions from a Workshop organized by WP3 in Metz in 1985:

- The need for surveillance, recording and publication was regarded as very high priority. This was the only epidemiological need recognized by the panel (on bacteria) for salmonellosis
- Surveillance of the use of sludge under the proposed Directive and recording of diseases associated with its use particularly with regard to bovine cysticercosis, ascariasis and also sarcocystosis, cryptosporidiosos and perhaps toxoplasmosis, Giardia and Entamoeba infections was identified as the first research priority by the panel on parasites.
- The panel on viruses concluded that the surveillance and good reporting of cases should be encouraged and organized internationally

These recommendations still are highly valid, unfortunately few initiatives have yet been taken.

It must be realized that surveillance of the overall incidence of infectious diseases will provide important background data for decisions on sludge management, but does not necessarily give direct information on the specific role of sludge. Highly complicated and expensive epidemiological studies would be necessary and these are generally regarded as not being cost-effective. Proper risk management of pathogens in sludge should be based on qualitative epidemiological data and quantitative knowledge on the behaviour of priority pathogens in sludge treatment and after application.

REFERENCES

1. BRUCE, A.M. and FISHER, W.J. Sludge stabilization - methods and measurement. In: Sewage sludge stabilization and disinfection (A.M. Bruce, ed.), pp. 23-47. Chicester: Ellis Horwood, 1984.
2. HAVELAAR, A.H. and BRUCE, A.M. Disinfection of sewage sludge. An enquiry among member countries of the EEC and some other countries. In: Disinfection of sewage sludge: technical, economic and microbiological aspects (A.M. Bruce, A.H. Havelaar and P. l'Hermite, eds), pp.245-252. Dordrecht, D. Reidel, 1982.

USE OF LIQUID AGRICULTURAL WASTE IN EAST EUROPEAN COUNTRIES; TOPICAL HYGIENIC PROBLEMS

J. HOJOVEC

University of Veterinary Science Brno, Czechoslovakia

Summary

In East European countries the risks involved in environmental pollution by agricultural wastes follow from the concentration of farm animals in individual localities. Litterless housing is the method mostly practised on the farms, the consequence being production of huge amounts of liquid waste. This, for the most part, is utilized for direct fertilization of fields and, when considered from the hygienic point of view, the quanta applied to the fields are not limited. Since a number of agricultural enterprises are lacking slurry storage tanks of sufficient capacity, they are compelled to practise land spreading of liquid manure in the course of the whole year. There is no legislative solution to this problem and hardly any efficient controlling system as well. In view of the social changes taking place in East European countries, both the public and the governments reveal intensified interests in the problems bearing on the environment. In response to these trends, agricultural enterprises are forced to solve this question, not only via construction of appropriate slurry storage tanks but also, primarily, in the form of installing equipment which provides for slurry processing with the use of biological aerobic or anaerobic methods. However, meeting these requirements is limited by shortage of the economic funds available.

The political systems in East European countries have been experiencing big changes most recently. These changes may be characterized as a return to the traditional democracy.

In the sphere of the national economy this process is relatively complicated; the change-over from so-called socialist but mostly state and centrally managed enterprises to private undertaking based on the principles of marketing suggests the necessity of facing and overcoming appreciable difficulties.

In the sector of agriculture these problems are very specific and delicate, for in the course of post-war period the private agricultural enterprises were mostly liquidated in the countries of East Europe (with the exception of Poland) to be replaced by the State farms, on the one hand, and by so-called Collective farms, on the other.

It is not our intention to deal with the political aspects of this phenomenon. Considered from the aspect of the problem observed in this paper the first point of material importance is seen in the appreciable concentration of farm animals. Concerning this point, the agricultural enterprises to provide housing for the highest numbers of farm animals have been established in the German Democratic Republic; the biggest of them have the capacity of housing as many as 4 000 dairy cows, 10 000 fattening steers and 100 000 pigs. Besides, there is a large number of farms where the concentration of housed animals is but one half of or even less than the figures mentioned above. In Czechoslovakia, and similarly in Hungary as well, the numbers of housed animals concentrated in one locality are lower varying, as a rule, between 600 and 1 000 dairy cows, about 10 000 pigs and between 1 000 and 2 000 calves. But from the views held in West European countries such high concentrations of farm animals are unusual.

Collective farms (Kolkhoz) and State farms (Sovkhoz) were established in the Soviet Union, but in view of the limited construction possibilities existing during the post-war period the agricultural enterprises distinguished by such high concentrations of farm animals have not become a common feature.

The construction of such big farms was reasoned by the necessity of raising efficiency of the production via specialization and the highest possible mechanization of all operations. But it has been evident for a number of years that the goals set were not fulfilled and the agricultural production in East European countries is appreciably lagging behind that in West European countries.

Thus, a comparison of the intensities of livestock production reveals clearly that the stocks of farm animals per 100 hectares (ha) in East European countries are essentially lower than the ones kept in West European countries, and the efficiency is appreciably lower too. To quote a few examples: in the sector of dairy cows the German Democratic Republic and Czechoslovakia rank first with some 30 head per 100 ha each, Hungary (15), USSR (7) and Bulgaria (11) following; whereas in Belgium, Germany, Switzerland and mainly in the Netherlands the counts are substantially higher ranging from 40 to 120 head per 100 ha.

It can be concluded from the figures given above that the weightiness of hygienic risks in East European countries follows primarily from the excessive concentration of farm animals in the locality and not from the intensity in livestock production. Intensity of the problem in question is underlined by the use of literless housing wherein liquid manure originates. Slurry utilization continues to be a serious problem also from the hygienic point of view, for a complete solution is still lacking in all of the countries mentioned above. At present, the fertilization with unprocessed slurry that has been stored in tanks for the required period of time remains practically the only way of its utilization. As the concentration of farm animals is high, capacity of the storage tanks must be large. They themselves represent an appreciable hygienic hazard (leakage), and from the aspect of disinfection during the occurrence of reportable diseases they become a problem whose solution is possible with much difficulty. A report on the results of our research work given over to these problems was read at the Meeting in Lelystad.

In this connection, however, the fact is of moment that a number of agricultural enterprises housing as many as 30 000 pigs have not been able to construct any storage tanks todate (from motives of economy) and are thus compelled to practise land spreading daily. It thus follows that no relevant directives are observed and that no effective legal measures can be taken to rule out for instance the application of slurry during periods particularly hazardous with respect to potential contamination of the ground or surface waters, for the simple reason that operation in such enterprises would have to be suspended. With the absence of some effective control system for environmental protection, there are frequent infringements of the directives and instructions compiled at various levels in individual countries and applicable to these problems. Only the cases of direct interferences with the environment are solved (such as contamination of the ground or surface waters) and the offenders penalized.

Insufficient legislation and absence of an effective control system is therefore another phenomenon affecting adversely the whole problem from our point of view. In most East European countries, efforts are used to solve the slurry problem by means of various methods of processing for which the high concentrations of farm animals in one locality create good conditions. Thus, for instance in Czechoslovakia, Hungary, and even in the Soviet Union, sporadic slurry-processing plants were put into operation in past years using the water-cleaning procedures based on the aerobic (TAS) or anaerobic principle with biogas production. However, further development of such facilities did not take place due to shortage of funds. The situation in Czechoslovakia was reported in more detail at the Meeting of our Working Group held in Dublin.

As a rule, the fertilization of plants is not based on their requirements for nutrients, and thus it often happens, especially because of a shortage of areas suitable for the purpose in the course of the year, that nitrogen and phosphorus are supplied in excess due to slurry spreading (mainly in the vicinity of big enterprises), which results in negative effects on the foodstuffs.

Under the new political climate of today, which is accompanied by intensified interests of the public in improvement of the environment, efforts are made towards solution of such problems. But all these things will depend upon further development in the sphere of agriculture and the national economy in individual countries.

However, it is to be pointed out that exposure of the environment to danger as discussed in this paper is not the only and the most important problem for the environment in East European countries. The German Democratic Republic and Czechoslovakia in particular, due to certain of their industries and the power plants burning fossil fuels, have become big polluting agents of the environment not only in the territories of their own but also in extensive areas of the other parts of Europe.

It is hoped that the facts mentioned above will exert influence on the process towards solution of the problem discussed here. The gigantic agricultural enterprises existing today are supposed to be experiencing the process resulting in the establishment of smaller farm units. Notwithstanding, the large-capacity sheds can be cancelled only on the assumption that new, smaller-sized sheds are constructed, which is not held for real from motives of economy. Similarly, purchasing larger quantities of agricultural products on the world markets is also held for unrealizable; accordingly, to secure at least the existing level of production of foodstuffs will remain a requirement in both economic and political respects. For this particular demand is a limiting factor in the process as a whole.

It is our opinion that solution to the problems outlined above will require a period of several years. The duty which follows therefrom for all countries in East Europe is seen in the necessity of minimizing the mentioned risks and solving the hygienic problems under existing conditions.

TAENIA SAGINATA AND BOVINE CYSTICERCOSIS IN DENMARK - A REVIEW OF THE PRESENT EPIDEMIOLOGICAL SITUATION WITH SPECIAL REGARD TO THE POTENTIAL RISK OF SPREADING TAPEWORM EGGS THROUGH SEWAGE SLUDGE.

B. ILSOE, N. KYVSGAARD, Sv.Aa. HENRIKSEN *) and P.NANSEN (DK).
Royal Veterinary and Agricultural University, 13, Bülowsvej, and *)National Veterinary Laboratory, 27, Bülowsvej, DK-1870 C, Copenhagen, Denmark.

SUMMARY

The risk of spreading bovine cysticercosis infections through tapeworm eggs in sewage sludge is considered a major obstacle in the promotion of sludge reuse in Danish agriculture.
In an epidemiological study including 288 Danish cattle herds with animals infected with cysticercosis, it was demonstrated that most of the affected animals only harbour very few cysts. The use of sewage sludge in agriculture could not be identified as a risk factor for low grade bovine cysticercosis.Infections most likely were acquired through contact with streams carrying sewage and effluent discharged from sewage treatment plants, thus dispersing Taenia saginata eggs in low concentrations over large distances.
Herds with heavily infected cattle in most cases had been into direct contact with septic tank sludge illegally spread or handled by farmers or private contractors, or had grazed close to small sewage works. Such events are apparently quite common, but will however, due to the very low prevalence of taeniasis, only on very few occasions lead to cysticercosis outbreaks. Agricultural use of sludge from municipal sewage works was not involved in any of the high grade cysticercosis outbreaks.

INTRODUCTION

Only about one third of the 210,000 tonnes of sewage sludge annually produced in Denmark, is reused in agriculture. The hygienic reservations towards using sewage sludge as a fertilizer are mainly based on the risk of accumulating heavy metals in the soil, but the risk of contaminating livestock or crops with infectious microorganisms present in the sludge is also a limiting factor in the public acceptance of sludge reuse (12, 13, 22).

The two pathogens of greatest concern for the livestock owners are Salmonella bacteria and the tapeworm Taenia saginata, as they can cause substantial production losses in individual herds.

Taenia saginata is now rarely affecting the Danish population, nevertheless detection of infections with the larval stage, bovine cysticercosis, is still the cause of downgrading and condemnation of beef. A number of cases of heavy infections in Danish cattle herds, the so-called cysticercosis storms, have occasionally been detected, causing mass condemnations of carcasses at slaughter and substantial economic losses for the herd owners. The source of these infections have in some cases been identified as the agricultural use, or handling, of sewage sludge (11, 22).

In Denmark, possibilities for breaking the Taenia saginata infection cycle is meat inspection for preventing bovine cysticercosis to infect the consumers, and limitation of the spread of Taenia eggs in the environment by restricting the use of sewage and sludge. Danish waste reuse legislation since 1984 includes the following hygienic restrictions: Sewage sludge can be spread on arable land only, and raw sludge must be deposited 10 centimeters below the soil surface. On farmland, within a year after spreading sludge, the area can only be used for grain or seed crops and grass for pelleting. Furthermore, all machinery, used for sludge handling, must be adequately cleaned before being used for other purposes (6). In 1989 the legislation was revised, and the crop restrictions were withdrawn for sludge subjected to "controlled hygienization" (heat at 70oC for one hour or equivalent treatment (7)).

With the purpose of evaluating the risk of spreading infections through the agricultural use of sewage sludge, a project was initiated in 1986, supported by the Ministry of the Environment and the National Association of Local Authorities. Taenia saginata was chosen as an indicator microorganism, and the aim of the project was to indicate the most prevalent infection routes of T. saginata cysticercosis under Danish conditions, in order to evaluate the relative importance of sewage sludge in this respect (13, 19).

In the following, a review of the collected epidemiological data will be presented, concluding in an assessment of the role of sewage sludge in the spreading of bovine cysticercosis in Denmark.

TAENIA SAGINATA AND SEWAGE TREATMENT

Based on information collected about the sale of Niclosamide, the only cestocide recommended for treating human tapeworm infections in Denmark, the prevalence of Taenia saginata infections is estimated to be 3:100,000 (13). Compared to reports about the sale of cestocide in other European countries, the number of diagnosed cases of human taeniasis is higher in European countries south of Denmark (2, 9, 12, 26).

The daily egg output into the faeces of a tapeworm carrier is enormous, and the Taenia saginata egg concentration in house-hold sewage therefore will be very high. But as the prevalence of infection is so low, the risk of the presence of Taenia saginata eggs in sewage from single houses or minor treatment plants is very small. It is much more probable that eggs will be present in sewage of plants, serving a large population, although the egg concentration then would be very low, due to the admixture of large amounts of non-infected sewage, rain-water and industrial wastewater.

In investigations of sewage sludge of large cities in some other European countries, low concentrations of Taenia eggs (up to a few eggs per litre) have been found (4, 23, 24), whereas an investigation in an area with a high prevalence of taeniasis revealed egg concentrations of several thousands per gramme of dewatered sewage sludge (2).
Concentrations of a few eggs per litre of raw sludge from Danish plants serving populations of more than a thousand persons have been calculated (13), and therefore routine examination for Taenia eggs in such sludge was not considered to be reliable with the available isolation methods of relatively low efficiency and small sample size.
The use of immunochemical methods, as described for investigating environmental samples for eggs of other Taenia species (5), could give a satisfactory egg recovery in samples with low egg concentrations, and at the same time make it possible to distinguish between eggs of Taenia saginata and of eggs of other Taeniid species, e.g. those infecting dogs and cats.

Due to the low settling velocities of Taenia saginata eggs, most of the eggs will not have the time to settle into the sludge during ordinary sewage treatment, and will subsequently be found in the effluent (4, 13, 25). Transmission of bovine cysticercosis to cattle grazing pasture irrigated with sewage effluents have been demonstrated (1, 16, 25).

In Denmark all effluent is disposed off to water recipients, and thus Taenia eggs in sewage effluent would be expected to be dispersed widely and in low concentrations with the water-courses.

EGG SURVIVAL IN THE ENVIRONMENT

If deposited on land, by flooding or irrigation from streams carrying wastewater or effluent, by sludge disposal or by defecation in the open, the Taenia saginata eggs can remain infective for extended periods of time. In an experiment, eggs were left on the soil surface for increasing time intervals, after which the infectivity was investigated by feeding the eggs to susceptible calves and determining the number of cysts which developed. This demonstrated that the infectivity was maintained in a minor part of the eggs for as long as 6 1/2 months during a spring-winter period and 5 1/2 months during an autumn-spring period, while no eggs remained infective for 9 1/2 and 8 1/2 months during the comparable seasons, respectively (13, 14).

Desiccation and exposure to high temperatures are reported to be important lethal factors for Taenia eggs (10, 20). Depositing below the soil surface could therefore, although it would decrease the direct availability of eggs, be expected to increase the survival of Taenia eggs in sludge, as it has been the case for eggs of other helminth species (3).

BOVINE CYSTICERCOSIS IN DENMARK

The prevalence of T. saginata cysticercosis in cattle slaughtered at Danish abattoirs has declined from a maximum of 1.35% in 1952 to 0.10% at present (8, 16). The highest prevalence is recorded to be 0.7% in South Jutland, compared to a Danish average of 0.11% (in 1986 (8)).

The slaughterhouse records will not present all the cysticercosis infections, as the traditional meat inspection methods seem to be of rather low efficiency. Thus recent studies on 23 experimentally infected calves demonstrated that only 4% of the total number of cysts were located in the surfaces inspected by the incision at meat inspection (18). Furthermore it was calculated that the average number of cysts per infected animal in Denmark was less than 4. From these figures it was concluded that less than 15% of naturally low grade infected animals would be detected by standard meat inspection.

EPIDEMIOLOGICAL FINDINGS OF HERDS WITH HIGH AND LOW GRADE CYSTICERCOSIS INFECTIONS

Low-grade infections (less than 10 cysts observed) constitute the major part of the naturally aquired cysticercosis infections. Thus only 1.8% of the infected animals harboured 10 cysts or more at the predilection sites, and consequently their carcasses were condemned (13).

Epidemiological investigations were carried out at all farms with a history of total condemnations at slaughter due to massive C. bovis infections in 1986 and 1987 (13, 15). Attempts were made to identify the sources of infection of 38 animals by on-site investigations of the 14 farms of origin. The most frequent infection route found was sludge from household septic tanks illegally applied on pasture or feed crops, in some cases after deliberate or accidentally having been mixed with animal slurry. Permanently housed herds had been infected through the feed or by contamination of their environment by handling of slurry mixed with septic tank sludge containing Taenia saginata eggs. Some herds had grazed in close proximity of sewage works of towns with a few thousand inhabitants, the cattle having access to streams carrying a high concentration of effluent. A few herds had grazed at fields likely to be used for indiscriminate human defecation or been fed fresh grass from such fields.

In contrast to observations prior to 1984, when the Danish sludge reuse was regulated (11, 22), none of the present farms had been in contact with sewage sludge from municipal treatment plants.

An additional epidemiological investigation included a case-control study of risks factors in cattle herds with and without cases of low-grade T. saginata cysticercosis infections detected at slaughter (19).
In 232 of the 274 case herds, only one animal had been affected, and most cases were recorded by the end of the grazing season.
The most important risk factor identified by this investigation was cattle having access to streams carrying effluent from sewage treatment plants or single households (odds ratio 3.6), explaining one third of the cases. No definite risk could be demonstrated by the agricultural use of septic tank or sewage sludge or proximity to sewage treatment plants, railways or camping sites. Illegal use of the same machinery for handling septic tank sludge and liquid manure was, common in both groups. Although this practice was significantly more often reported from the case group, this was regarded as being based on very subjective observations. There was no tendency towards geographical concentration of the affected herds, not even around sewage works or sludge dumps, thus vector transmission is considered to be of minor importance. Egg dispersal by vectors, eg. flies, are considered important for the spreading of infections with other tapeworm species over large areas (20, 21). However such transmission could not be confirmed in a Danish study involving eggs of T. saginata (17).

The large proportion of cases not possible to explain by the case-control investigation points towards the existence of other transmission routes. In this connection imported feed contaminated at the site of production or storing might be considered (19).

DISCUSSION

The epidemiological findings of the infection sources in herds affected with high and low grade cysticercosis infections can be explained by the infection pattern suggested in fig.1. Heavy infections require contact to high concentrations of Taenia saginata eggs, which is possible in septic tank sludge and concentrated effluent from small sewage works. The illegal emptying of septic tanks on pasture with or without mixture of animal slurry is apparently a common practice often carried out in the grazing season when summer cottage septic tanks run full. Following the rules would imply paying a fee for delivering the sludge for disposal or treatment at sewage works or depositing it at arable land, which is not available at that time of the year. The explanation for the relatively few cases of herds affected by cysticercosis storms through this illegal septic tank sludge use or handling, could be that the risk of this sludge containing Taenia saginata eggs is extremely low due to the low prevalence of taeniasis.

The fact that many of the lightly infected herds probably had aquired the infection through contact to streams carrying sewage effluent are consistent with the observation that a major portion of the eggs will leave the sewage treatment plants with the effluent, being widely dispersed in low concentrations; which is confirmed by the fact that these infections occurs single and geographically widespread. Carrington et al. (4) also explains the parallel British cysticercosis infection pattern by the existence of a well dispersed and diluted infection source.

Fig.1: INFECTION ROUTES FOR BOVINE CYSTICERCOSIS:

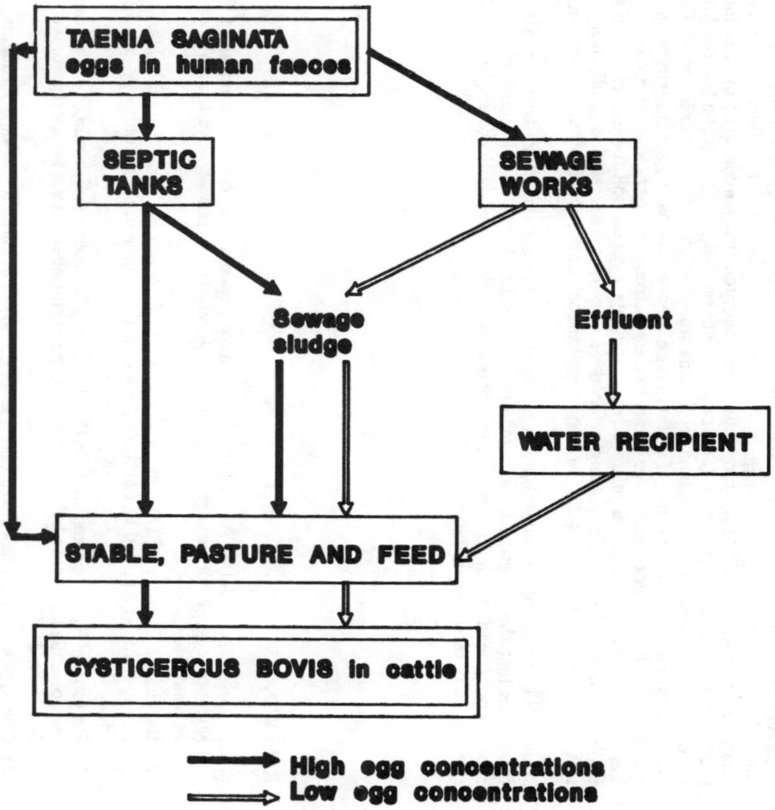

CONCLUSION

Agricultural use of sewage sludge apparently does not involve a significant risk of intensifying the infection cycle of Taenia saginata in Denmark, if the current sludge regulations are followed. In an epidemiological study it was demonstrated that a major part of infections was related to spreading of eggs through effluent and not by sludge. High level infections however were caused by spreading or handling septic tank sludge in contradiction to the regulations, or other practices implying exposure of cattle to high concentrations of Taenia saginata eggs. The relatively small number of such cases is due to the low frequency of tapeworm carriers.
The demonstration of survival of infective T. saginata eggs on soil for just over half a year implies that the required nograzing interval of one year after sludge application will be sufficient to prevent the spreading of bovine cysticercosis under Danish weather conditions.

Although taeniasis and cysticercosis can be regarded as being almost eradicated in Denmark, it appears that the infection cycle is maintained at a low level. The demonstrated infection routes could lead to an increase in the prevalence of bovine cysticercosis, if the Danish prevalence of taeniasis should increase.

REFERENCES

(1) Arundel JH, Adolph AJ (1980): Preliminary observations on the removal of Taenia saginata eggs from sewage using various treatment processes. Australian Veterinary Journal, 56, 492-495.

(2) Barbier D, Perrine D, Duhamel C, Doublet R, Georges P (1990): Parasitic Hazard with Sewage Sludge Applied to Land. Applied and Environmental Microbiology, 56 (5), 1420-1422.

(3) Bergstrøm K, Langeland G (1981): Overleving av spolormegg, salmonellabakterier og termostabile koliforme bakterier i jord og på grønnsaker dyrket i infisert jord. Nord.Vet.-Med., 33, 23-32.

(4) Carrington E G, Crewe W, Pike E B (1988): Parasites in Sewage Effluents. Water Research Centre Environmental Report, PRD 1403-M/2. WRc Medmenham Lab. Po Box 16, Marlow, Bucks, SL7 2HD England., 1-42.

(5) Craig P S, Macpherson C N L, Watson-Jones D L, Nelson G S (1988): Immunodetection of Echinococcus eggs from naturally infected dogs and from environmental contamination sites in settlements in Turkana, Kenya. Transactions of the Royal Society of Tropical Medicine and Hygiene, 82(2), 268-274.

(6) Danish Environmental Agency (1984): Bekendtgørelse nr 574 af 14 november 1984 - Bekendtgørelse om anvendelse af slam i landbruget.

(7) Danish Environmental Agency (1989): Bekendtgørelse nr 736 af 26 oktober 1989 - Bekendtgørelse om anvendelse af slam, spildevand og kompost m.v. til jordbrugsformål.

(8) Danish National Veterinary Services (1989). Personnal Communication to the Department of Meat Inspection. (9) Geerts S (1986): Academic session: "Tapeworms, meat and man, a complex interrelationship". From the International Colloquium held at the Instituut voor Tropische Geneeskunde, Antwerpen 11-12 december 1986, 1-19.

(10) Gemmell MA (1986): General epidemiology of Taenia saginata. In: Epidemiological Studies of Risks Associated with the Agricultural Use of Sewage Sludge: Knowledge and

Needs. (Block, Havelaar and L'Hermite eds.) Proc. of a CEC seminar, Metz, May 1985. Elsevier, 60-71.

(11) Haugaard P (1984): Cysticercosis and salmonellosis in cattle caused by sewage sludge on Danish farmland. Waste Management and Research, 2, 163-168.

(12) Havelaar A, Oosterom H, Notermans S and van Knapen F: Hygienic aspects of the application of sewage sludge to land. In: Proc. of the Int.Symp.on Biol. Reclamation and Land Util. of Urban Wastes. Napels 11-14/10 1983. 34pp.

(13) Ilsøe B, (ed.) (1988): Rapport om overvågningsprojekt vedr. anvendelse af slam i landbruget. National Veterinary Laboratory, Copenhagen. 157pp.,English summary.

(14) Ilsøe B, Kyvsgaard NC, Nansen P, Henriksen SA (1990a): A study of the survival of Taenia saginata eggs on soil in Denmark. Acta. vet. scand. 31 (2) 153-158.

(15) Ilsøe B, Kyvsgaard NC, Nansen P, Henriksen SA (1990b): Bovine cysticercosis in Denmark. A study on possible infection sources on farms with heavily infected animals. Acta. vet. scand. 31 (2) 159-168.

(16) Jepsen Aa, Roth H (1950): Parasitologiske og bakteriologiske problemer vedrørende spildevand, specielt i forbindelse med oversprøjtningsmetoden. Nordisk Veterinærmedicin, 2, 967-991.

(17) Kyvsgaard N, Ilsøe B, Henriksen SvAa, Nansen P (1988): An Attempt to Evaluate the Spreading of Taenia saginata Eggs in the Environment. Brief Communication to Acta Veterinaria Scandinavica, 29, 511-513.

(18) Kyvsgaard NC, Ilsøe B, Henriksen SA, Nansen P (1990): The distribution of Taenia saginata cysts in carcasses of experimentally infected calves and its significance for routine meat inspection. Research in Veterinary Science 49, 29-33.

(19) Kyvsgaard NC (1990): Taenia saginata cysticercose. Undersøgelser over epidemiologi, diagnose og kontrol. Licentiatafhandling til Den Kgl.Veterinærog Landbohøjskole. 112 pp.

(20) Lawson JR, Gemmell MA (1983): Hydatidosis and Cysticercosis: The Dynamics of Transmission. Advances in Parasitology, 22, 261-308.

(21) Lawson JR, Gemmell MA (1985): The potential role of blowflies in the transmission of taeniid tapeworm eggs. Parasitology, 91, 129-143.

(22) Nansen P, Henriksen Sv Aa (1986): The epidemiology of bovine cysticercosis (C.bovis) in relation to sewage and sludge application on farmland. In: Epidemiological Studies of Risks associated with the Agricultural Use of Sewage Sludge: Knowledge and Needs. (Block,Havelaar and L'Hermite eds.) Proc. of a CEC seminar, Metz, May 1985. Elsevier, 76-82.

(23) Schwartzbrod J, Thevenot MT, Collomb J, Baradel JM (1986): Parasitological study of waste-water sludge. Environmental Technology Letters, 7, 155-162.

(24) Watson DC, Satchwell M, Jones CE (1983): A Study of the Prevalence of Parasitic Helminth Eggs and Cysts in Sewage Sludges Disposed of to Agricultural Land. Water Pollution Control, 3, 285-289.

(25) Wilkens S (1981): Untersuchungen bei die Ansteckungsmöglichkeiten von Rindem mit Taenia Saginata und Sarcocystis spp. auf Abwasserverregnungsflächen und über das Absetzverhalten von Helmintheiern in vitro. Inaugural-dissertation, Tierärzliche Hochschule Hannover.

(26) Zimmermann G (1985): Kritische Betrachtung zur Verbreitung der Zystizercose beim Rind und Vorschläge zu verbesseren Bekämpfungsmassnahmen der Rinder-Bandwurm-Invasion beim Menschen. Tierärtzliche Umschau, 40, 257-264.

SEWAGE SLUDGE AND AGRICULTURAL WASTE HYGIENIZATION THROUGH AEROBIC STABILIZATION AND COMPOSTING

Marco de Bertoldi, Marcello Civilini and Marisa Manzano

Department of Industrial Microbiology, University of Udine, Udine, Italy

Summary

The use of not stabilized organic matter in agriculture may cause problems concerning competition for nutrients between plant roots and microorganisms, phytotoxicity, and pathogenic microorganism diffusion.

For these reasons sewage sludge and liquid agriculture waste should be treated by aerobic stabilization processes before land utilization. Aerobic stabilization processes give better final products, much more compatible with plant roots than those derived by anaerobic processes.

This paper describes different systems of haerobic stabilization and composting for sewage sludge and slurries; results obtained in independent experiments are discussed.

The problem of pathogenic microorganisms is deeply discussed with particular reference to pathogen reduction and process management, pathogen regrowth, indicator microorganisms, microbial specifications for disinfected sludge and compost. The importance of the following microorganisms both as pathogen or as indicator, is particularly analysed in this paper: Salmonella sp, faecal Streptococci, faecal Coliforms and Aspergillus fumigatus.

1. Introduction

Ecological, energy and sanitary requirements have drastically changed the management of sewage sludge and refuse of recent years. Uncontrolled landfill, tipping and incineration were the only means of disposal used until a few years ago, but ecological devastation caused by an invasion of waste in the environment and its high concentrations in some places has brought the question of these disposal methods on to the carpet.

The recent economic crisis in Europe has changed the concept of waste. It is no longer considered as whatever society has to eliminate; it has become a potential source of energy, which could be recovered. When the idea of recycling wastes for energy recovery caught on, it was discovered that the choice of this alternative meant a reduction in the amount of waste to be disposed of, which in turn meant lower costs.

The heterogeneus chemical composition of the organic fraction of sewage sludge is the ideal substrate for a whole series of microorganisms, so it is possible to use it in several, different microbial transformation processes, (production of: biomass, ethanol and organic fertilizers (compost)).

Low energy costs and good outlets make composting the most widely-used of all these processes over the world.

Several hundred industrial composting plants have been constructed and as many more are under construction or planned.

Unfortunately this process has two great drawbacks: 1) release of heavy metals into the environment; 2) the presence of animal and human pathogens.

This paper deals with the latter risk, with particular attention to the effectiveness of various systems used during composting to reduce pathogen content in the end-product (compost).

2. Pathogenic microorganisms present in waste-water sludge

Sludge may contain a huge number of pathogens-bacteria, virues, fungi and parassites (Tab. 1).

Waste is not the ideal habitat for many of them, so that their numbers tend to fall steeply in time.

Some pathogens, such as parassites and virues, are incapable of reproducing outside their host, so may survive for a long time in waste but without multipling their numbers.

Other pathogens, such as bacteria and fungi may reproduce, feeding on the assimilable organic matter in waste, so multiply in number in time.

The latter group of pathogens make it necessary to provide conditions for compost sanitization, such that these microorganisms not only are reduced in number, but also they are unable to grow and multiply in mature compost.

When pathogens present in sludge are spread on land, the homeostatic properties of soil and competition with native microorganisms usually have the effect of reducing their numbers. The fact remains however, that there is a risk of some pathogens reaching man and, if present in sufficiently high numbers of them causing diisease or food poisoning.

It is therefore extremely important for pathogens to be reduced to a minimum so that the health and hygiene risk be minimised too.

Technically speaking this could be done by sterilization, but this would not only be economically impratical, it would not eliminate the risk of re-growth in some pathogens (e.g. Salmonella) which, in absence of microbial competition, could grow beyond measure.

It must not be forgotten that the destination of this matter is the soil, which already harbours some pathogens such as Clostridium tetani and Aspergillus fumigatus, so complete sterilization is unneccessary. It would be superflous and a waste of money and energy to eliminate pathogens naturaly present in soil.

Composting must therefore guarantee the reduction to below a certain level of all those microorganisms which are foreign to soil and which, above certain concentrations result in the danger of contamination. There is a risk of direct contagion when pathogen-contamined vegetables are for human consumption and indirect contagion if consumed by animals for human consumption.

All this underlines the importance of reducing the number of pathogens present in waste before their use in agriculture.

The two main problems that emerge are first, how far to take sanitization and secondly, how to check the degree of sanitization once completed.

The former is simply a problem of economics; sludge, by definition, cannot carry heavy overheads for trasformation into compost which would make it non-competitive on the open market.

For this reason, pathogen reduction must of necessity be a compromise between cost of processing and benefit of a safe, sanitized end-product.

With regard to the second point, that is how to check the degree of sanitization, this can be done either during, or on completion of, the process.

Sanitizing occurs naturally during composting since the composting mass reaches a certain temperature level, which it maintains for some time. Temperature and time-span must be selected on the basis of a logarithmic reduction in the main pathogens; what degree of reduction there should be in compost has yet to be defined.

Nations have singly set temperature standards varying from 55°C to 65 °C covering a time-span of 3 days and 24 hours respectively.

Table 1 - Pathogens likely to be found in waste-water sludge.

PATHOGEN	DISEASE
Virus	
Enterovirus	gastro-enteritis, heart disease, meningitis
Rotavirus	gastro-enteritis
Parovirus	gastro-enteritis
Adenovirus	respiratory tract infections, conjunctivitis
Hepatitis A virus	viral hepatitis
Polio virus	poliomyelitis
Ecovirus	meningitis
Coxsachivirus	meningitis
Bacteria	
Salmonella (1700 types)	typhus, salmonella
Shigellae	sigellosis
Mycobacterium tuberculosis	tuberculosis
Vibrio cholerae	cholera
Escherichia coli	gastro-enteritis
Yersinia enterocolica	gastro-enteritis
Clostridium perfringens	gangrene
Clostridium botulinum	botulism
Listeria monocytogenes	meningo-encephalitis
Fungi	
Candida spp.	systemicand skin mycoses
Tricosporon cutaneum	skin mycosis
Aspergillus fumigatus	lung mycosis
Tricophyton spp.	skin mycosis
Epidemophyton spp.	skin mycosis
Microsporum spp.	skin mycosis
Protozoa	
Eutamoeba	amebiasis
Giardia lamblia	giardiasis
Balantidium coli	balantidiasis
Naegleria fowleri	meningo-cephalitis
A. canthamoebae	meningo-cephalitis
Helminths	
Ascaris lumbricoides	ascariosis
Ancylostoma spp.	ancilastomiosis
Necator amaricanus	necatoriasis
Enterobius vermicularis	enterobiasis
Strongyloides stercoralis	strongilaidiasis
Toxocara spp.	larvae in the viscera
THrichuris thrichuria	thricuriasis
Taenia saginata	tapeworm
Hymenolepsis nana	tapeworm
Echinococcus granulosus	echinococcosis
Echinococcus multilocularis	echinococcosis

Before market release, the end-product should undergo a whole series of complicated, lengthy and expensive analises to check its degree of sanitization. Some laboratories check the disappearance of only particular groups of pathogens, taking their absence or reduction to indicate that of other groups too. The use of indicator microorganisms too, has been considered by some laboratories, and experiments so far carried out indicate that faecal Coliforms, faecal Streptococci, Enterobacteriaceae, certain viruses and parassites eggs give reliable results.

For an indicator to give reliable results it must satisfay the following requisites:

1) It must be present constantly and in great quantities in the raw material before composting;

2) it must belong to a group with the same reactions to the treatment undergone (i.e. thermal changes);

3) it must have a higher resistance than pathogens to be eliminated, for a safety margin;

4) it must be determined by simple and cheap tests.

One of the most controversial problems under discussion is the quantitative determination of the reduction that indicators should undergo to guarantee hygienization of end-products.

In the present paper different systems of aerobic stabilization of sludges and their mixtures are discussed and their potential for pathogenic microorganism reduction, in particular referred to Salmonella sp. and two groups of indicators: faecal Coliforms and faecal Streptococci that seem to be the best candidates for such a purpose.

3. Factors that may influence the pathogen growth in sewage sludge

Only some pathogenic bacteria and fungi can multiply in sludge, but only under particular conditions.

Even viruses and parasites, may survive a long time on fresh sludge, they do not multiply outside the host.

Many factors which influence pathogen growth in sludge are here discussed.

a) organic matter

All pathogens present in sludge are heterothrophic and therefore they need a source of soluble organic matter to metabolize to growth. In sludge the organic matter is extremely heterogeneous (from short chain organic acids to high stable organic polimers like lignin).

A large number of microorganisms can growth in this heterogeneous habitat but they can utilize only the organic matter which are able to metabolize.

Bacteria and fungi are generally able to metabolize readily-assimilable organic matter, those which are the simplest, such as alcohol, organic acids, sugars etc. but are unable to multiply on complex compounds like cellulose, lignin and humic substances. When sludge and bulking agents undergo transformation into compost, a gamut of non-pathogenic microorganisms transform the organic matter through mineralization and humification. In cured compost the residual organic fraction is composed mainly of stable polymers, often combined together to make up the complex humic fraction. Easily soluble substances are transformed during the metabolism of the bacteria and fungi which bring about composting.

All this makes conditions unfavourable for the metabolism and growth of pathogens since they do not have organic matter on which to feed.

b) moisture

As with all other biological processes, growth in microorganisms depends on the presence of moisture. Moisture level in waste is usually at a level high enough to support pathogen growth. Even a very high moisture content, such as in sludge, does not impair pathogen development, since many -bacteria in particular- are facultative anaerobes, and can therefore grow even in the absence of oxygen.

Low moisture levels, below 25%, slow or halt all forms of microbial growth, pathogens included.

Cured, stabilized compost, with a moisture content of 25-30% should not therefore create problems concerning the re-growth of pathogens unless local meteorological conditions raise the moisture level.

c) <u>temperature</u>

All pathogen microorganisms have threshold resistence to temperature which varies from one microorganism to another and according to environmental conditions. All sporigenous and non-sporigenous thermophilous bacteria are able to resist temperatures above 100°C; this means that no composting process is capable of eliminating these microbial forms. Fortunately the most dangerous pathogens for the environment and agriculture in waste and sludge are neither sporigenous nor thermophilous and can therefore be eliminated through bland heat treatment.

The effectiveness of this treatment depends firstly on the temperature reached and the length of time it is mantained. Moisture conditions the technical effect by increasing the conductivity of the mass. Sanitization is the result of high temperatures (65°C) generated during composting and is more active if moisture content is high. This lethal effect would be much less marked in the dry end-product.

The central aspect in sanitization by the thermal effect is that a particular temperature must be reached in all parts of composting mass, at the same time. This is at times very difficult to achieve technically; in some parts of the mass temperatures may rise to above 80°C while others are not sanitized by the end of the process.

Later in this paper there will be a discussion of pratical composting systems, dealing with the disadventages of each in sanitation.

d) <u>oxygen supply</u>

We have said that temperature helps to produce sanitized mature compost. The temperature in the mass is the direct result of the accumulation of calories which are given off during the exothermic reactions during composting. Since these reactions are bio-oxydations which require a constant oxygen supply, it is easy to see that temperatures will rise only where there is a good supply of oxygen for oxidation. When there are parts of the mass where oxygen supply is insufficient, anaerobic processes take hold over the aerobic; much lower energy levels are liberated so temperature in the mass does not rise.

Ways of oxygenating the whole of the mass have enabled practical composting processes to be developed which by technical means, create aerobic conditions throughout the processes.

The technical side of the connection between oxygen supply and sanitation will be discussed below in the chapter "Practical composting systems".

e) <u>microbial competition and antagonism</u>

Microbial competition and antagonism are the second most important factors in pathogen control during composting. An enormous number of saprophytes are involved in composting.

This microflora shoul be considered as native or natural in composting matter. Waste contains a second microbial population, pathogens, which are numerically an insignificant fraction of the total microbial population. When population density is relatively high antagonism and competition for nutrition set in, in this heterogeneous microbial community. The native saprophitic population has a considerable advantage over the other population in this competition.

Composting matter is not the natural environment for pathogens, so in this ecosystem competition brings about the elimination of the weaker rival.

4. Practical Systems of Composting

I- Open systems

It is impossible to discuss composting without examining in detail the various systems in use today, with their respective operating requirements.

Table 2 reports briefly the practical systems for composting most widely in use all over the world.

We are now going to analyse the various composting systems with reference to sanitization. Results of analyses on different materials, obtained by the present Authors are reported for each system.

The microorganisms analyzed on all occasions were Salmonella sp., faecal Coliforms and faecal Streptococci as indicators for pathogenic bacteria.

a) Turned pile

Merely turning a pile does not guarantee constant or satisfactory oxygenation. One hour after turning, oxygen levels within the pile fall drastically, reducing microbial bio-oxydative activity. The pile needs to be turned frequently for sufficient oxygenation, but this gives rise to other problems of a technical and economic nature. Turning requires several precautions: the external layer of the pile which does not undergo a rise in temperature, and is therefore not hygienised, must be turned right into the center of the pile.

Pile size is an important factor in sanitization. Piles that are too high (above 3m) are difficult to aerate properly for sanitization.

The trials discussed here were carried out on piles 1.5 m in height, protected by roofing and turned twice- weekly.

Matter composted was the organic fraction of solid urban waste from the Dano plant in Pistoia (Italy) mixed with aerobically stabilized waste-water sludge.

The results of the five separate experiments were carried out are reported in Table 3

Salmonella persisted after 30 days composting in two experiments. None of the five experiments showed a consistent reduction in faecal Coliforms or faecal Streptococci. Repeated turning of the mass disturbed composting and altered the temperature profile, but oxygen requirements in the centre of the pile made turning at least twice a week essential.

Periodic turning for pathogen control is therefore not completely satisfactory. It must be added that, during the final stages of composting, when the material is nearly dry, dust containing a lot of Aspergillus fumigatus spores is released into the air.

b) static pile with bottom suction

Pile size is a critical factor in this method too. Piles over 2.5-3 m high make uniform aeration almost impossible. Usually the air-flow creates passages in the pile which leave large areas without air supply and therefore not aerated. The physical characteristics of the material also affect aeration: highly porous material, such as wood chips and sludge, can take a greater height than the organic fraction of S.U.W., which tends to settle.

When using static piles the whole pile must be blanketed before starting with a layer of cured, sanitized compost which will insulate the mass against heat loss, enabling high temperatures to be reached throughout its volume, right to the outermost layers to be composted.

Research using this method has shown fairly satisfactory results, as long as management of the process is correct - in other words, air-flow must always be carefully dosed in order not to cool the pile nor dry it out too quikly.

Our trials were carried out on piles of 20 T each, 1.5m high x 2.5 m wide. Trials were carried out on the organic fraction of M. S.W. mixed with aerobically stabilized waste-water sludge.

Results are reported in Tables 4, 5, 6 and 7.

c) static pile with bottom blowing

The same experimental conditions as under b) were set up and four experiments were carried out using different air flows.

In contrast with suction, this method tends to cool and dry the bottom of the pile while the upper portion is hotter and moister. With correct, careful processing the final product is well-sanitized. Results are reported in Tables 8, 9, 10 and 11.

d) static pile with inversion of flow

Experimental conditions identical to those under b) were set up. Temperature probes were inserted in the pile, one at 20 cm and the other at 1.20 m above the bottom. When the temperature probes revealed a difference of over 15°C air-flow was inverted automatically by a valve at the head of the fans.

Two independent trials were carried out using two different air quantities.

This process gave the best and most reliable results of all, in all respects (evolution and control of process and sanitization of end-product).

Periodically inverting air flow keeps temperatures and moisture content more even throughout the pile, giving reliably even sanitization. Results of these trials are shown in Table 13.

II- Closed systems

a) Continuous vertical reactors

There are several different types of continuous vertical reactor. They have in common the characteristic of processing a large quantity of material in bulk (up to 2000 m^3) inside reactors up to 9 m high. These reactors are usually charged from the top and discharged from the bottom by various means and oxygenation is carried out by forced air from the bottom.

Conditions inside these reactors are such that bottom ventilation is not sufficient to achieve the required effects. For optimal oxygenation levels the mass must have a particular air-flow per unit time. This means that the mass must be no higher than 2-3 m. Higher heaps require a proportionally higher quantity of air per unit surface area according to the height of the pile. This causes hyperventilation in the lower portion of the reactor, with all its drawbacks (cooling, drying, etc.), and insufficient oxygenation in the upper portion, where there is most need of it, since fresh material is in this part. As the forced air travels through the mass its composition changes because of microbial action; CO_2 content increases and O_2 falls. When it reaches the top most layers in the reactor the oxygen level is too low to support aerobic microbial metabolism. Partial anaerobiosis sets in with predominantly fermentations and anaerobic respiration, which does not cause a sufficiently high rise in temperature to guarantee a sanitized end-product.

This is one of the greatest drawbacks to continuous vertical reactors in which the composting mass is 7-10 m high. Under these conditions no type of ventilation whatsoever, whether from the bottom or the top, is able to guarantee uniform oxygenation throughout the mass. Analyses of resulting end-products confirmed our expectation that it is impossible to control the process : the conditions of the product are prevalently anaerobic and it is not sanitized. These partially anaerobic and mesophyllic conditions are ideal for the Enterobacteria and Streptococcaceae to multiply.

Tables 13, and 14 report data from analyses on material from a 100 m^3 vertical reactor charged with a mixture of sludge and the organic fraction of M. S.W.

Table 15 reports data from analyses on material (sludge+ compost+ sawdust) composted in an open vertical reactor for 15 days (300 m^3).

b) horizontal reactors

These reactors contain a mass of material, no more than 2-3 m in height, laid lengthwise. This means that the process is not much different from the aerated-pile method, but it has the advantage of making it easier to control the process, offensive odors and temperature.

When carefully controlled, these reactors give excellent results in quantity, quality and sanitization of the end-product.

It is possible to control temperature without difficulty inside these reactors,where oxygen is supplied either by turning or by aeration, and the mass can be uniformly oxygenated, which also means satisfactory sanitization of the end-product.

Results from trials carried out with horizontal reactors are reported in Tables 16, 17 and 18.

5. Conclusions

a) Pathogenic bacteria and indicators

One of the chief problems in marketing and agricultural use of compost is the presence of pathogenic microorganisms. The indiscriminate use of non-sanitized material could become the cause of a spread of these microorganisms in the environment, resulting in a serious sanitary hazard for man and animals. These products must therefore be treated before use in the environment and their processing must be able to guarantee a reduction in all pathogens present to avalue below risk level.

This boils down in practice to resolving two problems: 1) What is the threshold level below which pathogens present are no longer a hazard? 2) How shall we monitor compost to check on sanitization.

The former question does not have an easy answer. Threshold level varies from one organism to another and it must not be forgotten that there is no such thing as an economically feasable completely hazardless level.

On the latter point it has to be admitted that which pathogen microorganisms to monitor is a moot point. The present research is an attempt to make a contribution in this direction.

To analyse all composts produced, or available on the market, for hygiene would turn into an enormous economic problem - in view of the numbers of pathogens possibly present in waste used as starting material for compost (see Tab. 1).

It would be advisable to limit the choice to a few representative pathogens which are easy to determine and sufficiently frequently present in material for analysis.

Many scientists agree that one extremely suitable candidate is Salmonella genus.

Its presence in compost is probably a sure sign of non-sanitization. In order not to have to analyze every single other group of pathogens it would be useful to be able to identify indicators, but here too, the question is still not fully answered.

This research aimed at verifying whether the groups " faecal Coliforms " and " faecal Streptococci " were useful indicators for sanitization, perticularly where Salmonella is concerned.

Results obtained in this research indicate that these two groups behave well : they are always abundantly present in the starting material, whereas Salmonella is sometimes absent; they survive well in both aerobic and anaerobic conditions; their resistence to temperature is different

- Coliforms are more sensitive (similar to Salmonella), while Streptococci are more resistant (45-48°C). Both groups comprise species which are able to multiply in waste under favourable conditions so are also suitable as a sign of re-growth.

Faecal Coliforms and Streptococci seem to be highly suitable as indicators of pathogenic bacteria. In all composting trials carried out in this research, Salmonella disappeared regularly before Coliforms reached the $5x10^2$ level, or Streptococci the $3x10^3$ level.

At this point it would be interesting to extend research further in order to discover what limit to set for these indicators. They should also be compared with other groups of pathogens such as viruses and parassites which behave in a different way from bacteria during the process of composting.

The following emerges from research carried out on analysis of various types of waste (organic fraction of M. S.W., sludge, tree bark, sawdust) and different composting systems:

1) analysis on compost has never yet considered Aspergillus fumigatus, because although it is present in compost it does not constitute a health hazard; this fungus is in fact widespread in nature and is almost always present in soil so that even if it is introduced into soil by compost, its presence in the environment would not be altered;

2) when composting is properly carried out, i.e. when the whole mass has been subjected to a temperature of 65°C for at least two days, Salmonella spp. disappears completely.

3) under these working conditions, Coliform indicators are always inferior to $5x10^2$ and faecal Streptococci to $3x10^3$.

These values could therefore be taken as the limit for sanitization of compost, from the point of view of bacteria.

b) compost processing and sanitization

Composting and sanitization are not necessarily synonymous.

Sanitization depends on several factors which are not always present in all composting processes. In theory, when a process has been properly handled and the whole mass has been held at a sufficiently high temperature, the end product can be taken as having been sanitized, but this is not enough to guarantee stability. Pathogens may re-grow, so as well as drastically reducing pathogens, composting has also to transform waste in such a way as to prevent such an occurrence. This can be achieved by adequately stabilizing the organic matter, mineralizing all simple compounds, easily assimilated by pathogenic microorganisms and by humifying the other components. Moisture level too in the end product should be fairly low (25-30%), to avoid microbial activity.

Although several factors are concerned in the elimination of pathogens during composting, the only one which quantitatively can be measured and controlled is temperature, so it is this which is used to control the process.

Since temperature increase in mass is the direct result of microorganism bio-oxidation activity, anything affecting this activity, directly or indirectly, will have consequences on sanitization.

This means that from the practical aspect, satisfactory sanitization depends not just on the way composting is carried out, but also on conditions in during the process.

The analyses discussed above clearly showed the disadvantages of some types of process: mere periodic turning without ventilation does not guarantee good sanitization unless piles are low (1-1.30 m.). Continuous vertical reactors which accumulate material up to a height of 9 m do not esily permit the required optimal conditions controlling the process, and so give no guarantee of sanitized end-product. The other systems analyzed (piles ventilated by various means and horizontal reactors) are in theory suitable for controlling the process effectively, with good hygienization as a consequence.

The most important parameters to keep in mind for carrying out the process are the following:

the whole mass must be kept at at least 65°C for 2-3 days; composting should confer biological stability to the end-product, in order to reduce health hazards to a minimum by impeding pathogen re-growth. This requires the entire mass to be oxygenated throughout, because where anaerobiosis sets in, the temperature will not rise sufficiently for sanitization. Vertical reactors proved to be the least suitable of all in this respect, particularly when the mass is more than 3 m high.

In conclusion, a sanitized end-product is not the automatic result of any composting process; important factors to be considered are working conditions and how the process is runned and controlled.

* * * * * *

Table 2 - Practical systems of composting sewage sludge and waste.

OPEN SYSTEMS	
Turned pile	
Static pile	with air suction with air blowing with inversion of flow with air blowing in conjunction with feed back control (temperature or oxygen)
CLOSED SYSTEMS	
vertical reactors	continuous discontinuous
horizontal reactors	static with bottom ventilation with movement of material in conjunction with ventilation

Table 3 - Number of cells present per g dry weight in starting material (sewage sludge 40% and organic fraction of municipal solid waste 60%) and in compost obtained after 30 days processing on pile turned twice a week.

		COMPOST				
Microorganisms	**Starting material**	**1° trial**	**2° trial**	**3° trial**	**4° trial**	**5° trial**
Salmonella spp.	2.1×10^1	0	0.8×10^1	2.3×10^1	0	0
faecal Coliforms	4.8×10^6	5.8×10^5	4.1×10^6	2.3×10^5	3.8×10^4	6.5×10^4
faecal Streptococci	3.8×10^5	2.0×10^5	7.0×10^4	6.1×10^4	3.1×10^4	3.9×10^3

Table 4 - Number of cells present per g dry matter weight in starting material (sewage sludge and the organic fraction M.S.W.) and in compost obtained after 30 days processing in static pile with bottom ventilation (suction: 0.5m³/min/ton d.w.)

Microorganisms	Starting material	Compost
Salmonella spp.	$3.1x10^2$	0
faecal Coliforms	$8.3x10^5$	$2.9x10^2$
faecal Streptococci	$7.1x10^5$	$1.7x10^3$

Table 5 - Number of cells present per g dry weight in starting material (sewage sludge and the organic fraction M.S.W.) and in compost obtained after 25 days processing in static pile with bottom ventilation (suction: 1.0 m³/min/ton d.w.)

Microorganisms	Starting material	Compost
Salmonella spp.	absent	====
faecal Coliforms	$6.7x10^5$	$1.9x10^2$
faecal Streptococci	$4.2x10^4$	$2.0x10^2$

Table 6 - Number of cells present per g dry weight in starting material (organic fraction M.S.W. 60% + sewage sludge 40%) and in compost obtained after 30 days processing in static pile with bottom ventilation (suction: 0.5 m³/min/ton d.w.)

Microorganisms	Starting material	Compost
Salmonella spp.	$3.0x10^3$	====
faecal Coliforms	$4.2x10^7$	$1.9x10^2$
faecal Streptococci	$2.0x10^6$	$7.4x10^2$

Table 7 - Number of cells present per g dry weight in starting material (organic fraction M.S.W. 60% + sewage sludge 40%) and in compost obtained after 25 days processing in static pile with bottom ventilation (suction: 1 m^3/min/ton d.w.)

Microorganisms	Starting material	Compost
Salmonella spp.	$3.4x10^3$	====
faecal Coliforms	$7.8x10^7$	$5.4x10^2$
faecal Streptococci	$4.6x10^6$	$1.2x10^3$

Table 8 - Number of cells present per g dry weight in starting material (sewage sludge and the organic fraction M.S.W.) and in compost obtained after 30 days processing in static pile with bottom ventilation (blowing: 0.5 m^3/min/ton d.w.)

Microorganisms	Starting material	Compost
Salmonella spp.	absent	====
faecal Coliforms	3.8×10^5	1.2×10^2
faecal Streptococci	4.5×10^4	9.7×10^2

Table 9 - Number of cells present per g dry weight in starting material (sewage sludge and the organic fraction M.S.W.) and in compost obtained after 25 days processing in static pile with bottom ventilation (blowing: 1 m^3/min/ton d.w.)

Microorganisms	Starting material	Compost
Salmonella spp.	8.1×10^1	====
faecal Coliforms	7.4×10^6	2.1×10^2
faecal Streptococci	5.2×10^5	7.8×10^2

Table 10 - Number of cells present per g dry weight in starting material (organic fraction M.S.W. 60% + aerobic sludge 40%) and in compost obtained after 30 days processing in static pile with bottom ventilation (blowing: 0.5 m^3/min/ton d.w.)

Microorganisms	Starting material	Compost
Salmonella spp.	3.4×10^3	====
faecal Coliforms	4.5×10^7	2.9×10^2
faecal Streptococci	9.4×10^6	3.4×10^3

Table 11 - Number of cells present per g dry weight in starting material (organic fraction M.S.W. 60% + aerobic sludge 40%) and in compost obtained after 25 days processing in static pile with bottom ventilation (blowing: 1 m^3/min/ton d.w.)

Microorganisms	Starting material	Compost
Salmonella spp.	3.1×10^3	====
faecal Coliforms	3.9×10^7	2.1×10^2
faecal Streptococci	8.1×10^6	1.2×10^3

Table 12 - Number of cells present per g dry weight in starting material (organic fraction M.S.W. 60% + aerobic sludge 40%) and in compost obtained after 30 days processing in static pile with inversion of flow (suction and blowing: 0.5 m^3/min/ton d.w.)

Microorganisms	Starting material	Compost
Salmonella spp.	7.4×10^3	====
faecal Coliforms	5.4×10^6	1.3×10^1
faecal Streptococci	6.1×10^4	6.2×10^2

Table 13 - Number of cells present per g dry weight in starting material (organic fraction M.S.W. + sewage sludge) and in compost obtained after 14 days processing in closed vertical reactor (capacity 100 m^3)

Microorganisms	Starting material	Compost
Salmonella spp.	absent	====
faecal Coliforms	1.5×10^7	2.6×10^6
faecal Streptococci	3.5×10^5	9.1×10^6

Table 14 - Number of cells present per g dry weight in starting material (organic fraction M.S.W. + sewage sludge) and in compost obtained after 21 days processing in closed vertical reactor (capacity 100 m^3)

Microorganisms	Starting material	Compost
Salmonella spp.	1.4×10^3	2.3×10^1
faecal Coliforms	3.5×10^8	6.4×10^4
faecal Streptococci	6.8×10^6	1.4×10^9

Table 15 - Number of cells present per g dry weight in starting material (sewage sludge + compost + sawdust) and in compost obtained after 15 days processing in open vertical reactor (capacity 300 m^3)

Microorganisms	Starting material	Compost
Salmonella spp.	absent	
faecal Coliforms	3.0×10^8	2.4×10^5
faecal Streptococci	5.5×10^7	6.5×10^6

Table 16 - Number of cells present per g dry weight in starting material (sewage sludge + compost) and in compost obtained after processing in horizontal reactor .

Microorganisms	Starting material	Compost
Salmonella spp.	1.5×10^2	====
faecal Coliforms	5.8×10^4	2.3×10^1
faecal Streptococci	8.5×10^6	1.6×10^2

Table 17 - Number of cells present per g dry weight in starting material (sewage sludge + bark + sawdust) and in compost obtained after 14 days processing in horizontal reactor with static material aerated by bottom blowing .

Microorganisms	Starting material	Compost
Salmonella spp.	8.7×10^2	====
faecal Coliforms	6.8×10^8	9.5×10^1
faecal Streptococci	5.1×10^6	1.7×10^2

Table 18 - Number of cells present per g dry weight in starting material (sewage sludge + organic fraction of M.S.W.) and in compost obtained after 28 days processing in horizontal reactor with static material aerated by alternating bottom blowing and suction.

Microorganisms	Starting material	Compost
Salmonella spp.	6.3×10^3	====
faecal Coliforms	8.9×10^4	2.0×10^1
faecal Streptococci	9.1×10^5	1.2×10^2

Session V

IMPLEMENTATION OF REGULATIONS, PRACTICAL EXPERIENCES AND PUBLIC PERCEPTION

Chairman : A. KOUZELI–KATSIRI (GR)

Rapporteur : P.J. NEWMAN (UK)

SITUATION OF SEWAGE SLUDGE IN AUSTRIA - USE IN AGRICULTURE, NATIONAL GUIDELINES AND LAWS, FUTURE ASPECTS (A GENERAL VIEW)

K. AICHBERGER

Federal Institute of Agrobiology, Wieningerstr. 8, 4020 Linz, Austria

Summary

Presently in Austria there is produced about 4 million m^3 (≈ 200.000 t dry matter) sewage sludge per year. This quantity is disposed of in agriculture (25 %), used for land reclamation (3 %), for composting and incineration (37 %), or is simply stored in dumping areas (35 %). The quality of Austrian sludge is normally quite good for use in agriculture since the content of plant nutrients is acceptable and heavy metal levels are low on the average.

In the past years very strict regulations and laws on environment and soil protection came into force and caused a decrease in using sludge in agriculture. These laws generally require sludge and soil analyses, limit values for heavy metals in sludge and soil, a maximum rate of sludge per year, a delivery note for sludge, and a property attestation by the authority. For the future it is estimated that the disposal of sludge on agricultural land, landfills, and storage will further decrease and incineration and composting will become the main strategies for disposal of sewage sludge in Austria.

1. INTRODUCTION

Protection and purification of water are the main points of the environment protection programm of the Austrian Federal Government. The law of water rights from 1959 requires associations of water companies to build and manage waste water treatment plants. In the past 15 to 20 years the establishment of treatment plants increased, and more and more inhabitants and industrial plants were able to join into public waste water system. Thereby the water quality (rivers, lakes) was improved significantly but at the same time the quantity of sewage sludge increased tremendously. A correlation between treatment plant capacity (number of inhabitants equivalents) and sludge production could be established in Upper Austria for the years 1970 - 1990 (see Fig. 1).

2. USE OF SEWAGE SLUDGE IN AGRICULTURE, ITS CHEMICAL COMPOSITION, NATIONAL GUIDELINES AND LAWS

With about 11 million inhabitant equivalents in Austria presently there is a sewage sludge production of 11.000 m^3/day or 4 million m^3 wet sludge per year. This quantity - corresponding to about 200.000 t dry matter - has to be disposed of in agriculture, or used for land reclamation, landfilling, composting or incineration. Because of the saving of costs and the recycling of plant nutrients in Austria the utilization of sludge on agricultural land was a very frequent way of use (30 - 50 %) in former times. Today the use of sludge in agriculture varied between 20 and 90 percent dependent on the economical structur (agriculture/forestry/urban area) of the different federal countries (1). In Table 1 sludge production and the different ways of disposal in the nine federal countries of Austria are presented.

The sewage sludge produced in Austria is mainly anaerobic stabilized has a favorable content of plant nutrients and low values of heavy metals. Non dewatered sludge has a content of about 4 - 8 percent dry matter and slightly over 50 percent organic matter in dry matter.

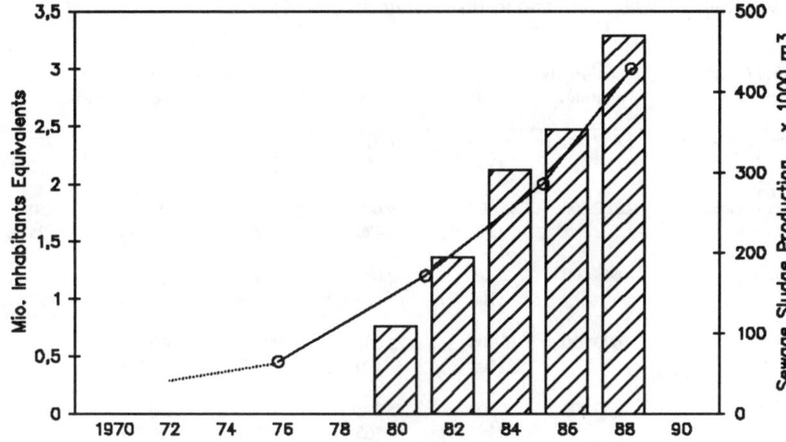

Fig. 1: Correlation between treatment plant capacity (inhabitants equivalents) and the production of sludge in Upper Austria from 1970–1990

The nitrogen values are in a quite good range (3.9 percent on average) with an ammonium component of 1/5. Content of phosphate and calcium-oxide vary considerable depending on type and plant process. (2, 3).
Table 2 presents the range and most frequent contents of nutrients and heavy metals in Austrian sludge. The statistics were calculated on data bases from Upper- and Lower Austria, Styria, Burgenland, Salzburg, and Tyrol and represent a summary of results of more than 10 years. In Austria the average heavy metal content of sewage sludge is remarkable low (see mode values in Tab. 2). This made it possible to introduce very strict guidelines for low limit values for heavy metal concentration (significantly lower than the council directive of the European Communities from June 1986). In addition to that we were able to recommend a higher dose of sludge on agricultural land which results in a stronger effect of the nutrients.

Therefore the main points of the Austrian guideline (4) for use of sewage sludge were:

- control and analysis of sludge before using in agriculture
- permission for using sludge only if the limit values of heavy metal concentration are not exceeded (2000 ppm Zn, 500 ppm Cu, Pb and Cr, 100 ppm Ni, 10 ppm Cd and Hg).
- the maximum application rate shouldn't exceed 100 m^3 or 5 tons dry matter/ha/year
- taking into account the charge of nitrogen and phosphorus per ha
- general bans for sludge application (e.g. in water protection areas; in horticulture and forestry; on grassland during the period of vegetation; on soils saturated with water, frozen or covered with snow).
- advice and supervision by water authorities and chambers of agriculture.

Table 1: Sludge production and its disposal in Austria (1)

Federal Country	Inhab. Equiv. (actual)	Sludge Prod. t dm/a	Disposal t dm/a Agricult.	Dump.areas	Land reclam.	Incin./Comp.
Lower Austria	1.487 000	29 550 (100 %)	14 775 (50 %)	13 002 (44 %)	-	1 773 (6 %)
Upper Austria	2.800 000	45 000 (100 %)	9 000 (20 %)	22 500 (50 %)	2 250 (5 %)	11 250 (25 %)
Salzburg	648 000	7 728 (100 %)	2 087 (27 %)	773 (10 %)	464 (6 %)	4 405 (57 %)
Tyrol	670 000	15 000 (100 %)	3 000 (20 %)	9 000 (60 %)	3 000 (20 %)	-
Vorarlberg	750 000	8 300 (100 %)	5 395 (65 %)	2 075 (25 %)	830 (10 %)	-
Burgenland	548 000	10 000 (100 %)	9 000 (90 %)	1 000 (10 %)	-	-
Styria	1.300 000	25 000 (100 %)	6 500 (26 %)	13 750 (55 %)	-	4 750 (19 %)
Carinthia	548 000	7 940 (100 %)	1 667 (21 %)	5 240 (66 %)	-	1 033 (13 %)
Vienna	2.600 000	51 100 (100 %)	-	-	-	51 100 (100 %)
Austria	11.351 000	199 618 (100 %)	51 424 (25,8 %)	67 340 (33,7 %)	6 544 (3,3 %)	74 311 (37,2 %)

Sources: Austrian Water Recources Association, 1989
Federal Ministry of Agriculture and Forestry, 1990

Table 2: Contents of nutrients and heavy metals in sewage sludge

Parameter		n	min	-	max	mode *
H_2O	%	1928	0,1	-	99,8	92,1
Dry Mat.	%	1928	0,2	-	99,9	7,9
Org.Mat.	% d.m.	1657	1,5	-	91,3	54,0
pH		468	2,1	-	12,3	6,8
el.Conduct.	mS	449	0,3	-	27,8	2,6
N	% d.m.	1665	0,3	-	38,5	3,9
NH_4	% d.m.	1465	0,01	-	32,5	0,7
P_2O_5	% d.m.	1665	0,2	-	24,2	3,0
K_2O	% d.m.	1413	0,04	-	6,5	0,6
CaO	% d.m.	1665	0,06	-	65,4	8,1
Mg	% d.m.	1414	0,1	-	5,6	1,0
Mn	mg/kg d.m.	419	42	-	2700	298
Co	mg/kg d.m.	1141	1	-	140	10,4
Mo	mg/kg d.m.	456	1	-	650	14,6
Cu	mg/kg d.m.	2393	12	-	4310	190
Zn	mg/kg d.m.	2402	18	-	14370	1320
Pb	mg/kg d.m.	2396	5	-	19150	145
Cr	mg/kg d.m.	2403	5	-	97600	64
Ni	mg/kg d.m.	2392	2	-	1840	37
As	mg/kg d.m.	394	0,1	-	54	3,6
Hg	mg/kg d.m.	2369	0,01	-	460	1,8
Cd	mg/kg d.m.	2374	0,1	-	285	3,0

* Most frequent values calculated on weighted median and mean values of sludge analyses from Upper - and Lower Austria, Styria, Burgenland, Salzburg and Tyrol during a period of more than 10 years.

Because of the comprehensive advice, the analytical control of sewage sludge and the taken measures against pollutioners it was possible to improve the quality of sewage sludge and to decrease the heavy metal levels continuously during the past ten years (see Fig. 2). The limit values for Cd, Ni, Hg, Pb and Cu are exceeded very seldom only in case of some local emittents. Sludge from tanneries gives frequently problems with chromium, but this is only of local importance. Zinc is the element which exceeds mostly the limit value and this is due to various diffuse, unknown sources (5,6). Other metals like Co, As, Mn probably reflects the natural geochemical background.

Although the quality of sludge is quite good generally it's use in agriculture is decreasing nowadays. There are two reasons for this decrease:

a) The new founded treatment plants are mostly planed for high capacity and therefore the produced sewage sludge is often enriched with heavy metals.

b) In some federal countries new laws and regulations on soil protection and sewage sludge application on agricultural soils came into force.

These enviromental laws require a lot of preventive measures so that farmers often lose interest in using sludge. In Austria the federal countries are responsible for the laws on agriculture and soil protection. This is a real disadventage to have not just one law for the whole of Austria. But the laws of the federal countries coincide in essential points and paragraphs (7):

- sampling and analysis of sewage sludge by official institutes and authorities before agricultural use (generally analysis is performed on nutrients and heavy metals, particularly on hygienic parameters, organic contaminants and biological tests).
- soil analysis before first sludge application (heavy metals, pH and some plant available nutrients; limit values of heavy metal concentrations (Tab. 3).

- property attestation for sludge and soil from the authority.
- regulation for maximum annual rate of sludge (2,5 tons d.m./ha for field crops, 1.25 tons d.m./ha for grassland)
- bans for sludge application (as in the guideline) taking in consideration continuous soil properties like texture, organic matter content, groundwater situation etc.
- delivery note for sludge with name and address of sewage plant, customer, field number, and the exact amount of sludge delivered.

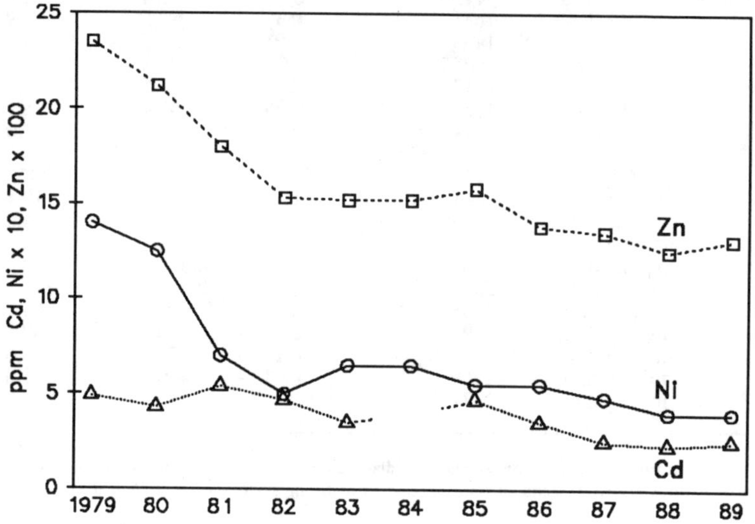

Fig. 2: Decrease of heavy metal contents in sludge from 1979 to 1989

Table 3: Limit values of heavy metal concentrations in sewage sludge and soil (7)

Metal	Sludge ppm	Soil ppm
Zinc	2000	300
Copper	500	100
Chromium	500	100
Lead	500	100
Cobalt	100	50
Nickel	100	50 - 60
Molybdenum	20	10
Cadmium	10	1 - 3
Mercury	10	1 - 2

3. FUTURE ASPECTS CONCERNING THE DISPOSAL OF SEWAGE SLUDGE

According to the latest survey from 1989 in Austria only 20 - 25 percent of sludge is utilized in agriculture today and for this quantity the environmental laws are in force. More than a third (35 percent) is used for landfills or stored in dumping areas, 40 percent for composting and incineration and the rest for land reclamation. Because of the trend nowadays it has to be accepted that the direct disposal of sludge on agricultural land will further decrease to 10 - 15 percent in the year 2000. The required measures by the authority like hygienic and analytical control of sludge and soil, property attestation, delivery note for each ton of sludge often are not in relation to the profit of sludge application. Therefore many farmers won't use sewage sludge anymore.

In Austria also storage and landfills will decrease, because no further land is available for new dumping areas due to the population density nowadays. Another important reason is that people refuse the establishment of new deposits since they are afraid of unpleasant odours and environmental effects. From the actual point of view it can be estimated that composting (together with municipal waste) and incineration (for energy production or simple volume reduction) will become the most common ways of sludge disposal in Austria in the future (see Fig. 3).

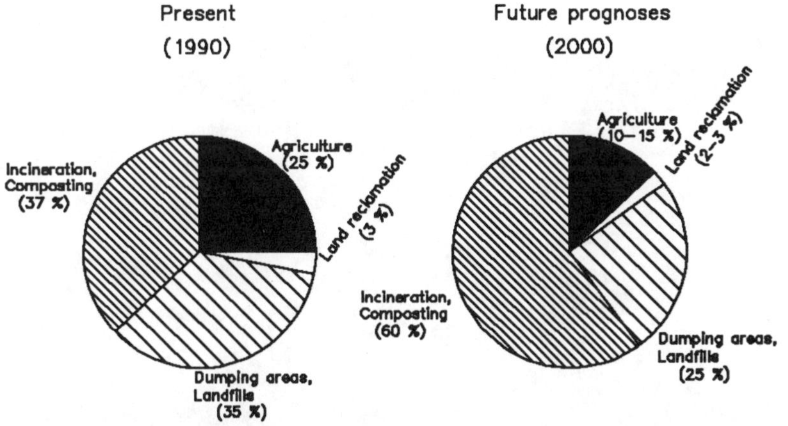

Fig. 3: Sludge disposal strategies in Austria

REFERENCES

(1) MINISTRY of AGRICULTURE and FORESTRY, and AUSTRIAN WATER RESOURCES ASSOCIATION (1990): Unpublished paper of an inquiry from 1989.

(2) AICHBERGER, K., MAYR, E., and SCHMOIGL, K. (1984): Systematical control and agricultural use of sewage sludge in Upper Austria. In Process and use of sewage sludge, 3 rd Intern. Sympos. Comm. Europ. Communities 158 ff.

(3) AICHBERGER, K., WIMMER, J., and HOFER, G. (1985): Wertgebende Inhaltsstoffe der Klärschlämme OÖ und deren Auswirkung auf den Boden. VDLUFA-Mitt. 2, 623-627.
(4) ÖSTERR. WASSERWIRTSCHAFTSVERBAND (1984): Landwirtschaftliche Verwertung von Klärschlämmen - Empfehlungen für Anlagenbetreiber. Regelblatt 17.
(5) MAYR, H.E. (1988): Klärschlammentsorgung in Oberösterreich. Öster. Wasserwirtschaft 40, 9/10, 246-255.
(6) AICHBERGER, K., and HOFER, G. (1990): Analytik von Siedlungsabfallstoffen. Jahresbericht der Bundesanstalt für Agrarbiologie 1989.
(7) LAWS and ORDERS concerning the utilization of sewage sludge on agricultural land (Vorarlberg 1987; Steiermark 1987; Niederösterreich 1989; Oberösterreich 1990).

CONSEQUENCES OF THE ORDINANCE ON SUBSTANCES ON SWISS AGRICULTURE

E. SPIESS

Swiss Federal Research Station
for Agricultural Chemistry and Hygiene of Environment
CH-3097 Liebefeld-Bern

Summary
In Switzerland the Ordinance relating to Environmentally Hazardous Substances of 1986 (1) prescribes, that fertilizer dressings must be adapted to the crop requirement and that mineral fertilizers should be applied only, if animal manure does not suffice for fertilization.
Nutrient balances for nitrogen, phosphorus and potassium were calculated to investigate how far animal manure can cover the crop requirement in Switzerland, provided that spatial distribution is optimal.
The N balance for the whole country is positive, but manure does not suffice to cover the crop requirement because of losses and the low efficiency of organic nitrogen. The phosphorus balance shows a small deficit. The average requirement of P in Switzerland amounts to 28 kg/ha. If the spatial and the quantitative distribution of animal manures were optimal, manure would cover the total potassium requirement.
Further calculations show that a surplus of 23000 t P and 54000 t K is annually imported into the nutrient cycle through feedstuffs and fertilizers. Recommendations for decreasing the nutrient surpluses are given. Farmers, advisory services and researchers must help to achieve a better distribution in time, space and quantity of the different kinds of fertilizer.

1. INTRODUCTION

In 1986 the Ordinance relating to Environmentally Hazardous Substances (1) was enacted in Switzerland. This ordinance prescribes, that fertilizer dressings must be adapted to the crop requirement and that mineral fertilizers have to be applied only, if the animal manure does not suffice or is not appropriate for fertilization.

The aim of the COST project "Management of animal manure in Switzerland" was to investigate how far animal manure can cover the crop requirement in Switzerland, provided that spatial distribution is optimal. For this reason nutrient balances for nitrogen (N), phosphorus (P) and potassium (K) were calculated for the whole country of Switzerland and for different regions on the basis of the national farm census of 1985.

2. MATERIAL AND METHODS

In this paper a nutrient balance is defined as a comparison of the nutrient content in manure with the crop requirement. The determination of the nutrient content in manure was based on model calculations (COST project "Production and composition of animal manures in Switzerland"; Table 1). The nutrient requirements of the arable crops was determined according to the official guidelines (2) (Table 2). The estimate of nutrient requirements of meadows took into account the net yields (without harvest and feeding losses) calculated on the basis of herbage consumption of ruminants.

As completion a nutrient cycle for nitrogen, phosphorus and potassium was computed for the country of Switzerland. If nutrient imports into the nutrient cycle exceed exports, nutrient losses (mainly for nitrogen) and soil enrichment (particularly for P and K) are likely to occur.

Table 1: Nutrient content in manure (in kg/year) (2, 3, 4)

Type of animal:	N	P	K
Dairy cow	106	18.5	148
Fattening pig	14.8	4.15	4.6
Mother sow	41.3	9.3	14.4
Laying hen	0.9	0.26	0.33

Table 2: Nutrient requirements of some important crops in Switzerland (in kg/ha)

Culture:	N	P	K
Wheat	100	30	100
Barley, oats and triticale	80	25	120
Silage and grain maize	110	45	220
Potatoes	100	45	250
Sugar and fodder beets	100	40	250
Rape	110	40	120
Fruit growing	110	30	150
Vine growing	50	25	120
Meadows with a net yield of			
12 t DM/ha	200	44	240
10	167	37	200
8	133	30	160
6	100	22	120
4	67	15	80

3. RESULTS AND DISCUSSION

3.1. Nutrient balances

Even though the nitrogen balance for the whole country is positive (Table 3), manure does not suffice to cover the nitrogen requirement of the plants because only about half of the total nitrogen is plant available. The nitrogen efficiency in the long term is likely to be lower than for mineral fertilizer as great losses by ammonia volatilization occur and the efficiency of organic nitrogen in animal manure is only low. Nevertheless a reduction in the nitrogen fertilizer consumption is desirable.

The most instructive balance is that of phosphorus because a fertilization strictly according to plant uptake was assumed for all crops. The efficiency of phosphorus in animal manure amounts in the long term to 100% (5, 6). Over the entire country animal manure would theoretically cover 95% of the crop requirement of P (Table 3). In the eastern and the central part of Switzerland there are great P surpluses (Figure 1) due to the high pig density in these areas (Figure 2). As cattle in Switzerland is mainly fed by roughage grown on the farm, manure surpluses are rarely caused by cattle. On the contrary pigs and chicken are fed in general by imported concentrate. Therefore on farms with a high percentage of pigs and chicken the nutrient imports with concentrates exceed the nutrient exports with meat and eggs.

The potassium content in manure exceeds the crop requirement by 17% (Table 3). If the spatial and the quantitative distribution of animal manure were optimal, the total potassium requirement could be covered. But it must be taken into account that the potassium offtake of meadows exceeds the crop requirement due to luxury consumption by the plants. For example the crop requirement of a meadow with a net yield of 10 t DM/ha amounts to

Table 3: Nitrogen, phosphorus and potassium balances (nutrient content in manure minus crop requirement) and sufficiency levels (nutrient content in manure : nutrient requirements x 100%) for the whole country of Switzerland and for different regions.

	Nutrient content in manure			Crop requirement				Balance			Sufficiency level		
	N	P	K	N	P	P	K	N	P	K	N	P	K
	t	t	t	t	t	kg/ha	t	t	t	t	%	%	%
Switzerland	153953	29469	184662	121304	30805	28.7	156612	32649	-1336	28050	126	95	117
Valley zone	79198	15505	91060	66555	17970	34.9	88796	12643	-2465	2264	118	86	102
Alpine foothill zone	24978	4779	29906	18193	4459	30.1	23030	6785	320	6876	137	107	129
Mountain zone 1	20326	3808	25308	14759	3444	26.6	18263	5567	364	7045	137	110	138
Mountain zone 2	19925	3676	25496	14445	3275	20.6	17602	5480	401	7894	137	112	144
Mountain zone 3	7590	1357	10243	5836	1316	14.9	7088	1754	41	3155	130	103	144
Mountain zone 4	1934	342	2647	1513	339	11.8	1831	421	3	816	127	101	144
Northern part	27717	5217	34340	24570	6573	31.7	32521	3147	-1356	1819	112	79	105
Western part	31995	5952	40694	31465	8402	28.1	41303	530	-2450	-609	101	70	98
Southern part	5601	1013	7421	4479	1090	15.1	5720	1122	-77	1701	125	92	129
Eastern part	30612	6047	34529	19705	4683	29.7	24703	10907	1364	9826	155	129	139
Central part	28546	5632	31910	17521	4087	29.1	21745	11025	1545	10165	162	137	146
Berne	29468	5596	35754	23551	5957	30.3	30605	5917	-361	5149	125	93	116

Figure 1: Sufficiency level of phosphorus (P content in manure : P requirement x 100%) in different regions.

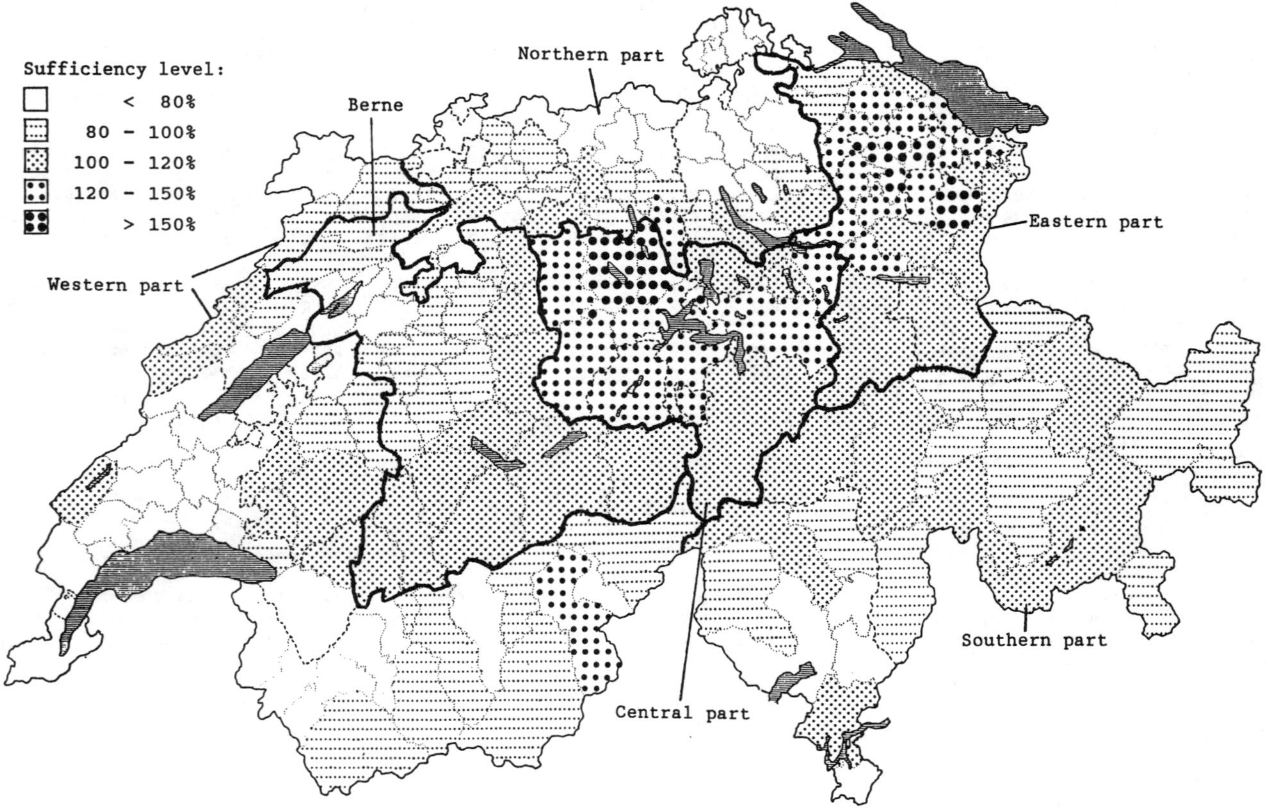

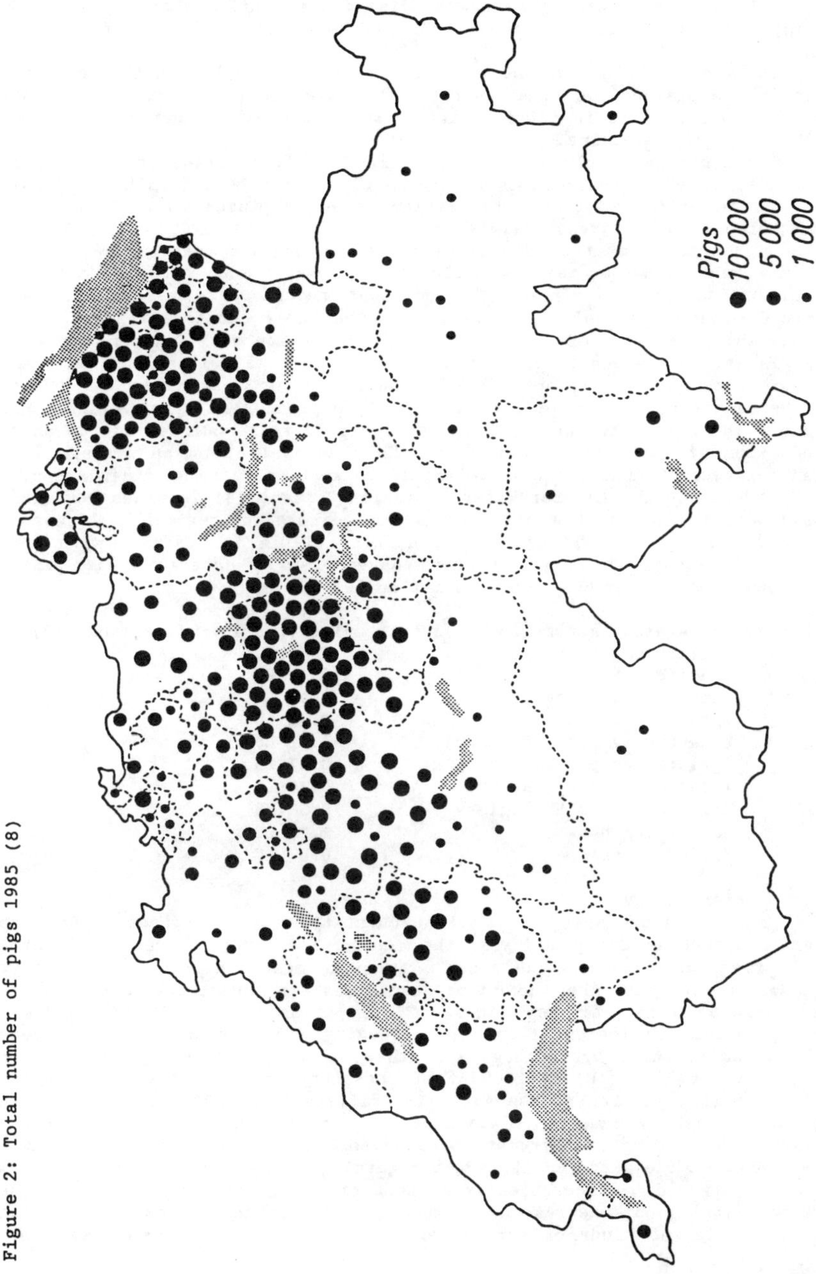

Figure 2: Total number of pigs 1985 (8)

200 kg K/ha, but according to new studies (7) the offtake averages to 300 kg K/ha. The difference is about 100 kg K/ha or 50% of the crop requirement.

It is remarkable that the sufficiency level is higher in the mountains than in the plains. Due to unfavorable climatic conditions yields of meadows and therefore their nutrient requirements are much lower in the mountains than on the plains.

Great manure surpluses in some regions of Switzerland (especially in the central and the eastern part) are mainly caused by the high pig density. For an efficient plant utilization these surpluses would have to be transfered to other areas (mainly to the western part).

On the basis of the crop requirement per hectare, the maximal number of livestock whose excrements could be applied without creating nutrient surpluses can be calculated. The nutrient requirements per hectare are quite low. The phosphorus requirement of the whole country averages to 28 kg/ha. This quantity corresponds to the P production of 1.5 dairy cows or 7 pig places. In the canton of Thurgau (eastern part of Switzerland) with the highest nutrient requirements, the P requirement of 35 kg/ha can be covered by the manure production of 2 cows or 8-9 pig places. In the mountain zones 3 and 4 the P requirement of 15 respectively 12 kg/ha corresponds to the P production of less than 1 dairy cow. This means that on an average less than 1 livestock unit per hectare should be kept in these mountain areas.

At present the Law for Water Pollution Control (9) is under revision. The new draft prescribes that the animal manure of maximal 3 livestock units is allowed to be applied per hectare (Table 4). Farmers with more than 3 livestock units per hectare have to make contracts with their neighbours or must reduce their animal number.

Table 4: Conversion factors for livestock units (LU) in Switzerland (10)

1 dairy cow	1.0 LU
1 heifer (> 2 year)	0.7 LU
1 heifer (1-2 year)	0.4 LU
1 calf (< 1 year)	0.25 LU
1 fattening pig (20-50 kg)	0.12 LU
1 fattening pig (50-80 kg)	0.17 LU
1 fattening pig (>80 kg)	0.20 LU
1 mother sow with piglets	0.38 LU
100 laying hens	1.0 LU
100 broilers	0.4 LU

3.2. Nutrient cycle

Figure 3 shows that the nutrient quantity in the roughage amounts to nearly as much as the quantity in the manure. The nutrient quantity imported by feedstuffs exceeds the exports by milk, meat and eggs considerably. In addition to this, great nutrient quantities are imported into the nutrient cycle with mineral fertilizers. The input into the soil (fertilizer, precipitation, biological N-fixation, feeding and harvest losses) exceeds the output (offtake by roughage and other feedstuffs, feeding and harvest losses, vegetable food) by 23000 t of phosphorus and 54000 t of potassium which are mainly fixed in the soil. The difference of 172000 t of nitrogen is mainly lost by ammonia volatilization, nitrate leaching and denitrification. The feeding and harvest losses also contain great nutrient quantities. Most of the nitrogen is probably getting lost.

There would be theoretically no need of any P and K mineral fertilizers in Switzerland. One reason for the surpluses is the uneven distribution of the different kinds of fertilizer (especially of animal manure). Some plots are overfertilized with slurry and farmyard manure whereas other plots are fertilized with mineral fertilizers. In addition to this only low manure quantities are applied in arable crops and the nutrient content of manure is often underestimated or not taken into account at all.

Figure 3: Nutrient cycle of Swiss agriculture 1985 (with alpine pastures; numbers in 1000t N/ 1000t P/ 1000t K per year)

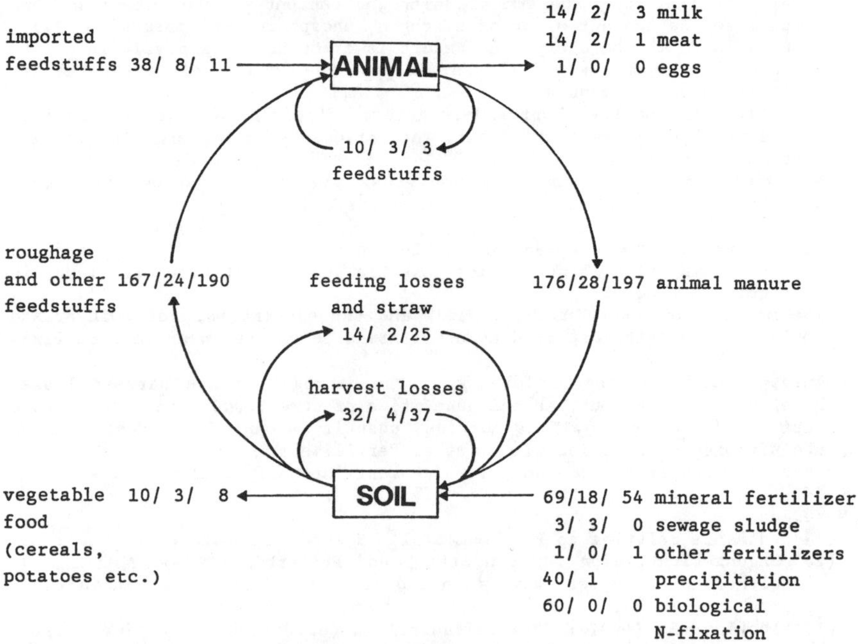

+ 395/56/314 input soil
- 223/33/260 output soil
= 172/23/ 54 balance soil

4. CONSEQUENCES

To reduce the considerable nutrient surpluses, a better distribution in time, space and quantity of the different kinds of fertilizer must be achieved by Swiss farmers:

- The consumption of mineral fertilizer must be reduced in the next years (especially for phosphorus and potassium) by a better fertilization management. The better the use of the different kinds of fertilizer, the higher is the efficiency of utilization of the nutrients (lower losses and less soil enrichment). Thus the fertilizer quantity per hectare can be reduced. Slurry and farmyard manure would suffice for the fertilization of a greater surface so that the consumption of mineral fertilizer could be reduced.

- An optimal use of animal manure is only possible, if farmers know the nutrient content of their manure. The content can be measured or estimated by different methods:
  . Official standard values which take into account the type of animal, the performance (e.g. milk production, number of piglets per mother sow), liveweight, feed ration and its mineral content.
  . Rapid tests (Quantofix, Gülle-Max and others). They are cheap and the result of analysis is quickly available. A disadvantage of the rapid tests is that only the ammonia nitrogen content can be determined and not also the content of total nitrogen, phosphorus and potassium (11).
  . Analysis in a laboratory: The result is exact but an analysis in a laboratory is expensive and the farmer gets the result only after the application of the manure.
- Investigations on the short and long term efficiency of nitrogen in animal manure have to be carried out. The efficiency of the organic nitrogen must not be neglected.
- Net yields of the meadows and not gross yields have to be taken into account for fertilization of grassland (net yield = gross yield - losses).
- More slurry has to be spread in arable crops, even though the application at the optimal time is not always possible and the nitrogen efficiency is difficult to predict.
- The manure surpluses in the central and the eastern part of Switzerland would have to be transported to other areas to ensure an optimal utilization of the manure.
- Investigations on the efficiency of the nitrogen in the harvest losses have to be carried out, as the quantities of some 30000 t in these losses must not be neglected. This nutrient quantity amounts to almost half of the nitrogen consumption with mineral fertilizers.
- Research workers and advisory service must intensify their work.

References

(1) Ordinance relating to Environmentally Hazardous Substances (1986).

(2) Düngungsrichtlinien für den Acker- und Futterbau (1987). Authors: U. Walther, J.-P. Ryser, R. Flisch and A. Siegenthaler. LBL Lindau (CH). 36pp.

(3) Flückiger E. (1987): Nährstoffanfall in den Hofdüngern - eine Modellrechnung. Landw. Jb. Schweiz <u>101</u>, 285-311.

(4) Flückiger E. (1988): Einfluss der Fütterung auf den Anfall an N, P, und K in Schweinegülle. Swiss Federal Research Station for Agricultural Chemistry and Hygiene of Environment, Liebefeld-Bern. 16pp.

(5) Amberger A., Wünsch A. und Gutser R. (1984): Kaliumwirkung verschiedener Güllen. Z. Pflanzenernährung und Bodenkunde <u>147</u>, 125-130.

(6) Steffens G. und Vetter H. (1985): Mittelfristige Nährstoffbilanz und Nährstoffausnutzung bei Gülledüngung. Kali-Briefe (Büntehof) <u>17</u>, 441-460.

(7) Kessler J. (1989: Mineralstoffgehalt im Wiesenfutter: Zusammenfassende Ergebnisse. Landwirtschaft Schweiz <u>2</u>, 523-526.

(8) Kipfer W. (1988): Swiss Agriculture. LMZ, Zollikofen (CH). 51pp.

(9) Law for Water Pollution Control (1971).

(10) Wegleitung für den Gewässerschutz in der Landwirtschaft (1979). Editor: Bundesamt für Landwirtschaft and Bundesamt für Umweltschutz. Bern. 89pp.

(11) Spiess E., Roulet M. und Schwab P. (1990): Gülle: Wie hoch ist der Nährstoffgehalt? UFA-Revue 3/90, 21-23.

REGULATIONS ON ADDITIONS OF SLUDGE-BORNE METALS TO SOIL AND THEIR ADAPTATION TO LOCAL CONDITIONS

J.H. WILLIAMS*

Soils Consultancy Services, 82 Redhouse Road,
Tettenhall, West Midlands, UK. WV6 8SP

Summary

The aim of the Council Directive (86/278/EEC) is to regulate the use of sewage sludge in agriculture in such a way as to prevent harmful effects on soil, vegetation, animals and man. Provided it is used correctly, its application as a source of N and P and as a soil ameliorant can be encouraged. If damage to crop growth or the risk of ingesting harmful amounts of sludge-borne contaminants by farm animals as a result of soil intake or excessive crop uptake is to be avoided, cumulative safe upper limits need to be observed for concentrations of all potentially toxic elements in soil. Having established these safe limiting soil concentrations, they must form the basis for calculating limit values for additions and concentrations of those elements in sludge as a guide to those persons responsible for the field application of sludge. Maximum tolerable additions and sludge concentration limits for 9 elements are presented based on the maximum tolerable increase in soil concentration when certain assumptions regarding soil density, cultivation practice and period of enrichment are made.

1. INTRODUCTION

Sludge, in particular liquid digested sludge, does have some very useful agronomic properties in terms of its nitrogen and phosphorus content, both of which are readily utilised by crops (1). De-watered sludge cake is a valuable source of organic matter, useful for improving soil physical characteristics in difficult soil situations, e.g., restoring disturbed or derelict land to agricultural use, improving humus content and water holding capacity of light-textured sandy soils and the workability of clayey soils (2)(3)(4). Its application in agriculture as a source of N, P and organic matter is to be encouraged provided it is used correctly strictly according to agronomic requirements and does not result in deterioration of soil quality, of the agricultural produce or the health of grazing livestock.

Sewage sludges often contain appreciable quantities of metalloids which are potentially injurious to crop plants and/or to human and animal health. Metals added to soil are firmly adsorbed onto organic matter and clay surfaces and, as such, are not subject to leaching. They remain in soil almost indefinitely. With few exceptions, metals are less available to plants at near neutral and alkaline pH's. Under acid soil conditions such as often occurs in permanent pasture situations, some metals are much more mobile and readily taken up by crop plants.

If industrial effluents enter the sewage treatment works, the resulting sludge may well contain a range of potentially toxic elements at such concentrations that limits its use in agriculture. The elements can be divided into two groups:

* Presented in Mr Williams' absence by J.E. Hall, UK WRc

(a) those which affect crop growth such as zinc, copper, nickel and chromium, and

(b) those which are not normally toxic to crop growth but may be harmful to animals which ingest the treated crop. The elements of greatest concern are copper, lead, molybdenum, fluorine, cadmium and mercury.

Before applying sludge to land, soils should be analysed to determine the background level and then periodically according to the frequency of application, to monitor the accumulation of the elements in soil. To avoid any harmful effects it is necessary to set and observe safe limiting concentrations for all potentially toxic elements.

2. LIMIT VALUES

Basis

The overriding controlling parameter must be the soil concentration of the element concerned since this is the only reliable measure of potential toxicity which can be checked from time to time over the long term. Calculations of additions of sludge-borne contaminants from sludge analysis and estimates of sludge quantities spread on land at intervals over a period of time can be unreliable. Calculations of sludge metal recovery by analysis of field soils are variable and often very low. Such poor recoveries are difficult to explain in practice but there are suggestions that there may be losses as organo-metalloid complexes through volatilisation into the atmosphere, by physical leaching through the soil profile or both. Having decided on the safe limiting concentrations of the elements concerned in soil, these must then form the basis for calculating limit values for additions and concentrations of these elements in sludge as a guide to those persons responsible for the field application of sludge. Knowing the increase in soil concentration which can safely be tolerated, subsequent additions of any element which may be safely applied and its concentration in sludge can be calculated.

In Annexes 1A, 1B and 1C of the Council Directive (86/278/EEC) there is no common relationship between the values for soil, sludge and the annual additions given in each of the three annexes respectively. Provided certain basic assumptions are made, there is no reason why this should not be the case. The basic premise to be made is that the maximum increase in toxic element concentration which can safely be tolerated is the difference between the typical level in an un-amended soil and the upper limiting soil concentration, as given for the 6 elements in the Directive (Annex 1A).

Typical trace element concentrations of un-amended soil free from other sources of contamination have been obtained from data published by Archer and Hodgson (5) and McGrath (6, 7).

Assessment of safe tolerable additions based on this difference will vary depending on the degree of contamination to which the soils have been subjected to in the past.

Other soil factors which need to be taken into account when calculating safe tolerable additions are field soil density and depth of cultivation or sludge incorporation. A field soil density of 1.20 and a depth of cultivation of 25 cm is adopted here. This gives a weight of air-dry soil of 3.0×10^6 kg per 25 cm hectare.

Concentration limits for metalloids in sludge are based on an enrichment period of 100 years assuming optimal agronomic dressings of

(a) 2 tonnes/hectare and (b) 4 tonnes/hectare of sludge dry matter per annum on average for arable crops. Based on the foregoing assumptions, the limit values calculated for potentially toxic elements in cultivated soils and in sludge are given in Table 1 for Cd, Cu, Ni, Pb, Zn, Hg and Cr.

Additional elements (Cr, Mo and F)

Chromium in sludge applied to agricultural land has not caused any real problems. The recommended upper soil limit and tolerable additions are based on the findings of a review study conducted on behalf of the European Commission (8). They are based on yield and uptake data.

Molybdenum, because of its interacting effect with copper, inducing hypocupraemia in grazing animals, needs to be controlled as it is readily taken up by arable crops, grass and clover. There is evidence that molybdenum present in a liquid digested sludge is readily taken up by grass and clover grown in pots, in the year of application and in the first residual year. The results suggested that sludge molybdenum is more effective than inorganic molybdate in maintaining elevated plant contents in the residual year, particularly of white clover (9). The results of the study suggest that molybdenum concentration in soil should not be raised above 2.0 mg/kg by the application of sludge. Very high levels of molybdenum were also recorded by Davis in turnip, kale, lucerne, red and white clover when these crops were grown in a soil to which 5 kg/ha Mo in sludge had been added. On a theoretical basis this addition is equivalent to an increased soil concentration of 1.5-2.0 mg/kg Mo, a level which should not be exceeded in the pasture/grazing animal situation (10). If in doubt, it is advisable to have the herbage analysed and advice sought if levels in excess of 2 mg/kg Mo in the dry matter are found. Unlike most other metals the uptake of molybdenum is enhanced by increasing soil pH. Liming, usually recommended to reduce the toxicity of sludge metals and maintain a soil pH of 6.5-7.0, increases the availability of molybdenum and hence its potential toxicity to grazing animals.

Fluorine contents of sludges are not usually high enough to cause problems in animals although isolated cases have occurred in the past. To avoid the risk of fluorosis in cattle, a limit of addition is recommended because of the risk of direct ingestion. For the same reason the content of F in sludge applied to pasture should not exceed 1000 mg/kg in the dry soils. It has been suggested by Underwood that intakes of fluorine greater than 30 mg/kg in the diet can be toxic to dairy cattle (11). In the UK there are statutory constraints on levels of fluoride in the whole feed for various classes of animals (Fertiliser & Feeding Stuffs Amendment 1976) - 30 mg/kg for cattle in milk and 50 mg/kg for other cattle. Direct uptake of fluoride via the roots is unlikely to be of any great significance in raising the fluoride content above the threshold value of 30 mg/kg F. Of much greater significance will be direct ingestion of sludge or sludge/soil residues in the shallow surface soil horizon. Assuming an average value of 5% for soil contamination of grazed herbage and an average food intake for cattle of 10 kg/day of dry matter, an upper soil limiting concentration of 500 mg/kg F ensures that the 30 mg/kg F safe threshold in the diet is not exceeded. Limit values for fluorine and molybdenum are given in Table 2 for the grazing situation together with those for copper and lead since they are the four elements which present the greatest risks of toxicity to the grazing animal.

3. DISCUSSION

The basis for the calculation of the limits on tolerable additions and on concentrations of the elements in sludge are given in the footnotes to the tables. Assuming an average field bulk density of 1.20 for arable soils and a depth of cultivation and sludge incorporation of 25 cm, an addition of 3 kg/ha of a potentially toxic element in sludge incorporated to this depth, produces a theoretical increase in soil concentration of 1 mg/kg. Thus the tolerable addition in kg/ha will be 3 times the tolerable increase in soil concentration in mg/kg. The factor will vary according to soil density and cultivation practice.

For arable crops, optimal agronomic dressings for sludge vary between 2 and 4 tonnes per hectare of dry matter. The calculated sludge concentration limits in Table 1 related to sludge dry matter dressings of 2 and 4 tonnes/ha per annum assume an enrichment period of 100 years. For Cd, Cu and Hg, the upper limiting values for sludge applied at 2 t/ha are very close to the upper limiting values given in Annex 1B of the Council Directive. The limit values for nickel and lead in the Council Directive are somewhat lower than those in Table 1 whilst those for zinc are slightly higher. In Annex 1C of the Council Directive are given the limit values for amounts of metals which may be added annually. If these amounts are related to the maximum tolerable addition (line 4 in Table 1), the calculated enrichment period varies from 22 years for Zn to 50 years for Cd, Ni and Pb with intermediate values for Cu and Hg.

To regulate the use of sludge and ensure that the accumulation of potentially toxic elements in the soil does not lead to the soil concentration limits being exceeded, from a practical standpoint it is probably more meaningful to observe sludge concentration limits related to the average quantity of sludge dry matter it is intended to use each year, than it is to observe the limit values on annual amounts of metalloid given in Annex 1C. Before sludge concentration limits related to quantities of sludge dry matter can be calculated it is obviously necessary to adopt a period of enrichment over which the sludge may be applied without exceeding the soil concentration limits. A period of 100 years has been adopted as this would provide an adequate margin for the protection of the soil environment and a continuing outlet for sludge.

In the permanent grass situation regularly grazed by cattle and sheep where liquid digested sludge may be applied to the surface at frequent intervals, the situation is quite different to that of annual cropping where there is much greater dilution of the applied sludge. In the permanent pasture situation there will be an accumulation of sludge contaminants, especially metals, in a shallow surface horizon. As such there is a very real risk of ingestion of sludge contaminated herbage, soil and sludge by the grazing animal especially if the stock graze closely. The risk is greatest in soils which are wet and poach easily. The criteria therefore need to be modified somewhat to guard against excessive ingestion by the grazing animal particularly of the elements Cu, Pb, Mo and F. Whilst there may be a very steep concentration gradient of elements from the soil surface downwards, it is not practical to sample soils to depths shallower than 5.0 cm and this depth has been adopted for the calculations made to arrive at the recommended

TABLE 1. Upper limiting values for potentially toxic elements in arable soils and sludge. (assuming a soil pH of 6.0-7.0)

			Cd	Cu	Ni	Pb	Zn	Hg	Cr
1.	Typical content of un-amended soil (mg/kg)		0.5	20	25	50	80	0.1	50
2.	Upper limiting soil concentration (mg/kg)		3.0	140	75	300	300	1.5	250
3.	Maximum tolerable increase in soil concentration (mg/kg)(2-1)		2.5	120	50	250	220	1.4	200
4.	Maximum tolerable addition to soil (kg/ha)		7.5	360	150	750	660	4.2	600
5.	Sludge concentration limit (mg/kg in dry matter)	(a)	40	1800	750	3750	3300	20	3000
		(b)	20	900	375	1875	1650	10	1500

Notes

4. The addition is based on the maximum tolerable increase in (3) assuming a field soil density of 1.2 and a depth of cultivation and sludge incorporation of 25 cm (3000 tonnes of air-dry soil per 25 cm.ha).

5. Based on an enrichment period of 100 years and optimal agronomic dressings of (a) 2 tonnes/hectare and (b) 4 tonnes/hectare of sludge dry matter per annum, on average or pro rata

TABLE 2. Upper limiting value for Cu, Pb, Mo and F in permanent pasture soils. (assuming a soil pH of 6.0 or greater)

		Cu	Pb	Mo	F
1. Typical content of un-amended soil (mg/kg)		20	50	1.0	200
2. Upper limiting soil concentration (mg/kg) (0-5.0 cm horizon)		140	300	2.0	500
3. Maximum tolerable increase in surface soil (0-5.0 cm) concentration (mg/kg)		120	250	1.0	300
4. Maximum tolerable addition to soil (kg/ha)		60	125	0.5	150
5. Sludge concentration limit (mg/kg in dry matter)	(a)	600	1250	5.0	1500
	(b)	300	625	2.5	750
	(c)	150	310	1.3	375

Notes

4. The addition is based on the maximum tolerable increase in (3) assuming a surface soil density of 1.0 and a depth of penetration of sludge solids of 5.0 cm (500 tonnes of air-dry soil per 5 cm.ha).

5. Based on an enrichment period of 100 years and sludge dressings of:
(a) 1 tonne/ha/annum of dry matter
(b) 2 " " " and
(c) 4 " " " on average
Recommended that sludges containing more than 1000 mg/kg Pb or F should not be surface-applied because of the risk of poisoning for more sensitive livestock.

limits given in Table 2 for pasture soils. Because of the high organic matter content of such soils at the surface, a soil density of 1.0 has been adopted giving a weight of air-dry soil of 500 tonnes per 5.0 cm ha. Thus an addition of 1.0 kg/ha of a potentially toxic element in sludge distributed through this depth produces a theoretical increase in soil concentration of 2.0 mg/kg. The sludge concentration limits have again been calculated for an enrichment period of 100 years for average sludge dry matter dressings of 1, 2 and 4 tonnes per hectare per year. Young stock are at greater risk from fluorosis or lead toxicity than adult animals and where such stock are present, it is recommended that sludges containing more than 1000 mg/kg Pb or F in the dry matter should not be surface-applied. Having established the soil concentration limit which should not be exceeded for any one of the potentially toxic elements, there is no reason why the same basic criteria cannot be used to calculate the maximum tolerable addition and sludge concentration limit for each element in turn. Slight changes my need to be made depending on local tillage practices, land use and soil type. Whilst it is necessary to regulate the application of sludge to land by controlling the quantities applied and ensuring that the sludge concentration limits are not exceeded, it is essential to keep a check on soil concentrations at regular intervals to ensure that limit values are not exceeded. If the concentration of one or more of the potentially toxic elements in soil exceeds the limit value, the use of sludge should be stopped or, if in the grassland situation it may be possible to dilute the residues and reduce the concentrations to safe levels by ploughing, mixing and reseeding.

REFERENCES

(1) Commission of the European Communities (1982). "Nitrogen and Phosphorus Value of Sewage Sludges", Document Nr. SL/82/82 - XII/ENV/35/82.

(2) Commission of the European Communities (1981). "The influence of Sewage Sludge Application on Physical and Biological Properties of Soils". Proceedings of an EEC Seminar held in Munich, pp 1-133 (Reidel Publishing Company).

(3) HALL, J.E. and WILLIAMS, J.H. (1983). The use of sewage sludge on arable and grassland. Proceedings of an EEC Seminar held in Uppsala, pp 22-35 (Reidel Publishing Company).

(4) HALL, J.E. and VIGERUST, E. (1983). The use of sewage sludge in restoring disturbed and derelict land to agricultural use. Proceedings of an EEC Seminar held in Uppsala, pp 91-102 (Reidel Publishing Company).

(5) ARCHER, F.C. and HODGSON, I.H. (1987). Total and extractable trace element contents of soils in England and Wales. Journal of Soil Science, 38, 3, 421-432.

(6) McGRATH, S.P. (1987). The range of metal concentrations in topsoils of England and Wales in relation to soil protection guidelines. In: Proceedings of 20th Conference on "Trace Substances in Environmental Health" held at University of Missouri - Columbia in 1986.

(7) McGRATH, S.P. (1987). Computerised quality control, statistics and regional mapping of the concentrations of trace and major elements in the soils of England and Wales. Soil Use and Management, 3,1, 31-38.

(8) WILLIAMS, J.H. (1988). Chromium in sewage sludge applied to agricultural land. Report prepared for the Commission of the European Communities. (Copyright ECSC-EEC-EAEC, Brussels - Luxembourg, 1988).

(9) WILLIAMS, J.H. and GOGNA, J.C. (1983). Molybdenum uptake - residual effect of sewage sludge application. In: Proceedings of Int. Conf. on Heavy Metals in the Environment, Vol. I, pp 483-486 (CEP Consultants Ltd).

(10) DAVIS, R.D. (1981). Uptake of molybdenum and copper by forage crops growing on sludge-treated soils and its implications for the health of grazing animals. In: Proceedings of Int. Conf. on Heavy Metals in the Environment, pp 194-197 (CEP Consultants Ltd).

(11) UNDERWOOD, E.J. (1971). In: Trace Elements in Human and Animal Nutrition. (London, Academic Press).

EXPERIENCE WITH A COMPUTER SUPPORTED ORGANISATION MODEL FOR AGRICULTURAL USE OF SEWAGE SLUDGE

H. WITTE, M. KEDING, University of Siegen,
Federal Republic of Germany

Summary
The paper describes in part A the actual situation of sludge application in the FRG from legal and practical aspects. In part B possibilities for an improvement of the sludge application situation and experiences with an organisational model are described.

A. General Evaluation of the Situation in Germany

1. Introduction

Having been asked to report on a more practical side of sewage sludge utilisation it seems necessary to start with a general information about the situation on the legal side in Germany which is in the process of changing. Necessary also seems to be a survey about the development of the amounts of sludge used in agriculture in the past up to the present. It must be regarded that regulations in comparison with other industrial countries seem to be quite stringent already and will be even more stringent in the near future. Following the rates of sludge, which go into agriculture, one gets the impression that too stringent regulations may kill whole systems of waste management - at least, if the ones who are meant, are not prepared for them. This will be subject of part A of this report. That this need not be true in all cases will be shown in the more practical part B.

2. Development of the actual situation of agricultural sludge reuse

The history of agricultural sludge reuse is connected with the progress of mechanical and biological wastewater treatment. In the FRG until 1960 sludge mostly was digested, dewatered on drying beds and wherever possible agriculturally reused. There was a balance between total sludge amount and the need for cheap fertilizers in agriculture. After 1960 this relation changed due to the increasing amount of sludge from wastewater treatment plants and due to positive developments of wastewater treatment as well as a saturation of the fertilizers market. Thus it was necessary to make way for other technical methods of discharge.

An analysis of the situation of sludge discharge for 1975 showed that the amount of agricultural reuse of sludge was about 30 to 43 % of the total sludge amount of about 1.7 million tonnes of dry substance per annum. Due to improved analysing methods and a rising sensitiveness of environmental problems amongst the population the discussion on dangerous substances in sludge started in the seventies. At the beginning the discussion focussed on the contents of heavy metals in sludge. The discussion caused a considerable decrease in sludge reuse rates. As a result of the discussion in 1981 the Government of the FRG published the "Klärschlammverordnung". Despite of the fact that this measure was meant to promote trust in consolidating the situation on resuse the initial effect caused by increased administration was, that not only farmers but also treatment plant operators became more critical of agricultural sludge application and preferred the relatively simple method of landfilling. This development is proven by a statistic of the Abwassertechnische Vereinigung (ATV = The German Association for Water Pollution Control) of 1984 and 1985 (Fig. 1).

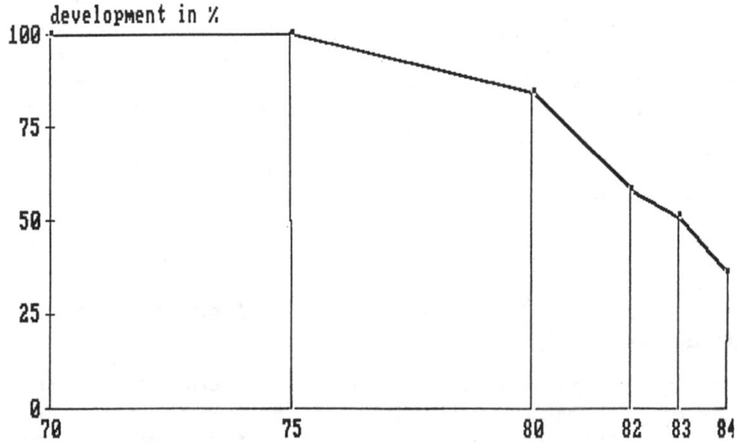

Fig. 1 - Development of agricultural use of sewage sludge in 104 communities in the FRG with approx. 22 million population and population equivalents since 1975 (1975 = 100%)

In total the statistic showed a decrease of agricultural sludge application to 15 - 22 % of the total sludge amount of approx. 2,1 million tonnes of dry substance (1985).

Over the next years sludge application stabilised on this level. Due to research activities to improve the organisational structure [1] a positve trend could be seen. In 1988 another considerable disturbance was caused by the beginning discussions on dioxins and a recommendation of the Federal Minister of the Environment not to use sludge on pastures because of its PCB and dioxin contents. This discussion is not yet finished but in the meantime a general consensus for the continuation of agricultural application under certain conditions can be noticed.

An overview of the actual discharge situation of sludge in the FRG in comparison to other European countries is shown in Fig. 2 [3]

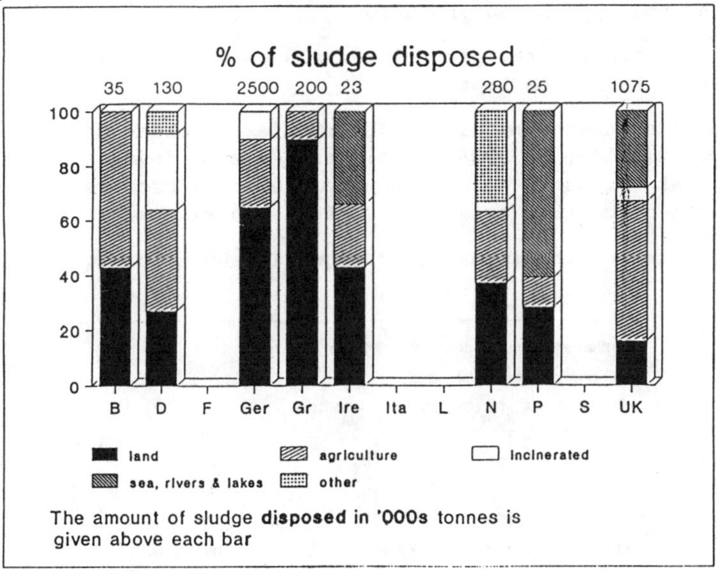

Fig. 2 - Overview of the actual discharge situation of sludge in the FRG in comparison to other European countries

3. Legal Situation

Based on the "Abfallgesetz" (Federal Waste Law) the Federal Government published the "Klärschlammverordnung" (Sewage Sludge Regulation) in 1982. The "Klärschlammverordnung" defines the legal requirements on agricultural sludge application in detail. Essential statements are as in other European countries similarly:

- sewage sludge may be agriculturally used only, if intervals of 6 months sludge samples have been analysed for certain defined heavy metals and nutrients
- sludge may be agriculturally used only if the soils of the land in concern have been examined beforehand for their contents of heavy metals
- decontamination of sludge is required before using it on pasture and fodder crop land
- the contents of heavy metals of sludge and soil must not exceed certain values
- over a period of 3 years more than 5 tonnes of dry substances per ha may be used
- a record for the path of the sludge amounts in application has to be kept

The drafted amendment of the "Klärschlammverordnung" considers the EC Guideline of 1986 as well as new experiences regarding the uptake of heavy metals and apprehensions because of the contents of certain organic pollutants in sludge.

An overview of the important changes is shown in Fig. 3

Fig. 3

- Implementation for plants < 5000 p.e.
- soil analyses every 10 years
- new limits for soil (mg/kg ds):old limits

	light soil	other	
Pb	100	100	100
Cd	1	2	3
Cr	100	100	100
Cu	60	100	100
Ni	30	50	50
Hg	1,0	1,5	2
Zn	150	300	300

- new limits for sludge (mg/kg ds): old limits

	light soil	other	
Pb	1000	1200	1200
Cd	6	10	20
Cr	1200	1200	1200
Cu	800	1200	1200
Ni	200	200	200
Hg	6	10	25
Zn	2200	3000	3000

- Further limits for PCB, Dioxin and AOX in sludge
- Reuse on greenland to be discussed

Fig 3 - Old and new legal requirements in FRG for agricultural sludge reuse

4. Critical remarks on former discussions and proposals

In order to end existing disturbances caused by the publically carried out discussion now lasting 2 years and the renunciation of decisions for investments the amendment of the Klärschlammverordnung should be finalised as soon as possible. Some comments on critical aspects, which may be interesting from a European aspect, shall be given here.

3.1 Polychlorinated biphenyls

The burden on the environment due to PCB is decreasing due to the limitation of production and use. This tendency is also scientifically shown in the contents of PCB in sludge which are today already below possible levels of concern regarding accumulation in soil and transfer to crops [4], [6]. A fixing of regulations for PCB values for application of sludge on arable land are not necessary in the authors' opinion .

On pasture land no hazards caused by agricultural use could be proved. There are certain apprehensions of the Federal Government that a direct uptake of plants covered by sludge particles might lead to an accumulation in milk fat. Therefore eventually a regulation of 0.2 mg per kg dry substance in sludge per singular substance PCB could be ascertained.

3.2 Dioxin

The Federal Minister of the Environment in 1988 recommended not to use sludge on pasture land. This recommendation was based on analyses which showed that the average sewage sludge contents was about 200 ng TE per kg/DS. In the meantime it was shown through more representative investigations that the actual average values are about 25 % of former assumptions (50 ng TE/kg/DS). From this one cannot derive any danger from accumulation in soil or uptake by plants.

However, locally higher sludge concentrations did occur. To avoid additional accumulation in soil a regulation of 200 ng TE per kg/DS could be fixed. Reusing sludge on pastures a standard of 100 ng TE per kg/DS could be feasible.

3.3 Further requirements

As control parameter including several singular substances the parameter AOX (Absorbable Organic Halogen) and EOX are discussed a standard of 500 mg AOX per kg/DS in sludge is favoured.

As is stated in the EC Guideline for Sludge Application as well as in the amendment of the Klärschlammverordnung the controlled use of fertilizers and, in connection with this, an extensive agricultural consultation for farmers is of special interest.

B. Experiences with a computer supported organisation model

During the second big break-down of sludge application, within the ATV offices (The German Association for Water Pollution Control) in the early eighties the idea was born to find out reasons for the decrease and possible measures for a better utilisation. In an enquiry of operators of sewage plants several replies after setting the regulation gave some first perceptions of that work [5]. Sludge could not be brought into agriculture because

1. limits of heavy metal concentrations in sludge had been exceeded
2. limits of heavy metal concentrations in soil had been exceeded
3. the daily volume of sludge was too great compared with the area necessary according to the new regulations
4. sewage sludge had to be in competition with liquid manure
5. the new regulations seemed too complicated to handle
6. it seemed more comfortable and safer to the increasing powers of public prosecutors to use landfills
7. farmers were influenced by negative press and public media
8. farmers were willing to take sludge, but were not able to judge the risks
9. farmers were asking for liability guarantees, which could not be given by sewage plant operators
10. there was a general fear that limiting standards could be lowered after some time so that either measurements taken by the operators could be useless or that the land of farmers could devalue.

Parallel to these results it was noticed that in certain regions of Germany sludge utilisation in agriculture was going on as ever before, where others reacted with a total collapse on the new regulations. The regions with constant success could be found as regional associations of communities mainly in rural parts of the country with some tradition and experience in this "technology". So in this area the key to further procedures had to be found.

It soon was clear that the modern sludge utilisation was no longer a "low" technology but needed well trained "managers" with experience in

- marketing
- operating sewage plants and their sewers
- growing plants and raising cattle
- solving logistic problems

to fight the ten points mentioned above. Such specialists had to be paid well and had to be supplied with offices, cars, hard- and software for data processing and controlling. No small community with one or two plants could economically afford such an institution. Consequently the next step of actions was to find a region where several communities joined together to pay for an organisational centre. This centre started its work with financial support of the German Ministry of Research (BMFT). The enterprise at first was established by hiring a civil and an agricultural engineer, who could fall back on a secretariat and a PC as well as other help necessary for running a business. The work started in 1983 with an analysis of

- annual amounts and qualities of sewage sludge in the region concerned. The first step was to measure the quality of sludge for its contents of heavy metals according to regulations
- the size of arable land and pastures available within the area joined and the soil qualities according to limiting standards

A general result of this analysis was that only about 10 % of the regions' sludges went into agriculture, although, according to the sludge quality it could have been much more.

In those parts of the area, where the sludge quality proved adequate, the next step could be taken. The agricultural engineer contacted the farmers to offer his goods. So in 1984 first negotiations were successful and controlled land application could start in 1985. The main work of the agricultural engineer was:

- to inform the farmer about pollutants and nutrients in the sludge he was to receive. In many cases it was necessary to convince him, that regulations for pollutants were set so low, that no disadvantages had to be feared
- to discuss farming plans of farmers, sequences of crops they wanted to grow
- analysing nutrients, mainly phosphorus in soils to be taken into application systems and calculating fertilizers necessary and delivered by sludge application in a fertilizing balance
- discussing other problems which might occur with his customers
- inform the sewage plant operator about amounts required, date and locality of delivery
- organise picking or delivery of sludges
- recording and collecting dates in the information centre
- reporting on results to farmers, plant operators and authorities according to the regulations

In the meantime the civil engineer was engaged to improve the quality of sludges were it was insufficient or even below regulation. This meant intense investigations into the sewage system. Large or small industrial plants with heavy metals emissions had to be found and emissions had to be stopped. Furthermore the necessity arose to store sludges when they could not be brought on to fields. So, plans and performance of sludge thickness, dewatering machinery and other consulting problems e.g. with the pretreatment of industrial sewage or others had to be initiated.

In the meantime a sewage sludge fund had been developed so that farmers who want more security can enter it by contract. Payment is made by the communities.

The result of this work can be taken from the following figure. It shows the increase of sludge application (Fig 4.) with the area under contract. In the meantime the system has been privatised and continues its work with great efficiency. This figure must be seen in contrast to Fig. 1 which shows the general tendencies in the FRG at that time.

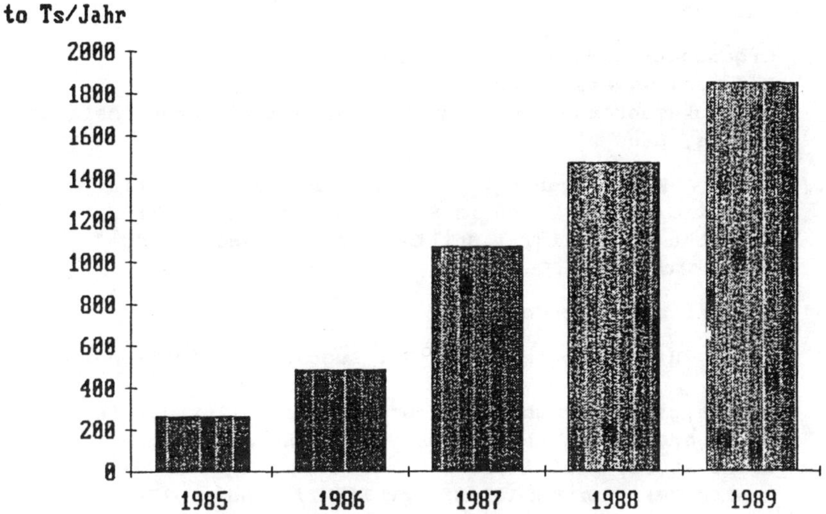

Fig. 4 - Development of agricultural use of sewage sludge in tonnes of sludge-DS per annum for operators of sewage plants controlled after by AGROTEC GmbH

The team is confident in the further development. At the moment the introduction of new limiting regulations, especially for organic micropollutants is being prepared. It can be seen that some sludges will no longer pass the examination. So, either improvements have to be made or sludges from the neighbourhood have to be "imported" to serve farmers who have already got accustomed to with a low cost fertilizer.

In conclusion it can be expected that with more stringent regulations the system of sludge utilisation will be more difficult to handle but it gives more security. This seems to be necessary because landfill of organic substances will be stopped sooner or later in Germany, so that the only alternative will be thermal treatment which, as everybody knows, is not less difficult to handle and, on the whole not less risky and creates higher costs.

This is not the only successful model in this area. One hopes that their number will increase, so that the system of sludge utilisation under the sceptical eyes of the the public can gain greater importance again.

Literature:

[1] Organisationsmodelle zur landwirtschaftlichen Klärschlammverwertung
ATV Dokumentation und Schriftenreihe aus Wissenschaft und Praxis, Band 21

[2] Keding, M., Langenohl, T., Witte, H.; St. Augustin:
PCB, Dioxin and Furan in Sewage Sludge and Their Effects with the Agricultural Utilisation of Sewage Sludge
KORRESPONDENZ ABWASSER 13/89, p. 134 ff.

[3] WSA Workshop, London 1990

[4] Witte, H., Langenohl, T.;, St. Augustin; Offenbächer, G.; Bonn
Investigation of the Entry of Organic Pollutants into Soils and Plants through the Use of Sewage Sludge in Agriculture, Part A and B
KORRESPONDENZ ABWASSER 131/88, p. 118 ff. and p. 126 ff.

[5] Stand der Klärschlammverwertung in der Bundesrepublik Deutschland
KORRESPONDENZ ABWASSER 11/84, S. 912 ff

[6] Witte, H.; St. Augustin:
Investigations of the Entry of Selected Organic Pollutants into Soils and Plants by the Use of Sewage Sludge in Agriculture in Proceedings "Organic Contaminants in Wastewater, Sludge and Sediment ...."
Edited by D. Quaghebeur, I. Temmerman and G. Angletti, Elsevier, London 1989

CONTROLS ON MANURE AND SLUDGE APPLICATIONS TO LIMIT NITRATE LEACHING DOES THE EVIDENCE JUSTIFY THE RESTRICTIONS WHICH ARE BEING PROPOSED?

R J UNWIN[(1)], M A SHEPHERD[(2)] AND K A SMITH[(3)]

Agricultural Development and Advisory Service
Ministry of Agriculture, Fisheries and Food
Bristol, (1) Gleadthorpe Experimental Husbandry Farm (2)
Wolverhampton (3) United Kingdom

Summary

Current and proposed legislation to restrict nitrate loss from agricultural land in the European Community is briefly reviewed. The potential effects of complying with restrictions on organic manure applications upon farmers and disposers of sewage sludge in the United Kingdom are discussed. These restrictions could be severe for a significant number of operations. Any regulation should take into account all the relevant evidence. The development of techniques for the direct measurement of nitrate leaching in the field now enables this evidence to be updated and clarified. Preliminary results from one such study are presented which highlight the differences between poultry manure and cattle farm yard manure and their time of application to arable land in autumn and winter.

1. INTRODUCTION

"Large areas of farm land could be pulled out of arable production as the Government introduce tight new controls to reduce nitrate leaching into drinking water" (1).

Statements such as these in the farming press in the United Kingdom have brought home to the agricultural industry the potential seriousness of the proposed European Community restrictions on agricultural activities to reduce the leaching of nitrate from agricultural land. Although challenged in official European circles this scenario of major change was supported following an enquiry by the Upper House of the British Parliament (2). This report also acknowledged the serious consequences to many livestock producers of proposed limits both in terms of stocking rates and possible limitations on time of manure application and other cultural practices. It is these aspects relating to livestock manures and by implication to sewage sludge disposal which form the basis of this paper. The European legislative framework is outlined and the steps taken to date in the United Kingdom are described. The practical problems faced by the agricultural and water industries are highlighted and relevant research work reviewed. Preliminary results from a new research project are presented.

2. CURRENT AND PROPOSED LEGISLATION

European Community

It is now 10 years since the European Community Drinking Water Directive (3) was published. Compliance with the maximum concentration of 50 mg/l NO_3 for all consumers at all times continues to provide problems for water undertakers in a number of member states. Nitrate concentrations in ground waters are predicted to continue to rise despite the uncertainties attached to our understanding of movement and denitrification within the unsaturated and saturated zones (4). In 1988 the Commission, in an attempt to reduce the problem at source, published their proposals to protect fresh coastal and marine waters against diffuse nitrate pollution (5). Although subject to amendment during negotiations the basic principles for control proposed a limit on manure to land equivalent to 170 kg/ha N or a stocking rate of approximately 2 LU/ha to be combined with additional timing and cultural rules as felt necessary by member states. Such an arbitrary limit based on livestock numbers was not however felt a prerequisite for environmentally safe slurry use by a joint working group of FAO and EC/COST 681 (6).

Member States of the European Community

Some member states such as Denmark and the Netherlands already have legislation in place to limit application of manures to land and thereby nitrate leaching as do various States of the Federal Republic of Germany. These and other controls on manure and sewage sludge applications were reviewed for the Commission up to 1987 (7).

In Denmark the primary concern is to ensure that runoff of all polluting material is avoided and to this extent all farm manures, solid or liquid, are subject to caveats on soil and weather conditions during and after spreading (8). To limit the leaching of nitrate, liquid manures can not be applied to bare ground between harvest and 1 November. All farms are subject to a stocking rate limitation of 2.3 LU/ha. In contrast the Netherlands has introduced manure limits based on the phosphorus content, although currently the limit varies for different crops it is planned to reduce to a maximum value of 100 kg/ha P_2O_5 for all situations. All manures are subject to the same limitations on time of spreading but differences are drawn between soil type and land use. On grassland spreading is prohibited from 1 October - 15 November and also 1 January - 15 February if the land is covered in snow. There are no restrictions on arable land on clay soils but on arable sandy soils applications may not be made between harvest and 1 November if there is no green cover or harvest and 1 October if there is green cover. In neither country do farmers receive direct compensation in respect of these restrictions. However in the Netherlands where more restrictive measures are required in localised water protection areas compensation is payable by the water undertaker. Also in certain states of the Federal Republic of Germany compensation is payable to farmers who follow certain rules of land management.

At first sight these restrictions not only contain major differences between countries but together or individually they suggest inconsistencies with the available scientific evidence on nitrate leaching. Discussions between the senior author and farmers representatives in both countries suggest that there has been a degree of practical compromise in reaching these restrictions. Solid manures are generally agreed to present less of a leaching risk due to a lower proportion of ammoniacal or ureic nitrogen content which results in a reduced availability of nitrate available for leaching. It has been suggested that as only 10% of Dutch manure is in a solid form it was of limited practical significance to treat it differently in the legislation. This avoided the problems of defining what is slurry and what is solid manure. This latter distinction is acknowledged to be a very real problem but such a view does nothing to encourage the development of manure handling systems which are potentially more nitrate friendly and have other benefits such as to improve animal welfare. Solid based systems might however make it more difficult to control emissions of ammonia should this become a high priority in future.

United Kingdom

The 1974 Control of Pollution Act introduced the concept of Good Agricultural Practice (GAP) (9) to reduce the risk of water pollution from farming activities. This Code of GAP is currently being updated. Following the 1989 Water Act a farmer is no longer able to plead that he had followed GAP as a defence against prosecution . In fact this defence provision was never used in court under the former Act. Neither were farmers prosecuted for having caused nitrate pollution.

The situation with regard to nitrate and water supplies in England was reviewed by a Working Party who reported in 1986 (10). Amongst their recommendations was a need to study the practicality of water protection policies and the costs involved. This led to a detailed study of a sandland catchment in Staffordshire and an appraisal of the various land use and water treatment options necessary to comply with the limit of nitrate in drinking water (11). The results showed that there could a wide range of costs involved with solving the problem according to local circumstances. By inference it was clear that every catchment is a unique case and justifies scrutiny to decide the least cost option to the community on meeting the nitrate limit. Solving the problem by agricultural restrictions alone could be economically possible according to the degree of severity of the required restrictions. However if it was necessary to convert arable land to low intensity grass production then there would be major implications for farmers in a given area. This Study was followed by a further exercise on another 10 groundwater catchments which confirmed the above points (12).

As a result the Ministry of Agriculture has launched in 1990 a Pilot Nitrate Sensitive Area Scheme in 10 areas of England designated by Statutory Instrument under the 1989 Water Act (13). This document details the compensation which will be paid to farmers who join this voluntary scheme to reduce nitrate leaching. A scale of compensation has been set for each area. For introducing a range of measures on all land within the designated area farmers will receive payments of between £55-95/ha per year for an initial 5 year period. This may be compared with payments of 310 DM/ha in Baden Wurtenburg designed to compensate farmers for an expected 10-15% loss in yield following the adoption of nitrate friendly practices. In addition the UK Scheme allows for payments of up to £380/ha for taking arable land out of production and converting it to low intensity grassland. There are specific restrictions relating to organic manures:-

1. Organic manure should be limited to 175 kg/ha total N per year.
2. No slurry, poultry manure or liquid sewage sludge to be applied between 31 August and 1 November to grass fields and between 30 June and 1 November to fields not in grass.
3. Storage of slurry and liquid sewage sludge must not be a point source of nitrate pollution.
4. An agreed manure plan must be prepared.
5. Application of organic manures to the NSA land on the farm may not be increased by more than 25% of current application unless the manure is produced within the NSA.

6. In most cases no manure can be applied to the low intensity grassland option.

For intensive pig/poultry producers payments may be made towards extra storage requirements or a transport subsidy may be made where it is necessary to move a proportion of their manure output from the area designated under the Scheme. The effects of these and all other measures will be monitored to determine their overall effect upon nitrate leaching. This Scheme has been devised to investigate the true costs and practicalities involved with meeting the Drinking Water Directive. It will also give valuable experience for the introduction of whatever overall measures are agreed as part of the proposed Nitrate Directive (5).

At the time of writing negotiations are continuing in the UK to finalise new regulations which will include controls over the collection and storage of manures and slurries. These will relate to the standard of construction, siting, control of liquid run-off and required storage periods.

3. THE PRACTICAL PROBLEMS

The quantities of nitrogen applied to agricultural land in the United Kingdom by way of livestock manures and sewage sludge have not changed substantially in recent years. In 1978 they were estimated as 1150 kt and 26 kt respectively (14). It has previously been estimated by the current authors that the quantity of nitrogen handled in livestock manures is approximately 75% of the total. Given that probably less than 1% of agricultural land receives sewage sludge in any one year whilst only approximately 31% of land receives livestock manures (15) the actual rate of nitrogen application from manures and sludge is therefore similar. Thus where sludge is applied to land in Nitrate Sensitive Areas it is likely that similar precautions will need to invoked as are brought to bear on livestock manures.

In the United Kingdom it is mainly the arable areas of eastern England where nitrate in surface waters may be above the 50 mg/l limit although ground waters over a wide area of the country are a cause for concern. Although predominantly an arable area, a significant proportion of both the national pig herd and poultry flock are concentrated in large intensive units in eastern regions. Although individual holdings exceed the suggested stocking rate norms, transport of a relatively short distance could ensure that the limitations were overcome. However it must be acknowledged that to date the rate of application on many farms has been considerably in excess of levels recommended as GAP. As the arable crops are predominately autumn sown winter cereals and oil seed rape, the most convenient spreading time has been to stubbles after harvest on many farms. This is true not only for livestock manures but also for the disposal of sewage sludge as this is the time when farmers are most likely to accept sludge onto their land.

To meet this relatively narrow time window storage facilities in many cases have had to be available to hold most of the manure for much of the year. Therefore on many farms restrictions on timing of applications need not necessitate large extra storage costs. However there will be a number of producers who have been able to spread manures for a greater part of the year, either on relatively small areas of grassland, on neighbouring farms or on areas of spring sown or spring planted crops. The spreading of animal slurries and poultry manure to growing crops in the spring has been practised in recent years but only by a few farmers. The reasons for this low uptake include lack of suitable machinery, particularly to coincide with the width of tramlines being used for other operations, the adverse affects on growth of uneven spreading and on heavier land the problems of access and damage to soils.

The options available to a farmer with pig slurry are less than poultry producers whose manure has a value which makes it easier to export to land elsewhere. Storage is also less of a problem with the more solid manure. Even the intensive dairy farmer may find himself exceeding the introduced limits. Although his grassland gives him more available time to apply manures, dealing safely with the different fractions at acceptable cost can pose problems.

Although they are often overstated, on a large number of holdings they are real logistical problems which present severe difficulties if they are to be overcome satisfactorily. In the main areas of concern the vast majority of the manure to be spread will be as slurry or poultry manures with their relatively high proportions of ammoniacal or ureic nitrogen. However with increasing concern about animal welfare issues, and the likely end to straw burning as a disposal option in 1992, more farmers are looking to the reintroduction of straw bedding systems. This desirable change could be complicated if autumn spreading of all manures were banned.

Similar comments apply to the disposal of sewage sludge. Although it is much easier to store high dry matter sludge cake this needs to be cultivated into the soil and therefore autumn application is a required option. Untreated liquid sludges have to be injected or cultivated into the ground as soon as practical after application (16). This together with the need to avoid odour problems from certain treated sludges and the problems of soil trafficability in winter and spring also restrict the times of application.

In can be seen that even if costs were not a limiting factor there are practical problems which would preclude farmers or sludge operators being able to comply with a complete ban on applications to land during late summer and autumn. In the extreme situation the alternative for farmers could mean an end to their business. The potential solution of industrial scale treatment as developed in the Netherlands is unlikely to be possible because of an insufficient concentration of stock in a given area. Sludge producers would be faced with finding alternative outlets all of which would themselves have environmental, practical and cost constraints.

4. NITRATE LEACHING FROM MANURES AND SLUDGES

This subject has been a regular topic at COST 681 (6) (17) (18) and other meetings (19) (20) (21) (22) during the 1980's. Space does not allow a detailed review of all the individual contributions to these meetings or of the many scientific papers written on the subject during the period under consideration. However for much of the time the techniques available for determining nitrate leaching have been restricted. Lysimeters offer a direct measure of nitrate loss but cannot be guaranteed to reproduce field conditions, certainly not over a wide range of soils types. Mineral nitrogen measurements to depth in the soil profile at best give only an indication of leaching risk. The modelling of nitrogen transformations in soil after the application of organic manures and subsequent nitrate leaching has not yet progressed as far as our understanding of processes in unamended soils. In more recent years the use of porous ceramic cups to extract soil profile water have been used together with hydrological balances to calculate nitrate flux from the profile. Currently considerable attention is being paid in the United Kingdom to establish a satisfactory methodology for this technique (Lord & Sheppard Private Communication). Now that reliable field technique appears to be available, at least for some soil types, it is likely that earlier work will need to be repeated to establish actual rather than inferred measurements of leaching under field conditions. Nothing less is likely to be acceptable on which to base policies which could have series consequences for individual farmers, and the agricultural and water treatment industries in general.

Also of concern is the potential for increased nitrate loss following repeated application of manure to the same area of ground in successive years. A livestock unit is likely to be present on a given site for at least 20 years; yet few trials concern themselves with more than one or two years of study. Monitoring below operational farm units is one of the few measures we have of such practises and these commonly show high accumulations of nitrate in soil and in subsoil water below livestock farms (eg 23).

It is well established that manures applied to arable land in the late summer and autumn have a much reduced fertiliser efficiency than applications made in the spring. This reduced efficiency has been shown by lysimeter studies to coincide with an increased nitrate leaching loss. Crop cover over winter generally reduces leaching, with grassland being particularly effective, but there is a limit to how efficient a cover crop can be in recovering manure applications made in late summer and autumn (24). Soil cultivation itself can release extra nitrate which is available for leaching so that the incorporation of manure or the establishment of cover crops can have significant indirect effects.

It may be concluded that the effects of organic manures on leaching vary according to the type of manure, the time of application and the utilisation of the land. It has therefore been considered necessary in the United Kingdom to undertake a research project to more clearly establish actual nitrate losses from a number of alternative practices. Preliminary results are presented below from one trial concerned with the application of poultry manure and cattle farmyard manure (FYM) to an arable sandy soil.

5. NITRATE LEACHING AFTER MANURE APPLICATION - EXPERIMENTAL RESULTS

Method

A small plot experiment commenced in October 1989 on a loamy medium sand soil at the ADAS Gleadthorpe Experimental Husbandry Farm in Nottinghamshire. All treatments were replicated 3 times in randomised blocks. Following a crop of spring barley from which the straw had been baled and removed poultry manure or strawy cattle FYM was applied each month from October - February and the plots rotavated within 48 hours. The poultry manure with a C:N ratio of 7(range 6-8) was applied at 6 t/ha and the FYM at 40 t/ha with an average C:N ratio of 41(31-61). The total

N applied by the poultry manure varied from 215-360 kg/ha and by the FYM from 193-432 kg/ha. The October and November applications of poultry manure supplied 228 and 215 kg/ha N respectively. Plots were kept weed free throughout the winter and the soil cultivated and sugar beet drilled in April. Soil samples in 30 cm increments to 90 cm were taken for mineral N determinations in October and from each plot, bulked by treatment, in March. Five ceramic cups were installed per plot at 100 cm from the surface prior to manure application and soil water samples extracted at 2 weekly intervals until the end of the drainage season. Monitoring will recommence after the sugar beet crop to detect any residual effects. Leachate volumes for the calculation of nitrate loss were obtained from monolyth lysimeters extracted from, and reinstated on plots adjacent to the current experiment.

Results and Discussion

The early part of the winter was dry with only 41% of long term average rainfall between November and January however a total leachate volume during the monitoring period of 225 mm was near to average for the site. In October the soil contained 20, 13, 15 kg/M/ha N in each 30 cm depth of soil. The calculated nitrate leaching losses are given in Table 1 and selected treatments are shown graphically in Figure 1 as nitrate concentrations in soil water.

TABLE 1 Nitrate leaching after application of poultry manure and cattle FYM.

Application Date	Poultry Manure N Leached kg/ha	Poultry Manure Leached as proportion of N applied %	Cattle FYM N Leached kg/ha	Cattle FYM Leached as proportion of N applied %
6 October	123	39	41	3
9 November	139	49	59	8
19 December	52	7	42	5
15 January	36	1	33	0
7 February	34	0	35	1
Control	33	-	33	-

SED (CV%) N Leached kg/ha± 9 (19.4%) Proportion N Leached ± (40.4%)

Nitrate leaching losses from October and November applied poultry manure were 40-50% of the total N applied whilst applications made in January and February gave no increased loss above that from the control. Losses from FYM were much less with an average 5% loss from October - December applications. Soil mineral nitrogen data for March are given in Figure 2. These results clearly indicate an almost complete loss of mineral nitrogen residues from the October poultry manure. By contrast applications from November to January have left appreciable residues with an obvious movement to the lower part of the profile following applications in November and December. FYM showed a lower residual mineral nitrogen content from the October dressing compared to other months. There had been only limited mineralisation of the nitrogen

Figure 1. N Leaching from Poultry Manure (PM) and Cattle FYM (FYM), Winter 1989/90

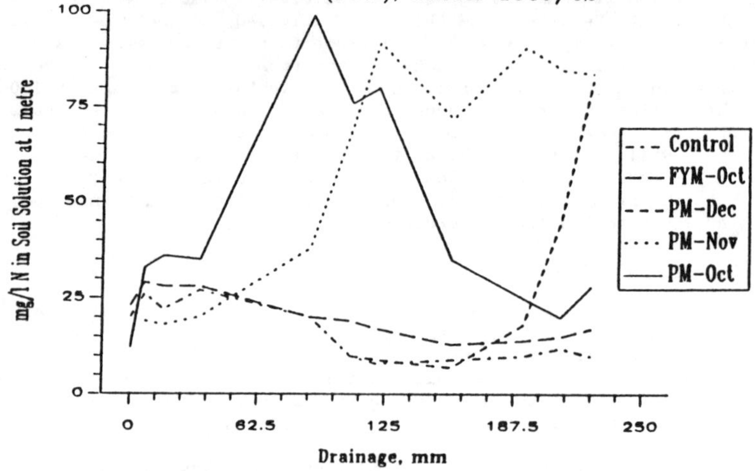

Figure 2. Soil Mineral N After Manure Applications Sampled 27 March, 1990

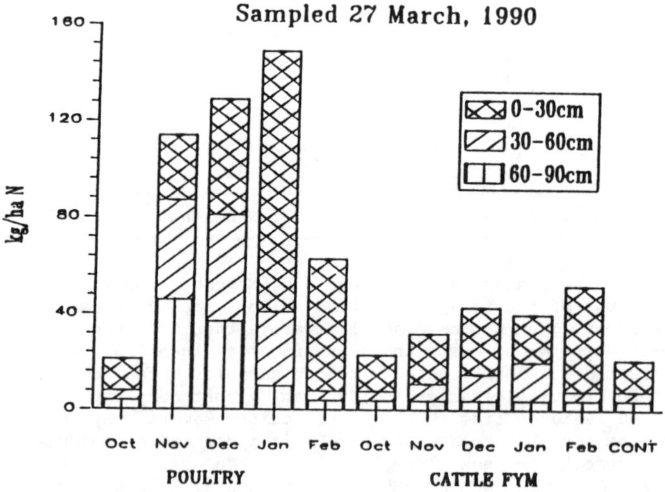

reserves and no evidence of nitrate movement to depth in the soil profile from any treatment.

Conclusions

The results to date confirm the general impressions from earlier studies and demonstrate the ability of current techniques to estimate actual leaching losses and to satisfactorily explain soil mineral nitrogen analyses. In 1990 the experimental programme will include earlier manure treatments, cover crops and other manure types hopefully including sewage sludges. A similar programme will commence on soils under grass to determine if increased leaching losses can be detected from this land use.

REFERENCES

(1) HARVEY, J 1988. Farmers Weekly 30 Sept p. 14.
(2) HOUSE OF LORDS 1989. Nitrate in Water. Select Committee on The European Communities. 16th Report. HMSO London.
(3) EEC 1980. Directive relating to the Quality of Water intended for Human Consumption. 80/778/EEC. Brussels.
(4) OWEN, T R and JURGENS-GSCHWIND, S 1986. Nitrates in drinking water; a review. Fertiliser Research 10 3-25.
(5) EEC 1988. Proposal for a Council Directive concerning the protection of fresh, coastal and marine waters against pollution caused by nitrate from diffuse sources. Com 88/708 Final. CEC Brussels.
(6) VETTER, H and STEFFENS, G (1989). Guidelines for an economic and environmentally safe use of slurry on agricultural land. In Safe and efficient slurry utilisation. Ed. H Vetter, G Steffens and P L'Hermite. SL/131/89. XII/ENV/3/89. CEC Brussels.
(7) WILLIAMS, J H 1988. Guidelines, recommendations, rules and regulations for spreading manures, slurries and sludge on arable and grassland. SL/124/88. XII/ENV/14/88. CEC Brussels.
(8) Anon 1988. Statutory order from the Ministry of Environment no 568 of Sept 22 1988 on farm yard manure and silage. In pursuance of Environmental Protection Act 1985 as amended 1987. Copenhagen.
(9) MAFF (1985). Code of good agricultural practice. Ministry of Agriculture, Fisheries and Food/Welsh Office Agriculture Department. HMSO London.
(10) DOE (1986). Nitrate in water. Pollution paper No. 26 Department of Environment. HMSO London.
(11) SEVERN TRENT WATER (1988). The Hatton catchment nitrate study. Severn Trent Water, Birmingham.
(12) DOE (1988). The Nitrate issue. Department of the Environment, HMSO London.
(13) SI (1990). The nitrate sensitive area (Designation) order 1990. Statutory Instrument 1990 No. 1013. HMSO London.
(14) ROYAL SOCIETY (1983). The nitrogen cycle of the United Kingdom. Royal Society, London.
(15) ANON (1990). Fertiliser use on farm crops in England and Wales 1989. Survey of fertiliser practice. ADAS, MAFF London.
(16) SI (1989). The sludge (use in agriculture) regulations 1989. Statutory Instrument 1989 No. 1263. HMSO London.
(17) BROGAN, J C (1981) Ed. Nitrogen losses and surface runoff from land spreading of manures. Nijoff/Junk. The Hague.
(18) DAM KOFOED, A WILLIAMS, J H and L'HERMITE, P (1986) Eds. Efficient landuse of sludge and manure. Elsevier, Barking.

(19) VAN DER MEER, H G, UNWIN, R J, VAN DIJK, T A and ENNIK G C (1987) Eds. Animal manures on grassland and fodder crops. Nijoff, Dordrecht.

(20) WELTE, E and SZABOLCS, I (1988) Eds. Agricultural waste management and environmental protection. International Scientific Centre of Fertilisers (CIEC) Belgrade.

(21) HANSEN, J A and HENRIKSEN, K (1989) Eds. Nitrogen in organic wastes applied to soils. Academic Press, London.

(22) GERMON, J (1989) Ed. Management systems to reduce the impact of nitrates. Elsevier, Barking.

(23) RITTER, W F and CHIRNSIDE, A E M (1987). Influences of agricultural practices on nitrates in the water table aquifer. Biological Wastes 19 165-178.

(24) LANDMAN, A (in press). Use of green manures to prevent nitrate leaching after slurry spreading. In Fertilisation and the Environment. Proceedings of a symposium held at Catholic University of Leuven, 27-30 August 1989.

AN ENVIRONMENTALISTS VIEW OF THE USE OF SEWAGE SLUDGE

BY G A GLEGG AND P V HORSMAN

GREENPEACE UK,
30-31 Islington Green,
London N1 8XE

SUMMARY

Sewage sludge is being generated in ever increasing amounts and the dilemma centres on how to find an environmentally acceptable way to utilise sewage. The problems encountered are commonly those of contamination with pathogenic organisms and chemicals and those of a practical, logistic nature. This paper considers these problems and suggests some ways in which they may be overcome. It is essential that sewage is treated appropriately so that sludge can be used beneficially in the environment.

1. INTRODUCTION

The UK government has recently announced that it will end dumping sewage sludge at sea by 1998. It also announced that it will end the pumping of untreated sewage effluent into coastal waters which will increase the amount of sewage sludge generated. However, as yet little is known about the government future plans for dealing with sewage sludge.

Currently, 40% of UK sludge is applied to agricultural land, 15% in landfill, 30% is dumped at sea and 15% is disposed of in various ways including incineration. Sewage sludge is a valuable resource which should be used for its organic, nitrogen and phosphorous content. However, the practical restraints and the presence of trace contaminants and pathogens can impede its use. This paper will discuss these constraints and suggest the factors to be considered when making decisions about the use of sludge. Throughout, the beneficial use of the components of sewage sludge is assumed essential.

2. THE CHEMICAL CONTAMINATION OF SEWAGE SLUDGE

Chemical contaminants arise from two main sources: industrial and domestic effluents; others also arise from urban runoff. Assessing the relative proportions which arise from these sources is difficult particularly in the UK where information concerning industrial discharges to sewers is currently not in the public domain. It is likely that the industrial sources outweigh the domestic (1). Very little detailed work on this problem has been done and although such research would be interesting the real solution lies in preventing the contaminants reaching the sewage works. The levels of contaminants can be reduced most effectively by reduction at source (2,3). Removal of contaminants once they have reached the sewage sludge is neither economically feasible (4) nor likely to be practically possible.

Trace metals

Historically sludge and/or trace metal application limits were developed on the basis of phytotoxicity but more recently these were lowered to take into account the possibility of accumulation in the food chain (3). High levels of trace metals applied to land with sludge can lead to increased concentrations in vegetation (5). Following concerns about the contamination of sewage sludge discussed in European fora, an EC Directive has been adopted which details acceptable levels of trace metals in sludge and in the land to which it is to be applied (6). While this is a necessary step in the right direction it does not address the real problem of limiting the contamination of sludge.

Cadmium presents a significant risk to human health (5,7) and accumulates in many plants such as soyabeans (8), wheat (9), corn (10) and vegetable crops (11). However, recently (1970s), sludge containing high levels of cadmium as a result of discharges from a lightbulb factory, was being applied to soil in the US (12). Attempts to correct this have involved a long term programme of soil replacement, liming and ploughing. However, it is unlikely that all soils contaminated by this source will ever be identified and remediation carried out (12).

A more insidious problem may be the effect that elevated trace metal concentrations have on soil microbial populations (7). In extensive experiments carried out on the same site for over 40 years workers have found that the nitrogen fixing activity of the bacterium Rhizobia, is seriously affected by moderate levels of trace metals in the soil. The root nodules of clover grown in soil with a high trace metal content were found to be abnormally small and contain a strain of Rhizobia which tolerate high levels of trace metals but are not effective at fixing nitrogen (13). Other soil microbes are also affected by high trace metal concentrations in the soil (7).

Awareness of problems with trace metal contamination has resulted in a considerable reduction of levels in sewage sludge and a concomitant lowering of the regulated acceptable levels (3,14). This fall in trace metals is laudable but could be considered to be too late in view of the areas of land which are presently contaminated. Trace metals applied to soils have long residence times with offtake in crops and leaching (unless the soil is very acidic) being very low (15). Around 40% of applied trace metal was recoverable after 40 years (20 years after the last sludge application) but two thirds of the 'losses' were accounted for by lateral transport (15). The levels of trace metals which do affect microbial action are very close to those permissable under the recently adopted EC guidelines (6,16) and considering the long residence times levels of trace metals reaching soils must be minimised and industrial wastes excluded from domestic sewage as a matter of urgency (7).

Trace organics

Although the problem with trace metals has been acknowledged, the problems with contamination by trace organic materials have not been addressed in the same comprehensive way. In a survey carried out in Canada of 144 priority organic contaminants analysed 106 were found to be present in sewage or sewage sludge (17) and the authors noted that there were a further 80 organic compounds on the Canadian Environmental Monitoring Priority Pollutant List which were not determined. A review of the literature by Rogers (18) also noted a very wide range of organic contaminants in sewage.

At present, few organic contaminants are routinely monitored in UK sludges. The EC has a list of 129 dangerous substances but the procedure to initiate the Directives is very slow (18). Moreover, the inherent flaw in any procedure to list prohibited chemicals is that contaminants have to be seen to cause damage before action is taken. A more prudent philosophy, given the range of chemicals available and the rate at which new ones are developed, would be to adopt the precautionary approach and prevent contamination at source.

The treatment that sewage requires is dependent on its end use. However, in general, it must be stabilised by a digestion or composting procedure which will remove some of the organic contaminants added by industrial or domestic users. However a considerable amount of work has shown that this is not the case for all contaminants (19,20). An example of this is chlorinated organic compounds which in many cases persist on land for considerable lengths of time (more than 10 years) (19).

Polychlorinated biphenyls (PCBs) in the environment have received considerable attention because of their toxicity, persistence and tendency to bioaccumulate. During sewage treatment chlorinated organics accumulate in sludge with accumulation being greater with higher degrees of chlorination (20). Work by Fairbanks *et al.* (21) found that after 240 days between 8 and 33% of PCBs added to soil with sludge had been removed. However, only 1-11% of this loss was due to degradation the rest being lost by volatilisation. Thus, although calculated 95% disappearance times ranged between 6.1 and 26.3 years, breakdown rates would be very much slower. Results from another study indicated no loss of PCBs, within the error of analysis (about 20%), from soil over a period of a year (22).

A review by Jacobs *et al.* (19) discusses differing levels of trace organic uptake by plants depending on a variety of factors such as soil, plant and compound type. A recent study showed little uptake of PCBs from sludge amended soil into lettuce and fescue but did observe some uptake into the skins of carrots (23). However, the authors conclusion that this is of minimal concern because the contamination would be removed by normal culinary practices' is unacceptable.

3. THE BIOLOGICAL CONTAMINATION OF SEWAGE SLUDGE

The contamination of sewage with pathogens has long been recognised widely as a potential health problem. Reductions of certain pathogens can be achieved 'at source' by disease prevention schemes (24) but generally contamination by pathogenic organisms cannot be avoided and treatment is required to produce a microbiologically acceptable sludge for the proposed use. The traditional indicator of pathogenic contamination has been coliform bacteria, however it has been shown that in media such as seawater these are not always appropriate as indicators as their decay does not mirror that of other pathogens (25).

Work has been carried out on the survival of a wide range of pathogenic organisms in sludges. This has shown that the application of untreated sludge may result in risks to the health of humans, animals and crops due to a variety of pathogens (26,27,28,29). Numerous viruses are found in sewage sludge after traditional cold digestion treatments (30). However, the health hazard from viruses can be minimised by appropriate waste water treatment (27). The most important factors in the destruction of pathogens are thermal treatment, loss of moisture through evaporation, microbial antagonism, exposure to high pH and irradiation (27).

Common methods of eradicating pathogens generally involve heat treatment such as thermophilic digestion, composting and other technologies discussed at this conference. However, problems can occur if temperature regimes are not adhered to during processing as microbes may persist at temperatures a few degrees lower than recommended microbes (31). In addition, it is essential that there is total separation within the plant of the treated sludge from incoming wastes and sources of contamination (32,33).

4. THE PRACTICAL ASPECTS OF SEWAGE SLUDGE USE

The last broad area of concern is the practical aspects of sewage sludge transport to where it can be utilised and its acceptability. These points are often raised especially when considering sludge from urban areas.

Transport

Concern about carbon dioxide emission is often raised in discussions about transport but decisions are generally based on economics. Analyses of environmental impacts must include the implications of all elements of the equation. For example in a discussion of the greenhouse gas emissions from different sewage sludge treatment and disposal routes, Frost and Bruce (34) conclude that dumping at sea involves the least generation of greenhouse gases whilst application to agricultural land ranks middle and landfill dumping last, generating the greatest amount of greenhouse gases. However, no consideration is given to the saving in greenhouse gas generation from reduced fertilizer production and transport which ought to have been included in a comprehensive survey.

That there is a market for properly treated sewage sludge is demonstrated by the importation of 20,000 tonnes of pasteurised filter cake from Switzerland to Swansea in Wales which was sold for agricultural use and land reclamation schemes (35). Whilst not wanting inefficient long distance hauling of sludge it is very important to utilise the benefits of sludge. The volume and weight of sludge can be reduced by various digestion and dewatering processes. In the US sludge has been dried before transport and then rewetted to enable application. In addition, at present the only transport commonly discussed is by road, but other means need to be investigated, such as sea (possibly in old dump ships), by rail and pipeline.

Agriculture

At present 40% of the UK sludge arisings, and 34% of EC sludge is used on agricultural land (36). It is in this use that chemical and biological contamination, transport and acceptability are most commonly highlighted as limiting factors.

In the UK sewage sludge is supplied free to farmers and the authorities regard the farmers as doing them a favour by providing an outlet for the sludge. Thus sludge is being viewed as a disposal problem, rather than as a resource, though it is unlikely that farmers would use untreated sludge if asked to pay for it. Hence, sludge should be treated and converted to a useful product rather than simply dumped on land in the hope that nothing will go wrong. Perhaps the sludge is given free because the authorities feel that they are dealing with a substance of suspect quality which they are glad to 'get rid' of. This could lead to the authorities being nervous of increases in the quantity of sludge applied to land.

The product is valuable to farmers as a good, cheap fertilizer and soil conditioner. The value of the nitrogen and phosphorous content has been estimated to be £5.8 million and could contribute 2% of the total farming fertilizer requirement (36) which, with animal wastes, could make a real contribution to these needs in the UK. Sludge acts as a slow release fertilizer which reduces land runoff of fertilizers (36,37) and with the return of interest in sustainable and organic farming there must be a role for these wastes (38,39).

Agricultural land is not available for sludge application throughout the year due to crop cycles and the weather. Nonetheless, as part of an integrated scheme it can play an important part in any programme of sewage sludge utilisation. The key to the use of sludge in agriculture lies in the knowledge that the sludge is of guaranteed chemical and pathological quality. This guarantee can only be supplied if there is no industrial input to the domestic sewage system and appropriate treatment is undertaken by the authorities.

Composting

Composting sludge reduces biological contamination of sewage sludge and makes it acceptable for application to land. This treatment has the advantage of being able to produce a variety of different media, increasing the scope of markets available (40). It may be used to prepare agricultural fertilizer and soil conditioner, materials for use in land reclamation, forestry, and for the horticultural markets (41,42,43). However, this application, as with all the others, requires sludge free from chemical contamination.

Composting has been used in the United States since the 1970s and is now of growing interest there and in Europe (43,44). Large composting facilities have recently been built to serve major population centres, the largest being in Philadelphia which has a maximum capacity of 367,400 dry kg per day (41). Composting can also utilise other waste materials such as straw, urban leaves and grass cuttings, and the organic component of domestic refuse (43,45). In the UK one technique has been developed which can use untreated liquid sewage and straw to produce compost (46) which may be widely used in the future when straw burning is banned.

The limitation on the use of composting is generally considered to be the size of the market however work suggests that for a satisfactory product there is a readily available market (41,43,46). It is also reported that there are few problems with public perception once the product has been seen and used. Since composting already requires some handling during processing it is amenable to manipulation to create the required product. Work is being carried out to try to produce an acceptable sewage based peat substitute (47). At present in the UK damage is caused by peat extraction in large quantities from bogs a considerable distance from the population centres where it is required.

Forestry

In the UK the potential for sewage sludge as a forest fertilizer has been estimated to be between 6 and 11% depending on the transport distance accepted (48). It is already being used extensively in some US states and is being investigated in the Netherlands and France (49,50,51).

It has much to offer forestry because it fertilizes and conditions the often poor soil used in forestry and requires less frequent application than conventional fertilization systems by inducing natural nutrient cycling (52,53). Significant improvements in growth have been observed following application of sludge and scheduling application according to season is less restrictive than agriculture (50,54). Forestry offers a long term outlet which can be used for the bulk of the year.

However the key to any land application programme including forestry is high quality sludge (50). Pollution problems of water runoff and the build up of contaminants in wildlife needs to be guarded against (55).

Land reclamation

A wide variety of land types, including derelict mined land and colliery spoil, may be reclaimed to permanent vegetation by the addition of sludge (56,57,58). Large quantities of sludge can be applied (ensuring a long term outlet) and experiments confirm that sludge is superior to more traditional methods at enabling a self-sustaining vegetation cover to be established (56,59).

Transport is less likely to be a barrier to this use because many sites are situated around industrialised centres thus enabling short hauling distances (59). Although reclaimed land is seldom used for arable pasture it is not acceptable to use contaminated sludge as this may again lead to build up in wildlife, grazing farm animals and ground water.

Novel uses

A variety of other options have been discussed; two of these, thermal conversion and use in building materials, seem promising (60,61). There has been some interest in the use of sludge as a feedstuff for a variety of animals and fish (61). However, in the light of the recent food scares in the UK it would seem very unlikely that this would now be acceptable to the public (supposing it ever would have been).

Thermal conversion of sewage sludge to fuel oil and char has been researched in Canada where a full-scale demonstration plant is being built (62). Sludge disposal costs by this route are estimated to be considerably less (around half) than existing methods. In the UK an analysis comparing low temperature conversion and incineration found that at 1985 fuel costs thermal conversion would be 40% more economic than incineration (60). However, no extensive environmental analysis appears to have been carried out and this must be a prerequisite of any implementation programme. Only uncontaminated sludges should be used as the resulting fuels or wastes would contain any contaminants which had been present in the original sludge and any which were synthesised during the processing.

The use of sludge in building materials has received considerable attention in the US where sludge has been used successfully to make 'Biobricks' (61). It has been estimated that 15% of municipal sludge production in the US could be incorporated in bricks (60).

5. THE DEVELOPMENT OF A SLUDGE USE STRATEGY

The essential prerequisite for any sludge use programme is that it is not contaminated with industrial wastes or domestic products which may lead to adverse effects or a build up of contaminants in the environment. Industrial pollutants can be removed by disconnecting their effluents from the domestic sewage system. Industries will then have to treat their own effluents thus forcing them to accept responsibility for their wastes.

It is very difficult to control domestic pollutants once they are available on the open market. Certainly, public education should be able to prevent some materials reaching the sewage system and the European Commission 'eco-labelling' scheme may help. The working paper suggests that products should only be considered after a full 'cradle to grave' assessment of the environmental (63). This should involve studies of the effects on sewage and sludge quality.

Of course, wide ranging questions should be raised about products that damage the environment. Using the precautionary approach, producers should prove that their products do no harm to the environment. Industry should develop clean production which ensures that production does not cause pollution and that the product is not polluting either during or at the end of its life.

Sewage sludge must be treated to remove all pathogens which will enable the general public and the water workers to have confidence in its safety in handling and use. A variety of treatment methods are available which can be used to develop a wide spectrum of products for different applications.

In developing any sludge use scheme all aspects of the environmental equation must be considered even those which do not have a direct economic influence but have an environmental benefit. The US Environmental Protection Agency has emphasised the importance of beneficial reuse of sewage sludge and assessments by certain states have come to the conclusion that this outweighs certain economic arguments (64). In The UK House of Commons Environment Committee Third report it is stressed that 'disposal routes which convert sludge into useful products .. seem more desirable and their viability should be researched and reviewed thoroughly' (65).

The key to the whole problem must lie in the water authorities being willing to develop a wide range of sludge outlets and not expecting a single strategy to fulfil all their requirements. The present thinking is exemplified in the discussion by North West Water of their options for sewage sludge disposal (66). They discuss only agricultural land disposal, landfill and incineration as alternatives to sea disposal. The other options discussed in this paper, such as forestry and composting, are grouped in a separate section where it is noted that none have the 'capability to provide substantial and continuous outlet capacity for the quantities of sludge' and that 'it would be imprudent to place early reliance on an untried method' (66).

More creative thinking is essential if we are to realise the benefits of sewage sludge and establish a long term environmentally secure programme. Strategies cannot be entirely dependent on one disposal method but must be able to incorporate a range of uses to enable the practical difficulties to be overcome.

6. CONCLUSIONS

Sewage sludge is a resource which if treated properly can be of use in a variety of different products. When developing a strategy for using sewage sludge effectively, it is essential that assessments are not made solely on economic grounds but also recognise the environmental factors. The various options for using sewage sludge must be fully investigated in the decision making process. In a world of diminishing resources it is vital that we make maximum use of those resources available to us.

REFERENCES

(1) HATHAWAY, S.W. (1980). Sources of Toxic Compounds in Household Wastewater. United States Environmental Protection Agency Report No. EPA-600/2-80-128. Municipal Environmental Research Laboratory, Cincinnati, Ohio. pp 83.

(2) MOLLER, U. (1986). Overview Of Sewage Sludge Treatment Processes: An Outlook. In 'Processing And Use Of Organic Sludge And Liquid Agricultural Wastes' (Ed L'Hermite, P.) Commission of European Communities 4th International Symposium, D. Rendel Publishing Co., Dordrecht, p 2-24.

(3) TJELL, J.C. (1986). Trace Metal Regulations For Sludge Utilization In Agriculture: A Critical Review. In 'Processing And Use Of Organic Sludge And Liquid Agricultural Wastes' (Ed L'Hermite, P.) Commission of European Communities 4th International Symposium, D. Rendel Publishing Co., Dordrecht, p 348-361.

(4) BASTIAN, R.K. (1986). Overview On Sludge Utilization. In 'The Forest Alternative For Treatment And Utilisation Of Municipal And Industrial Wastes' (Eds Cole, D.W., Henry, C.L. and Nutter, W.L.) University of Washington Press, p 7-25.

(5) CHANG, A.C., HINESLY, T.D., BATES, T.E., DONER, H.E., DOWDY, R.H. and RYAN, J.A. (1987). Effects Of Long-term Sludge Application On Accumulation Of Trace Elements By Crops. In 'Land Application Of Sludge' (Eds Page, A.L., Logan, T.J. and Ryan, J.A.) Lewis Publishers Inc, p 53-66.

(6) EC (1986). Council Directive (86/278/EEC) on the Protection of the Environment, and in Particular Soil, when Sewage Sludge is Used in Agriculture. Official Journal of the European Communities, No L 181/6 - 181/12.

(7) GILLER, K. and McGRATH, S. (1989). Muck, Metals And Microbes. New Scientist, 4 November 1989, p 31-32.

(8) HECKMAN, J.R., ANGLE, J.S. and CHANEY, R.L. (1987). Residual Effects of Sewage Sludge on Soyabean: I. Accumulation of Heavy Metals. J. Environ. Qual. 16,113-117.

(9) LUBBEN, S. and SAUERBECK, D. (1989). The Uptake of Heavy Metals by Spring Wheat and their Distribution in Different Plant Parts. Presented to 'Alternative Uses for Sewage Sludge' held at York, September 1989.

(10) RAPPAPORT, B.D., MARTENS, D.C., RENEAU Jnr, R.B. and SIMPSON, T.W. (1988). Metal Availability in Sludge-amended Soils with Elevated Metal Levels. J. Environ. Qual. 17,42-47.

(11) KEEFER, R.F., SINGH, R.N. AND HORVATH, D.V. (1986). Chemical Composition of Vegetables Grown on an Agricultural Soil Amended with Sewage

Sludges. J. Environ. Qual. 15,146-152.

(12) LAGAN, T.J. and CASSLER, D.E. (1989). Correcting Widespread Cadmium Contamination. Wat. Environ. Technol. 1,312-315.

(13) GILLER, K.E., McGRATH, S.P. and HIRSCH, P.R. (1989). Absence Of Nitrogen Fixation In Clover Grown On Soil Subject To Long-Term Contamination With Heavy Metals Is Due To Survival Of Only Ineffective Rhizobium. Soil Biol. Biochem. 21 841-848.

(14) CANDINAS, T., GUPTA, S.K., ZAUGG, W., BESSON, J.M. and LISCHER, P. (1989). Changes in Heavy Metal Contents of Swiss Sewage Sludge and Future Prospectives. Presented to 'Alternative Uses for Sewage Sludge' held at York, September 1989.

(15) McGRATH, S.P. and LANE, P.W. (1989). An Explanation For The Apparent Losses Of Metals In A Long-Term Field Experiment With Sewage Sludge. Environ. Pollut. 60 235-256.

(16) McGRATH, S.P., BROOKES, P.C. AND GILLER, K.E. (1988). Effects of Potentially Toxic Metals in Soil Derived from Past Applications of Sewage Sludge on Nitrogen Fixation By Trifolium Repens L. Soil Biol. Biochem. 20,415-424.

(17) ONTARIO MINISTRY OF THE ENVIRONMENT (1988). Thirty Seven Municipal Water Pollution Control Plants, Pilot Monitoring Study. Volume I, Interim Report, Queen's Printer for Ontario.

(18) ROGERS, H.R. (1987). Organic Contaminants In Sewage Sludge (EC 9322 SLD): Occurrence And Fate Of Synthetic Organic Compounds In Sewage And Sewage Sludge - A Review. WRc Environment, Medmenham Laboratory, Henley Road, Medmenham, Bucks SL7 2HD.

(19) Jacobs, L.W., O'Connor, G.A., Overcash, M.A. and Rygiewicz, P. (1987). Effects Of Trace Organics In Sewage Sludges On Soil-Plant Systems and Assessing Their Risk To Humans. In 'Land Application Of Sludge' (Eds Page, A.L., Logan, T.J. and Ryan, J.A.) Lewis Publishers Inc p 101-143.

(20) KIRK, P.W.W., and LESTER, J.N. (1988). The Behaviour of Chlorinated Organics During Activated Sludge Treatment and Anaerobic Digestion. Wat. Sci. Tech. 20,353-359.

(21) FAIRBANKS, B.C., O'CONNOR, G.A. and SMITH, S.E. (1987). Mineralsiation and Volatilisation of Polychlorinated Biphenyls in Sludge-amended Soils. J. Environ. Qual. 16,18-25.

(22) MARCOMINI, A., CAPEL, P.D., LICHTENSTEIGER, TH., BRUNNER, P.H. and GIGER, W. (1989). Behaviour of Aromatic Surfactants and PCBs in Sludge-treated Soil and Landfills. J. Environ. Qual. 18,523-528.

(23) O'CONNOR, G.A., KIEHL, D., EICEMAN, G.A. and RYAN, J.A. (1990). Plant Uptake of Sludge-borne PCBs. J. Environ. Qual. 19,113-118.

(24) POYRY, T., STENVIK, M. and HOVI, T. (1988). Viruses in Sewage Waters during and after a Poliomyelitis Outbreak and Subsequent Nationwide Oral Poliovirus Vaccination Campaign in Finland. App. Environ. Microbiol. 54,371-374.

(25) WHEELER, D. (1990). The Pollution of Beaches by Viruses. In: House of Commons Committee Enquiry into the Pollution of Beaches. Appendices to the Minutes of Evidence 11,24-39. HMSO, London.

(26) WATKINS, J. and SLEATH, K.P. (1981). Isolation and Enumeration of Listeria monocytogenes from Sewage, Sewage Sludge and River Water. J. App. Bacteriol. 50,1-9.

(27) HURST, C.J. (1989). Fate of Viruses during Wastewater Sludge Treatment Procedures. CRC Critical Reviews in Environmental Control. 18,317-343.

(28) THARALDSEN, J. and HELLE, O. (1989). Survival of Parasite Eggs in Livestock Slurry Utilised for Compost Heat. Acta Agric. Scand. 39,381-387.

(29) SPAULL, A.M., McCORMACK, D.M. and PIKE, E.B. (1989). Effects of Various Sewage Sludge Treatment Processes on the Survival of Potato Cyst-nematodes (Globodera spp.) and the Implications for Disposal. Wat. Sci. Tech. 21,909-916.

(30) NASHIDA, M., SHIMOHARA, E., SUGISHIMA, S. and KANEKO, M. (1989).

Viruses Isolated from Activated Sludges, 1984-1986. Wat. Sci. Tech. 21,279-282.
(31) PIKE, E.B., CARRINGTON, E.G. and HARMAN, S.A. (1988). Destruction of Salmonellas, Enteroviruses and Ova of Parasites in Wastewater Sludge by Pasteurisation and Anaerobic Digestion. Wat. Sci. Tech. 20,337-343.
(32) FUJIOKA, R.S., HIRANO, W.M. and LOH, P.C. (1988). Disinfection, Recontamination and Regrowth of Indicator Bacteria in a Wastewater Sludge Heat Treatment System. Wat. Sci. Tech. 20,329-335.
(33) BURGE, W.D., ENKIRI, N.K. and HUSSONG, D. (1987). Salmonella Regrowth in Compost as Influenced by Substrate. Microb. Ecol. 14,243-253.
(34) FROST, R.C. and BRUCE, A.M. (1989). Energy from Sludge. Presented to 'Alternative Uses for Sewage Sludge' held at York, September 1989.
(35) HOUSE OF COMMONS (1990). Toxic Waste Disposal in Wales. Minutes of Evidence taken before the Welsh Affairs Committee, Session 1989-1990, 6 December 1989. HMSO, London.
(36) DAVIS, R.D. (1989). Agricultural Utilisation of Sewage Sludge: A Review. J. IWEM, 3, 351-355.
(37) WRC. Soil Injection of Sewage Sludge, A Manual of Good Practice (2nd edition). Ref No. FR 0008. Foundation for Water Research, Water Research Centre, Medmenham, Bucks. pp15.
(38) REGANOLD, J.P., PAPENDICK, R.I. and PARR, J.F. (1990). Sustainable Agriculture. Sci. Am. June, 72-78.
(39) STANHILL, G. (1990). The Comparative Productivity of Organic Agriculture. Agric. Ecosystems Environ. 30,1-26.
(40) BORDER, D., COOMBES, C. and SHELLENS, M. (1988). Composting Straw with Untreated Liquid Sludge. Biocycle. July, 54-55.
(41) KUCHENRITHER, R.D., DIEMER, D.M., MARTIN, W.J. and SENSKE, F.J. (1987). Composting's Emerging Role in Sludge Management: A National Perspective. J. WPCF, 59,125-131.
(42) LOPEZ-REAL, J.M., WITTER, E., MIDMER, F.N. and HEWETT, B.A.O. (1989). Evaluation of Composted Sewage Sludge/Straw Mixture for Horticultural Utilisation. Wat. Sci. Tech. 21,889-897.
(43) GOLDSTEIN, N. (1988). Steady Growth for Sludge Composting. Biocycle, November-December,27-32.
(44) MARTEL, J.L. (1989). Economics and Marketing of Urban Sludge Composts in the EEC. Presented to 'Alternative Uses for Sewage Sludge' held at York, September 1989.
(45) PEREIRA, J.T., STENTIFORD, E.I. and SMITH, D.V. (1986). Survival of Faecal Indicator Micro-organisms in Refuse/Sludge Composting Using the Aerated Static Pile System. Waste Management and Res. 4,397-406.
(46) MATTHEWS, P.J. and BORDER, D. (1989). Compost: A Sewage Sludge Resource For The Future. Presented to 'Alternative Uses for Sewage Sludge' held at York, September 1989.
(47) BRAGG, N. (1990). Ministry of Agriculture Fisheries and Food, personnel communication.
(48) TAYLOR, C.M.A. and MOFFAT, A.J. (1989). The Potential For Utilising Sewage Sludge In Forestry In Great Britain. Presented to 'Alternative Uses for Sewage Sludge' held at York, September 1989.
(49) LAVERGNE, G. (1989). The Evaluation Of Dewatered Sludge From The Marseille Treatment Plant For Use In Forestry. Presented to 'Alternative Uses for Sewage Sludge' held at York, September 1989.
(50) NICHOLS, C.G. (1989). US Forestry Uses Of Municipal Sewage Sludge. Presented to 'Alternative Uses for Sewage Sludge' held at York, September 1989.
(51) OLESEN, S.E. and MARK, H.S. (1989). Long-term Effects Of Sewage Sludge Application In Conifer Plantation On Sandy Soils. Presented to 'Alternative Uses for Sewage Sludge' held at York, September 1989.
(52) MOFFAT, A.J. and BIRD, D. (1989). The Potential for Using Sewage Sludge in Forestry in England and Wales. Forestry, 62,1-17.

(53) BAYES, C.D., TAYLOR, C.M.A. and MOFFAT, A.J. (1989). Sewage Sludge Utilisation In Forestry: The UK Research Programme. Presented to 'Alternative Uses for Sewage Sludge' held at York, September 1989.

(54) BURD, R.S. (1986). Forest Land Applications Of Sludge And Wastewater. In 'The Forest Alternative For Treatment And Utilisation Of Municipal And Industrial Wastes' (Eds Cole, D.W., Henry, C.L. and Nutter, W.L.) University of Washington Press, p 3-6.

(55) HAUFLER, J.B. and WEST, S.D. (1986). Wildlife Responses To Forest Application Of Sewage Sludge. In 'The Forest Alternative For Treatment And Utilisation Of Municipal And Industrial Wastes' (Eds Cole, D.W., Henry, C.L. and Nutter, W.L.) University of Washington Press, p 110-116.

(56) METCALF, B. and LAVIN, J.C. (1989). Consolidated Sewage Sludge as Soil Substitutes in Colliery Spoil Reclamation. Presented to 'Alternative Uses for Sewage Sludge' held at York, September 1989.

(57) SOPPER, W.E. and SEAKER, E.M. (1983). 'A Guide For Revegetation Of Mined Land In Eastern United States Using Municipal Sludge', School Of Forest Resources And Institute For Research On Land And Water Resources, The Pennsylvania State University, March 1983.

(58) WERNER, W. and SCHERER, H.W. (1989). Experiences of the Usage of Heavy Amounts of Sewage Sludge for Reclaiming Opencast Mining Areas and Amelioration of Very Steep and Stony Vineyards. Presented to 'Alternative Uses for Sewage Sludge' held at York, September 1989.

(59) BYROM, K. and BRADSHAW, A.D. (1989). The Potential Value of Sewage Sludge in Land Reclamation. Presented to 'Alternative Uses for Sewage Sludge' held at York, September 1989.

(60) FROST, R.C. and CAMPBELL, H.W. (1986). In 'Processing and Use of Organic Sludge and Liquid Agricultural Wastes' (Ed L'Hermite, P.) Commission of European Communities 4th International Symposium, D.Reidel Publ. Co., Dordrecht. p94-109.

(61) WEBBER, M.D. (1989). Resource Recovery Through Unconventional Uses of Sludge. Presented to 'Alternative Uses for Sewage Sludge' held at York, September 1989.

(62) CAMPBELL, H.W. and BRIDLE, T.R. (1989). Conversion of Sludge to Oil: A Novel Approach to Sludge Management. Wat. Sci. Tech. 21,1467-1475.

(63) ENDS REPORT. (1990). Brussels Reveals Eco-labelling Plans. ENDS Report 182,23.

(64) Sludge Management Study Blue Plains Wastewater Treatment Plant -Washington D.C. Final Environmental Impact Statement, USEPA March, 1990.

(65) HOUSE OF COMMONS. (1987). Third Report from the Environment Committee. Session 1986-7. Pollution of Rivers and Estuaries. Volume 1.

(66) MINISTRY OF AGRICULTURE FISHERIES AND FOOD (1990). Report on the Disposal of Waste at Sea 1986 and 1987. HMSO, London.

SESSION REPORT

P.J. NEWMAN, WRc Environment, Marlow, Bucks, UK

A total of six papers were presented during the final Session of the Symposium. The first five dealt with various aspects of sludge regulations in several European countries whilst the last considered an "environmentalist's view of the use of sewage sludge". To include such a paper in the Symposium was an interesting innovation and one that should be repeated at other Symposia.

The first paper by K Aichberger of the Federal Institute of Agrobiology, Linz, Austria dealt with the "Situation of sewage sludge in Austria - Use in Agriculture. National Guidelines and Laws, Future Aspects (A General View)". This showed that in Austria some 3.5 million m^3 (~ 200000 tds) sewage sludge were produced each year. 25% of this was disposed in agriculture, 3% in land reclamation, 37% in composting and incineration with 35% being stored in dumping areas.

Even though Austrian sewage sludge is of good quality because of the control of effluents from industry to the public sewer and can be used widely in agriculture, the Austrian authorities are considering stricter regulations to obtain an even greater degree of safety. This will greatly reduce the amount that can be used in agriculture and excess may have to be incinerated.

Concern was expressed during the discussion of the paper that as different standards apply in the individual federal states in Austria, it could result in sludge being transported for disposal in the state with the least stringent regulations. It was pointed out, though, that in reality the differences in the regulations set in the individual states do not vary very much and so transportation of sludges between states is unlikely.

The second paper dealt with "Consequences of the Ordinance on Substances on Swiss Agriculture". It was presented by E Spiess of the Swiss Federal Research Station for Agricultural Chemistry and Hygiene of Environment in Liebefeld. This described the Swiss Ordinance relating to Environmental Hazardous Substances introduced in 1986 which prescribes that fertiliser dressings must be adapted to the crop requirement and that mineral fertilisers should only be applied if animal manures are not sufficient for the purpose. Nutrient balances for nitrogen, phosphorus and potassium were calculated to investigate how far animal manure can provide crop requirements in Switzerland.

During the discussion of the paper Dr Spiess commented that although taxes had not yet been imposed on farmers in order to reduce the use of fertilisers, they could be introduced in the future. Furthermore the situation may exist where more animal manure may be produced in Switzerland than the total that could be disposed of safely in agriculture. If this proves to be the case it may be necessary to reduce, say by 10% of the total, the number of animals kept in the country.

The next paper was by J H Williams of the Soil Consultancy Services in Tettenhall, UK and dealt with "Regulations on Additions of Sludge-Borne Metals and their Adaption to Local Conditions". It was presented by J E Hall of UK WRc in Mr Williams' absence.

It stressed that provided sludge is used correctly in agriculture, it can be a useful source of N and P and a good soil ameliorant. Even so, damage to crops or farm animals due to sludge-borne contaminants must be avoided by setting cumulative safe upper limits for all potentially toxic elements in the soil treated with sludge. Having established these safe limiting soil concentrations, they must form the basis for calculating limiting values for additions and concentrations of these elements in sludge as a guide for those persons responsible for the field application of sludge.

Maximum tolerable additions and sludge concentration limits for 9 elements (Cd, Cu, Ni, Pb, Zn, Hg, Cr, Mo and Fr) were then presented in the paper.

There followed a detailed discussion about the application of sludge regulations in the UK. Concern was expressed particularly that the levels of nickel permitted may lead to phytotoxic effects. Mr Hall was able, though, to demonstrate the validity of the value chosen and to show that much higher levels of nickel could be tolerated before phytotoxic effects are observed.

Mr Ramsay of the UK Department of the Environment also pointed out that the values set in the UK were the total heavy metal content permitted by the EC. Even so, only some 2.5% of the soils in the UK would have zinc levels above 2-300 mg/kg within 100 years. Lower values could be set but these would only cause unnecessary problems for the farmers. It was possible for some sludges to have a lower heavy metal content than for some soils and they still cannot be used in agriculture.

The following paper was by H Witte and M Keding of the University of Siegen in the Federal Republic of Germany. It described their "Experience with a computer supported organisation model for agricultural use of Sewage Sludge". It was divided into two parts. The first examined sludge application in the Federal Republic from the legal and practical aspects. It considered the need to amend the Federal Republic's Sewage Sludge Regulation in line with the EC Sewage Sludge to Agricultural Land Directive. It also discussed the risks due to PCBs, dioxin and other potential contaminants in sludge. The second part of the paper considered the opportunities for an improvement of the sludge application situation. Experiences with an organisational model were also described.

During the discussion of the paper, Dr Witte claimed that although the organisational model worked well and that the sludges in the Federal Republic were of good quality because of the restriction of discharges of heavy metals from industry into the public sewerage system, farmers were still uncertain how to consider sludge. They were prepared to accept it being spread on their fields. However, as they cannot judge the extent of the risk, it is with some apprehension. They had even asked for liability guarantees from the plant operators, which could not be accepted.

The fifth and final paper in the first part of the session was presented by R J Unwin of the Agricultural Development and Advisory Service of the UK Ministry of Agriculture, Fisheries and Food. This paper, written jointly by M A Shepherd and K A Smith, dealt with "Controls on manure and sludge applications to limit nitrate leaching - does the evidence justify the restrictions which are being proposed?"

This reviewed current and proposed legislation to restrict nitrate loss from agricultural land in the European Community. Also discussed were the potential effects of complying with restrictions on organic manure applications upon farmers and disposers of sewage sludge in the UK. Here it was shown that these restrictions could be severe for a significant number of operations. Finally preliminary results from one study were presented which highlighted the differences between poultry manure and cattle farm yard manure and their time of application to arable land in Autumn and Winter.

Great interest was expressed during the discussion of the paper about how the regulation's would be applied in practice. Question's were asked concerning the compensation paid to farmers to take their land out of arable production. Mr Unwin indicated that although the scheme was only voluntary, some 50% of the farmers had taken up the offer. It had been decided to adopt a procedure by which farmers could receive a general compensation/hectare for that region rather than a specific amount for the particular farm.

The second part of the Session consisted of a paper by G A Glegg and P V Horsman of Greenpeace UK which dealt with "An environmentalists view of the use of sewage sludge". It was presented by Gillian Glegg who said that she considered that it was essential to treat sludge properly, ie by the removal of potentially pathogenic organisms and chemicals, so that it could be used beneficially in the environment. In particular, she stated that discharges from industry should be prevented from entering public sewers so that the resulting sludge would have a much lower chemical, and especially heavy metal, content.

During the discussion of the paper, Ms Clegg said that there was not enough information available about the quantity and quality of sludge produced at present to advise how it should be disposed. In the circumstances she felt it should be landfilled until further information became available.

Many delegates doubted that it was practical, as proposed by Ms Glegg, to disconnect all industrial discharges from sewers. In urban areas where there would not be sufficient land to build a treatment work, the liquid waste would need to be tankered to separate facilities at the municipalities wastewater treatment work. Even wastewaters from laundrettes would need to be tankered in this way. The result would be additional road tanker journeys and a duplication of treatment facilities.

The experiment of including in the Symposium a paper expressing the views of the environmentalist groups worked successfully and was well received by the delegates. Similar experiments should be considered for other Symposia.

EVALUATION OF THE CONFERENCE AND CLOSURE

by

R C RAMSAY
Vice-President
COST 681 Programme

Department of the Environment
43 Marsham Street
LONDON, UK

I want to say how sorry I am that Mario Santori, President of the COST 681 Programme, has not been able to be with us this week for the final Symposium of the Concerted Action Programme. He has been associated with this project since it began in 1972, and should have had the honour of making this closing statement. However I am sure all of us here would like his colleagues, Mr Spinosa and Mr Minini, to take back to him our best wishes for a speedy return to good health.

I would also like to say how pleased I am that Dr Heinrich Ott, who has also been involved with COST 681 since the beginning and is now fully recovered, from his recent illness and is able to attend the Symposium.

I have some general points to make which seem to me to have emerged from the papers and discussions this week, but all of you will be able to come to your own conclusions when you receive your copies of the final proceedings - to be published later this year.

Firstly:

There is general agreement that we must aim by all practicable means to reduce the concentrations of heavy metals and other contaminants in sewage sludge so that it can be disposed of with least adverse effect on the environment. We have heard from several speakers here that in many countries our efforts in this direction are showing significant signs of success - certainly in respect of heavy metals. Information on organic contaminants is not yet sufficient for us to see how far we have managed to contain them, but a great deal of work is under way to establish what is the degree of risk and what needs to be done to make it acceptable.

However, results of research in recent years are reassuring in that the concentrations in the environment of such contaminants as dioxins, PCB's, PAH's etc are relatively low and are falling. Also, as we heard from Dr Leschber, nature can in certain circumstances degrade such materials to safe levels over acceptable timescales. We also know that these materials are not transferred from soils to arable crops in significant quantities, although there is need to be particularly cautious when using sludge on grassland in order to protect the health of grazing animals and thus the quality of the human food chain.

Secondly:

In respect of the risks from the pathogenic content of sewage sludges, we have had confirmation from practical experience in many countries that regulation of sludge use by specification of minimum standards of treatment coupled with constraints on subsequent planting and harvesting of certain food crops, and restrictions on grazing by animals is successfully achieving adequate safety levels.

Nevertheless, we must continue to be vigilant and maintain our research effort to protect against new strains of pathogens which are causing occasional public health problems and whose spread may be, in part related to sludge disposal operations.

Thirdly:

We have heard this week on how the problems of disposal of surplus slurries and manures from intensive stock-rearing operations in some countries are being solved by the development of practicable treatment methods. These processes are of course expensive, but in those countries where they are needed, the extra cost is considered acceptable in order to secure a safer environment. There may be a wider problem to be dealt with in future, caused by the effect of intensive agricultural operations on the levels of nitrate in surface and groundwaters. This may have significant effects on the reuse of animal slurries on cropland as well as on the general use of inorganic fertilisers in agriculture. A great deal of research and development work in this direction is still needed.

Finally:

Although the scientific community agrees to differ about some of the precautions which we should take when utilising sewage sludge, there appears to be a general consensus that there is a wealth of scientific backing for the procedures now in use, varying of course in different countries because of differences in geography, climate and farming practice.

There is however, the continuing problem on how to convince politicians and articulate members of the public that what we have recommended on scientific evidence are indeed the best practicable options taking all environmental/economic factors into account. Gillian Clegg's well presented paper today indicated for us the extent of that problem. This is the first of our conferences at which such a paper has been presented and I hope that we can continue to explore this more formal dialogue with the general public so that a compromise approach to the problem of sludge treatment and disposal might be evolved to satisfy all parties in the future.

In conclusion, and on behalf of all delegates, I would like to thank the Scientific Committee for the Symposium for arranging a very interesting scientific programme.

Many thanks are also due to session chairmen, speakers, participants to the discussions, authors of the posters and to the interpreters, whose contribution has been particularly appreciated.

Special thanks must be paid to Dr Alexandra Kouzeli-Katsiri of the National Technical University of Athens and to Dr Pierre L'Hermite of DGXII of the European Commission and Secretary of the COST 681 Programme for their invaluable contribution which made the Symposium possible.

Finally I would like to take this last opportunity to give my good wishes for the future to all my colleagues involved in the work of COST 681.

Poster Session

ABILITY OF A THERMIC SLUDGE TO SUPPLY CROPS WITH PHOSPHORUS

B. POMMEL(1) , R. BILLIET(2) , A. GANDILHON(2) , R. ROCHE(2)

(1)INRA Laboratory of Agronomy (F)78850 THIVERVAL-GRIGNON
(2) Students, Agronomy, INA-PG,
16, rue Claude Bernard (F)75231 Paris Cedex 05

SUMMARY

The sludges resulting from thermic conditioning in the Acheres-Paris treatment plant were tested as phosphorus source in comparison with monocalcium phosphate.

Ray-grass was used as P extractant,in such conditions that P is always the limiting factor,others beeing optimized, and a relative efficiency was calculated for each date of cutting on the basis of P cumulative uptake for all supplied doses.

Three kinds of experiments were carried on, in a greenhouse.

1. Evaluation of the *per se* activity of the sludge, measured as a kinetic. The relative efficiency of the sludge rise from 13% to 32% when the time pass from 900 degree.day after sowing to 2500 degree.day.

2. Determination of the behavior of the sludge and fertilizer when applied on soils with different P fixing capacities depending on their organic matter content. In the very short term (300 degree.day), the relative efficiency of the sludge is 88% on a loamy soil, and 44% on the same soil supplied with wheat straw 3 months before. Then, 2000 degree.day after sowing, both of them have stabilized, respectively at 50% and 35%.

3. Evaluation of the value as P source of soils whose P fertility has been built up for a long time by repeated sludges application.

1. INTRODUCTION

The plant of Achères near Paris, treats 2.1 millions m^3 of waste water in a day, and produces yearly 100 to 120 000 t of sludges at 50% of dry matter.

The specificity of this plant results in a thermic conditioning of digested sludges at 200°C of temperature and 18 bars of pressure during 45 minutes, according to the "Porteous" process. Afterwards, the sludges are filtered on press-filter.

These sludges have a low nitrogen (1.27% d.m. basis) but a heavy phosphorus content (3.1%). Their heavy metal content and the evolution of these elements after landspreading is presently the subject of researchs (1).

The S.I.A.A.P. (Syndicat Interdépartemental pour l`Assainissement de l`Agglomération Parisienne), responsible of the treatment of water and sludges and removal of sludges organized their utilisation in agriculture in the Vexin area, N.W. of Paris, on the basis of phosphorus valorisation (2). To this end, it was necessary to obtain a correct evaluation of the availability of their phosphorus, intrinsically or in conditions more representative of agricultural situations.

So, we appreciated the offer of phosphorus by the sludge to the plant :
- directly and immediatly on quartz sand,
- in competition with the fixation sites on an agricultural soil,
- from an agricultural soil anciently fertilized.

2. MATERIAL AND METHODS

We evaluate the sludge P availability by the means of a test based on P uptake by ray-grass, comparatively from sludge or from monocalcium phosphate. The plant, grown in a greenhouse in 700 g capacity pots, is in optimum conditions regarding every factor excepted phosphorus : this element is supplied in growing but always limiting doses as each source. The cumulative phosphorus uptake throughout the experiment is a linear function of the supplied dose : the sludge P relative efficiency is calculated as the ratio between slopes of regression lines for the sludge on the one hand, and for the fertiliser on the other (3).

The repetition of cuttings give informations about the variations of this relative efficiency with time.

3. RESULTS

1/ Specific value of the sludges.

It was determined after mixing the sludge in a quartz sand substrate. The relative efficiency of the sludge rise from 13% to 32% when the time pass from 900 degree.day after sowing to 2500 degree.day, and continue to grow afterwards, as suggested by the reading of the curve A (fig 1). Therefore this product is a slow-release phosphorus fertiliser (4).These results are beeing specified.

2/ Determination of the behavior of the sludge and fertilizer when applied on soils with different P fixing capacities depending on organic matter content.

In order to avoid the introduction of a parasite P source, we use a loamy soil, originating from Grignon (Yvelines), which has been depleted with phosphorus by 40 years of cultivation without P fertilisation (5). Two process are carried on :

1 - after receiving the P sources, the soil is amended with pulverized wheat straw as organic matter source, then fertilized, watered, and incubated at 18°C for 3 months. After this period of time,it is sowed ;

2 - the soil is sowed as soon as receiving the P source without any addition of organic matter.

The evolution of the cumulative uptake according to the treatments is shown on the figure 2 : the curve A represents the references in which the P originating from the soil and the straw is very low in comparison with the one from the seeds. The difference between the B and the A curves represents the P taken up from the sludge in the amended soil. The one between the D and the C curves represents the decrease of soil fixation of the soluble fertilizer due to the organic

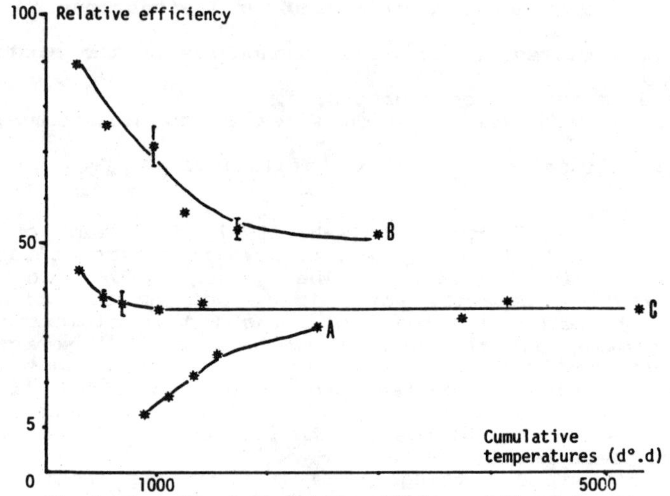

Fig. 1 : Evolution of the relative efficiency of the sludge, as percent, of monocalcium phosphate. A : On quartz sand ; B : In an agricultural soil ; C : In the same soil after application of pulverized straw and incubation

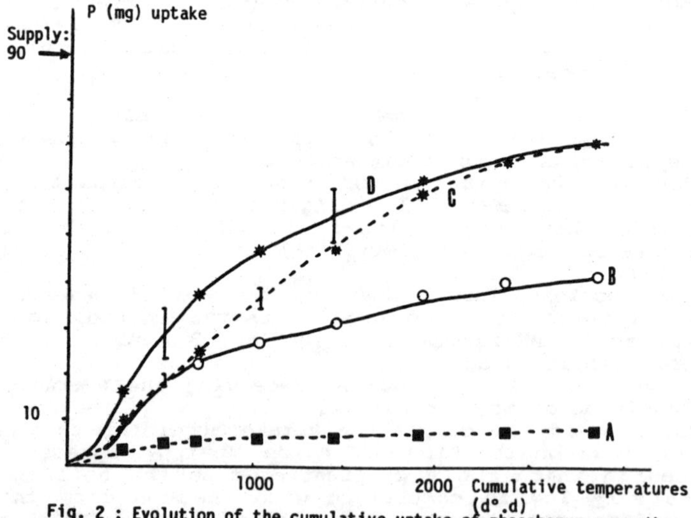

Fig. 2 : Evolution of the cumulative uptake of phosphorus according to the treatment. A : Soil alone ;
With 90 mg P supplied as : B : Sludge on amended soil ; C : monocalcium phosphate alone ; D : Monocalcium phosphate on amended soil

amendment.

In the fig 1, the B curve represents the relative efficiency of the sludge without incubation, and the C curve the same datas after incubation. In the short term the relative efficiency of the sludge in the non-amended soil is extremly high, due to the important fraction of the soluble fertilizer that is adsorbed on the soil, out of reach of the root extraction (B curve). The insoluble phosphorus of the sludge is not concerned in the same extent by this soil fixation.

But the adsorption of the monocalcium phosphate on the soil is mainly reversible, and for this reason, the relative efficiency of the sludge decrease with time, to reach an equilibrium at fifty per cent.

When the supply of a carbon source and incubation in the soil take place, the relative efficiency of the sludge is 45% in the short term and, later on, decrease slowly to reach an equilibrium (C curve). This value much lower than previously could be explained as a consequence of the stimulation of microbial life, resulting in the production of organic compounds able to combine with phosphorus in low energy bonds where phosphorus remain root accessible (6).

The divergence remaining at the end of the experiment between the B and the C curves represents the excess of irreversible fixation of soluble phosphorus in the case of direct application into the soil.

By itself,the thermic sludge is a better source of phosphorus when its application is not associated with organic amendment and incubation of the receiving soil.

3/ Evaluation of the effects of ageing at field of the sludge P.

The purpose of this experiment is to appreciate the consequences of ageing of the thermic sludges and of the superphosphate, in different soil conditions, and after different lengths of time. To this end, we sampled, in the Vexin area, soils in 10 situations where the P fertilization as thermic sludge application ceased between 1984 for the oldest one and 1989 for the most recent one. Concurrently, we sampled soils in 3 situations, homologous with the previous one, where the P fertilization as superphosphate ceased in 1975, 1985 or 1988.

These soil samples are presently under examination.

4/ Co-evolution of sludge nitrogen and phosphorus.

In parallel with a field experiment,we determine concurrently the evolution of N mineralisation and P availability at different stages of incubation of the soil-sludge mixture (with S. Bourgeois and M. Sbih). This experiment is in progress.

4. CONCLUSION

Specifically, the thermic sludge is a slow release phosphorus fertilizer. As a consequence of this, the phosphorus in this product is efficiently protected against soil fixation, and its short term efficiency relatively to the soluble fertilizer is, in the loamy soil used, very high. Later, beyond 2000 degree.day, this parameter stabilize at about 50%.

In other respects, the organic amendment of the soil receiving the sludge and its incubation does not improve the availability of the sludge P. Beyond 2000 degree.day, the relative efficiency of the sludge in the amended soil is stable at about 35%.

Presently, we try to determine the effects of the ageing of the sludge P at field on its plant availability.

ACKNOWLEDGMENTS

The authors would like to thank Christine Bouchard and Josiane Jean-Jacques for their technical assistance. This work was supported by the "Syndicat Interdépartemental d`Assainissement de l`Agglomération Parisienne".

REFERENCES

(1) BERMOND, A., IRELAND-RIPERT, J., and DUCAUZE, C.J. (1986 and 1987). Etude complémentaire des sols du Vexin soumis à des épandages de boues d`Achères. INA-PG, Laboratoire de Chimie analytique.

(2) E.V.S. (1988). Recueil de données ; valorisation agricole des boues d`Achères, 65p.

(3) POMMEL, B. (1984). Aptitude des déchets organiques à alimenter les cultures en phosphore. Thèse, Univ. Paris VI, 126p.

(4) GANDHILON, A., GARCIA, F. (1988). Etude de la valeur fertilisante des boues d`Achères. A.S.E./INA-PG, 60p.

(5) FOURBET, J.F. (1983). Essai permanent de fertilisation de Grignon 1958-1983, 7p.

(6) BILLIET, R. (1989). Essai d`évaluation de la valeur fertilisante en phosphore d`un apport de boues thermiques d`Achères. D.A.A. d`Agronomie, INA-PG, Paris.

TREATMENT OF TABLE OLIVE INDUSTRY WASTEWATER IN GREECE

G. C. KOPSIDAS

ELEOURGIKI, CENTRAL COOPERATIVE UNION OF OLIVE PRODUCTS, ATHENS

There are many table olive industries in Greece today. All these produce wastewater, which is disposed untreated in the environment. A bibliographic review shows that there are various methods, which can be used for the treatment of the previous referred wastewater. This work gives a solution to the wastewater treatment problem of a Greek table olive industry. According to this solution the wastewater is seperated in two portions. The one portion, which contains sodium chlorine, is treated at the following stages: (i) flow equalization, (ii) sedimentation, (iii) oxidation (rotaring biological contactors), (iv) clarification, (v) activated carbon treatment and (vi) recycling of the wastewater as brine. The other portion, which doesn't contain sodium chloride, is treatment at the following stages: (i) sedimentation and (ii) disposal in the environment.

1. INTRODUCTION

The wastewater, which is produced by the Greek table olive industry, is disposed untreated in the environment. In this work, the wastewater treatment problem is examined and a solution to it is given.

2. METHODS FOR TREATMENT OF TABLE OLIVE INDUSTRY WASTEWATER

A bibliographic review has shown that there are various methods, which can be used for the wastewater treatment. The use of these methods for the treatment of brine only gives better results, than the use of them for the treatment of the total amount of wastewater. These methods are given below.

1) Wastewater elimination

1.1) Disposal in the sea (1). The sea is a very good receiver for wastewater which contains sodium chloride. The disposal of wastewater in the sea doesn't create special problems. In California the wastewater disposal in the sea is strengthened by the state.

1.2) Underground disposal (2). The underground disposal of wastewater requires a water proof cesspool and an absorbent cesspool with large depth and adequate volume. If the subsurface disposal field water is used for irrigation, then this method of wastewater disposal will create problems.

1.3) Natural evaporation in lagoons (3). This method uses the solar energy for the wastewater evaporation in proper lagoons, as in the salt-pits. The residue, which remain in the lagoons, can be used for the sodium chloride recovery. This method requires proper climatological conditions and large area.

2) Methods which permit the reuse or the recovery of the subproducts.

2.1) Methods which eliminate the organic compounds. If the wastewater recycling is desirable, then the organic compounds elimination is required. The organic compounds elimination is necessary because: (i) perhaps it is the singular treatment which is required for the brine recycling, (ii) the organic compound concentrations in the wastewater must be lower than some supreme limits and (iii) high organic compounds concentrations create problems in some wastewater treatments.

The following methods are used:

2.1.1) Wastewater oxidation (4). This is the easiest and most tested method for the reduction of the organic compounds concentration. The biological oxidation reduces the BOD5 and the COD and increases the suspended solids. The aeration reduces the acids and the sugars and increases the microorganisms. The extension and the velocity of the oxidation don't depend from the sodium chloride concentration (<11.6 g/l). The following questions must be answered: (i) which are the organic compounds that remain after the oxidation and (ii) if these compounds can be removed by activated carbon treatment. A type of organic compounds (phenols), which are contained in the wastewater, is oxidized very hard.

2.1.2) Activated carbon treatment (5). A fixed-bed column is used as a means of contacting wastewater with granular carbon. Activated cardon with active surface 1.000-2.000 m2/g adsorb the 60-70% of the organic compounds from the wastewater. The activated carbon treatment increases the PH and the acidity, reduces the COD and the sodium chloride (little) and eliminates the colour of the wastewater. The carbon regeneration, at 100°C requires an amount of water equal to the one tenth of the amount of wastester which was treated by this carbon. The residue of the carbon regeneration has high concentration of the organic compounds, wich can be recovered, and low concentration of the sodium chloride.

2.1.3) Chemical and mechanical treatment. This method is a combination of the flocculation (6) and the filtration. The stages of the wastewater treatment are: (i) the NaOH addition (for PH=12), (ii) the proper additive addition, (iii) the agitation (30 min), (iv) the filtration (filter press is used), (v) the HCl addiditon (for desirable PH) and (vi) the activated carbon treatment. This method permits the wastewater recycling.

2.1.4) Reverse osmosis (7). Reverse osmosis is a process in which wastewater is separated from dissolved sodium chloride and organic compounds by filtering through a semipermeable membran, at a pressure greater than the osmotic pressure caused by the dissolved salt in the wastewater. This method deducts the 75-99% of the sodium chloride from the wastewater and produces water for irrigation.

2.1.5) Hyperfiltration (8). The hyperfiltration is a basic type of the reverse osmosis and uses a membran which permits the passing of the sodium chloride and the small molecular weight organic compounds through it. This method deducts the large molecular weight organic compounds (the 67% of the organic compounds) from the wastewater and can be used before the activated carbon treatment of this wastewater.

2.2) Methods which eliminate the inorganic compounds. These methods seperate the sodium chloride from the wastewater.
The following methods are used:

2.2.1) Methods without phase change. They are:

2.2.1.1) Dialysis and electrodialysis (9). Dialysis is a process in which the sodium chloride of the wastewater is separated through the use of semipermeable ion-selective membrane (celofan). Electrodialysis is a process in which the ions of the sodium chloride are separated through the use semipermeable ion-selective membranes, with an application of an electrical potential between two ellectrodes (accelerated dialysis).

2.2.1.2) Ion exchange (10). Ion exchange is a unit process by which the ions of the sodium chloride are displaced from the wastewater by ions of a different species in solution. For this purpose, both anionic and cationic exchange resins, which contain Ca^{++} and OH^-, must be used.

2.2.2) Methods with phase change. There are various methods (evaporation, boiling etc), which use the phase change of the water for the separation of the water from the wastewater.

3. THE TABLE OLIVE INDUSTRY IN GREECE

The annual Greek table olive production is about 65.000 tons. The 70% of the olives are black and the 30% of the olives are green. For the treatment of the table olives, there are many small domestic units and a small number of factories. There are about 75 factories with capacities between 200 and 7.000 tons table olives per year. The 50% of these factories have capacities near to 500 and 2.000 tons table olives per year. From these factories, the 75% of them elaborate green and black olives, the 20% of them elaborate black olives and the 5% of them elaborate green olives.
The treatment stages of the olives in the above factories are the choice, the cleaning, the fermentation and the packing for the black olives and the choice, the cleaning, the bitterness reduction (October-November), the cleaning, the fermentation and the packing for the green olives.
The table olive factories have been established in regions which produce this agricultural product. The 60% of the factories only are near to the sea. The 90% of the factories are near to villages and small towns.

4. THE QUANTITY AND THE CHARACTERISTICS OF TABLE OLIVE INDUSTRY WASTEWATER

The wastewater flow rate in a table olive factory is depended by the type, the quantity and the treatment stage of the olives, which are elaborated in it. A very large variation in the wastewater flow rate at all the factories (especially when they elaborate green and black olives) are observed.

The mean wastewater flow rate q (m3/h) in a factory (when it operates) is given by the following relation:

$q = 0,012Qg + 0.0016Qb$ (October-November)

$q = 0.0016(Qg+Ob)$ (rest of the year)

where Qg and Qb are the quantities (tons) of the green and black olives which are elaborated in the factory per year.

In considering usage rates of water supplies in a factory, it is important to note the distinction between the amount q_s (m3/h) of wastewater which contains sodium cheoride (brine) and the amount q_o(m3/h) of wastewater which doesn't contain sodium chloride. For the means prices of the qs and qo the following relations are given:

$qs = 0.0008\ (Qg+Qb)$

$qo = q - qs$

There are not independet collection systems of above two wastewater portions in the various factories. A collection system only collects all the amount of the wastewater.

Measuring in two factories (November) gives the following mean prices of the wastewater characteristics:

PH: 6-6,5
Temperature: environment temperature
Colour: light brown-black
Suspended solids: 100 mg/l
BOD5: 800 mg/l
COD: 1400 mg/l
Oils and fats: 10 mg/l
Sodium chloride: 12 g/l

5. A SOLUTION TO THE PROBLEM OF TABLE OLIVE INDUSTRY WASTEWATER

There is not wastewater treatment plant in the various Greek table olive factories today.

The wastewater treatment problem of a Greek table olive factory can be solved as follows:

1) Separation of the wastewater in two portions. One portion which contains sodium chloride (brine) and another portion which doesn't contain sodium chloride. Each portion is treated independently from the other. If such separation is not possible, then all the amount of the wastewater will be treated as brine.

2) Treatment of the wastewater which contains sodium chloride as follows:

2.1) Flow equalization. Variations are observed in the influent-wastewater flow rate. Flow equalization is used to overcome the operational problems caused by these variations and to improve the performance of the downstream process. An equalization basin, with mixing system and pump must be used. The volume V(M3) of this basin must be, at least, the 67% of the maximum wastewater volume which can be produced in a day (8 working hours.)

2.2) Sedimentation. The objective of treatment by sedimentation is to reduce the suspended solids content. A rectangular tank can be used. The determination of the dimensions of this tank must be based on the following date: (i) detention time >1.5 h, (ii) surface loading rate < 1m/h, and (iii) horizontal velocity < 36m/h.

2.3) Oxidation. One or more rotaring biological contactors can be used for the oxidation of the wastewater. A rotaring biological contactor consists of a series of closely spaced circular disks of polystyrene chloride. The disks are partially submerged in wastewater and rotated slowly through it. In operation, biological growths become attached to the surfaces of the disks and eventually form a slime layer over the entire wetted surface area of the disks. The rotation of the disks alternately contacts the biomass with the organic material in the wastewater and then with the atmosphere for adsorption of oxygen. The disk rotation affects oxygen tranfer and maintains the biomass in an aerobic condition. The determination of the number and the dimensions of the required rotaring biological contractors is based on their active surface and efficiency.

2.4) Clarification. Clarification is used for removal of the sludge, which is produced from the wastewater oxidation. As in the sedimentation, a rectangular tank can be used. The determination of the dimensions of this tank must be based on the following data: (i) recirculation ratio about 0.2, (ii) detention time >2, (iii) surface loading rate <1m/h and (iv) horizontal velocity < 36m/h.
The characteristics of the produced activated sludge are unknown.
A sludge-drying bed can be used for the dewatering of the sludge which is produced by the sedimentation and the clarification.
After clarification and chlorination (legal demand), the wastewater can be disposed in the environment (the sea is preferred), otherwise the following stages are proposed:

2.5) Activated carbon treatment. For this treatment one or two fixed-bed columns with granular carbon can be used.
The wastewater is applied to the top of the column and withdrawn at the bottom. The carbon is held in place with an underdrain system at the bottom of the column. The design of the fixed-bed columns must be based on the carbon adsorption isotherm for the under study wastewater, which is not known now. A Langmuir isotherm equation can be used at this time.

2.6) Recycling of the wastewater. After the activated carbon treatment, sodium chloride is added in the wastewater and it is used in the production as brine.

3) Treatment of the wastewater which dosen't contain sodium chloride as follows:

3.1) Sedimentation. As for the previous wastewater portion.

3.2) Disposal in the environment. This wastewater portion can be disposed without problems.

6. CONCLUSIONS

The proposed solution to the wastewater treatment problem of a Greek Table olive factory will reduce the problems, which are ctreated by the under study wastewater and will permit the reuse of the dissolved sodium chloride.

The operation of the first proposed wastewater treatment unit will give more data for a better design and operation of all this type wastewater treatment units, which will be constructed in future.

REFERENCES

(1) BURCHETT, M.E.,TCHOBANOGLOUS, G.and BURDOIN, A.J.(1967). A practical aproach to submarine outfall calculations. Public Works, Vol. 98, No. 5.

(2) THOMAS, R.E (1973). Land disposal II: An overview of treatment methods . J. Water Pollut. Control Fed., Vol. 45 pp 1476-1484.

(3) BARTSCH, E.H., and RANDALL, C.J. (1971). Aerated lagoons- A report on the state of the art. J. Water Pollut. Control Fed., Vol 43, No. 4.

(4) MEHTA D.S., DAVIS, H.H.and KINGSBURG, R.P. (1972). Oxygen theory in biological treatment plant design. J. San. Eng. Div.,ASCE, Vol 98,No. SA3.

(5) ALLEN, J.B., JOYCE, R.S.and KASCH,R.H.(1967). Process design calculations for adsorption from liquid in fixed beds of granular activated carbon. J. Water Pollut. Control Fed., Vol. 39, No. 2.

(6) Mc KEON, W.R.and MULDOWNEV., J.J. (1987). Evaluating alternative coagulants to determine efficiency and cost effectiveness. J. Am. Water works assoc., Vol 79.

(7) CULP, R.L., and CULP, G.L. (1971). Advanced wastewater treatment. Van Nostrand Reinhold, New York

(8) SCHOEDER, E.D (1977). Water and wastewater treatment. Mc Graw-Hill, New York.

(9) TCHOBANOGLOUS,G. (1979). Wastewater engineering: Treatment, disposal, reuse. Mc Graw-Hill, New Delhi.

(100 ELIASSEN, R., WYCKOFF, B.M.and TONKIN,C.D (1965). Ion exchange for reclamation of reusable supplies. J. AWWA, Vol 59, No. 57.

A TRHEE YEARS STUDY OF AGRICULTURAL USE OF SEWAGE SLUDGE

A. MANRIQUE; I. ARROYO; A. MORE; A. SANZ
Departamento de Análisis Ambiental
Consejería de Medio Ambiente
Junta de Castilla y León
Burgos (Spain)

Abstract

A trhee years field study concerning the agricultural use of sewage - sludge produced in the city of Burgos (Spain) has been carried out. In this contex the application effect of increasing amounts of sludge with a low concentration of heavy metals, except chromium, on a barquey crop, without addition of a mineral fertilizer was assessed. The sludge application rates were 18, 36, 54 and 72 Tons/ha, year (dry matter bassis).
A decrease in barley yield was observed in connection with chemical conditioning (lime and chloride) of sewage sludge.
The quality of soil and plant material was controlled. Nitrogen concentration in the grain were lowly dependent upon treatment. Phosphorus and potassium were no dependent upon treatment. Soil organic matther increased slightly with sewage sludge application rate. Heavy metals (Cd, Cr, Cu, Ni, Pb and Zn) were also controlled in soil and plant material. An increase of Cr, Zn and Cu was noted in soil by the application of greater amounts of sludge. Only Zn and slightly Cu showed an increase in barley grain.

1. INTRODUCTION

The Castilla y Leon Regional Governement and The Town Hall of Burgos have set up a research projet designed to investigate the possibility - of using sewage sludges, that are being produced in the treatment work plant of Burgos, in agriculture. During three years and experiment has - been developed about sewage sludge application on a cereal crop to obtain information on the following points:

- The productivity of crops treated with sewage sludge in comparison with ordinary mineral fertilizer application.
- Changes in crop quality, with particular reference to the absoption of heavy metals.
- Changes in soil characteristics following treatment with sewage sludge.

This paper presents preliminary results of a field experiment of barley crop growing on an alluvial soil near of the Burgos wastewater -- treatment plant.

2. METHODS AND PROCEDURES

In 1985, the Departamento de Análisis Ambiental and the Servicio Muni cipal de Aguas de Burgos (Spain) established a field test site to evaluate the use of municipal sludges of treatment work of Burgos city on cropland.

The soil at the disposal site is a loam textured. The soil contained 23,2 % clay, 42,1 % silt, and 34,7 % sand. The cation exchange capacity was 9,4 meq/100 gm. The others main soil characteristics are reflected in Table I.

From 1985 through 1987, chemicaly deshydrated anaerobically digested sludges were applied into experimental plots (6 by 5 m) at rates ranging from 18 t/ha per year to 72 t/ha per year. Each treatment was replicated four times. Sludges were spread on the soil surface and incorporated into the soil to a depth of aproximately 15 cm. A barley crop was planted each year following sludges application and, except for fertilizer control, no fertilizer was used. The main sludge characteristics are reflected in Table I.

Barley yields were determined and plant tissue samples (grain and leaf) were obtained for laboratory analysis. The soil was sampled at depth of 0-20 cm.

Grain and leaf samples were dried at 70º C.and ground. The samples were wet-ashed with a HNO_3 - $HClO_4$ (2:1) mixture.The digest were analyzed for heavy metals using flame of flameless atomic absorption spectroscopy. Nitrogen was determined by mean of Kjeldahl method. Phosphorus was determined in ashed solution by colorimetry.

Soils samples were air-dried and ground to pass a 1.0 mm opening - screen. Dry soil was refluxed in "aqua regia". After diluting to a known volume, the aqueous phase was analyzed for heavy metals by atomic absorption. Nitrogen and phosphorus were determined by Kjeldahl and colorimetry method respectively. Oxidable carbon was determined oxiding with $Cr_2O_7K_2$. Potassium was determined by emission spectroscopy in the solution of heavy metals.

3. RESULTS AND DISCUSSION

Crop yield varied considerably beween the different years, as reflected in Table II by consequence of drought during the second year of experience. A crop yield decrease has been detected between the mineral fertilization used as control and the sewage sludge treatment. Statistical analysis shows a significative negative interaction effect (p 0,001) - between the control and sewage sludge treatments.

This negative effect of application of sewage sludge may be in connectiontion with chemically conditioning (lime and chloride) of sludge. Has been admitted by several authors (1 and 2) that lime and chloride addition has a negative effect on phosphorus availability. In the other hand, sewage sludge used in this experiment had a low potassium content and it is an additional reason to depress yield crop.

The nutrient contents of barley grains and leafs are reflected in Table III. We can observe a nitrogen increase when sludge rates are encreased. No adverse effects associated with nitrogen have been observed in the experience. Nitrogen content is similar to the results obtained by other authors (3 and 4) and far of nitogen toxicity. In connection with phosphorus content a slight decrease has been observed. This in contrast with sludge application, because sludge introduce importants amounts of phosphorus in soils are not reflected in grains and leafs. This phenomenon can be explained as a blockade as consequence of chemical sludge conditioning.

Substantial amounts of heavy metals had been added to the soil trough

TABLE I MAIN CHARACTERISTICS OF SOIL AND SEWAGE SLUDGE

	SOIL	SEWAGE SLUDGE
pH 1:2,5	7,72	9,45
oxid. C. %	1,18	27,54
N %	0,15	2,36
P_2O_5 %	0,19	2,47
K_2O %	0,03	0,11
Ca %	2,53	12,53
Mg %	0,26	0,69
Cu ppm	22,4	269,00
Cr ppm	16,1	2300,00
Ni ppm	25,6	69,00
Zn ppm	50,7	941,00
Pb ppm	26,5	268,00
Cd ppm	0,07	1,55

the sludge application. In Table IV and V the heavy metal concentration of grains and leafs are reflected. Compared whith the control, there were slight, but consistent elevations of Zn and Cu, a very slight elevation for Cr and no modification for the others elements analyzed. If the metal content of plant tissue is used for diagnosis phytotoxicity, all values are withing the range concentrations considered normal for barley (5 and 6).

The plant tissue might accumulat e sufficients amounts of toxic elements to become a potential health hazard to consumers. In this respect Zn and Cu poses a serious danger (7), but by the moment the accumulation in our experiments do not pose any problem. In our situation, the increase of Zn and Cu content is higher in grain than leaf.

Soils present important modifications in their characteristics in - connection with repeated applications of sewage sludge. These variations are reflected in Tables VI and VII.

The most important modifications are refered to the organic matter content (expressed by oxidable carbon) and heavy metals.

The increase of organic matter han been important and similar to the increase obtained by other authors when sludges were applicated at the same magnitud order (8 and 9). This aspect is very important specially when sludge in used in soils with a very organic matter content, as such occurs in the area when this experience has been carried out.

At the 0-20 cm depth of the soil profile, all sludge treated soils showed elevated Cu, Cr and Zn values, as is reflected in Table VII Only the increase observed for these heavy metals were statisticaly significative between soil concentration in control and sewage sludge trearments. Increassing in heavy metals concentration in soils have been refered for several authors (10, 11 and 12).

Althought the heavy metals increase in soil has been important, however, the maximun value found for the highest rate of sludge application was within of the C.E.C. Directive (1986) limits. The main limitation to use this sludge in agriculture is chromium because its concentration in sludge is high and can pose serious problems as result of repeated additions. The other elements analyzed during this field experience do not present significant problems of heavy metal accumulation in soils.

TABLE II CROP YIELD OF BARLEY in Kg/ha

sewage sludge treatment	1 st year	2 nd year	3 rd year	mean
Control	5333	4666	5133	4845
18 t/ha	4111	2904	4766	3910
36 t/ha	4166	2841	4583	3836
54 t/ha	4499	3023	4033	3867
72 t/ha	4249	2867	3658	3594

TABLE III CONCENTRATION OF N,P,K AND Ca IN GRAINS AND LEAFS OF BARLEY (expressed in % of d.m.)

sewage sludge	N		P_2O_5		K_2O		Ca	
treatment	g.	l.	g.	l.	g.	l.	g.	l.
Control	2,95	2,28	0,54	0,71	1,69	0,50	1,36	0,042
18 t/ha	3,74	2,59	0,56	0,64	1,75	0,48	1,49	0,051
36 t/ha	3,54	2,88	0,56	0,69	1,86	0,50	1,58	0,043
54 t/ha	3,69	3,21	0,53	0,63	1,56	0,50	1,56	0,069
72 t/ha	4,21	2,91	0,52	0,60	1,65	0,52	1,68	0,044

TABLE IV AVERAGE CONCENTRATION OF HEAVY METALS IN BARLEY LEAFS (expressed in mg/kg of d.m.)

sewage sludge treatment	Cu	Cr	Ni	Zn	Pb	Cd
Control	11,81	6,42	4,49	21,70	0,89	0,02
18 t/ha	11,77	6,10	3,38	22,81	0,68	0,03
36 t/ha	12,82	5,22	2,75	27,65	0,79	0,03
54 t/ha	13,00	5,33	3,67	23,11	0,88	0,03
72 t/ha	13,33	7,13	3,83	26,25	0,79	0,02

TABLE V AVERAGE CONCENTRATION OF HEAVY METALS IN BARLEY GRAINS (expressed in mg/kg of d.m.)

sewage sludge treatment	Cu	Cr	Ni	Zn	Pb	Cd
Control	5,10	0,66	0,30	25,65	0,31	0,04
18 t/ha	6,61	1,31	0,61	34,18	0,45	0,09
36 t/ha	7,01	0,94	0,43	36,47	0,50	0,03
54 t/ha	7,14	0,66	0,48	37,52	0,48	0,02
72 t/ha	7,12	0,84	0,51	38,89	0,31	0,01

TABLE VI CONCENTRATION OF NUTRIENTS, pH AND OXIDABLE CARBON IN SOIL AFTER TREATMENTS

sewage sludge treatment	N	P_2O_5	K_2O	pH	oxid. C.
	%	%	%	1:2,5	%
control	0,14	0,18	0,27	7,74	1,24
18 t/ha	0,15	0,16	0,23	7,85	1,25
36 t/ha	0,17	0,22	0,26	7,78	1,42
54 t/ha	0,18	0,20	0,26	7,82	1,53
72 t/ha	0,19	0,25	0,26	7,92	1,56

TABLE VII AVERAGE CONCENTRATION OF HEAVY METALS IN SOILS AFTER TREATMENT (in mg/kg of d.m.)

sewage sludge treatment	Cu	Cr	Ni	Zn	Pb	Cd
control	21,77	17,37	22,27	53,75	29,25	0,06
18 t/ha	24,64	28,58	21,16	57,20	28,41	0,07
36 t/ha	33,93	44,12	23,87	75,79	30,68	0,09
54 t/ha	32,43	55,16	23,58	77,43	31,45	0,08
72 t/ha	40,79	69,43	24,10	80,83	31,54	0,07

4. BIBLIOGRAPHY

(1) DE HAAN, S. (1981) - Sewage sludges as a phosphate fertilizer. In: -- Phosphorus in Sewage Sludge and Animal Waste Slurries. T.W.G. Hucker G Catroux Eds., D. Reidel Publishing Company. Dordrecht. 149-162 pp.

(2) DE FRAJA FRANGIPANE, E.; VISMARA, R.; MEZZANOTE, V. (1986) - Environmental impact assesment of agricultural use of sewage sludge. In: - - Processing and Use of Organic Sludge and Liquid Agricultural Wastes. P. L'Hermite Ed., D. Reidel Publishing Company. Dordrecht. 414-426 pp.

(3) SILLAMPAA, M. (1985) - Comparison of microelement contens of different plant species grown side by side. In: Newsletter from the F.A.O. European Cooperative Network on Trace elements.

(4) BATTEN, G.D.; KHAN, M.A. (1987) - Uptake and utilisation of phosphorus and nitrogen by bread wheats grown under natural rainfall. Australian/ Journal Experimental Agriculture 27. 405-410 pp.

(5) CHANEY, R.L. (1982) - Fate of toxic substances in sludge applied to -- cropland. In: Proceding of an International Symposium on Land Aplica-- tion of Sewage Sludge. Tokyo 1982.

(6) WILLIANS, J.H.; SMITH, K.A.; JONES, R.J. (1986) - Varietal tolerance - in cereals to metal contamination in sewage a sludge treated soil. In: Proceedings the Fourth International Symposium Processing and Use of - Sewage Sludge and Liquid Agricultural Wastes. P. L'Hermite Ed. D. Reidel Publishing Company. Dordrecht. 537-542 pp.

(7) CAST (1976) - Council for Agricultural Science and Technology. AMES. Application of Sewage Sludge to cropland. Appraisal of Potencial Ha- - zards of the heavy metals to plants and animals. U.S. Depart of Comer- ce. NTIS Report PB-264015.

(8) KLADIVKO, E.J.; NELSON, D.N. (1979) - Changes in soil properties from/ application of anaerobic sludge. Journal Water Pollution Control Fede- ration. 51. 325-332 pp.

(9) EASTON, J.S. (1983) - Utilization and effects of anaerobically diges-- ted sludge on a red sandy soil of Natal. Water SA, 9-2, 71-78 pp.

(10) CHANG, A.G.; PAGE, A.L.; FOSTER, K.V.; JONES, T.E. (1982 a) Comparison of cadmium and zinc accumulation by four cultivars of barley grown in sludge amended-soils. Journal Environmental Quality. 11-3, 409-412 pp.

(11) McGRATH, S.P. (1983) - Metal concentrations in sludges and soil from a long-term field trial. J. Agric. Science Comb. 103. 25-35 pp.

(12) TACKETT, S.L.; YADVISH, A.M. (1987) - Heavy metals in soils from land/ reclamation with sludge. Journal Water Pollutio Control Federation. 59 5. 294-298 pp.

(13) C.E.C. (1986) - Council Directive of 12 June 1986 on the protection/ of the environment, and in particular of the soil, when sewage sludge is used in agriculture. Oficial Journal of the European Communities -- nº L 18,16 of 4-7-86

EFFECT OF SEWAGE SLUDGE USED AS FERTILIZER IN THE CENTRAL REGION OF SPAIN

I.WALTER, R.MIRALLES, E.FUNES, M.J.GOROSPE and M.BIGERIEGO
Dpto. Produccion y Tecnología de Alimentos.CIT-INIA.
Madrid-Spain.

Summary

Winter wheat was grown in clay-basic soil treated with sewage sludges in a long term field experiment, to study their value as fertilizer and their contamination risk.The soil was treated, every year,with 50 or 100 tm/ha of sludges from two different treatment plants of Madrid. Grain average yield, heavy metal accumulation in soil and uptake by the plant, from 1983 to 1989 in all treatments were examined. Grain yield through the years did not give a uniform response to sludge application.
The sludge incorporated into the soil causes an increase in the total content and DTPA-soluble fraction of all the heavy metals studied. This increase is greater in the DTPA-soluble fraction than in the total content of all the heavy metals studied.
The amount of Cr and Cd was undetectable in grain and straw even after the six year period, but a significant increase of the level of Zn, Ni, Pb and Cu in plant tissues was detected as a result of the treatment with sludge.

1. Introduction

The use of sewage sludge in soil as organic-mineral fertilizer is perhaps the best way to dispose of these residuals C.E.Clapp et al. 1986 (1).

However there are risks associated with the use of sludge on agricultural land; sludges present larger heavy metal concentration than soils, in particular Cd, Zn, Cu, Ni,Pb and Cr. Consequently an increase of the content of these metals in the top soil is produced when sewage sludge is applied year after year A. Andersson and K. Nilsson 1973(2), A. Bidwell and R. Dowdy 1987(3).

These levels may also increase in the crops,which is undesirable in the case of toxic elements A. Page 1974(4), A. Chaney and S. Hornick 1978(5).

This paper reports the results from a long-term field experiment, started in 1983 in the Central Region of Spain. Succesive incorporations of sewage sludge from two treatment plants of Madrid were added to the soil at two different rates each year,during six consecutive years, winter wheat being the crop under study.

2. Materials and Methods.

Winter wheat was produced on sludge-amended soil following a 6 yr continuous application of anaerobic sludges (S and V) from two different treatment plants of Madrid. Physicochemical properties of the soil and the sewage sludges are found on table 1.

TABLE 1

Characteristics of soil and sewage sludges

	Soil	Sewage sludges S (Sur)	Sewage sludges V (Viveros)
pH 1:2,5 H_2O	8,4	8,6	8,8
% oxidable C	0,47	24,21	25,27
% N	0,04	2,65	3,96
Texture	>10% clay loamy soil	–	–

Heavy metals mg/Kg	Soil HNO_3-$HClO_4$	Soil DTPA	S (Sur) HNO_3-$HClO_4$	V (Viveros) HNO_3-$HClO_4$
Cu	13,00	1,10	1639	510
Zn	42,50	1,17	2028	1451
Cr	18,14	0,20	275	390
Ni	12,60	0,27	81	84
Pb	80,78	2,57	733	623
Cd	0,30	0,03	82	37

The treatment consisted of three replicates with two application rates of 50 and 100 Tm/ha per year of both sewage sludges. A control without treatment was maintained for the whole period. Another control, with a basic inorganic fertilizer, was applied every year, similar in N, P and K to the lower sludge rate. The two controls had both three replicates. The sludge and mineral fertilizer was incorporated into the top soil before sowing with a moldboard plow in one annual application and always at the same rate in each plot. At the end of each cropping cycle (July) wheat was harvested and grain yield was recorded every year; Cd, Zn, Cu, Ni, Pb and Cr content was measured on grain and straw separately I. Walter and M. Bigeriego 1987 (6).

Soil samples were taken after the harvest from the surface layer (0-20 cm) each year during the 1984-1989 period. The soil samples were processed to determine the total (HNO_3-$HClO_4$) and available metals (DTPA-extraction) W. Lindsay and N. Norwell 1978 (7).

Heavy metals in solution were determined by atomic emission spectroscopy (DCP).

All data were analyzed statistically using the split-plot ANOVA procedure and Duncan test was used for means comparison. R. Waller and B. Duncan 1969 (8).

3. <u>Results and Discussion</u>

a). Crop yields

Sewage sludge application had no significant effect on grain yield compared with mineral fertilizer.

Table 2 shows that grain yield fluctuated annually, being higher in years 5 and 1 and lowest in year 3 where the minimum crop yield of the experiment (405 Kg ha) can be found. Lack of rain at critical growth stages probably reduced overal grain yield in this year and also in year 6.

Nitrogen levels were suitable for normal growth in all treatments, as indicated by the analysis (not shown) undertaken in grain and straw.

Total dry matter yield, expressed as the sum of grain and straw, did not differ significatively among treatments either.

TABLE 2

Average grain yield through the 6 years

TREATMENTS	RATES	YEAR 1	2	3	4	5	6	X
		(Kg/ha)						
Control		1315	860	700	1230	1300	540	990
Mineral	D_1	1380	1650	560	1230	2040	680	1257 A
S	D_1	1877	1200	650	1280	2100	800	1318 A
S	D_2	2100	1190	405	1130	2150	780	1293 A
V	D_1	1890	1510	650	1300	2070	840	1377 A
V	D_2	1620	930	560	1150	2100	710	1227 A
ANOVA		ns	ns	ns	ns	ns	ns	

Means followed by the same character are not different at $p=0.05$

b). Heavy metals in plant tissues

Zn,Ni, Pb and Cu content increased consistently in grain and straw following sludge applications.

Total metal uptake, defined as metal removed by harvest (aerial part), depends on the specific metal with some fluctuations between years.

The grain and straw Zn concentration (table 3) varied from 36,5 to 46,1 mg/Kg and from 39,1 to 66,2 mg/Kg when sludge was incorporated in year 6. These values are significantly higher than the control, but they are below the tolerance levels recomended by the National Research Council (1980).

Table 3 shows that sludge application causes a significant increase in Ni and Pb content only in straw,though there is no correspondence between sludge rate and metal uptake. This increase ranges from 0,21 to 0,30 mg/Kg for Ni and 0,10 to 0,37 mg/Kg for Pb. Both elements are also below the recomended levels, as in the case of Zn.

The accumulation of Cu in wheat grain was high when sludge S was applied but there are no significant differences between values with sludge V. The concentration of Cu in wheat straw did not increase with the incorporation of any sludge.

The Cd and Cr concentration in grain and straw in control was undetectable, under our conditions, and they remained the same after 6 years of sludge input.

TABLE 3

Average concentration of heavy metals of straw and grain in the last year

TREATMENTS	RATES	Cu	Ni	Pb	Zn
		Grain mg/Kg			
Control		7,29 B	0,00	0,00 B	14,27
Mineral	D1	8,26 B	0,27A	0,18AB	23,64 C
S	D1	15,91A	0,25A	0,24A	36,48 BC
S	D2	17,29A	0,22A	0,22A	46,15A
V	D1	7,62 B	0,20A	0,23A	44,74AB
V	D2	8,53 B	0,17A	0,21AB	37,87ABC
		Straw mg/Kg			
Control		5,64A	0,00	0,00 B	13,84
Mineral	D1	7,31A	0,12 BC	0,10 B	32,18
S	D1	7,62A	0,23AB	0,17AB	49,76
S	D2	7,62A	0,30A	0,18AB	66,24
V	D1	7,64A	0,25AB	0,37A	39,07
V	D2	7,61A	0,21AB	0,10 B	54,06

Means followed by the same character are not different at p=0.05.

c). Heavy metals content of soil.

With respect to mineral fertilizer, heavy metals measured by HNO_3-$HClO_4$ (as total amount of metals) and by DTPA-extraction (as metal availability) show a consistent increase yearly in all the treatments with sludge. This increase is more accused in the available fraction (DTPA-extract) than in the total content in all the metals studied except in Cr where the increase is the same in both measurements.

Total soil metal content in the last year of the experiment was greater than the metal content of the mineral control (table 4);however the levels are below the guidelines of the CEE.

The extractable metal content (table 5) does not give a uniform response to sludge type or rate although these values are always higher than controls.

The amount of Cd and Cr added with the sludges were sufficient to increase the concentration of these elements in the soil. None the less their availability for the plant seems to be small since they are not detected in wheat grain or straw.

This effect could be caused by the high pH level of the soil (above 8).

The constant use of sludge as fertilizer causes an increase in the Zn,Cu,Pb and Ni content in soil as well as in the wheat grain and straw

TABLE 4

Average concentration in soil of total metals in the last year (HNO_3-HClO_4)

TREATMENTS	RATES	Cu	Cr	Ni	Pb	Zn	Cd
				mg/Kg			
Control		11,07 B	20,71	16,65 C	71,25	32,26A	0,29
Mineral	D1	15,55 B	35,58	17,13 C	102,93	40,82A	0,50
S	D1	90,77	65,18	28,85 B	140,11A	145,51	2,00
S	D2	148,92	78,14A	35,11A	144,15A	195,83	2,17
V	D1	70,67A	53,51	34,66A	141,09A	162,19	1,67
V	D2	79,39A	81,63A	31,85A	169,52	176,48	1,89

Means followed by the same character are not different at p=0.05

TABLE 5

Average concentration in soil of available metals in the last year (DTPA)

TREATMENTS	RA.	Cu	Cr	Ni	Pb	Zn	Cd
				mg/Kg			
Control		1,09 B	0,17 CD	0,29 B	2,74 B	0,82 B	0,02B
Mineral	D1	1,45 B	0,14 D	0,33 B	2,41 B	1,79 B	0,04B
S	D1	18,17A	0,19 BCD	0,78	10,39A	27,03	0,83A
S	D2	37,29	0,24 BC	1,25	12,07A	43,93A	1,34
V	D2	18,93A	0,27AB	1,67A	16,24	41,83A	0,87A
V	D2	25,20	0,34A	1,66A	18,41	42,51A	0,88A

Means foloowed by the same character are not different at p=0.05.

References

(1) Clapp,C.et al.(1986). The role of organic matter in modern agriculture. Martinus Nijhoff Publishers. The Netherlands 209-253.

(2) Andersson,A. and Nilsson,K.(1976). Influence on the levels of heavy metals in soil and plant from sewage sludge used as fertilizer.Swedish J. Agric. Res.6 151-159.

(3). Bidwell,A. and Dowdy,R, (1987). Cd and Zn availability to corn following termination of sewage sludge applications. J. Environ. Qual. 16,4 438-442.

(4). Page,A. (1974). Fate and effects of sewage sludge when applied to agricultural lands EPA 670/2-74-005 USEPA. Cincinnati OH.
(5). Chaney,R. and Hornick,S. (1978). Accumulation and effects of Cd on crops p.125-140 In Proc. 1st Cadmium Conf. San Francisco.C.A. 30 Jan-2 Feb.1977. Metal Bulletin Ltd. London.
(6). Walter, I. and Bigeriego, M. (1987). Evaluation of sewage sludge from treatment plants as a fertilizer. 4th. Int. CIEC Symp. Agricultural Waste Management on Environmental Protection 11-14 May vol II 379-382.
(7). Lindsay, W. and Norwell, W. (1969). Development of a DTPA DTPA soil test for Zn, Fe, Mg and Cu. Soil Sci. Soc. Am.J. 42 421-428.
(8). Waller, R. and Duncan, D. (1969). A Bayes rule for the symmetric multiple comparisons problem. J.A.S.A. 64 1484-1503.

MEASUREMENT OF LABORATORY VANE SHEAR STRENGTH AS COMMUNITY STANDARD OR REFERENCE METHOD FOR TESTING THE SUITABILITY OF DE-WATERED SLUDGES FOR USE AS LANDFILL

U. LOLL, Darmstadt/Wiesbaden, FRG

1. INTRODUCTION

In the context of the activities of Working Party 1 of the Community's COST 681 Programme "Concerted action on treatment and use of organic sludges and liquid agricultural wastes", it was decided at the first meeting in Nancy, France, in 1987 to set up a study group to deal with the testing of the suitability of sludges as landfill.

The following members were involved:
U. Loll, Germany (Spokesman)
P. Ockier, Belgium
P. Geuzens, Belgium
W. Berndsen, Netherlands
J. van den Berg, Netherlands
H. Siegrist, Switzerland
Th. Lichtensteiger, Switzerland
R. Otte-Witte, Germany.

Following joint experiments with various sludges, thorough documentary evaluation and ongoing consultations, the members of the group reached the conclusion that the measuring procedure described below can be recommended for introduction as a Community standard or reference method.

2. BACKGROUND

Given the requirements of dumping techniques and tip stability, the suitability of sludges, especially sewage sludges, as landfill must not be decides merely on the basis of a minimum value for the dry solid matter which remains from the dewatered sludge, as is often still the case today. Various experiments, such as the cooperative experiments lasting several years, have shown beyond any doubt that parameters from the field of soil mechanics are more appropriate for defining the suitability of sludges as landfill. The abovementioned work and other publications mentioned in the bibliography demonstrate that, according to the latest knowledge, measurement of laboratory vane shear strength is to be recommended as a standard or reference method for determining suitability as landfill.

It is extremely important that a uniform measuring method be introduced because in many project decisions there is confusion over which parameters are appropriate for testing suitability as landfill or for setting limits in this field.

As no standardized measuring technique has yet been introduced in any country, Community Member State or otherwise, participating in the COST 681 Programme, there should be nothing to prevent the described measuring procedure being introduced, particularly if limits for determining suitability as landfill continue to be lacking.

3.0. MEASUREMENT OF VANE SHEAR STRENGTH USING THE LABORATORY VANE APPARATUS AS STANDARD OF REFERENCE METHOD

In order to obtain initial values for calculating the stability of tips and assessing the suitability of sludges as landfill, it is absolutely essential to measure shear strength. An appropriate quick mehtod is to use a laboratory vane apparatus.

The measured value is to be represented by τ_{VA}, where:

	τ	=	shear strength
and	VA	=	vane apparatus.

Its dimensions are kN/m^2.

3.1. PREPARATION OF SPECIMEN FOR TESTING LABORATORY VANE SHEAR STRENGTH

When testing vane shear strength, uniform conditions for the preparation of the specimen and performance of the experiment are essential to obtain precise and comparable results. The following parameters are of particular importance: particle size of the sludge, dimensions of the specimen, compaction of the substance and its density.

The de-watered sludge is broken up into particles no larger than about 10 mm, before it is placed in the sample container.

The sample container (a small Proctor mould - Fig. 1) has an internal diameter of 100 mm and a depth of 120 mm. An extension ring ensures that the specimen can be compacted in the prescribed manner up to the top edge of the test cylinder.

The sludge is placed in the container in three equal layers, each approximately 100 mm thick before compaction, which is performed by the small Proctor rammer with a drop weight of 2,5 kg, drop height 300 mm and diameter of face 50 mm. A correction plate (diameter 99.5 mm) placed on top of the sample ensures that compaction takes place evenly over the whole cross-sectional area. Each layer is compacted with 10 strokes of the rammer.

After compaction of the sample, the extension ring is removed and any sludge protruding above the top edge of the test cylinder ist cut off. Density is determined before performing the experiment.

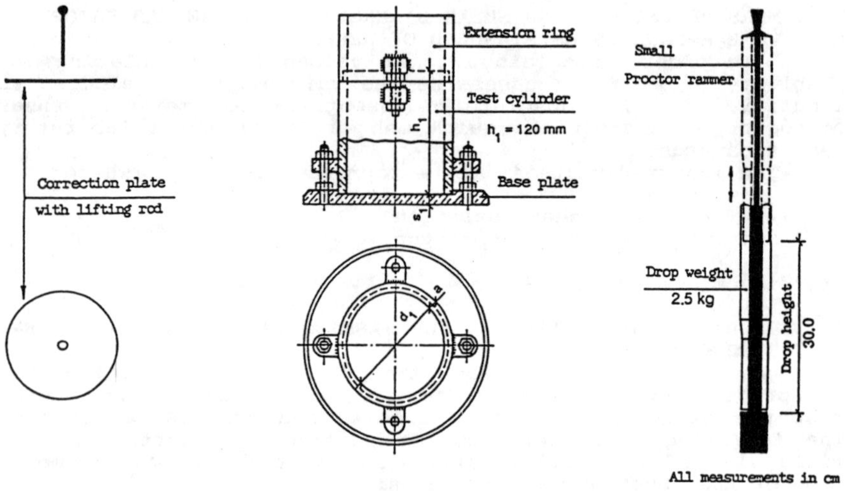

Fig. 1 : Proctor apparatus with Proctor rammer; Proctor mould with base plate and extension ring

3.2. DESCRIPTION OF LABORATORY VANE APPARATUS

In tests using the vane apparatus, the horizontal torsional moment which results in the shearing of the sludge sample is established by means of a vane inserted into the specimen. The vane consists of four plates set at right-angles to each other. As it rotates in the specimen, it carves out a cylindrical mass of sludge.

Experiments performed hitherto have used a laboratory vane apparatus which is produced by a UK manufacturer and commonly flund throughout Europe (Fig. 2). The vane, which is connected to a torsion spring with a display, is turned by a small electric motor or a manually operated crank. The angle of deflection is shown on a 360° scale with trailing pointers. The torsional moment is calculated from the angle of deflection and the spring constants. It increases as the angle of torsion of the vane increases. When the specimen fails, the angle of deflection, and thus the maximum torsional moment applied at that point, continues to be indicated on the scale. Three tests are carried out on the same specimen to obtain a definitive reading and eliminate random and "freak" results.

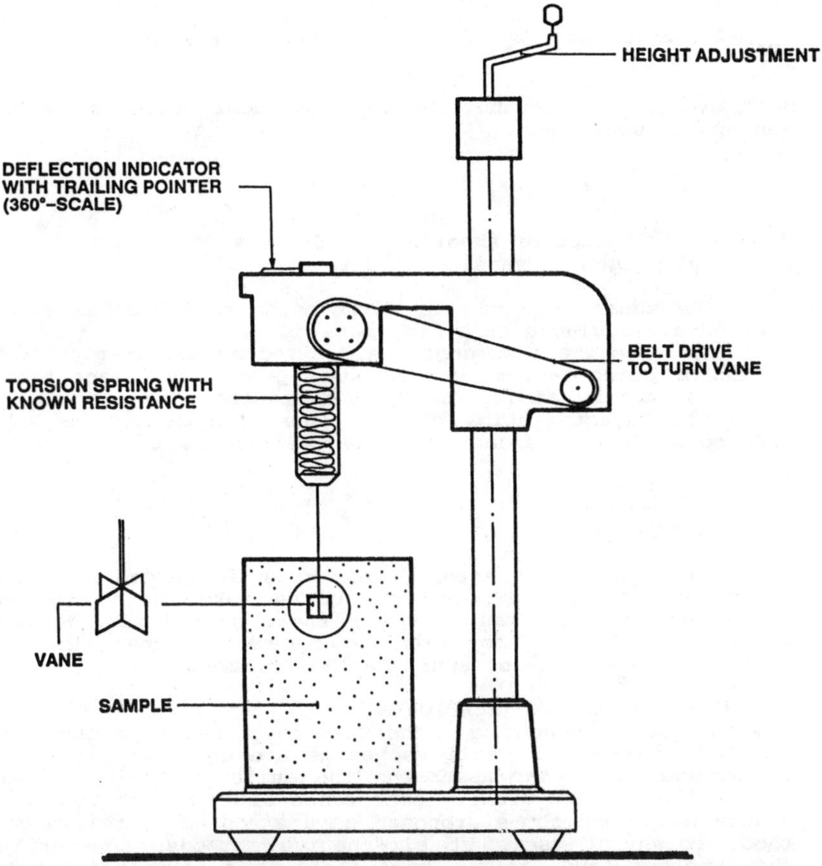

Fig. 2: Laboratory vane apparatus for measuring shear strength

The laboratory vane apparatus is a fully-developed divice which meets the same handling requirements as other laboratory equipment. It is easily operated by trained staff.

The vane consists of four fixed intersecting rectangular plates on a vertical shaft (Fig. 2).

The rotation of the vane in the sample produces a cylindrical shearing surface within which the volume remains constant. The shear strength is derived from the maximum torsional moment M_f. The shear stress opposes rotation at the casing and the vane faces. The shear resistance τ_{VA} resulting from the equilibrium of forces around the rotational axis is the total shear strength under undrained conditions:

$$\tau_{VA} = \frac{M_f}{K} \text{ [kN/m}^2\text{]} \quad M_f = \text{maximum torsional moment in [kN m]}$$

With even load distribution over the shear surfaces, K is given by the equation:

$$K = \frac{\pi}{2} \cdot d^2 \cdot h \left(1 + \frac{d}{3h}\right) [m^3]$$

d = diameter of sheared cylinder = width of vane
h = height of vane.

The shear strength τ_{VA} corresponds to the cohesion c_u of the undrained sludge at a high rate of shear.

The vane shear strength is derived by mathematical calculation or using tables for various types of equipment from the torsional moment at failure of the specimen and from the geometrical characteristics of the vane. The dimensions of the vane correspond to standards values, where h = d.

$$\tau_{VA} = \frac{3 \cdot m_f}{2 \cdot \pi \cdot d^3}$$

As the measured shear strength is influenced by the rate of shear, all measurements should be taken using the motor drive and at a constant rate of shear of 10°/min. With vane measurement d = 12.7 mm, this gives a rate of shear in the cylindrical shear area of approximately 1 mm/min.

4. OTHER MEASURING TECHNIQUES

As other measuring techniques concerned with the simulation of processes in soil mechanics are certainly appropriate in general terms for assessing the suitability of sludges as landfill, they are usually acceptable alternatives to the procedure described above proposed as a standard or reference method. Potential users of alternative methods, however, must show whether (and, if so, how) there is a correlation between their parameters and the reference values.

5. PRACTICAL IMPLICATIONS OF THE MEASURING TECHNIQUE

According to current knowledge, the suitability of sludges, especially sewage sludges, as landfill can be described with sufficient accuracy and reliability by measuring laboratory vane shear strength.

This measuring technique has long been used as a standard in soil mechanics and enables conclusions to be drawn regarding specific properties of matter or possible treatment methods.

The relevant equipment is fully developed, standardized and used on an international basis. The measurement procedure is suitable for implementation by staff in the sewage treatment and waste disposal sectors and produces meaningful results in a relatively short time (about 20 minutes).

6. SCOPE AND POSSIBLE LIMITS VALUES

According to current knowledge, laboratory vane shear strength limits of approximately 20 kN/m^2 (minimum 15 kN/m^2) should be observed for the depositing of de-watered sludges on domestic refuse tips if satisfactory compaction is to take place.

Higher limits are necessary for sludge-only tips of for dumps of special categories of sludge.

The stability of a whole tip must be calculated through separate investigations.

7. REFERENCES

[1] BS 1377: Methods of Test for soils for civil engineering purposes, Part 2. Determination of the liquid limit, Test 2a (1975).

[2] Cadling/Odenstad: THE VANE BORER, An Apparatus for Determining the Shear Strength of Clay Soils directly in the Ground. Royal Swedish Institute. Proc. No. 2, Stockholm (1950).

[3] Demberg, W. u. W. Tischer: Erfahrungen über die Ablagerung von gekalktem Klärschlamm am Beispiel einer Monedeponie. Schriftenreihe des Fachgebietes Siedlungswasserwirtschaft Universität - Gesamthochschule Kasse, Heft 2, S. 390 (1988).

[4] DIN 4096 Baugrund-Flügelsondierungen, Maße des Gerätes, Arbeitsweise, Auswertung (1980).

[5] Möller, U., G.C.W. Gay, W. Kassner, D. Köhlhoff, U. Loll, R. Otte-Witte: Neudefinition von Klärschlämmen, Korrespondenz Abwasser 31, H. 11, S. 928-933 (1984).

[6] Möller, U., G.C.W. Gay, W. Kassner, D. Köhlhoff, U. Loll u. R. Otte-Witte: Neudefinition der Deponierfähigkeit von Abwasserschlämmen. Teil 1, Schriftenreihe Siedlungswasserwirtschaft Bochum, Band 6, Bochum (1985).

EVALUATION OF ENTEROVIRUS CONTAMINATION OF SEWAGE SLUDGE

M. ALBERT, C. BERIL and L. SCHWARTZBROD

Faculté de Pharmacie
5, rue Albert Lebrun - 54000 NANCY France

Summary

In the present study , evaluation of viral contamination of primary sludge is performed with sludge coming from two sewage treatment plants : the first one is located in Nancy (France) and treats 300,000 equivalent inhabitants sewage, the second one is located in Barcelona (Spain) and treats 2,000,000 equivalent inhabitants sewage.
Virological analysis is performed from 40 g sub-samples of sludge which are submitted to an extraction-concentration procedure : extraction is realized using different eluents (glycine, beef extract, skim milk solutions) at alkaline pH. Then the eluate is concentrate either by organic flocculation according to Katzenelson et al., (8), either by two precipitation with Polyethyleneglycol 6000 according to Lewis and Metcalf (9).
The concentrates are inoculated on BGM or MA 104 cells cultures and enterovirus are enumerated using plaque assay. Results are expressed as PFU/100 ml^{-1} or 100 g^{-1}.
The results show that sludge are nearly always contaminated with enterovirus. But there is variations in virus concentration in the different samples :
- 100 % sludge samples from Nancy contain enterovirus : from 10 to 262 UFP 100 ml^{-1} (mean 69).
- 90 % sludge samples from Barcelona are polluted with enterovirus : from 0 to 347 PFU 100 g^{-1} (mean 135).

1. INTRODUCTION

Urban wastewater epuration treatments often lead to the production of large quantities of sludge, in which many micro-organisms such as the enteroviruses are gathered.

As a matter of fact, more than 130 virus types have been detected in wastewater (10) among which enteroviruses, hepatitis virus and also agents causing gastro-enteritis.

A suggestion is put forward : using this sludge for enrichment purposes in the agricultural field. This solution can be envisaged, provided that this agricultural use does not involve risks for human health. It is thus necessary to be in a position to determine the corresponding load of parasite eggs, bacteria and also viruses.

Considering that viruses are adsorbed on sludge solid particles (6), the sludge virological analysis usually comprises a phase during which viruses are extracted by elution followed by a phase during which viral particles are concentrated (5, 7, 12).

This work aims at determining the viral concentration of the primary sewage sludge using various extraction-concentration methods.

Only the enteroviruses with a cytopathic effect on "in vitro" cell cultures have been quantified.

2. MATERIAL AND METHODS

Sludge

Samples from two sewage treatment plants have been analysed.

Most of them have been collected from the treatment plant of Nancy (France) which treats 150,000 m3 water every day for a population of 300,000 equivalent inhabitants.

The other samples have been obtained from the San Adrian treatment plant located in Barcelona (Spain) which concerns a population of 2,000,000 inhabitants.

Virological analysis have been performed on primary sludge samples ; sludge samples that were mostly liquid have been blended and centrifuged.

Viral analysis was then performed on the pellet.

Elution

Viruses are eluted from an aliquot of 40 g of sludge by homogenezing the sludge in a solution of alkaline pH. Four different eluents have been used :

- 10 % beef-extract solution pH 9
- 1 % skim milk solution in tris buffer pH 9
- 0.05 M glycine solution pH 9
- 3 % beef extract solution in borate buffer pH 9

The sludge fraction was mixed either with 160 ml (20 %) or with 360 ml (10 %) of eluent.

After centrifugation the pellet is discarded and the supernatant constitutes the eluate.

Concentration

Two concentration procedures have been used :

- organic flocculation (OF) (8)
- polyethyleneglycol precipitation (PEG) (9).

The organic flocculation was performed by acidification of the eluate.

For 10 % beef extract eluent, pH was directly brought down to 3.5.

For 0.05 M glycine eluent, 3 % beef extract was added to the eluate before acidification to pH 3.5.

For 1 % skim milk eluent pH was adjusted to 4.

The polyethyleneglycol procedure consisted in two successive precipitations with PEG 6,000 according to Lewis and Metcalf (9) and Biziagos (3).

Virus recovery

After disinfection with chloroform and addition of antibiotics and antifungi (4), the concentrates were assayed on cell cultures. Quantification was performed using the agar overlay plaque technique on BGM or MA 104 cells grown in 100 mm Petri dishes. After incubation at 37° C, plaques were counted every day and viral concentration was reported in terms of PFU 100 g^{-1} or 100 ml^{-1} sludge.

3. RESULTS

Viral recovery from Nancy treatment plant sludge

Results are gathered in histograms and take into account the viral concentrations obtained with each analysis procedure (figures n° 1, 2, 3 et 4).

Viral recovery from Barcelona treatment plant sludge

The primary sludge coming from the Barcelona treatment plant has been analysed using a single technique combining the elution by a solution at 10 % beef extract pH 9 with a concentration by organic flocculation. Results are gathered in figure 5.

Fig 1 : Virus recovery with procedure using 10% beef extract solution / organic flocculation

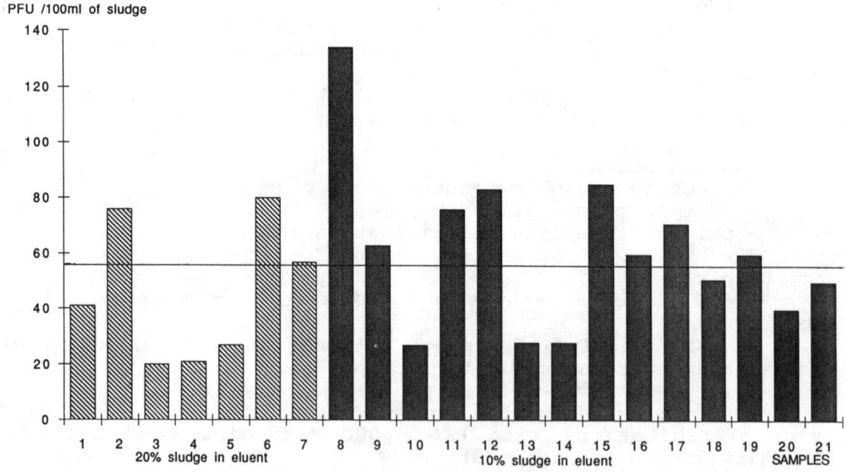

Fig 2 : Virus recovery with procedure using 1% skim milk solution /organic flocculation

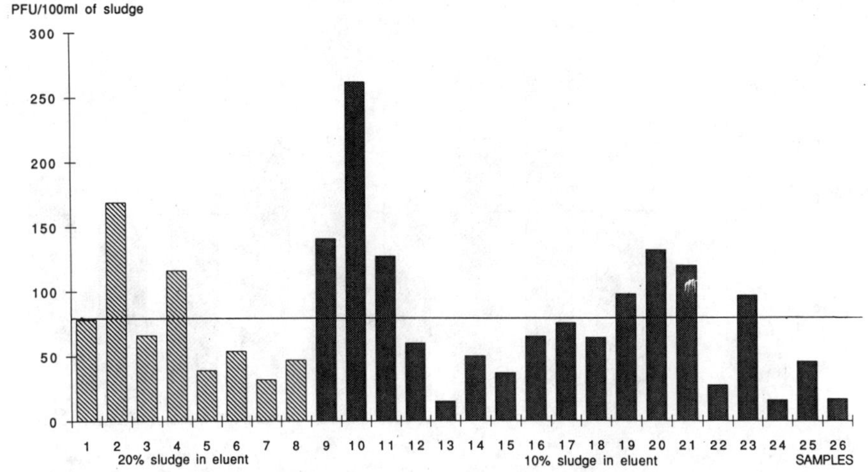

Fig 3 :Virus recovery with procedure using 0.05M glycine solution /organic flocculation

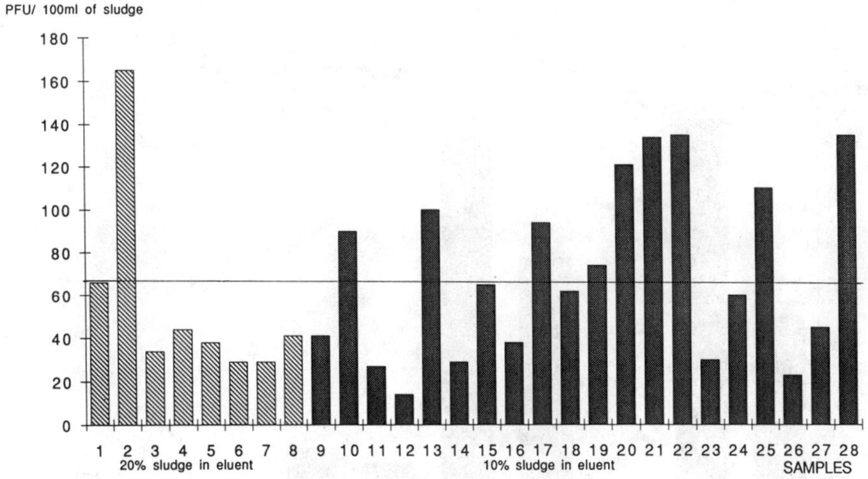

Fig . 4 :Virus recovery with procedure using 3% beef extract solution / PEG 6000 precipitation

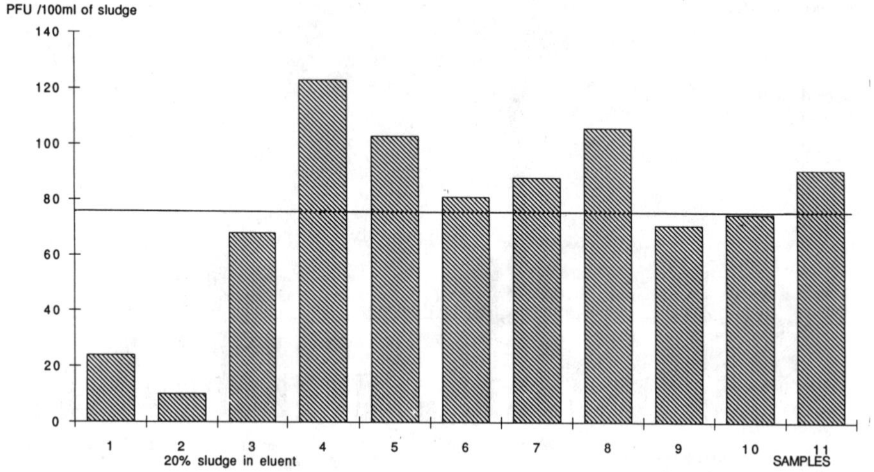

Fig.5 : Viral contamination of sludge from Barcelona treatment plants

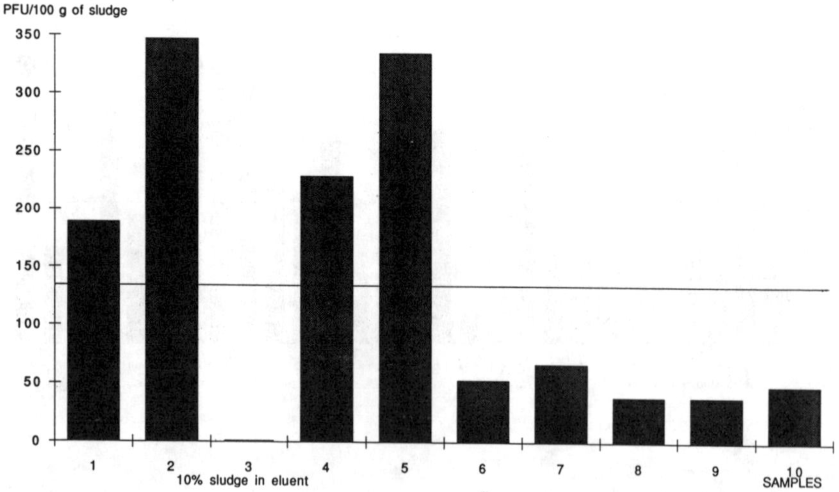

4. DISCUSSION CONCLUSION

When reading the histograms, you can notice that, regarding the primary sludge from Nancy

- on the one hand all analysed samples contained infectious enteroviruses,
- on the other hand, the viral concentration varies from 10 to 262 PFU 100 ml^{-1} with a mean value of 69 PFU 100 ml^{-1}.

Considering the results obtained with each analysis procedure (fig. 1, 2, 3 and 4) the mean values of viral density respectively amounting to 56, 79, 67 and 76 PFU 100 ml^{-1} are very closed to the general mean value of 69 PFU.

Very significant variations can be observed in the virus content depending on the samples tested.

Thus, concentrations vary :

- from 20 to 134 PFU 100 ml^{-1} fig. 1
- from 15 to 262 PFU 100 ml^{-1} fig. 2
- from 14 to 165 PFU 100 ml^{-1} fig. 3
- from 10 to 123 PFU 100 ml^{-1} fig. 4

Results obtaided in this way do not significantly differ whatever the extraction-concentration procedure may be.

As far as the results obtained with sludge from Barcelona are concerned, one sample out of 10 proved to be negative and the concentration in infectious virus varies from 0 to 347 PFU 100 g^{-1} with a mean value of 135 PFU 100 g^{-1}.

As a conclusion, the viral concentration in the sludge is much more significant in Barcelona than in Nancy.(135 as against 69 PFU).

However, the results obtained, be it in Barcelona or in Nancy, are in accordance with the values mentioned for similar sludge by :

. Bermann et al. (1) : from 27 to 2320 PFU 100 ml^{-1}
. Goyal et al. (6) : from 13 to 930 PFU 100 ml^{-1}
or . Scheuermann (11) : from 6 to 75 PFU 100 ml^{-1}

On the contrary, values are significantly lower than those reported by Brashear and Ward (5), who detect 1,000 to 14,000 PFU 100 ml^{-1} or by Berg and Bermann (2) with concentrations varying from 380 to 11,600 PFU ml^{-1}.

It thus appears that viral contamination in primary sludge is pratically permanent but with very significant variations in concentration from one sample to another.

REFERENCES

(1) BERMAN, D., BERG, G. and SAFFERMANN, R.S. (1981). A method for recovering viruses from sludges. J. Virol. Methods, 3, 283-291.

(2) BERG, G. and BERMANN, D. (1980). Destruction by anaerobic mesophilic and thermophilic digestion of viruses and indicator bacteria indigenous to domestic sludges. Appl. Environ. Microbiol., 39, 361-368.

3) BIZIAGOS, E., PASSAGOT, J., CRANCE, J.M., and DELOINCE R. (1989). Hepatitis A virus concentration from experimentally contaminated distilled, tap, waste and seawater. Wat. Sci. Tech., 21, 255-258.

(4) BLOCK, J.C. and SCHWARTZBROD, L. (1989). Viruses in water systems : detection and identification. V.C.H. publishers, New York, 136 p.

(5) BRASHEAR, D.A. and WARD, R.L. (1982). Comparison of methods for recovering indigenous viruses from raw wastewater sludge. Appl. Environ. Microbiol., 43 1413-1418.

(6) GERBA, C.P. (1984). Applied and theoritical aspects of virus adsorption to surface. Adv. Appl. Microbiol. 30, 133-168.

(7) GOYAL, S.M., SCHAUB, S.A., WELLINGS, F.M., BERMANN, D. GLASS, J.S., HURST, C.J., BRASHEAR, D.A., SORBER, C.A. MOORE, B.E., BITTON, G., GIBBS, P.H., and FARRAH, S.R. (1984) b. Round robin investigation of methods for recovering human enteric viruses from sludge. Appl. Environ. Microbiol., 48, 531-538.

(8) KATZENELSON, E., FATTAL, B. and HOSTOVESKY, T. (1976) Organic flocculation : an efficient second step concentration method for the detection of viruses in tap water. Appl. Environ. Microbiol., 32, 638-639.

(9) LEWIS, G.D. and METCALF, T.C. (1988). Polyethylene-glycol precipitation for recovery of pathogenic viruses.

(10) RAO, V.C., METCALF, T.G. and MELNICK, J.L. (1988). Virus humains dans les sédiments, les boues et les sols. Bull. O.M.S., 64, 341-356.

(11) SHEUERMANN, P.R., FARRAH, S.R. and BITTON, G. (1986). Development of a method for the recovery of enteroviruses from aerobically digested wastewater sludges. Water Res., 1, 15-19.

(12) SCHWARTZBROD, L. and MATHIEU, C. (1986). Virus recovery from wastewater treatment plant sludge. Water Res., 8, 1011-1013.

SIMULATION OF NITROGEN TRANSPORT IN SOILS DURING WASTEWATER APPLICATIONS

VASSILIS Z. ANTONOPOULOS

Dept. of Hydraulic, Soil Science and Agricultural Engng., Aristotle University
Thessaloniki 54006, Greece.

Summary

A mathematical model was developed to simulate water movement, mass transport and nitrogen transformations in soils. This model was used to evaluate the application rate and the variations of the soil solution in the unsaturated zone, as well as the groundwater quality, when nitrogen wastewater is applied to soils covered by plants. The model is one-dimensional and based on the Galerkin finite element method. The submodel of mass transport and transformations of nitrogen incorporates mechanisms and processes of mass transport by dispersion and convection of ammonium and nitrate nitrogen, nitrification of ammonium to nitrates in one step, denitrification of nitrates, ammonium exchange and uptake of NH_4 and NO_3 ions.

1. INTRODUCTION

Application of sewage water to land is a good method of preventing pollution of water resources and also enables wastewater to be reused. Sewage water applied to land can be used immediately for crop production, it can be renovated by soil filtration or it can cause groundwater contamination. Most of the nutrients must be removed from sewage water as it moves through the soil if land application is to be environmentally acceptable.

Nitrogen contributes to eutrophication of surface water and when excess nitrogen reaches the ground water, can be a health hazard. The extent of nitrogen removal in the land treatment systems is affected by numerous physical, chemical and biological processes in soil and by the dynamic state of the soil moisture.

A method of application of wastewater to land is the low-rate or irrigation system (5). With low-rate systems, the application of wastewater takes place by irrigation of lands where crops are grown. Nitrogen is uptaken by plant roots and removed from the field in the harvested plant material.

The knowledge of nitrogen mass transport and transformations in soils, when nitrogen is applied by fertilizer or by wastewater, nitrogen or organic, is obtained by mathematical modeling and computer simulation. Among the mathematical models which have been developed in recent years to describe the nitrogen transport and transformations in the soil there are some models which assume steady-state water flow conditions (2,6,9) and there are other assume nonsteady or transient water flow in the soil profiles (3,8). All are based on analytical solutions, on solutions of equations by finite difference analyses, or more recently on finite element analyses (4).

In this paper a mathematical model is developed to simulate water movement, mass transport and nitrogen transformations in variably saturated soils under transient flow conditions. The model is one-dimensional and based on the Galerkin finite element method.

2. MODEL DESCRIPTION

Water flow in variably saturated soil.

The one-dimensional vertical flow of water in variably saturated porous media with water uptake by plant roots, is described by the equation

$$L_w(h)=\frac{\partial}{\partial z}\left(K\frac{\partial h}{\partial z} - K\right)- C_h \frac{\partial h}{\partial t} + A =0 \qquad (1)$$

where h is the pressure head, K is the hydraulic conductivity, C_h is the specific moisture capacity, A is the sink term for root water extraction, z is the vertical coordinate (positive down) and t is the time. In the present mathematical model the macroscopic approach of water uptake by plants is described by the Molz and Remson model (7), as modified by Selim and Iskandar (1981). This model is represented as

$$A(z,h)=T\ R(z)\ K(h)/ \int_0^d R(z)\ K(h)dz \qquad (2)$$

where d is the maximum depth of the root zone in the soil and T is the evapotranspiration rate per unit area of soil surface, R(z) is the root distribution as a function of depth in the soil profile. This model was successfully used when high soil water contents (low suctions) were maintained in the soil root zone, such as in land treatment-slow infiltration systems.

Soil nitrogen transport and transformations

When sewage effluent is applied to a land surface after treatment, nitrogen may be removed by a variety of chemical and biological reactions. These reactions can be divided into two main groups. In the first group nitrogen is removed from the soil system by denitrification (biological), volatilization of ammonia, chemodenitrification and uptake by plants. In the second group, are the adsorption of NH_4-N on the soil cation exchange sites, adsorption by organic matter and incorporation into microbial tissue. Figure 1 shows a diagrammatic representation of a simplified model of nitrogen transformations in the soil, where the nitrification process is considered as a single step and the organic N phase is not incorporated in the model.

The one-dimensional vertical mass transport and transformations of NH_4-N and NO_3-N under transient flow and variably saturated soil conditions, are described by the equations

$$L_A(C) = \frac{\partial \theta C}{\partial t} + \frac{\partial \rho S}{\partial t} - \frac{\partial}{\partial z}\left(\theta D\frac{\partial C}{\partial z}\right) + \frac{\partial qC}{\partial z} + \Phi_1 - Q_{NH_4} = 0 \qquad (3)$$

$$L_N(Y) = \frac{\partial \theta Y}{\partial t} - \frac{\partial}{\partial z}(\theta D\frac{\partial Y}{\partial z}) + \frac{\partial qY}{\partial z} + \Phi_2 - Q_{NO_3} = 0 \quad (4)$$

where C and Y are the concentrations of NH_4-N and NO_3-N in the soil solution, respectively, D is the dispersion coefficient, θ is the volumetric moisture content, q is the Darcy velocity, ρ is the bulk density, S is the amount of NH_4-N in the absorbed phase per gram of soil, Φ_i is the rate of NH_4-N (i=1) or NO_3-N (i=2) transformations per unit soil volume and Q_{NH_4}, Q_{NO_3} are the rates of plant uptake of NH_4-N and NO_3-N per unit soil volume, respectively.

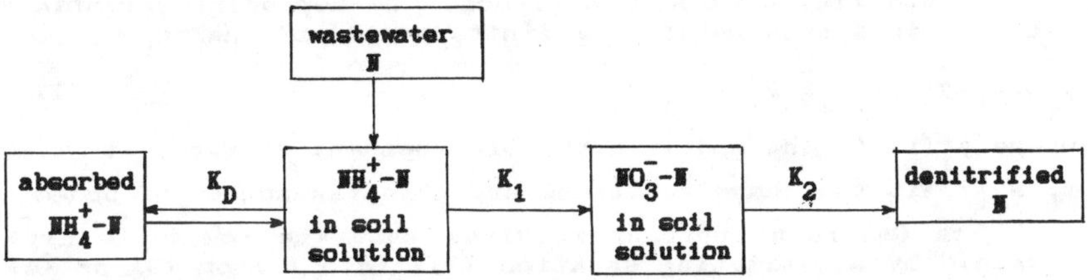

Fig. 1 Nitrogen transformation processes.

The transformation terms Φ_1 and Φ_2 were represented as follows (Fig. 1)

$$\Phi_1 = - K_1\theta C \quad (5)$$

$$\Phi_2 = K_1\theta C - K_2\theta Y \quad (6)$$

Equations (5) and (6) describe the nitrification of NH_4-N and denitrification of NO_3-N. These N transformation processes are assumed to follow a first order kinetic type reaction. The rate coefficients K_1 and K_2 are considered as variables dependent upon environmental and other factors, such as temperature, pH and aeration (8). In order to incorporate these factors, the coefficients are expressed as

$$K_i = k_i f_i \quad (7)$$

where k_i is considered constant and represents the maximum value and f_i is empirical function. For the nitrification rate K_1 the function f_1 is expressed as a function of pressure head, while the denitrification function f_2 is considered as a function of the degree of water saturation in the soil.

The adsorption of NH_4-N is considered instantaneous and reversible with equilibrium defined by a linear Freundlich type isotherm of the form

$$S = K_D C \quad (8)$$

where K_D is the distribution coefficient for the reaction.

The plant uptake terms Q_{NH_4} and Q_{NO_3} are described by a macroscopic model, in which the Michaelis-Menten approach is used to determine the rate of N uptake as a function of root density and concentration of ammonium and nitrate in the soil solution

$$Q_{NH_4} = I_{max} R(z,t)\ C / [K_m + C + Y] \qquad (9)$$

$$Q_{NO_3} = I_{max} R(z,t)\ C / [K_m + C + Y] \qquad (10)$$

where I_{max} is the maximum rate of N uptake per unit root length and K_m is the Michaelis constant which represents the concentration of N at $0.5I_{max}$.

Numerical solution by the finite element method.

In the finite element approach, the dependent variable y (=h,C,Y) is approximated by a finite series of the form

$$y(z,t) \approx \hat{y}(z,t) = \sum_{j=1}^{n} N_j(z) y_j(t) \qquad (11)$$

where $y_j(t)$ is the nodal values of dependent variable at time t, $N_j(z)$ is the shape functions and n is the number of nodes.

The Galerkin approach requires that the residuals $L(\hat{y})$, obtained by substituting equation (11) into (1) or (3) or (4), be orthogonal to the selected weighting functions

$$\int_0^d L(\hat{y}) N_i(z) dz = 0 \qquad (12)$$

where N_i is the weighting functions and d is the soil depth.

Substituting equations (1), (3) and (4) respectively, into (12), using Green´s theorem to remove the second derivative and applying integration by parts to the spatial derivatives, leads to:

$$[A_{ij}]\{y_j\} + [B_{ij}]\{\frac{dy_j}{dt}\} = \{F_i\} \qquad (13)$$

where, in the case of water flow:

$$[A_{ij}] = \sum_e \int_e K \frac{dN_i}{dz}\frac{dN_j}{dz} dz \qquad (14a)$$

$$[B_{ij}] = \sum_e \int_e C_h N_i N_j dz \qquad (14b)$$

$$\{F_i\} = q_0 - q_d + \sum_e \int_e K \frac{dN_i}{dz} dz - \sum_e \int_e Q_{NH_4} N_i dz \qquad (14c)$$

In the case of equation of NH_4-N transport:

$$[A_{ij}] = \sum_e \{ \int_e \theta D \frac{dN_i}{dz}\frac{dN_j}{dz} dz + \int_e q N_i \frac{dN_j}{dz} dz + \int_e \frac{dq}{dz} N_i N_j dz +$$

$$\int_e K_1 \theta N_i N_j dz + \int_e R \frac{d\theta}{dt} N_i N_j dz \} \qquad (15a)$$

$$[B_{ij}] = \sum_e \int_e R \theta N_i N_j dz \qquad (15b)$$

$$\{F_i\} = q_d C_d - J_{Cd} - q_0 C_0 + J_{C0} - \sum_e \int_e Q_{NH_4} N_i dz \qquad (15c)$$

and R is the retardation factor ($= 1+\rho K_D/\theta$)

and in the case of equation of NO_3-N transport:

$$[A_{ij}] = \sum_e \{ \int_e \theta D \frac{dN_i}{dz}\frac{dN_j}{dz} dz + \int_e q N_i \frac{dN_j}{dz} dz + \int_e \frac{dq}{dz} N_i N_j dz +$$

$$\int_e K_2 \theta N_i N_j dz + \int_e \frac{d\theta}{dt} N_i N_j dz \} \qquad (16a)$$

$$[B_{ij}] = \sum_e \int_e \theta N_i N_j dz \qquad (16b)$$

$$\{F_i\} = q_d Y_d - J_{Yd} - q_0 Y_0 + J_{Y0} + \sum_e \{ \int_e K_1 \theta C N_i N_j dz - \int_e Q_{NO_3} N_i dz \} \qquad (16c)$$

The time derivatives in equation (13) are approximated by a central difference formula which achieves as second-order accuracy (1). The results of these substitutions are three systems of linear algebraic equations which have tridiagonal matrices of unknowns.

3. RESULTS AND DISCUSSION

The mathematical model is applied to simulate problems of nitrogen wastewater application in a homogeneous and layered soil. The purpose of this analysis is to verify the accuracy of the model by comparison with other numerical solutions, and to determine the sensitivity of the numerical model to various parameters and the applicability of the model to field conditions.

The following applications of mathematical model presented here are based on the experimental data of Selim and Iskandar (1981). For all cases studied, wastewater which contained 25 mg/l NH_4-N and zero nitrate content was applied weekly at a rate of 0.5 cm/h and duration of 10 hrs, to a sandy loam soil. The physical and chemical properties of this soil are given by Selim and Iskandar (1981). On this soil, a two-year-old mixture of reed canary-grass and tall fescue was grown, for which the root distribution function is

$$R(z,t) = 226\exp(-0.1z) \qquad (17)$$

In the first application, wastewater was applied to a three-layer soil. The used values of coefficients of transformation reactions assumed were: K_D=0.25 cm^3/g, K_m=1.0 ppm. Imax=0.001 µg N/h cm of roots, k_1=0.1 and k_2=0.01 h^{-1}. The coefficients K_1 and K_2 were assumed dependent respectively from pressure head and the degree of water saturation (8).

Figure 2 shows the distributions of pressure head, NH_4-N and NO_3-N respectively, for the first week following the application of 5 cm wastewater. In this figure the results of the finite element model are compared to the results of the finite difference model of Selim and Iskandar. Pressure head and NH_4-N distributions predicted by the finite element model are almost identical to those of the finite difference model. The results of NO_3-N distributions, however, are not in close agreement with those of the finite difference model. The NO_3-N distribution on the 7th day presents a peak like that of the initial distribution, while the distribution of finite difference are more dispersed.

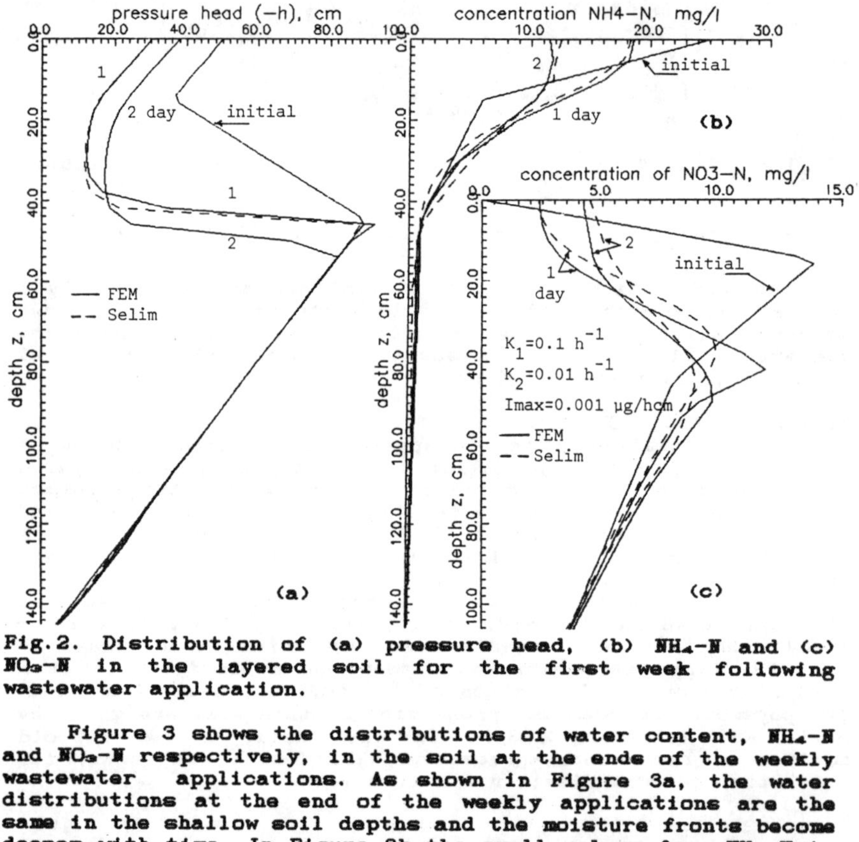

Fig. 2. Distribution of (a) pressure head, (b) NH_4-N and (c) NO_3-N in the layered soil for the first week following wastewater application.

Figure 3 shows the distributions of water content, NH_4-N and NO_3-N respectively, in the soil at the ends of the weekly wastewater applications. As shown in Figure 3a, the water distributions at the end of the weekly applications are the same in the shallow soil depths and the moisture fronts become deeper with time. In Figure 3b, the small values for NH_4-N in the soil indicate that approximately all the applied NH_4-N in the soil is removed or transformed. The NO_3-N distributions become deeper with time and the maximum concentrations decrease as the time increases.

Another set of simulations examine the influence of a range of Imax values on the fate of wastewater nitrogen in the soil. Simulations were performed for three different values of Imax: 0.0005, 0.001 and 0.0015 µg N/h cm, while all other parameters remained constant. The comparison results for NH_4-N and NO_3-N at the end of the first weekly wastewater application are given in Figure 4. As expected, for small values of Imax, the NH_4-N and NO_3-N, distributions show

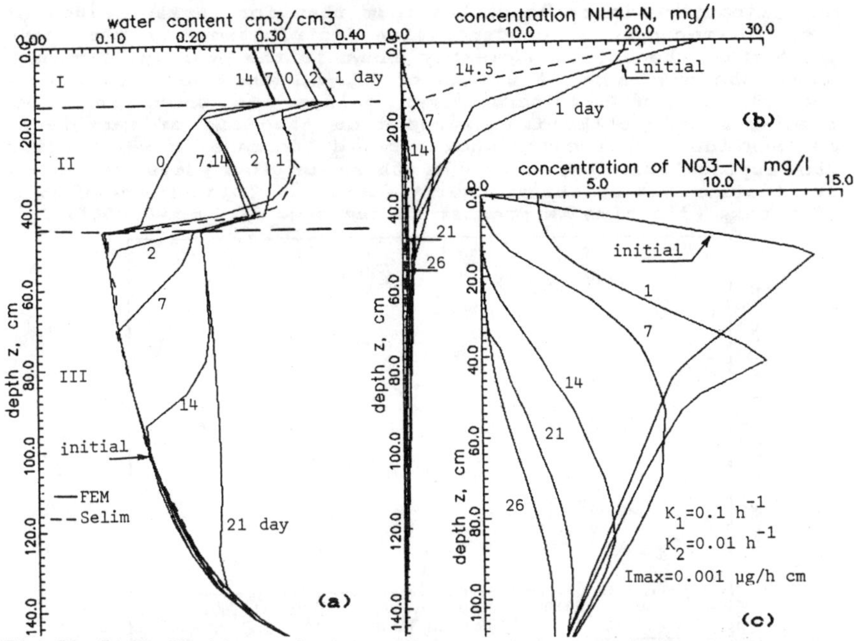

Fig.3. Distribution of (a) water content, (b) NH_4-N and (c) NO_3-N in the layered soil following weekly wastewater applications.

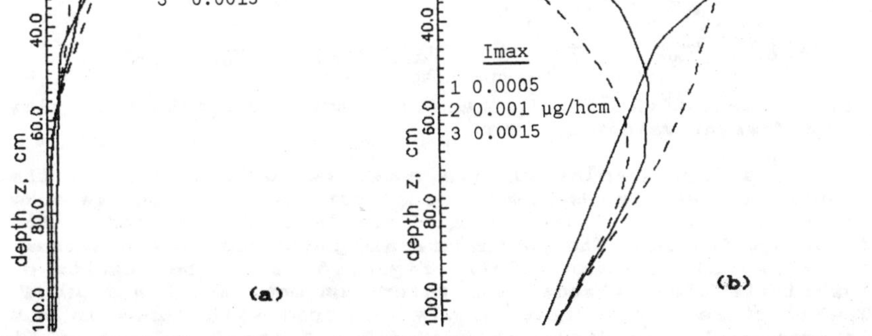

Fig.4. Distribution of (a) NH_4-N and (b) NO_3-N with depth at day 7 for different Imax values.

significantly higher concentrations than for larger values of Imax. Differences in concentration distributions in the upper portion of the soil are clearly shown in the profile. Figure 5 shows the cumulative N uptake of NH_4-N, NO_3-N and total-N by the plants for the three Imax values. As shown in these figures, NH_4-N uptake is dependent on the time of wastewater application. The results show a rapid increase in NH_4-N after the application, but over the three or four days following application the uptake gradually decrease. This change of rate of uptake with time is greater in the case of Imax=0.0015.

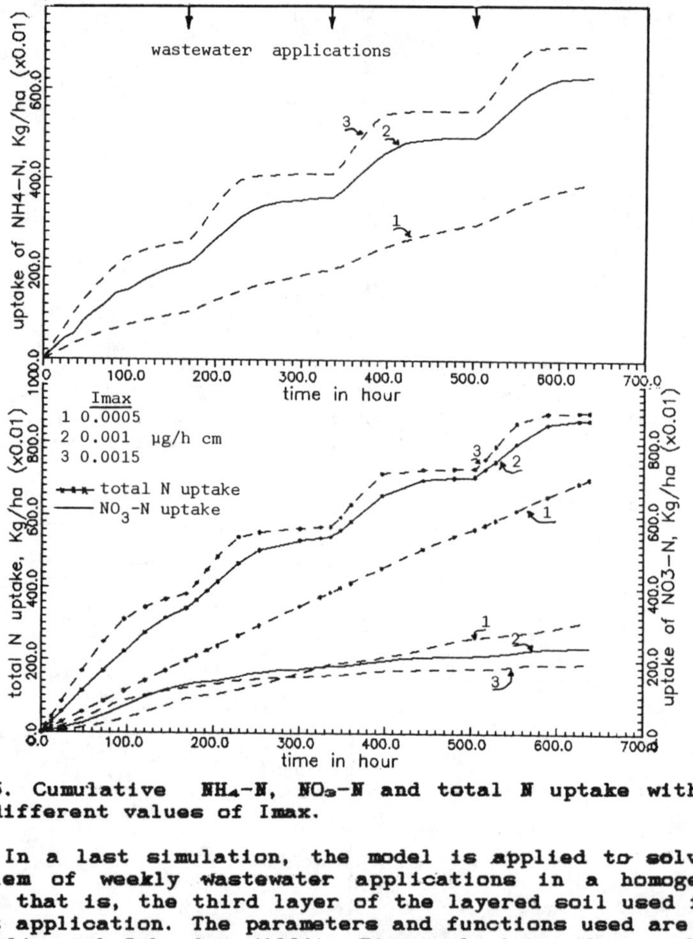

Fig.5. Cumulative NH_4-N, NO_3-N and total N uptake with time for different values of Imax.

In a last simulation, the model is applied to solve the problem of weekly wastewater applications in a homogeneous soil; that is, the third layer of the layered soil used in the first application. The parameters and functions used are those of Selim and Iskandar (1981). Figure 6 shows the results of the finite element model for water content, NH_4-N and NO_3-N. The moisture fronts move slowly compared with those in the layered soil. The distributions of NH_4-N and NO_3-N after 14 days from application are identical to those in the layered soil.

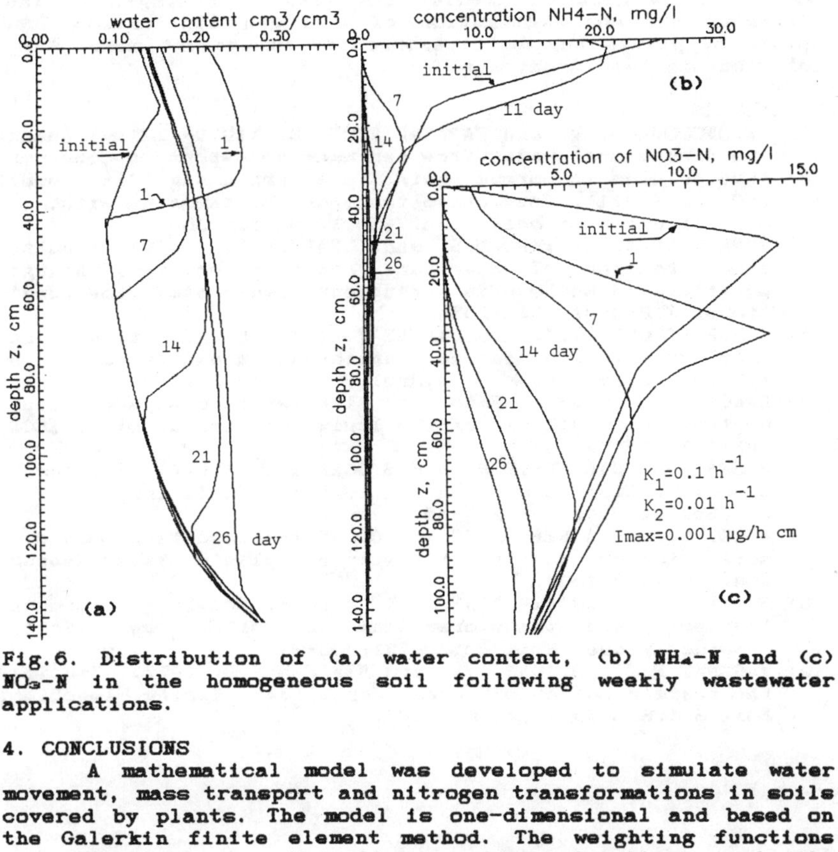

Fig.6. Distribution of (a) water content, (b) NH_4-N and (c) NO_3-N in the homogeneous soil following weekly wastewater applications.

4. CONCLUSIONS

A mathematical model was developed to simulate water movement, mass transport and nitrogen transformations in soils covered by plants. The model is one-dimensional and based on the Galerkin finite element method. The weighting functions are identical to the linear shape functions and the integration in time is approximated by a second order scheme, type Crank-Nicolson.

The model incorporates the plant uptake of water, the mass transport by dispersion and convection of ammonium and nitrate nitrogen, the nitrification of ammonium to nitrates in one step, the denitrification of nitrates, the ammonium exchange and the uptake of NH_4 and NO_3 ions. The nitrogen transformations are represented by first order kinetics. The ammonium-ion exchange is described by a Freundlich type linear reversible reaction. The plant uptake of water is approached by the Molz and Remson extraction model and the uptake of NH_4-N and NO_3-N is approached by the empirical Michaelis-Menten type model.

The accuracy of the proposed model was examined by comparison with other mathematical model results and available field data. The sensitivity of the model to changes in rate of

N uptake by plants and the applications of the model to a multileyered and a homogeneous soil were investigated. The model can be used to predict the fate of nitrogen in land treatment systems, estimation of nitrogen application load, uptake by plants and as a research or management model for N behaviour in land treatment.

REFERENCES

(1) ANTONOPOULOS, V. and PAPAZAFIRIOU, Z. (1990). Simultaneous one dimensional water flow and mass transport of conservative solutes in porous media. Water Res.Manag. (in press).
(2) CHO, C. (1971). Convective transport of ammonium with nitrification in soil. Can.J.Soil. 52:339-350.
(3) GUREGHIAN,A.B., WARD,D.S. and CLEARY,R.W. (1979). Simultaneous transport of water and reacting solutes through multilayered soils under transient unsaturated flow conditions. J.Hydrol., 41:253-278.
(4) KALUARACHCHI, J.J. and PARKER J.C. (1988). Finite element model of nitrogen species transformation and transport in the unsaturated zone. J.Hydrol., 103:249-274.
(5) LANCE, J.C. (1975). Fate of nitrogen in sewage effluent applied to soil, J. of the Irrig. and Drain. Div., ASCE, 101:131-144.
(6) MISRA,C.D.,NIELSEN,D.R. and BIGGAR,J.W. (1974). Nitrogen transformation in soil during leaching: I,II and III. Soil Sci.Soc.Am.Proc.,38:289-304.
(7) MOLZ, F.J. and REMSON I. (1970). Extraction term models of soil moisture use by transpiring plants. Water Resour. Res., 6:1346-1356.
(8) SELIM, H.M. and ISKANDAR, I.K. (1981). Modeling nitrogen transport and transformations in soils: Theoretical considerations. Soil Sci., 131:233-241.
(9) WAGENET,R.J., BIGGAR,J.W. and NIELSEN,R.D.(1977). Tracing the transformation of urea fertilizer during leaching. Soil Sci.Soc.Am.J.,41:896-902.

TECHNICO-ECONOMIC STUDY OF THE APPLICATION OF ANAEROBIC DIGESTION FOR ENERGY CONSERVATION IN SMALL-SIZED AGRO-INDUSTRIES

E. DIAMADOPOULOS, A. TATSI and X. DABOU, Chemical Process Engineering Research Institute, Thessaloniki, Greece

Summary

This work investigated the application of anaerobic digestion in five type of agro-industries which are characterized by their small size. These were cheese and dairy products manufacturing, fruit and vegetable canning, cow breeding for milk production, cattle breeding for meat and swine breeding for meat. It was found that significant conservation in the usual energy forms (fuel oil and electricity) can be achieved. These savings, however, were not sufficient to recover the capital needed for the anaerobic digestion unit, unless the savings of the investment for aerobic treatment were considered. In the latter case, the cost analysis indicated capital investment recovery in a reasonable payback period, with the exception of the fruit and vegetable canning industry.

1. INTRODUCTION

Anaerobic digestion has been successfully used for the treatment of agricultural wastes, for several decades. Along with the treatment of wastes, the biogas produced during the process can be used as an energy source. This results in conservation of the most commonly used forms of energy, such as fuel oil and electricity. Recently, the development of advanced anaerobic reactors has led to the construction of efficient treatment systems requiring relatively small bioreactor volumes or, equivalently, low reactor retention times.

The process of anaerobic digestion, however, finds very limited application in Greece, despite the favourable climatic conditions and the abundance of agro-industries. Most of them lack wastewater treatment facilities and, even where such facilities exist, they involve aerobic treatment with high energy consumption (such as extended aeration activated sludge). Two important reasons that limit the applicability and the cost effectiveness of anaerobic digestion are the size of the agro-industrial installations and the fact that anaerobic digestion alone cannot achieve complete wastewater treatment. The majority of the agro-industries in Greece is of small size and many are family operations.

The objectives of this work were to evaluate, in a technico-economic sense, the process of anaerobic digestion as a means to achieve wastewater treatment and substitute fuel oil and/or electricity in small-sized agro-industries: The expected benefits of such study include the specification of guidelines (such as capacity and type of agro-industry) which may result in significant energy conservation, as well as reasonable pay-back periods for the application of anaerobic digestion in the agro-industries under local conditions. It is hoped that this study may encourage agro-industries, consulting firms and government for the implementation of anaerobic digestion in Greece. In order to pursue the objectives of this project, we considered the agricultural industries in Central Macedonia (Northern Greece).

2. METHODOLOGY

The agro-industries under study were grouped in five categories. These include: a) cheese and dairy products manufacturing, b) fruit and vegetable canning, c) cow breeding for milk production, d) cattle breeding for meat; and e) swine breeding for meat. Other agricultural and food industries also exist (such as breweries; beet sugar manufacturing, wineries, etc.), but they represent a smaller fraction of the total number of agro-industries.

At first, actual capacity and energy data were obtained. The energy consumption reflected the use of both fuel oil and electricity. Waste volumes and characteristics were not possible to obtain, since they are not measured. In this case, correlation from national and international data were used (1, 2). In order to obtain comparable results in the design of the anaerobic digestion system, a common flow diagram was employed (Figure 1). The unit operations included wastewater pumping, heating by means of a external steam heat exchanger, anaerobic digestion in a continuous stirred tank reactor without recycle, compression and storage of biogas, a flare system and a biogas utilization unit. Two alternatives were evaluated for the latter: a boiler, for steam generation, and a gas generator (shown in Figure 1), for electricity production.

The bioreactor was a high-rate single-phase digester with mechanical stirring. Its sizing was based on the consideration of appropriate loadings for each type of wastewaters (Table 1). The quantities of methane produced are given by (3,4):

$$V_{CH4} = (0,35\ m^3/kg)\{[EQS_0(10^3\ g/kg)^{-1}]-1.42P_x\}$$

where : Vch4 = Volume of methane produced, m^3/d

0.35 = Stoichiometric coefficient for the conversion of ultimate BOD removed to methane produced

Table 1. Design variables for the digesters

Type of agro-industry	Capacity range tn/yr	Waste flowrate m^3/d	Organic load $KgBOD_5/d$	Reactor Loading KgBOD/M3/d
Milk and dairy products	1100-91000	11-610	110-6600	9.8
Canning	1600-38700	180-1200	200-1400	7
Cow breeding for milk production	100-340	1-4	26-90	15.9
Cattle breeding for meat production	12-180	1-18	26-400	15.9
Swine breeding for meat production	80-1270	5-80	85-1420	39.2

E = BOD removal, taken 80%

Q = Volumetric flowrate, m^3/d

So= Influent ultimate BOD, $mg/L=g/m^3$

1.42 = Factor converting biomass to ultimate BOD

Px = Biomass produced, kg/d, and

Ultimate BOD = $0.68(BOD_5)$

The biomass produced, Px, is given by:

$$Px = \frac{YQESo*(10^3\ g/kg)^{-1}}{1 + K_d 8_c}$$

where: Y = Yield coefficient, taken 0.05 mg/mg
Kd = Endogenous respiration coefficient, taken 0.02 d^{-1}
8c = Mean cell residence time, d. Since no recycle was considered, the mean cell residence time equals the hydraulic residence time, which, in turn, equals the reactor volume divided by the volumetric flowrate.

The equipment capital cost (Fixed Investment) was estimated from the data provided by manufacturers. The origin of the equipment, however, affected significantly the cost. Equipment manufactured in Greece was, in general, less expensive than imported. Therefore, equipment manufactured locally was preferred over imported. Furthermore, the total capital cost was calculated by the summation of all individual costs which were estimated as a fixed percentage of the equipment capital cost (Table 2).

The operational cost consists mainly of the cost of utilities, maintenance and labour costs. The cost of utilities (such as water, steam and electricity) was estimated for each case based on the demands for each item of equipment. The other two costs were taken as a fixed percentage of the total equipment capital cost (Table 2).

Table 2. Cost estimates for the treatment system

Type of cost	Estimate	Remarks
	Anaerobic Digestion	
Total capital	413% of Fixed Investment	Ref. 5
Labour+Maintenance	4.5% of Fixed Investment	Ref. 6
Utilities	Calculated for each case	
	Aerobic Treatment (Activated Sludge)	
Total capital	150.000 dr/KgBOD	Estmd. from Ref.7
Total operational	40.000 dr/KgBOD	Estmd. from Ref.7

Finally, in the estimation of the cost benefits derived from the application of anaerobic digestion, two alternatives were considered. The first consisted of the utilization of biogas alone, while the second included the reduction in the required cost, both capital and operational, for complete wastewater treatment by means of aerobic treatment (activated sludge). Cost estimates for the aerobic treatment of wastewaters are also presented in Table 2.

3. RESULTS AND DISCUSSION

The electric energy conservation for the various agro-industries is presented in Figure 2. This figure exhibits the percentage of electricity consumed that can be produced on the plant by means of a biogas generator, as a function of the relative plant capacity. This latter term refers to the plant capacities (and equivalently to waste flowrates and BOD massrates) as presented in Table 1. The energy conservation depends on the type of agro-industry and the relative capacity. The largest conservation seems to occur for the swine and cattle breeding cases. Similar results were obtained when the use of a boiler was considered for fuel oil substitution.

The results of the economic analysis are presented in Figures 3 and 4 for the use of boiler and gas generator, respectively. These figures show the time required for investment capital recovery as a function of relative capacity. Two remarks should be made about Figures 3 and 4. First, the

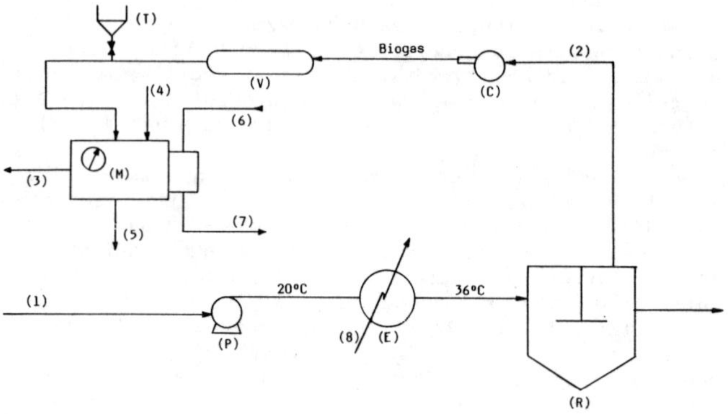

(R) Digester with mechanical stirring
(E) Steam heat exchanger
(P) Pump
(C) Blower
(V) Biogas storage
(1) Feed
(2) Biogas
(3) Electric energy produced
(4) Air
(5) Flue gases
(6) Cooling water
(7) Hot water
(8) Steam

Figure 1 - Flow diagram of the proposed anaerobic digestion unit

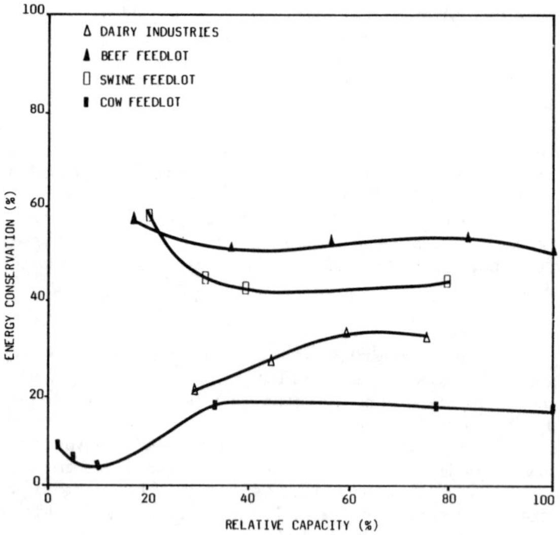

Figure 2 - Electric energy conservation

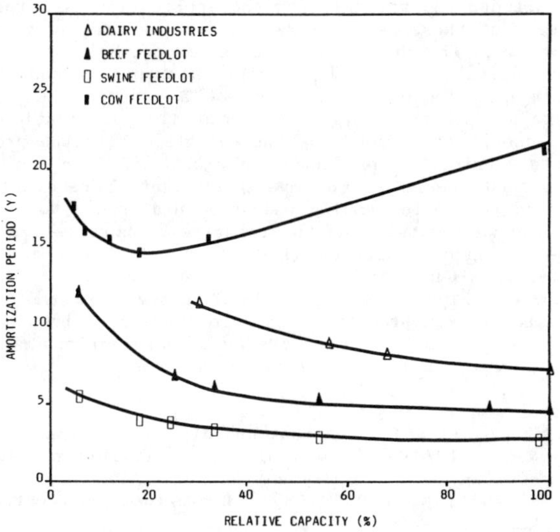

Figure 3 - The amortization period as a function of plant capacity considering the use of boiler

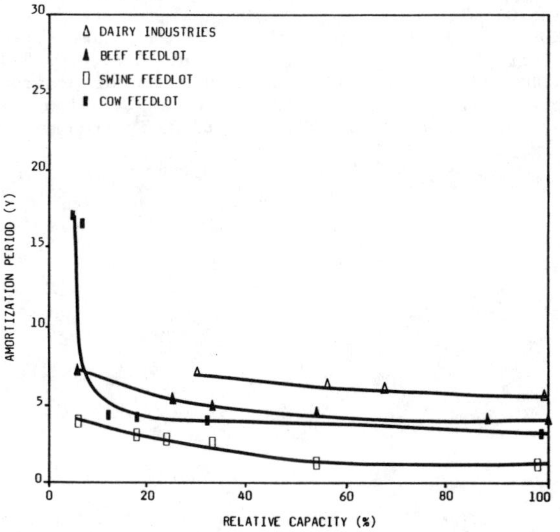

Figure 4 - The amortization period as a function of plant capacity considering the use of gas generator

total profit includes the savings from the application of aerobic treatment for the treatment of these wastewaters and for the same degree of treatment (80% BOD removal). If only the profits from the use of biogas are considered, no capital recovery is possible within a reasonable time. This agrees with similar findings in a study of a canadian beef feedlot (8), where the major profits originated from the utilization of protein extraction and use of it as food for the animals, while the profits for the production of electricity represented only a small fraction of the total profits. The second remark is no investment capital recovery is possible for the fruit and vegetable canning industy. This is due to the low organic load (BOD) of the wastewaters and the relatively short period of operation for this type of agro-industries (four months for the majority of the canning industries in Greece).

Comparing the Figures 3 and 4, we can see that the use of a gas generator for electricity production is more cost-effective than the use of a boiler for steam production. From the agro-installations considered, the swine breeding gave the best results for both cases of biogas utilization.

REFERENCES

(1) MARKANTONATOS G. (1986). Treatment and disposal of wastewater, Athens.

(2) NEMEROW, N.L. (1978). Industrial water pollution, Addison-Wesley Publishing Company.

(3) GRADY, C.P. and LIM, H.C. (1980). Biological Wastewater Treatment, Marcel Dekker Inc.

(4) METCALF and EDDY, Inc. (1979). Wastewater Engineering: Treatment Disposal Reuse, 2nd Edition, McGraw-Hill.

(5) PETERS, M. and TIMMERHAUS K. (1980). Plant design and economics for chemical engineers, 3rd Edition, McGraw-Hill.

(6) WOODS D. R. et al. (1978). Cost estimation techniques for sewage sludge disposal, Proc. Conf. on Sludge Utilization and Disposal, Environment Canada.

(7) GRAMMATIKOS, D. (1989). Cost analysis of wastewater treatment units, 1st Ntl. Conf. Envir. Sci. Techn., Mytilini, Greece, 378-390.

(8) CANVIRO CONSULTANTS LTD (1983). Demonstration of Anaerobic Digestion and Single Cell Protein Recovery at a Canadian Beef Feedlot, Technology Development Report EPS 4-EP-83-3, Environment Canada

The Use of Fermentation Sludge as a Fertilizer in Agriculture

A.B. Larsen*, F.H. Funch**

* Waste Water Dept., Novo Nordisk A/S
Hallas Allé, 4400 Kalundborg, Denmark

** Analytic and Environmental Services, Novo Nordisk A/S
Hallas Allé, 4400 Kalundborg, Denmark

ABSTRACT

Novo Nordisk is one of the world's largest producers of enzymes, insulin and biopharmaceuticals. This production is based mainly on biotechnology and fermentation processes. The waste from the fermentation process is non-toxic, but due to the high content of nitrogen and phosphorus the waste can cause eutrophication if it is discharged in the marine environment. Novo Nordisk has developed a process in which the fermentation sludge - which constitutes a large part of the waste - is use as an organic fertilizer - NOVOsludge - in agriculture.

The total amount of sludge is 500,000 m^3/year, corresponding to 800,000 kg of nitrogen and 300,000 kg of phosphorus. Around 10,000 hectares of farmland benefit from this. Since a great proportion of the raw materials used in fermentation are based on agricultural products, this use is an example of re-cycling of nutrients.

Investigations of the uptake of the nutrients in different types of crops are made in order to develop still more efficient uses. The system which has been developed has also contributed to new rules for the utilization of organic waste material.

KEYWORDS

Fermentation sludge, fertilizer, recycling of nutrients, uptake of nutrients, spreading techniques, regulations.

INTRODUCTION

Novo Nordisk is the largest single enzyme manufacturer in the world and one of the two largest insulin producers.

The company has sales offices all over the world and production facilities in Denmark, USA, Switzerland and Japan, as well as in Brazil.

The enzymes produced are used in a wide variety of industrial processes and in consumers' daily lives.

Detergent enzymes for use in washing powders, liquefaction enzymes for the starch industry, carbohydratedegrading enzymes for the alcohol-producing industries and so on.

The production processes are based mainly on fermentation using bacteria, yeast and moulds. The sludge coming from the recovery of the products consists of the cell material of the organisms, unused substrate and some process helping aids. So the potential impact on the environment is due not to the toxicity of the sludge but to the heavy amount of organically degradable material from the process.

The total amount of sludge from Novo Nordisk plants all over the world is around 5.000 tons/day.

So it is obvious that to deal with these amounts we had to develop an environmentally sound and economical way of treating the sludge. And as you will see, our answer is not final disposal but reuse of a valuable resource.

ORIGIN OF THE SLUDGE

As mentioned, most of the production is based on fermentation, followed by recovery and purification of the product.

The environmental influence of this production is shown schematically in the diagram, Fig. 1.

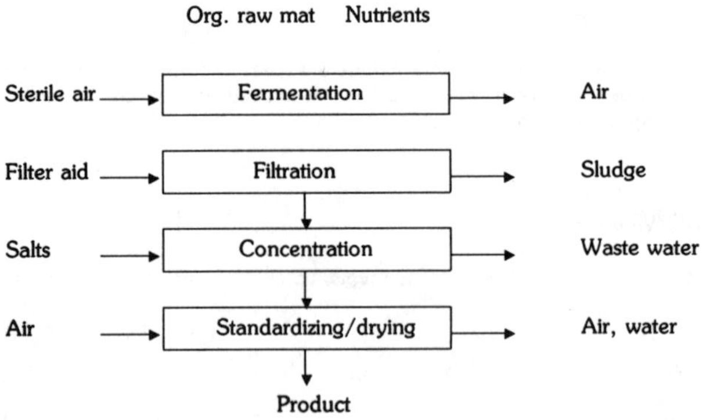

Fig. 1. Production process and environmental influence.

The fermentation process has several similarities to waste water treatment, such as the use of microorganisms, aerobic conditions, etc., but also several differences. Fermentation uses monocultures, the microorganisms do not aggregate in flocks and the microorganisms are well-fed and not competing for food as in an activated sludge process.

In the downstream processing several of the same unit operations as in waste treatment are in use, e.g. precipitation, flocculation, filtration and centrifugation. However, the most significant similarity is the fact that after the process the microorganisms become a waste product, which you need to get rid of in a proper manner. This article describes how the sludge from the biggest fermentation plant at Novo Nordisk, situated in Kalundborg, Denmark, is no longer regarded as waste, but rather as a valuable by-product for use as a fertilizer.

CHARACTERISTICS OF THE SLUDGE

The "sludge" consists of the fermentation broth after recovery of the product. This means that the content will be the microorganisms used, the nutrient surplus and the filter aid used. Two kinds of filter aid are used, namely inorganic and chemicals for precipitation and flocculation of organic matter, etc., and inert substances involved in the filtration processes like diatomaceous earth and activated carbon.

The composition varies according to the different fermentation processes, i.e. the microorganisms used and the products recovered, but generally the total mixture from the production plant in Kalundborg has a composition like the one described in Table 1. All figures in Table 1 relate to the dry solids in the mixture, which normally vary between 4 and 6%. It should be noted that approximately 55% of the dry solids is organic matter, while the rest is either salts, inorganic nutrients or inert filter aid.

TABLE 1. Average Composition of NOVOsludge (mg/kg DS)

Total nitrogen	45,000
Organic nitrogen	41,500
Ammonia-nitrogen	3,500
Total phosphorus	20,500
Potassium	9,000
Calcium	154,000
Magnesium	4,800
Manganese	100
Sodium	20,400
Chloride	23,200

The content of heavy metals is much lower than in sludge from waste water and therefore the product differs considerably from municipal sludge. Table 2 shows the typical content of heavy metals in NOVOsludge and municipal sludge, together with the limits for agricultural use of NOVOsludge given by the Danish National Agency of Environmental Protection (NAEP).

Table 2. Metal Content of NOVOsludge Compared with Municipal Sludge (mg/kg/ DS)

	Municipal sludge	NOVOsludge	
	Anaerobic stab.	Average conc.	NAEP permit
Cd	3 - 7	0.66	2.0
Co	3 - 8	<2	
Cr	30 - 90	2.4	
Cu	100 - 400	16	
Hg	2 - 7	<0.02	0.5
Mn	200 - 500	101	
Ni	20 - 60	3.8	
Pb	150 - 400	<4	20.0
Zn	1000 - 2500	96	

As shown in Table 2, the levels in NOVOsludge are 10-50 times lower than the content in municipal sludge. This fact is of course a great advantage for the NOVOsludge. However, this is not a coincidence, but the result of a very precise selection process, which has been going on for many years, where those buying raw materials for production have demanded low contents of heavy metals, both with regard to the actual products and the by-products.

TREATMENT OF THE SLUDGE

Before transportation to the fields, the sludge is inactivated and stabilized by addition of lime and by heat treatment at 90°C for at least one hour.

The purpose of stabilization is to achieve a product which is biologically stable during storage and transportation to the fields, which has no detrimental effect on soil and crops, and does not inhibit the mineralization of the sludge when applied to the soil. Several possibilities were considered, but lime was chosen because it has a positive influence on the inactivation process and makes it possible to maintain stability of the product. In addition, it has a very positive influence on smell, and it improves the soil by counteracting acidification. A rise in pH to above 11 on average requires approximately 5 kg $Ca(OH)_2$ per m^3 sludge.

When the sludge is stored for long periods, a separation into two phases takes place, and the water phase is gradually neutralized due to sedimentation of the surplus lime and carbonization of the soluble lime. It is possible to use the sedimented lime to prolong the stability of sludge by a simple mixing operation.

After the heat treatment, where it is kept for at least 1 hour at 90°C, the sludge is cooled to below 40°C. There are three factors which make the cooling necessary: soft tubes are used for the spreading, the crops are sensitive to excessive heat, and recovery of the heat reduces costs. Of these the two first considerations are of greatest importance, since the cooling makes it possible to use the equipment described below, and it allows the use of sludge on growing crops.

The heat treatment is carried out by a combination of heat exchangers and steam injection. Two types of heat exchanger are used: the spiral type and concentric tubes. The heat transfer and viscosity vary according to the production at the plant, but normally the sludge behaves more or less like water. The processes for inactivation and stabilization are shown in Fig. 2.

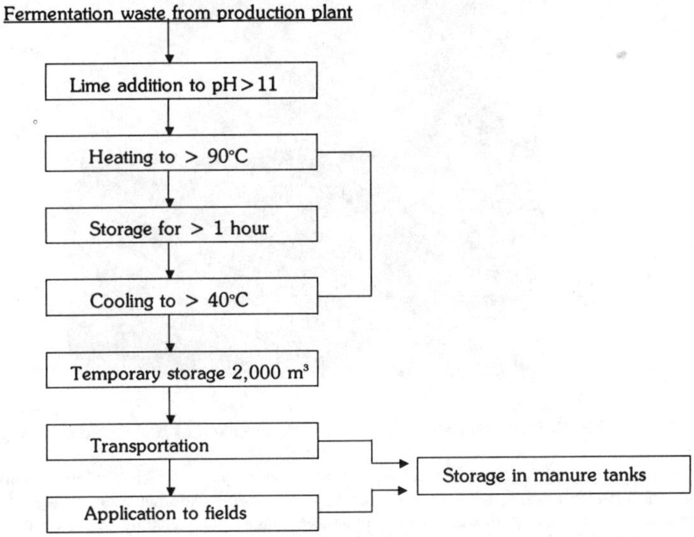

Fig. 2. Process flow-chart for treatment and handling of NOVOsludge.

SPREADING

From the sludge treatment plant the NOVOsludge is transported to application areas situated in the range of 40 km from the production plant.

Application to the fields is made by means of modified watering equipment which uses a spreading boom with an application width of 48 m for distribution - see Fig. 3.

Fig. 3. Spreading equipment used for application of NOVOsludge to fields.

This equipment was developed to achieve an application system which is gentle towards fields and crops, and is also flexible, efficient and useful all year round. Another requirement of the system is a very precise and uniform distribution of the sludge, since only a volume corresponding to a precipitation of 5-10 mm is used, and this has been achieved by installation of a flowmeter in combination with a small computer system controlling the drum on the watering machine.

During periods with "bad" weather, which means snowy periods and periods alternating with frost and thaw, the sludge is stored in manure tanks at present constructed on approximately 20 large farms.

As a full-scale experiment, a 4-km tube system from the production plant to a farming area was constructed in 1989. It seems that this will successfully avoid the expensive tank-truck transportation and, in addition, will reduce the environmental impact, since noise and vibration from the heavy tank-trucks are avoided. It is expected that this method of transportation will gradually take over.

The potential customers for the NOVOsludge are all the farms in the County of West Zealand, but in the selection of interested sludge recipients the nature (size, regularity, inclination, soil type, etc.) of the area is naturally taken into consideration, together with its position in relation to Novo Nordisk. In 1989, approximately 180 farms received sludge from Novo Nordisk.

Amounts of 1,500-2,000 m^3 sludge are brought out daily from the factory in Kalundborg. The applied dosage of 50 m^3/ha gives a daily area use of approximately 30-40 ha. In 1989 the annual sludge production was approximately 325,000 m^3. In 1990 the sludge production is expected to rise to 500,000 m^3, with a corresponding area use of approximately 10,000 ha.

Shortly after the conclusion of the delivery, the farmers receive a delivery note stating the amounts supplied and the dosage of nitrogen and phosphorus, as well as a map with marked areas. This enables the farmers to include the nutrient inputs in their fertilizer plans.

For the individual user of NOVOsludge, the saving in fertilizer consumption will bee around DKK 750 per ha (USD 115/ha), since Novo Nordisk pays all costs in connection with treatment, transport and spreading of the sludge.

FERTILIZER VALUE

The nutrient needs of the microorganisms are not very different from the needs of plants in general. However, some modification of the composition of nutrients takes place during the downstream processing, where different salts and organic chemicals are added during the precipitation processes.

The typical ratio between the major plant nutrients N, P and K is 10:4:3, which is different from commercial inorganic fertilizers, especially in respect of the relatively low potassium content.

Another difference is the fact that 90% of the nitrogen is part of the organic material, which means that this part of the nitrogen has to be mineralized by microbial activity in the soil before it becomes useful to the crop. Investigations has shown that, in general, 80% of the organic matter will degrade during the first year, and that will make 40-50% of the nitrogen available to the plant during the first year.

The phosphorus is partly organic and partly bound in inorganic salts. The organic part will be mineralized along with the breakdown of organic matter, while the inorganic part will be liberated slowly during the course of a few years.

All the potassium is water-soluble and therefore available to plants.

In addition to the fertilizer value of the macro-nutrients, the farmers appreciate the content of micronutrients like magnesium, manganese and copper and the high content of calcium.

It has been found that, when used for a few years, the sludge improves soil structure and water capacity, especially in sandy soils.

Although the nutrient content differs from that of commercial inorganic fertilizers, farmers have learned to use this product on many different crops. Especially grass, corn and sugar beet can benefit from the slowlyreleased nitrogen, but the fertilizer is also used for wheat, barley and other crops.

In Denmark, the climate in the winter season is cold and rainy, and because of this it can be rather problematic to spread organic fertilizer on the fields during this time.

Precipitation (mm), County of West Zealand

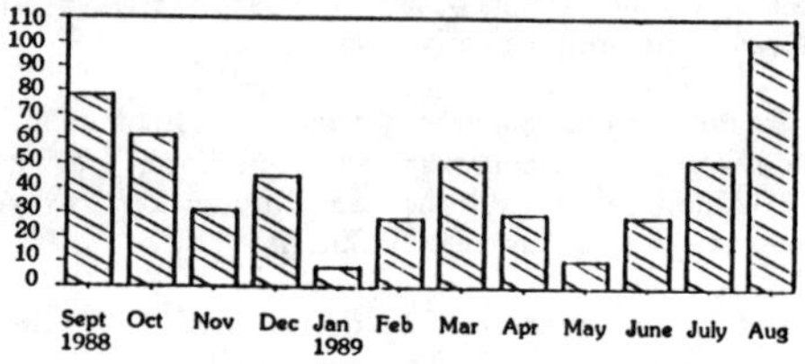

Fig. 4

Mean temperature (°C), County of West Zealand

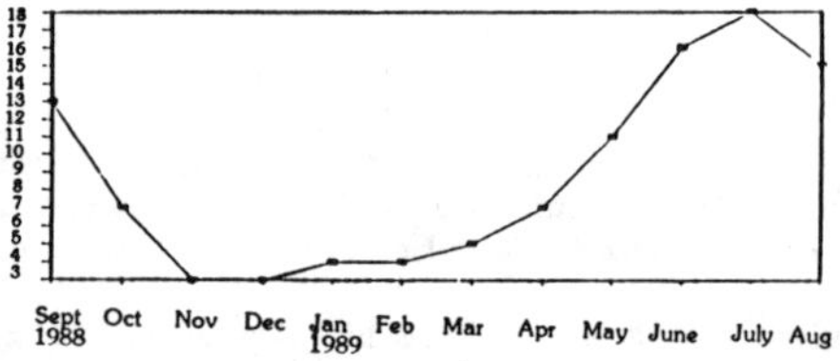

Fig. 5

However, the fertilizer tests with the NOVOsludge have shown that, depending on the type of crop, it is possible to achieve a good efficiency of use of NOVOsludge during the winter time.

Efficiency of NOVOsludge from N uptake

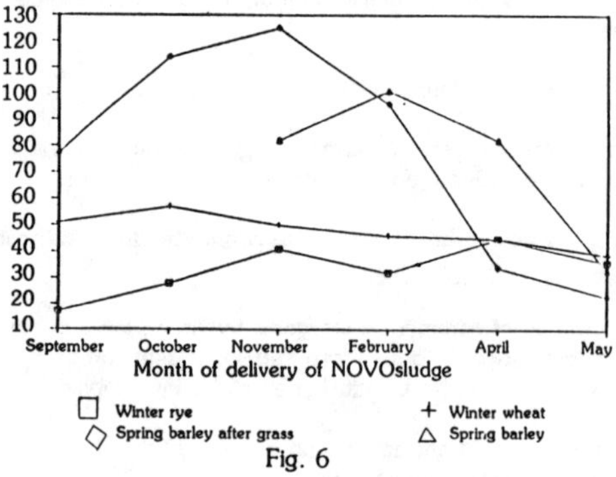

Fig. 6

For winter rye the efficiency in the early autumn months is rather low, but this is due to the fact that winter rye is used as a crop on very sandy fields where the risk of downwashing of the nitrogen is greater. Furthermore, the nitrogen uptake of winter rye is only 10-20 kg N in the winter months.

The other crops, spring barley and winter wheat, have an efficiency of over 40%. For spring barley and spring barley after grass, the efficiency in the autumn is very high.

For municipal sludge the average nitrogen uptake is only 25%.

LEGISLATION

The permit for application of NOVOsludge on farmland was given by the Danish National Agency of Environmental Protection (NAEP) in 1988. The permit states that the sludge is to be used as a fertilizer. The most essential conditions, which are very similar to those in the general permits for the use of municipal sludge in agriculture, are given below:

NOVOsludge Rules

On September 7, 1988, the Danish National Agency for Environmental Protection's Agricultural Office granted a permit for spreading production sludge from Novo Nordisk A/S in Kalundborg on farmland. On December 15, 1989, the Danish National Agency of Environmental Protection granted modification of the permit concerning crops.

The production sludge is approved and registered by the Danish Ministry of Agriculture's Fertilizer Control Agency as a fertilizer.

The Danish Ministry of Agriculture's Department of Plants has granted a permit to use NOVOsludge in ecological agricultural production. The NOVOsludge may be supplied in amounts of up to 25% of the ecological farm's total nitrogen need.

Generally

- Maximum dosage:

grain, seed crops:	170 kg N/ha per year, 60 kg P/ha per 2 years
grass:	250 kg N/ha per year, 80 kg P/ha per 2 years

- The sludge may not flow to lakes or streams (including drains in hollows)
- Manure, silage liquid and other types of sludge may not be spread in the same year on areas on which NOVOsludge has been spread.
- No more than a maximum of 80 kg chloride/ha per 2 years may be supplied (average for a property).

Distance

- A distance of at least 50 metres must be observed to residences, potable water borings and wells.

Ploughing into the Soil

- When applied on crops which are directly consumable (e.g. potatoes), the sludge must be ploughed into the soil before sowing.

Autumn

- In the period from harvest to November 1, sludge may not be spread in sandy soil areas (types 1, 2 and 3) unless the crop is a continuous grass crop.

- In the period from harvest to January 15, sludge may only be spread on green areas or on areas which are sown with a winter crop.

CONCLUSION

The techniques developed and knowledge gathered at Novo Nordisk have shown that use of sludge from fermentation processes as a fertilizer is a good and environmentally sound way of solving a major waste problem.

Amounts of 500,000 m^3/year are managed in this way, corresponding to the reuse of 800,000 kg of nitrogen and 300,000 kg of phosphorus.

The exact knowledge of the content of nutrients in the sludge and the precise way of spreading it makes it easy for the farmer to include nutrient inputs in the fertilizer plans, saving the cost of commercial fertilizers.

WASTEWATER SEWAGE SLUDGE TRANSPORT THROUGH LONG DISTANCE PIPELINES

O. G. MANOLIADIS
Assistant Professor, Tecnological Educational Institutes-Athens

Summary

An experimental investigation was made to determine the effect of the parameters suspended solids concentration and temperature that affect the phenomenon in the laminar flow region. The flow parameters were derived from shear stress - shear rate diagrams determined by a Stormer viscometer. Experiments were performed on samples at varying total concentration of solids by weight from 2% to 6.64% for raw digested and secondary sludge. The non Newtonian models Plastic and Pseudoplastic were used to determine the rheological parameters of sewage sludge.

The results of this investigation revealed that temperature and solids concentration have a significant effect on the flow characteristics of the wastewater sludges tested. The energy losses of the wastewater sludge have been calculated in comparison to those of the water. The ratio head loss of the wastewater sludge over the head loss for the water is estimated as a function of the rheological parameters concentration of solids and temperature and the type of sludge (raw, activated and digested).

1.INTRODUCTION

The use of pipelines as a means of transportation for wastewater sludge presents a solution for the remote disposal/storage of wastewater, especially for large cities and long distances. The transportation of sludge through pipelines is used, in general within treatment works or between treatment works and a storage or a disposal area. Usually, location, space limitations, land. values and foundation conditions at the treatment works are the limiting factors for the transportation of wastewater sludge over long distances. Cleveland, Chicago in USA, and Mogden and Birmingham in England have for many years processed sludge at sites widely separated from the sources of generation. Mixed primary and digested or activated sludge, reduced to an average of 60 percent volatile matter, is used for pipeline transportation.

It has long been reported that for concentrations above 3%, the wastewater sludge exhibits non-Newtonian behavior, i.e., the relationship between shear stress and shear rate is no longer linear. Because of the long conveyance distance of the pipelines used for transportation of wastewater sludge the accurate determination of the head loss becomes important. A proper understanding of the flow behavior of sludge is necessary for adequate hydraulic design of pumping systems.

In the literature, two types of non-Newtonian fluids ; the Bingham plastic and the pseudoplastic fluids- have been assumed and received special attention. Based on these two classifications, the rheological models are derived for the design of the pipelines . Extensive experimental studies of design procedures have been conducted to date for the plastic and the pseudoplastic model . Although studies on the design of pipelines have been under investigation for many years, there is little information concerning the role of concentration of solids on the rheological properties of wastewater sludge.

The role of temperature is also not well known. Babbitt and Caldwell conducted one of the most interesting studies on wastewater sludge characterization. This investigation includes study of the effect of temperature change on the rheological properties of sludges, measured by the rotational viscometer. No attempt was made to formulate the effect of temperature other than to show an increase of temperature resulted in a lowering of the yield value for the plastic model. Part of the Kenny et al study measured sludge slurry properties at different temperatures. Temperature was reported to be important but there is insufficient data to substantiate the claim.

Investigators in the field of rheology (Weltmann; Metzner; Skelland)(3) that temperature has a marked effect on polymer solutions, paper pulps, etc. It is common practice to assume no significant changes in the rheological behavior of sewage sludge and to ignore the effect of temperature and solids concentration on the rheological parameters. In this respect a new model is introduced in order to improve the accuracy of the design of sludge flow pumping systems. An experimental investigation was conducted to examine the effects of temperature and concentration of solids by the use of a rotational viscometer.

2. THE RHEOLOGICAL MODELS

Sludge tansported by pipelines is in general raw, digested or activated and has a solids concentration that averages 4.5% . It has long been recognised that for concentrations above 3% the wastewater sludge exhibits non-Newtonian behavior. In the literature two types of non-Newtonian models, the Bingham plastic and pseudoplastic, have been given special attention because of experimental evidence that wastewater sludge may be classified by them. The rheological equations for the plastic model reported by Babitt and Caldwell:

$$S_w = (D\Delta P/4L) = So + b(V/4L) \qquad (1)$$

here S_w is the shear stress at the pipe wall, D is the diameter, L the length, b the coefficient of rigidity, and V velocity. The energy consumption computed by analogy to the Poinseuille's results in:

$$H_L = 16S_OL/3\gamma D + 32bVL/\gamma D \qquad (2)$$

where H_L the head loss and γ the specific gravity.

b. For the pseudoplastic model the rheological equations were reported by Behn (2):

$$S_w = C(8V/D)^n \qquad (3)$$

$$H_L = 4C/\gamma D \; [(3n+1/n)(2V/D)]^n \qquad (4)$$

were C' and n' are the consistency index and H_L is the head loss.

3. EXPERIMENTAL APPARATUS

The basic component of the experimental set up was the Stormer viscometer in which tests were performed. The Stormer rotational viscometer has been used as a rheometer by Babbitt and Caldwell(1) and Kenny et al(2).The bob and cup used in testing the sludges had radii of 1.25in and 1.75 in respectively. A Neslab temperature control led bath circulator consisting of an Exacal 200 heating unit and an Endocal 150 refrigerating unit was used to adjust the temperature of sludges. Temperature was controlled continuously from a memory ring thermoregulator dial. Temperatures were continuously maintained with an accuracy of + 0.02^0 of the desired temperature. The total solids concentration was measured by the mass of dry oven solids at a temperature of 105^0C.

4. EXPERIMENTAL PROCEDURE

The sludge samples used were obtained from wastewater treatment plants in Durham, Somersworth and Newmarket New Hampshire in the United States.

Solids concentration ranged from 6.64% to 2% by weight. The total residue was measured by drying at 103^0 to 105^0C. A well mixed sample evaporated in a tared dish and dried to constant weight in an oven for 5 hours.

Before the flow properties of the sludge could be determined, calibration of the rotational viscometer was performed.

The temperature of the samples was adjusted by using the temperature control bath from 10^0C to 25^0C with increments of 5^0C. Operation of the rotational viscometer was performed as follows. A known weight to act as a driving force was attached to the string wound of the bob. The rotation of the modified bob caused by the imposed torque was recorded.

Stabilization of the temperature of the sample was possible by the surrounding water bath and was ascertained by an attached thermometer. Shear rates ranged from 20 to 200 sec^{-1} close to the prototype conditions.

5. RESULTS

Shear stress- shear rate curves were obtained for each sludge at varying concentrations, each type of sludge and varying temperature. Knowing the shear stress, S, for a given angular velocity Ω, and the radius of the bob and the cup, R1 and R2 respectively, it was possible to calculate the shear rate utilizing the equations (1) and (3).

Statistical regression analysis was used to determine the rheological coefficients yield stress S_o,and coefficient of rigidity b, for the plastic model and consistency index C and flow behavior index n of the pseudoplastic model are a function of temperature and concentration of solids.

The results of this analysis are the following:

For the plastic model the coefficient of rigidity was found to be essentially independent of temperature within the range of sludge concentration tested.

In contrast, the yield stess, S_o, was shown to be an exponential function of the temperature.

$$S_o = A e^{Bt}$$

where A and B are constants evaluated experimentally.

For the pseudoplastic model the flow behavior index may be assumed to be independent of the temperature based on the experimental data. The consistency index was found to vary with temperature in a similar way to the continuous phase:

$$A = a \ TS^b$$

where TS is the total solids concentration and a and b coefficients resulting from the regression analysis.

For the pseudoplastic model the consistency index C is an exponential function of the temperature and solids concentration:

$$\alpha) \ C = A \ e^{(bTS-Bt)}$$

where t is the temperature and A and B are coefficients resulting from the regression analysis. The flow behavior index, n, is expressed as an experimental function of total solids concentration:

$$n = a2exp(b2TS)$$

(4) The resulting rheological equations for the plastic and pseudoplastic models:

$$S = Ae^{Bt} + b(du/dy) \qquad \text{(Plastic)}$$

$$S = a1exp(bTS-Bt)(du/dy)^{a2exp(b2TS)} \quad \text{(Pseudoplastic)}$$

The coefficients of these models are tabulated in Table 1.

The ratio head loss of the wastewater sludge over the head loss for the water, r, is estimated in terms of comparing the Darcy Weisbach coefficient f for the two models. According to equations f is now a function of the rheological parameters concentration of solids and temperature and the type of sludge (raw, digested, secondary). If f_w and f_s are the Darcy Weisbach coefficients for the water and the sewage sludge the head loss ratio may be derived as follows:

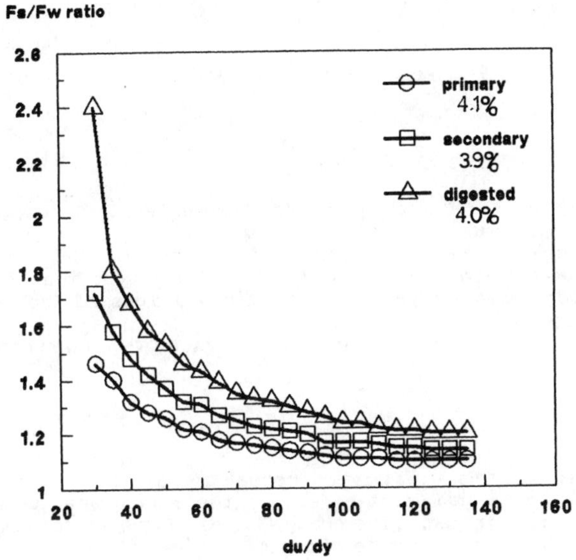

TABLE 1

PLASTIC MODEL			
	A	B	b
PRIMARY	-1.75	0.19	-1.99
ACTIVATED	-3.62	0.49	-1.56
DIGESTED	-3.59	0.66	-1.40

PSEUDOPLASTIC MODEL					
	a1	b	B	a2	b2
PRIMARY	-2.58	3.67	-0.11	-0.40	-2.44
ACTIVATED	-2.38	3.33	-0.21	-0.22	-1.63
DIGESTED	-1.98	3.05	-0.33	-0.13	-1.54

$$r = f_s / f_w$$

where

$$f_s = 16\ a1exp(-BT)\ (8V/D)^{1-(expb2TS)}\ /(\rho\ V\ D)$$

and

$$f_w = 16\ C^w\ /\ \rho\ V\ D$$

Prototype data of full scale pipelines documented by Mulbarger et al (5) were used for comparison.

For shear rates 8V/D from 20 to 100(sec $)^{-1}$ the relative rate of head losses was rated from 1.2 to 3.0. It also noticed that for the three types of sludge examined primary, activated and digested the head loss was increased accordingly:

$$H_L\ (primary) < H_L\ (activated) < H_L\ (digested)$$

6.CONCLUSIONS

The results of this investigation revealed that temperature and solids concentration has a significant effect on the flow characteristics of the wastewater sludges tested. The energy losses of the wastewater sludge have been calculated in comparison to those of the water.

The computer regression analysis of the rhelogical parameters lead to the determination of the flow parameters of the sewage sludge grouped to graphs for the accurate selection of pumps and the determination of the energy requirement in the design of sewage works.

REFERENCES

(1) Babbitt, H. E. and D.H. Caldwell, "Laminar Flow of Sludges in Sludges in Pipes with Special Reference to Sewage Sludge" Bulletin Series No. 319, Engineering Experimental Station, University of Illinois, 1939.

(2) Behn, V.C.,"Experimental Determination of Sludge Flow Parameters" Journal of Sanitary Engineering Division, Proceedings of the American Society of Civil Engineering, 88, SA3, 1962.

(3) Kenney J.P., E.J., Wasp, and T.L. Thompson, "A Design Model for Pipeline Flow for Solid Wastes", Water 1970, Chemical Engineering Progress Series 107, Vol. 67, 1971.

(4) Manoliadis O. and P.L. Bishop, "Temperature effect on the Rheology of Sludges" Journal of Enviromental Engineering Division, Procee-dings of the American Society of Civil Engineering, SA2, 109, 1983.

(5) Mulbarger, M.C., S.R. Copas, J.R. Kordic and F.M. Cash, "Pipeline Friction Losses for Wastewater Sludges", Journal of Water Pollution Control Federation,Vol. 58, No 8, August 1981.

SLUDGE DEWATERING USING FILTER PRESS

A.I. ZOUBOULIS* and A.G. GUITONAS **

*Lab. of Gen. and Inorg. Chem. Techn., Dept. of Chemistry
**Faculty of Engng., Dept. of Rural Engineering
Aristotelian University, GR-54006 Thessaloniki, GREECE

Summary
Sludges of different origin (domestic-industrial) were dewatered using a pilot-scale fitration press unit. Several additives were examined as conditioning agents, i.e. flocculants ($Fe_2(SO_4)_3$ and $Al_2(SO_4)_3$), clay (caolin) and flocculant aid (cationic polyelectrolyte). Comparisons between them were made in terms of specific resistance of sludge cake,resistance of filter medium solids content of the filter cake (totally and volatiles) and filtration rate. General comments are presented and recommandations are suggested.

1. INTRODUCTION

1.1. Sludge dewarering generalities

Today in Greece , a number of sewage treatment plants is in the stage of construction or of initial operation, operating usually with the activated sludge process. That means, in the today's relative small sludge quantities, produced in some industrial plants from their own wastewater treatment units, would be added soon huge amounts of sewage sludge. This sludge has a mean solid concentration 2-5% w/v or even less (after sedimentation and thickening). The problems of sludge management and disposal, considered as a solid waste, have to find sooner or later their own solution. A possible solution could be succeeded through the removal of the higher percentage of the slugde moisture, using the well known filtration techniques, thus through the subsequent volume reduction, it could be made easier the final disposal, (landfilling, incineration, etc.).

Even if it is not the purpose of this paper to expound upon the theory of filtration, or specifically pressure filtration, because there are volumes of literature published on the theory and practice of sewage sludge dewatering, the authors consider, that a short review would be useful for the better understanding of the problem.

Typical purposes of sludge dewatering are to prepare sludge for landfilling or spreading, to reduce transportation costs, or to increase the fuel value and to reach the solids levels required for incineration without supplemental fuel. Laboratory tests and engineering and cost studies are necessary to determine the optimum process of sludge handling, to achieve minimum construction and operation costs, inclunding labor, power and chemicals.

Considering thickening as the first step in the separation of solids from liquids,with filtration the second and drying the third step, the costs of these three steps are going to increase from the first to the last. Particularly with dilute solutions, it means it is usually cheaper to remove 95% of the initial liquid by thickening, 4% by filtration and the remaining 1% by drying (1).

1.2. Sludge dewatering practice

The dewatering of sewage sludge depends upon the varying factors in the solids make-up such as source, organic matter, age, pretreatment, concentration, etc. Usually sewage sludge contains solids of lower mineral and higher organic content (average: 70% organic - 30% mineral) (2). The moisture content ranges from about 80% in some digested sludges to about 99% in some activated sludges (3). The net effect of natural mineral matter (with specific gravity 2,5-2,65 g/m³) in any sludge, is to increase its specific gravity, and the percentage of solids concentrated in the sludge and filter cake and to make more stable the structure. From another point of view, the net effect of organic matter fraction present in a sludge, is practically equivalent to a corresponding increase in sludge moisture.

The selection between the different available dewatering devices, is controlled from the sludge properties, the available area and the methods for ultimate cake disposal. For smaller plants where land availability is not a problem, drying beds or lagoons are most frequently used. Controversely, for plants situated near to the constricted sites, mechanical dewatering devices are generally chosen.

The available mechanical dewatering processes include vaccumn filters, centrifuges, filter presses, horizontal belt filters, etc. It is worth noticed here, that not all the sludges are amenable to mechanical dewatering. Some certain sludges, particularly the aerobically digested, can be dewatered also on sand beds with good results.

The mainly influencing variables of the dewatering process are solids concentration, sludge and filtrate viscosity, sludge compressibility, chemical composition, the nature of sludge particles (size, shape, water content, etc.), and the extent of aggregation (4,5). Knowledge of the influence of these or other variables on the ease with which water can be removed from sludge, has not progressed to the extent that filterability predictions can be based on direct measurement of basic properties. Instead, it is necessary to measure performence parameters, which refflect the combined influence of the variables. These descriptive parameters most usually used are the <u>specific resistance</u> (r) and the <u>coefficient of compressibility</u> (s^0) of a sludge (6).

The basic theory of the filtration process was developped by Carman and extended by Coakley, by application of Poiseuille's and Darcy's laws (7,8). The basic equation is,

$$dV/dt = (P A^2)/[\mu(rWV + R_m A)],$$

where, V: volume of filtrate, (m^3), t: time (s), P: pressure, (N/m^2), A: filter area,(m^2), μ: viscosity of filtrate, (Ns/m^2), r: specific resistance of sludge cake, (m/kg), W: mass of dry solids per unit volume of filtrate, (kg/m^3), R_m: resistance of filter medium, (m^{-1}).

The specific resistance (r) is a measure of the filtrerability of the sludge and is numerically equall to the pressure difference required to produced a unit rate of filtrate flow, of unit viscosity through a unit weight of cake, so it provides a valuable tool for the relative filterability of sludges. High specific resistance values indicate poor dewaterability.

The high organic fraction account for the high compressibility of sewage solids, which even with some chemically bound mineral matter, are largely capillary porous or gelatinous. In other words, compressibility in sewage sludges and their filter cakes correlates closely and varies directly with the volatile solids content of the sludge.

As large volumes of water containing a small percentage of solids must be handled at low costs in sewage clarification, and as the solids are to a considerable extent of a colloidal nature, direct mechanical filtration is difficult.

Sludge conditionning refers to chemical and physical methods for altering the properties of the sludge to permit water to be removed more readily (9). The clear result is to enhance filterability and to reduce the specific resistance. In general, the goal is to transform the amorphous gel - like sludge mass into a porous material, which will relase water to pass through. It must be admitted, that conditioning technology has been not advanced to the extent, that it could be possible to select apriori the exact conditioning technique, required to filter a particular sludge with a particular process. The efficiency of alternate conditioning methods, could be then evaluated by use of parmeters such as r, s^0, etc.

Sludges are most frequently conditioned by chemical means. Chemical treatment usually involves coagulation-flocculation of sludge solids with polymerized hydrolysis products of multivalent metal ions and/or natural or synthetic organic polymeric materials to form an agglomerated, expanded structure, thus improving the sludge ability to be dewatered (10,11). Other solid industrial wastes like fly ash, has been also used as conditioning agents (12, 13). Chemical conditioning with either inorganic or organic coagulants is complicated by variables such as the sludge age the dilution and age of the conditioning chemical, and the degree and time of mixing of the conditioned sludge prior to be dewatered. It has been indicated, that the results of laboratory conditioning and filtration tests cannot by duplicated exactly in the field, because mixing conditions, following addition of conditioning chemicals cannot also be duplicated.

Chemical conditioning of the sludge, prior to filtation, is usually practiced in order to control the solids content and to improve the dewatering characteristics. In general, the optimum solids content for filtration is about 8 - 10%. (1). Higher solids make the sludge difficult to distribute and to be conditioned for dewatering, while lower solids content requires the use of larger installations.

1.3. <u>Sludge dewatering using filter press.</u>

In a filter press, dewatering is achieved by forcing the water from the sludge under high pressure. Biggest <u>advantages</u> of this method include :

1. This method is simple and without complicated auxilliaries, which results in low initial cost, low installation cost and low power requirements.

2. It can be operated with unskilled labor and provides large filter areas in a compact space, while high pressures are obtainable. Highly automated filter systems are easy to manage and operate unattended, expect during initial and final stage of the filter cycle.

3. Higher concentrations of cake solids (40% - 50%), are being produced.

4. Filtrate clarity is being improved (less than 100 mg/l particles are contained in filtrate), and

5. Denser cake is easier to incinerate, lighter for handling and relatively odorless if appropriate filter aids will be used for conditioning.

Main <u>disadvatages</u> include :

1. Higher labor costs, relating to the periodic discharge of dewatered cake, because filter pressing is mainly a batch operation.

2. Limitations on filter cloth life, due to the plate squeezing and the rough scraping to remove adherent cake. There maybe difficulty at time in obtaining efficient washing due to compression in cake formation and to channeling, with resultant bypassing of wash water through the line of least resistance, and

3. There are certain hazards, by filtering corrosive liquors.

It is well pointed out also, that the term "press" is not correct, because no mechanical press is involved. Pressures up to several hundreds psi, imported by the sludge pump, provide the gradient pressure, to pass filtrate through the sludge cake. Pressing time is dependent on the nature of the sludge and may be less than an hour, or more than a day. Filter cake solids concentrations in the order of 50 to 60% may be achieved even with organic sludges (15).

Although some generalities regarding filter press performance can be made, in practice the obtainable results are extremely variable and very difficult reproducable. In general, sludge conditioning, as already mentioned, is necessary to achieve high performance within short operation time. The results are controlled by a delicate balance between particle-particle and particle-filter media interactions. Electrical double layer theory predicts, that coulombic repulsion will be decreased by increasing the ionic concentration of the system or by decreasing the surface charge. Both can be achieved through the proper use of conditioning agents.

Partial decomposition produces a sludge, which is extremely difficult to be conditioned and it is almost unfilterable. Instead, fresh, raw or chemically precipitated sludges flow readily, filter most easily and yield high capacities and low moisture cakes. Digestion of sludge decomposes the organic matter in the solid, usually reducing the volatiles 50 - 70% and producing a sludge of high mineral content and low compressibility (2).

2. EXPERIMENTAL PART

2.1. Pilot Plant Description

The sludge dewatering experiments were performed using a pilot - scale, batch operated filter press (Fig. 1), mainly consisted from plates and frames placed alternatively, with a filter cloth between them, which it was consisted from natural cotton fibers; a total filtration area of 19,3 x 10^{-2} m^2 was used.

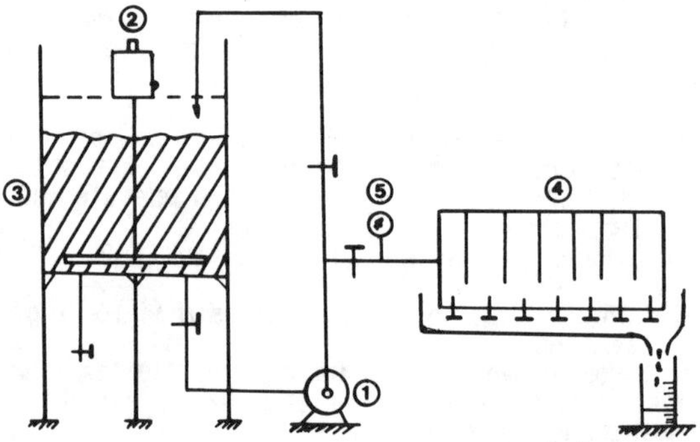

Fig 1. Filter press pilot plant, were: (1) centrifugal pump, (2) mechanical mixer, (3) feed tank (volume 100 1), (4) filter press section and (5) manometer.

The liquid carrying the suspended solids is forced into the filter chambers, which are designed so, that the liquid cannot escape except by passing through the filter medium and depositing the suspended solids on it. The flow through the filter press is not from chamber to chamber consecutively, but to all chambers simultaneously, so that each chamber fills up at the same rate, retains the same quantity of filtered out solids, and delivers the same quantity of filtrate. The number of plates, i.e. the number of filters used, are based mainly on the sludge characteristics and especially on the solids content of the sludge and the number of operating hours. A screw or hydraulic ram, fitted with a closing mechanism for forcing the plates and frames together, is situated on the one end of the filter press and operated manual, against a movable head.

2.2. Materials used

Two types of sludges from different origin, were examined:

1. Thickened secondary sewage sludge from a trickling filter unit.
2. Thickened secondary industrial sludge, aerobically digested, from a wastewater treatment plant operating with the activated sludge process.

As sludge conditioners, several additives were examined, like the flocculants, $Fe_2(SO_4)_3$, $Al_2(SO_4)_3$, and clay (caolin, argillopyritic mineral, fine disintegrated) and as flocculant aid, a cationic polyelectrolyte (solid Separan CP-400).

2.3. Experimental Description and Results Evaluation

A measured amount of sludge with known quantities of additives is added to the feed tank under stirring for few min. The applied pressure was 220 kN/m^2, while the duration of each experiment was 20 - 180 min.

For constant pressure, the basic filtration equation gives after integration:

$$t/V = (\mu W/2PA^2)V + \mu R_m/PA \qquad (y = ax + b)$$

The sludge specific resistance (r) can be determined from laboratory data, by plotting t/V versus V. If the value of the slope of the line passed through the data points is α, then from the previous equation results :

$$r = 2PA^2\alpha/\mu W,$$

while if the value of the abscissa in y-axis is b, then the resistance of the filter medium R_m can be calculated as. $R_m=(P\cdot A/\mu)b$.

Filtration rate is defined as the filtrate production (kg), per m^2 of filter area and per hour of filtration time.

It has to be pointed out, that differences in daily quality of the used sludges, make necessary the characterization and examination of all the used sludge samples. The solids content of the filter cake, was measured by heating at 105° C up to constant weight and expressed % w/w. The volatile solids content was then measured by incinerated the dry cake to 550° C for 2 h, and expressed also as % of the total solids. The used sludges had a viscosity around 1cp, (as this was measured with an Ostwald viscometer).

At the beginning of the procedure the filter permeability and the filtration rate was high. As on the filter the creation of the cake was started and the solids were rendering his permability and filtration rate were decreasing, but the filtrate quality was improved significantly and became very clear, without any observalbe solid. Usually in practice this filtrate is not suitable for final disposal, but it can be returned to the beginning of the treatment plant.

3. RESULTS AND DISCUSSION

3.1. Flocculants Addition

The first comparison was made, between the two types of examined sludge, the so called industrial and sewage, using clay. From the results, shown as Fig. 2, becomes clear, that the influence of applied clay was much more remarkable to the sewage sludge, in terms of r lowering and Rm increasing, while for the case of industrial sludge the differences were

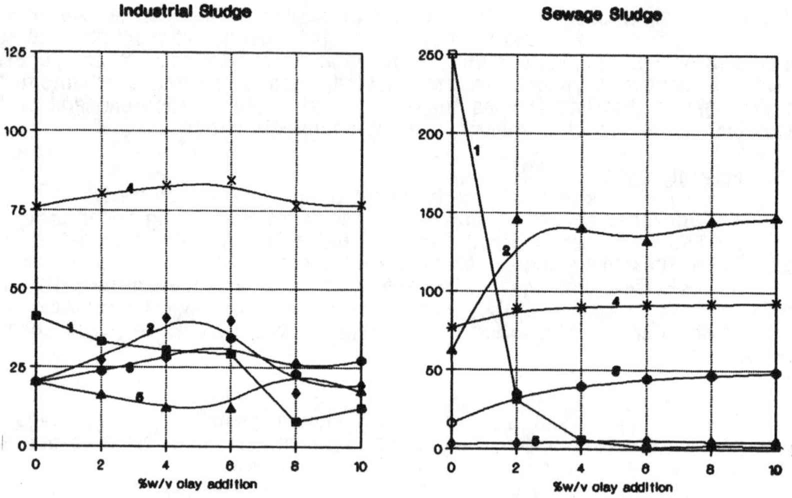

Fig. 2. Effect of clay addition on filter press performance,where (1) : r, $(x10^{-12})$, m/kg, (2) : Rm, $(x10^{-12})$, 1/m, (3) cake solidscontent, %,(4) : volatile solids content, %, (5) : filtration rate, $kg/m^2 \cdot h$.

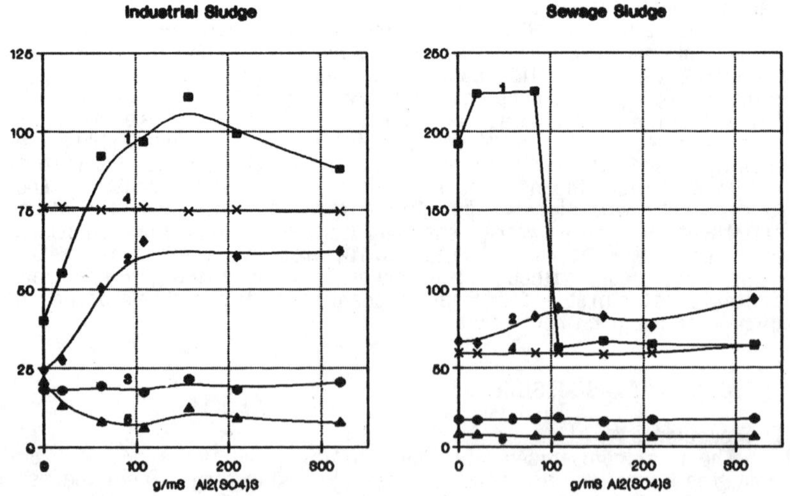

Fig. 3. Effect of $Al_2(SO_4)_3$ concentration on filter press performance. The notation remains the same.

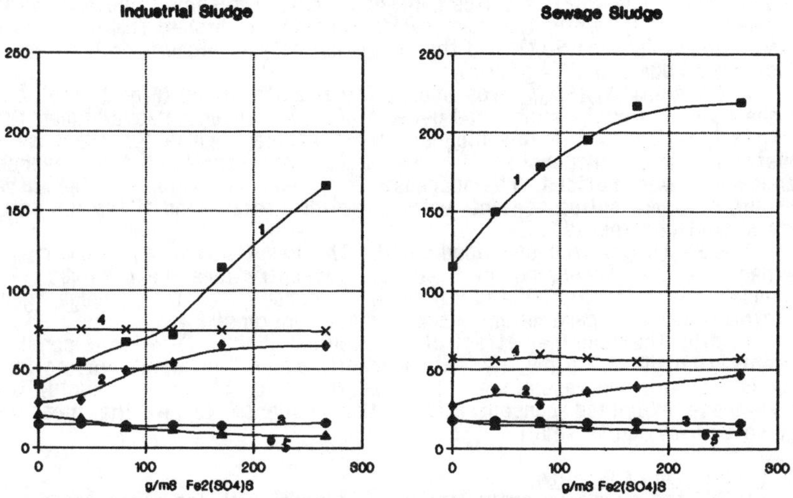

Fig. 4. Effect of $Fe_2(SO_4)_3$ concentrations on filter press performance. The notation remains the same.

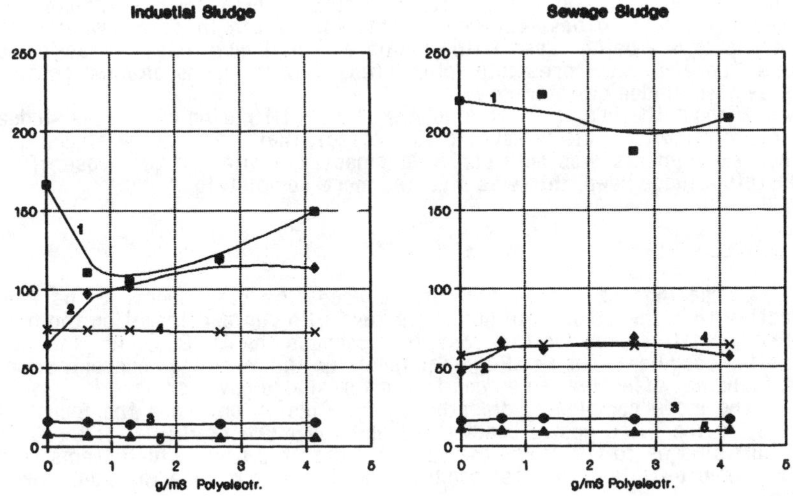

Fig. 5. Effect of cationic polyelectrolyte concentrations on filter press performance. The notation remains the same.

much smaller. The volatile solids content is about the same in both cases (in the range of 80%), while the cake solids contents is higher for the case of sewage sludge (around 50%), and the filtration rate is higher for the case of industrial sludge (15-25 kg/m^2·h).

After that, $Al_2(SO_4)_3$ was used. The results show. (Fig. 3) that only for the case of sewage sludge and using $Al_2(SO_4)_3$ concentrations higher than 100 g/m^3, a remarkable lowering of r was noticed, while for the case of industrial sludge increasing the $Al_2(SO_4)_3$ concentrations the reversed phenomenon was noticed, i.e. increase in r and Rm values. The other parameters (cake solids content, volatile solids content and filtration rate), were almost unaffected.

The $Fe_2(SO_4)_3$ was examined third. The results (Fig. 4.), were rather unexpectable, i.e. increasing the $Fe_2(SO_4)_3$ concentrations the r values were increased respectively, instead of being decreased, for both sludge types. The other measured parameters were almost constanst.

Finally, the combined effect of a flocculant ($Fe_2(SO_4)_3$ with a constant concentration of 260 g/m^3, i.e. rather small) and a flocculant aid (cationic polyelectrolyte) was examined. The results, (Fig. 5), show clearly, that under these examined concentrations, the influence to all the measured parameters was quite small.

3.2. Discussion of Results

In all the examined cases, the used industrial sludge appeared to have lower initial r values, but higher Rm values than the corresponding sewage sludge. Generally speaking, the filtration rate and the volatile solids contents are also higher for the industrial sludge case, while the cake solids content seems to be almost the same for the both cases.

The lowering of r values, because of the conditioning agents (or filter aids), is devoted to the phenomenon, that filter aids build up a porous, permeable and rigid lattice structure, which retains solids particles and allows the liquid to pass through. This is truth in cases, where the filter aids are being used with relative small quantities of solids, or where there is a compressible solids slime, like it hapens, in most cases of wastewater sludge dewatering.

Without the use of filter aids, the direct filtration of sewage sludges is very difficult. This is devoted to the fact, that the sewage sludge used in our experiments was only partially digested, while for the case of the industrial sludge used, this was digested more completely.

4. CONCLUSIONS

Comparing the three flocculants used, the clay seems to be more effective than the other two, but if we take into our consideration the much higher quantities used for this case, then perhaps the $Al_2(SO_4)_3$ is the most effective coagulant. The $Fe_2(SO_4)_3$, at least for the examined concentrations, was found not effective as a conditioner for sludge dewatering.

The more completely digested sludge, like in our case the industrial sludge, is the more easily conditioned and the more readily it filters. The difficult sludges to filter are the partially digested ones, which seems to be highly compressible even after conditioning, like in our case the used sewage sludge.

REFERENCES

(1) Randall, C. W., Turpin, J.K. and King, P.H., (1971). Activated sludge dewatering: factors affecting drainability. J. W.P.C.F., 43 (1), 102-122.

(2) Smollen, M., (1988). Moisture retention characteristics and volume reduction of municipal sludges. Water S.A., 14 (1), 25-8.(2)

(3) Heiskanen, Niemi, H. and Vesanto, A., (1988). The effect of filtration variables on the residual moisture of a filter cake. Proc., XVI Intern., Min. Proc. Congress., E. Forssberg, (ed.), Elsevier, Amsterdam, p.p. 337-350.

(4) Tosun, I. and Willis, M.S., (1987). Cake filtration theory : an Overview. Biotech - Forum, 4 (3), 136-41.

(5) Sürügü, G. and Cetin, F.D., (1989). Effect oı temperature, pH and D.O. concentration on filterability and compressibility of activated sludge. Wat. Res., 23 (11), 1389-95.

(6) Agunwamba, J.C., Egbuniwe, N. and Ademiluyi, J.O., (1989). Mass balance filtration Equation. Waste Management, 9, 141-9.

(7) Carman, P.C., (1933). Fluid flow through granular beds. Trans. Inst. Chem. Engng., (London), 15, 150.

(8) Coackley, P.,(1958). Laboratory scale filtration experiments and their application to sewage sludge dewatering. Biol. Treatm. Sew. Industr. Wastes, Vol. II. W.I. McCabe, and W.W. Eckenfelder, (eds.), Reinhold Publ., New York, p. 270.

(9) Dentel, S., Resta, J.J., Shetty, P.V. and Todd, A., (1988). Selecting coagulant, filtration, and sludge-conditioning aids. J. A.W.W.A., Jan., 72-84.

(10) Knocke, W.R., Hamon, J.R. and Dulin, B.E., (1987). Effects of coagulation on sludge thickenning and dewatering, J. A.W.W.A., June, 89-98.

(11) Novak, J.T., and Bandak, N., (1989). Chemical conditioning and the resistance of sludges to shear. J. W.P.C.F., 61 (3), 327-32.

(12) Papachristou, E., Guitonas, A. and Zouboulis, A., (1988). Fly ash employment in conditioning biological sludges for dewatering. 2nd Intern. Conf. on Wat. Poll. Contr., I.A.W.P.R.C., Bangkok, 9-11 Nov., p. 405-9.

(13) Zouboulis, A.I. and Tzimou-Tsitouridou, R., (1990). Fly ash utilization in environmental engineering : the case of Greece. 3rd Intern. Symp. on Reclamation, Treatment and Utilization of Coal Mining Wastes. Minestone Serv., Glasgow, 3-7 Sept.

(14) Rebhun, M. Zall, J. and Galil, N., (1989). Net sludge solids yield as an expression of filterability for conditioner optimization. J. W.P.C.F., 61 (1), 52-4.

(15) Thomas, C.M., (1971). The use of filter presses for the dewatering of sludges. J. W.P.C.F., 43 (1), 93-101.

SLUDGE PARASITOLOGICAL CONTAMINATION

J. SCHWARTZBROD, J.L. STIEN, M.T. THEVENOT and S. STRAUSS
Department of Microbiology
W.H.O. Collaborating Centre
Faculté de Pharmacie - 5, rue Albert Lebrun - 54000 NANCY France

Summary

The parasitological contamination of sludge has been carried through different types of samples.

. The study of 116 samples of primary, digested and semidehydrated sludge has revealed that all the samples contained helminth eggs at concentrations ranging from 113 to 135/100 g of sludge.
. Concerning the sludge at a dump, the number of eggs recovered varied in a great way according to the storage time. No eggs were recovered from samples stored for 4 years, on the other hand in sludges stored for 1-3 years, mean values of 12 eggs/100 g were found.
. The marine sediments from the sludge effluent zone at 4000 m offshore contained helminth eggs.
. Finally 60 % of compost samples with sludge from wastewater treatment plant and garbage contained helminth eggs after 120 days of compostage.

1. INTRODUCTION

In developed countries, wastewater treatments often lead to the production of large quantities of sludge. The various uses of sludge are summarized in table 1.

Although many studies have been done on viral and bacteriological pollution of the sludge, very little is known about parasitological contamination of this important resource. This investigation addressess the problem of sludge parasitological contamination at various treatment stage at the sewage plant ; after storage in a sewage disposal after transport offshore through a deep conduit and at last in compost composed of a mixture of garbage and sewage sludge.

Table 1 – Sludge disposal in developed countries (10)

COUNTRY	SLUDGE PRODUCTION (1000 tds/year)	AGRICULTURAL LAND	OTHER LAND	INCINERATION	SEA	UNSPECIFIED
UK	1200	39	27	4	30	0
BELGIUM	70	15	83	2	0	0
DENMARK	130	45	45	10	0	0
FRANCE	840	30	50	20	0	0
GERMANY	2200	39	49	8	2	2
GREECE	3	0	100	0	0	0
IRELAND	20	4	51	0	45	0
ITALY	1200	20	55	0	0	20
LUXEMBOURG	11	90	0	10	0	0
NETHERLANDS	230	60	27	2	11	0
AUSTRIA	140	small	large	30	0	0
FINLAND	130	40	45	0	0	15
NORWAY	55	18	82	0	0	0
SPAIN	45	60	——20——		20	0
SWEDEN	210	60	——30——		0	10
SWITZERLAND	50	80	10	10	0	0
USA	8000	42	15	27	4	12

2. STUDY OF SEWAGE SLUDGE

Sludge samples were collected at the treatment plant of Nancy which treats 150.000 m^3 of wastewater daily. The analysis were carried on primary, digested and semidehydrated sludge. For helminth eggs, the method of Janeckso Urbanyi (5) with potassium iodomercurate solution (d = 1.44) was used. Giardia cysts concentration was obtained with the method of Teleman-Rivas according to Bailenger (11). 116 sludge samples were analysed (table 2).

Table 2 – Helminth eggs concentration in sludges

	NUMBER OF SAMPLES	NEMATODES*	CESTODES*
Primary sludge	34	100	35
Digested sludge	34	103	22
Semidehydrated sludge	48	105	8

*** Eggs number/100 g of sludge (mean value)**

Helminth eggs (Cestodes and Nematodes) were found in all sludge samples. The number of eggs recovered varied from 113 to 135 eggs/100 g wet matter depending on sample type. The predominance of Nematodes eggs (Ascaris, Trichuris, Toxocara) in all samples must be noticed. Giardia cysts analyses were realized on 62 samples of primary and semidehydrated sludge (Table 3).

Table 3 - Giardia cysts in sludge

SAMPLES	SAMPLES NUMBER	% POSITIVE SAMPLES	MEAN VALUE /100 g SLUDGE
Primary sludge	34	67.6 %	160
Semidehydrated sludge	28	3.5 %	5

Giardia cysts were found in 67.6 % of sample with mean value 160/100 g whereas in semidehydrated sludge only 3.5 % of samples contained very low concentration of Giardia cysts.

3. ANALYSE OF SLUDGE AT A DUMP

77 sludge samples taken from the Nancy treatment plant and put in disposal during time intervals ranging from 1 to 4 years have been analysed (Table 4).

Table 4 - Helminth eggs concentration in sludge disposal

TIMES STORAGE	NUMBER OF SAMPLES	EGGS/100 G SLUDGE NEMATODES	CESTODES
For 1 year	27	13	0.4
For 2-3 Years	22	11	0.1
For 4 years	28	0	0

All eggs have disappeared in the 28 sludge samples put in a disposal 4 years ago. On the other hand, all sludge samples stored during 1 to 3 years contained helminth eggs with mean values of 13.4 and 11.1 eggs/100 g respectively. Comparing these concentrations with those of samples taken from the treatment plant, there is a reduction of about 90 %.

4. HELMINTH EGGS IN MARINE SEDIMENT

Wastewater from Barcelona were treated at San Adrian treatment plant and sewage sludge was transported offshore through a deep conduit at 4.5 km. Marine sediments were collected from offshore at depths varying from 50 to 52 m. The exact location of sampling point was obtained from the Transit satellite (table 5).

Table 5 - Location of sampling points

	LATITUDE N	LONGITUDE E	DEPTH (METER)
Sample A	41°23.22'	02°15.15'	50
Sample B	41°23.40'	02°15.40'	50
Sample C	41°23.30'	02°15.35'	50
Sample D	41°23.10'	02°15.40'	52

3 sampling campaings have been performed in March, June and November (Table 6).

Table 6 - Helminth eggs concentration in marine sediment

SAMPLING LOCATION	FIRST CAMPAIGN (MARCH)	SECOND CAMPAIGN (JUNE)	THIRD CAMPAIGN (NOVEMBER)
Sample A	256*	41	15
Sample B	88	ND**	45
Sample C	90	88	0
Sample D	132	83	120

* Eggs number/100 g wet sediment
** Not determined

Globally, 10 (i.e. 90.9 %) out of the 11 analysed samples contained helminth eggs belonging to Nematodes and/or Cestodes family. The predominance of Nematodes eggs (Ascaris, Toxocara, Trichuris) detected in all the positive samples (10/10) must be underlined whereas Cestodes eggs (Hymenolepis ; Taenia) have been detected in 6 samples only. The egg type detected are those that are usually described by different authors (4,7). In a previous paper, we did the parasitological analysis on sludges from San Adrian treatment plant (8). This sludge proved to be very rich in helminth eggs as 70 % of the samples contained at least 3 different egg types. Cestodes eggs were pratically represented in the same proportion as Nematode eggs since 80 % of samples contained Cestodes and 90 % Nematodes respectively. The marine environment therefore seems to be particularly hostile to Cestodes eggs which are only detected in 54.5 % of the samples.

A comparison between the various sampling campaigns shows that the helminths eggs concentration are the lowest in November whatever the sampling point. This observation might be explained by variations in the climatic conditions. The analysis carried out according to the sampling points shows that A and D samples were always strongly contaminated with concentrations in helminth eggs ranging from 15 to 256/100 g sediments, the C sample always having a lower contamination level. Theses discrepancies might be related to the existence of sea currents which might convey helminth eggs from the outfall ditch to open sea according to a route including sampling points A, B and D. A further analysis with additional sampling points made it possible to confirm this hypothesis (9).

5. PARASITOLOGICAL CONTAMINATION OF COMPOSTS

The analysed composts contained 65 % of garbage and 35 % of semidehydrated sludge. The composting method calls for windrowing.

Test samples of 20 kg taken at the upper and the lower third were used to form samples by quatering The follow up of the sanitary quality has been performed in carrying out parasitological analyses of the raw materials and the composts (table 7).

Table 7 - Parasitological contamination of raw materials and composts

SAMPLES	SAMPLES NUMBER	NUMBER OF POSITIVE SAMPLES (%)	
Garbage	16	1 (6.2 %)	Nematodes
Semidehydrated sludge	20	16 (80 %)	Nematodes Cestodes
Compost	56	32 (57.1 %)	Nematodes Cestodes

Concerning raw materials the percentage of positive samples is much higher (16/20) in the sludge than in the garbage being only slightly contaminated as only one sample contained helminth eggs.

We detected Nematodes and/or Cestodes eggs in 57.1 % of mixed compost samples after 120 days of compost preparation.

6. CONCLUSION

The analysis of various type of sewage sludge shows that parasitological contamination is rather significant in the sludge samples from Nancy treatment plant since all samples proved to be positive. Concentrations vary from 113 to 135 eggs/100 g depending on the sludge type which underlines the low impact of the treatment on helminth eggs. Detected eggs mainly belong to Nematodes family and more rarely to Cestodes family. This observation is in accordance with literature informations as Kabrick and Jewell (6) detected 10 eggs/100 g of digested sludge, Bergström (1) 51 eggs/100 g of primary sludge, Graham (3) 70 eggs/100 g of raw sludge and 54 eggs/100 g of digested sludge.

For Giardia cysts, the treatment impact is significant since 67.6 % of primary sludge samples and only 3.5 % of semidehydrated sludge samples proved to be positive.

The analyses of sludges stored at a dump confirm the great resistance to environmental conditions of helminth eggs as shown by different searchers (2,12).

In the marine sediments, a parasitological contamination is demonstrated in 90.5 % of the sample with very low concentration of Cestodes eggs. This observation seems to prove that this type is more labile in a marine environment.

At last, the analysis of the composts consisting of a mixture of garbage and semidehydrated sludge showed the presence of Nematodes and/or Cestodes eggs in 57.1 % of samples after 120 days of compostage.

In view of these results it appears that sludge has a rather significant and persistent parasitological load. Under these conditions it would be necessary to take this parasitological contamination into account when using to preserve human and animal health.

REFERENCES

(1) BERGSTROM K. (1981). Recovery of parasite eggs from sewage sludge, from sewage plants and septic tanks in Norway. Norsh. Vet., 93, 323-330.

(2) FEACHEM R.G., BRADLEY D.J., GARELICK M. and MARA D.D. (1983). Sanitation and disease health aspects of excreta and wastewater management. John Wiley ed., 380-393.

(3) GRAHAM H.J. (1981). Parasites and the land application of sewage sludge. Report Ontario Min. Environment n° 110.

(4) HANNAN J. (1985). Parasitological problems associated with land application of sewage sludge. In "Epidemiological studies of riskes associated with agricultural use of sewage sludge : knowledge and needs". Ed. Block, Havelaar, L'hermite, 102-108.

(5) JANECKSO A. et URBANYI L. (1931). Methode d'enrichissement coprologique. Rev. Gen. Med. Vet., 41, 496-497.

(6) KABRICK R.M. and JEWELL W.J. (1982). Fate of pathogenes in thermophilic aerobic sludge digestion. Wat. Res., 16, 1051-1060.

(7) PANICKER P.V.R.C. and KRISHNAMOORTHI M.P. (1981). Parasite egg and cyst reduction in oxydation ditches and aerated lagoons. Wat. Pollut. Control Fed., 53, 1413-1419.

(8) SCHWARTZBROD J., LUCENA F., BOSCH A., BERIL C., SCHWARTZBROD L. et PARES R. (1987). Influence des eaux continentales et des eaux usées sur la contamination microbiologique des sédiments marins. J. Fr. Hydrol., 18, 142-156.

(9) SCHWARTZBROD J., THEVENOT M.T. and STIEN J.L. (1989). Helminth eggs in marine and river sediments. Marine Poll. Bull., 20, 269-271.

(10) STIEN J.L. (1989). Oeufs d'helminthes et environnement : le modèle d'oeufs d'*Ascaris*. Thèse Université Metz.

(11) TELEMAN-RIVAS modifié BAILENGER (1973). In BAILENGER "Coprologie parasitaire et fonctionnelle". Ed. Drouillard.

(12) WALLIS P.M. and LEHMANN D.L. (1983). Biological health risks of sludge disposal to land in cold climates. In "Processing use of sewage sludge". Ed. L'hermite, Ott, 237-239.

ENVIRONMENTAL ASPECTS OF LIQUID PHASE OXIDATION OF SEWAGE SLUDGE

Pieter H.A.M.J. DE BEKKER

Grontmij n.v. De Bilt (Netherlands)
Consulting Engineers
De Holle Bilt 22, 3732 HM DE BILT
Netherlands

Summary

Liquid phase oxidation (LPO) or wet oxidation of sewage sludge is a relatively new method of sludge treatment, which may be compared with incineration of sewage sludge.In liquid oxidation, organic pollutants in the water phase are made to oxidize under high pressure and temperatures. The extent to which complete oxidation occurs depends on temperature, processing time and the nature of the organic pollutants. In the VerTech process the high pressure requires for liquid oxidation is achieved in a reactor with a length of approximately 1225 m, built vertically into the earth.

1. INTRODUCTION

The VerTech process for liquid oxidation has been described extendedly earlier [lit. 1]. It operates as follows: the reactor will be built in a deep concrete shaft. The reactor consists of a pipe system with two concentric pipes: the 'downcomer' and the 'upcomer' and a heat exchange system (fig. 1).

Sludge to be oxidized and oxygen will enter the reactor by way of the central pipe, the downcomer. Temperature and pressure will increase as the influent flows down. At a temperature of approximately 175 °C the oxidation process starts and the temperature of the reaction mixture in the bottom part of the reactor will rise as a result of exothermal oxidation. On account of the liquid column on top, pressure will increase and evaporation of the reaction mixture is prevented. Oxidized sludge will leave the reactor by way of the upcomer. In the upcomer, pressure gradually diminishes and part of the heat generated is transferred to the influent in the downcomer. The concentric downcomer and upcomer provide an efficient transfer of heat between cold influent and hot effluent.

The heat released in the oxidation process will not only be conducted to the influent, but also to a heat exchanger, which envelopes the reactor. The coolant of the heat exchanger, called heat exchange medium, is pumped down into the surrounding pipe and returns to the surface via an extremely insulated pipe. From this heated water, energy can be recovered by the production of steam. The heat exchange system is also used for starting up the installation.

Using an external heat source to warm the heat exchange medium and temporarily reverse the flow of the heat exchanger, so that the heat exchange medium flows down through the insulated

Center Tube Downcomer Vessel (CTD) Cross-Sectional Schematic

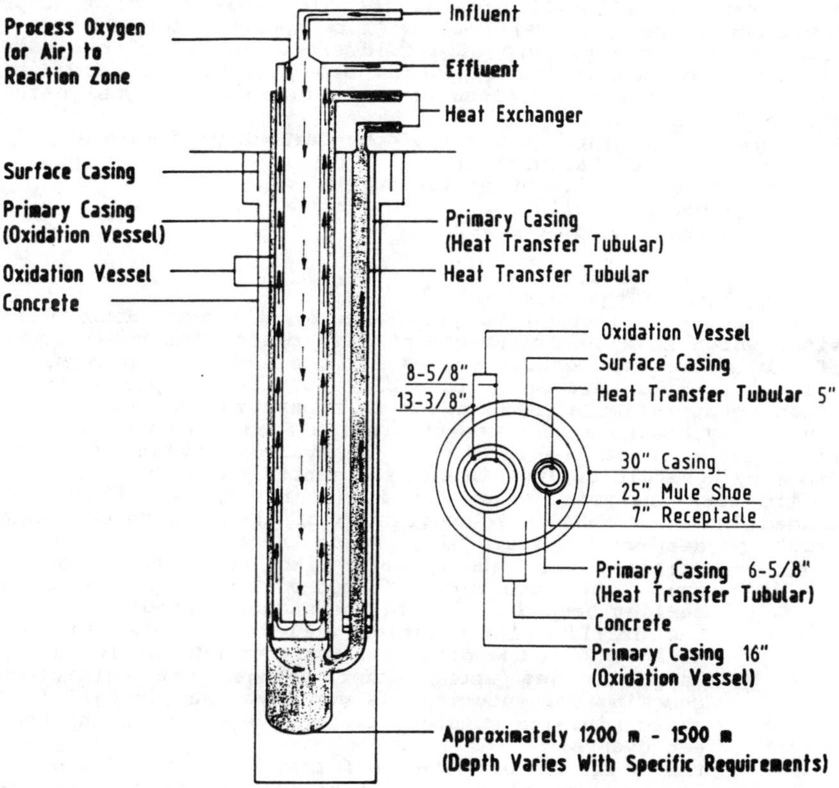

pipe, enables proper heating of the mixture at the bottom of the reactor. Once the oxidation process is in progress, the external heat source is switched off, the flow direction of the heat exchange medium reversed again and normal operation of the heat exchanger will resume. Near the bottom of the reactor, the temperature is controlled at approximately 277 °C. There the weight of the gas/liquid column creates a pressure of approximately 85-110 bar. At the discharge of the reactor the temperature of the effluent will be approximately 50 °C. After oxidation a mixture of liquid effluent, in which sludge ashes are suspended and off-gases are present, will remain. The oxidation process will convert the organic nitrogen in the sludge into ammonium.
Possible environmental aspects of the VerTech process are:

- soil contamination;
- liquid effluent pollution;
- off-gas pollution;
- leaching of the ashes.

2. SOIL CONTAMINATION

Construction of the reactor

Soil contamination is prevented by a secure double casing, which is also double cemented in place. The total depth of the subsurface reactor varies between 1200-1500 meters, of which the upper part where possible aquifers are present will reach to approximately 250 m. In this part the casing will be doubled to ensure a complete separation from the earth layers. The other part of the reactor is separated from the surroundings by a single casing, which is cemented in place.
In the risk analysis, which has been carried out, it is concluded that the chances of pollution of the environment as a result of deep-drilling are nil, seeing that:

- drilling will mostly take place in a formation of which the structure is well-known as a result of earlier deep-drilling in the neighbourhood;
- the drilling installation will be equipped in compliance with the direction and regulations in force, which implies among other things that sufficient checking and security systems have been installed so as to minimize consequences in the event of unforeseen events;
- the drilling will be performed in a way known to involve few risks, as is often applied in the extraction of natural gas and mineral oil.

Earthquakes

Soil contamination of the worst case which can theoretically arise, when an earthquake occurs and the reactor is filled with sewage sludge to be processed. An earthquake of such force causes the surrounding earth formations to move, the concrete mantling of the surface casing to disappear and the surface casing to break. In that case the heat exchange medium as well as the sludge would flow into the ground. Should the reactor crack as well, then the content of the reactor - approximately 120 m^3 of sludge with a suspended solids content of 5% in a 16" reactor - would be absorbed by the soil. The consequences of such an earthquake in Western Europe are hard to predict. If an earthquake of such magnitude were to occur, the disturbances of soil and soil hydrology would be so enormous that the quantities of coolant and sludge released as a result of a crack could end up both in other strata and immediately at the newly formed surface of the earth.

Leakage

The reactor will function only if the heat generated during liquid oxidation can be adequately removed. If the reactor or the heat exchanger do not function properly, the reactor will have to be shut down. With respect to various forms of irregular interactions between heat exchanger and reactor, a risk analysis has been drawn up.

* Casing leakage

If a change in the quantity of heat exchange medium is signalled, the reactor can be pumped dry and the reactor valves closed, after which repairs can be carried out.

* Leakage between upcomer and downcomer

A leak between upcomer and downcomer implies that a short circuit flow will occur in the reactor. Because of this, reactor output will diminish, which can be noticed on the effluent side of the reactor. In case of a possible short circuit flow, the reactor will have to be shut down, so as to enable repairs to be performed.

* Leakage between reactor and heat exchanger

This kind of leakage will cause oxygen to enter the heat exchanger, which will not lead to a potentially dangerous situation however. The presence of oxygen in the heat exchange medium will be detected. To have a leak of this type repaired, the reactor will have to be shut down.

3. LIQUID EFFLUENT POLLUTION

Liquid effluent is produced and it is to be expected that soluble components of the sludge will be hydrolysed during the process. The oxidation reaction will not reach to a full oxidation, but a certain part of the organics will be present in the effluent, mostly as lower fatty acids. Nitrogen originating from cell proteins is transformed into ammonia-nitrogen, which together with the fatty acids forms a strong chemical buffer against Ph-changes. Nitrates in the liquid effluent originate from reactor washings with nitric acid once per 14 days period and after equalization/storage dosed to the main flow. Phosphorus and metals will almost not be found in the liquid effluent, but will be present in the ash. A typical composition of the liquid effluent looks as follows:

COD	15000 mg/l
BOD	7500 mg/l
TKN	2250 mg/l
NH_4-N	1700 mg/l
NO_3-N	700 mg/l
Total-N	2950 mg/l

In most cases the effluent has to be treated to a high standard for BOD, TKN, NO_3-N, P and suspended solids.

There are several methods available for such treatment:

a. ammonia stripping followed by anaerobic and aerobic treatment;

b. ammonia precipitation by addition of magnesia oxide and phosphoric acid;

c. denitrification followed by nitrification.

The first process under a is rejected because of the high cost for chemicals to increase pH to 11. Besides this after stripping the ammonia gas has to be absorbed by an acid like sulphuric acid to form ammonia sulphate. In many countries this chemical in a relatively diluted form (30%) has no economic value. So it has to be evaporated and dried before it can be used as a fertilizer. This causes an intensive processing apart from the main stream, which will cost far more than may be earned back through selling of the ammonia sulphate.

The second process uses crystallisation of ammonia salt by adding magnesia oxide and phosphoric acid forming crystals of magnesia-ammonia-phosphate (MAP). In this process the cost for chemicals (magnesia oxide and phosphoric acid) are much more than can be earned back by selling the crystals as a fertilizer. Thus this process scheme also is rejected.

The third process has been chosen for the first commercial VerTech plant in Apeldoorn (The Netherlands) due to the fact that experimental affairs only will be considered among the liquid phase oxidation process. Therefore a conventional process has been taken for the treatment of the liquid effluent. Although the process may be conventional, the process factors are somewhat uncommon i.e. the high concentrations of components to be removed and the severe effluent standards. A pilot plant has been built at the Dordrecht Test Centre to verify process parameters for the effluent treatment.

Results of the test work indicate that:

- a sequencing batch reactor (fill and draw) is a type of treatment, which can generate a well-treated effluent. Samples however taken after the nitrification period show low TKN and NH_4-N levels, but relatively low NO_3-N levels, whereas samples taken after the denitrification period show the opposite effect. Complete denitrification showed to be possible even with BOD_5/N-ratios of below 2,86.
- effluent concentrations may be expected of:

COD	<	900	mg/l	reduction	93 %
BOD_5	<	15	mg/l	reduction	99,8 %
TKN	<	130	mg/l	reduction	94 %
NH_4-N	<	50	mg/l	reduction	97 %
NO_3-N	<	70	mg/l	reduction	90 %
Total-N	<	180	mg/l	reduction	94 %

- trace metals have to be added as the VerTech process converts these components into insoluble metal oxides.
- because of the low chemical buffering capacity of the liquid effluent lime or caustic soda have to be added to control pH.
- a biological system may operate successfully under the following circumstances:

F/M ratios	0,06	kg BOD_5/(kg MLSS.d)
	0,027	kg TKN/(kg MLSS.d)
	0,13	kg NO_3-N/(kg MLSS.d)
Temperature	33	°C

4. OFF GAS POLLUTION

The off-gases of the process will consist mostly of carbon dioxide. The distribution of components is represented in the next table.

TABLE I: Off-gas composition

Component	Volumetric percentage
Nitrogen	2,92%
Oxygen	2,86%
Carbon dioxide	90,23%
Carbon monoxide	3,97%

It was found that the effluent gases also contain traces of organic components. The measured concentrations of organic components in the effluent gases without any treatment are shown in the table below.

TABLE II: Off-gas organic components (without gas treatment)

Component	Concentration (ppm)	MAC/TLV (ppm)	
Acetone	5-30	1000	MAC
Furan	2-3	50	TLV
2-Butanone	3-4	200	MAC
2-Pentanone	1-2	200	MAC
1-Hexene	1-2	100	TLV
1-Heptene	1	400	TLV
Benzaldehyde	4,78	100	TLV
Furfural	0,34	5	MAC

MAC = Maximum Acceptable Concentration
TLV = Threshold Limit Value

5. ASH LEACHING

The ashes produced by the VerTech process will be dewatered in a filter press without addition of poly-electrolytes. The solids content of the remaining cake will be at least 40%. If the ash is to be dumped a solids content of 40% generally will be sufficient, as the material will be fairly stable from a ground-mechanical point of view. Solids contents of more than 70% might result in the ashes blowing away after dumping. During pilot tests with dewatering of the ashes, ash samples were analyzed for leaching action. The ash composition of heavy metals can be computed by using weighed average concentrations in the unprocessed sewage sludge and the theoretical thickening factor. In the table below the theoretical and analyzed composition of VerTech ash and the theoretical composition of sludge incineration ash is given.

TABLE III: Heavy metals content of VerTech ash [mg/kg]

Component	Theoretical VerTech ash	Analyzed VerTech ash	Theoretical incineration ash
As	12-15	29	15
Cd	6-7,5	4,5	19
Cr(III)	500-625	200	450
Cu	450-565	520	1600
Hg[1)]	3,5-4,4	3,0	<3
Pb	700-875	425	1000
Ni	70-88	75	300
Zn	2250-2815	2200	4200

1) In dry combustion processes, mercury will due to its volatile character not concentrate in the ashes, but in the fume gases. Therefore concentrations of mercury in the theoretical incineration ash is low.

Except for the metal arsenic, the ash quality is equal or better than the theoretical quality. The VerTech ash is not a residual substance, which must be marked as chemical waste referred to the Dutch Chemical Waste Act.

6. CONCLUSIONS

Wet oxidation of sewage sludge has compared to other techniques very few environmental impacts. In other words the discharge of gas, ash and liquid effluent are well controllable, so that pollution may be prevented in all cases.

ACKNOWLEDGEMENT

This paper has been written in cooperation with VerTech Treatment Systems b.v. in Baarn (Holland). All data mentioned are extracted from the Environmental Impact Assessment Report for the liquid phase oxidation installation to be built in Apeldoorn (Holland).

REFERENCES

1. DE BEKKER, P.H.A.M.J. and VAN DEN BERG, J.J. (1988), Wet Oxidation as the Alternative for Sewage Sludge Treatment, Proceedings of the EC Conference Sewage Sludge Treatment and Use by Elsevier Applied Science, ISBN 1-85166-418-1

EFFECT OF FREQUENT USE OF SEWAGE SLUDGE ON THE COMPOSITION OF ARABLE LAND AND CROPS

P.J. van Erp*, P. van Lune* and A. de Jong**.
* Institute for Soil Fertility Research,
P.O. Box 30003, 9750 RA Haren (Gr.), The Netherlands
** Research Station for Arable Farming and Field Production of Vegetables, P.O. Box 430, 8200 AK Lelystad, The Netherlands

Summary

In field experiments, carried out from 1976-1986, large amounts of sewage sludge were applied repeatedly to a reclaimed peat soil and a clay soil. Potatoes, sugar beets and wheat were grown in the experiments. Soil loads with Zn, Cu, Pb, Cr, Ni, Cd, Hg and As via sewage sludge were lower than permitted according to the Dutch guidelines for the use of sewage sludge. Repeated use of sewage sludge increased the content of some elements in the top layer of the soils and in the crops. However, the increase was moderate and the (suggested) maximum permissible levels for heavy metals in soils and crops were not exceeded. The results of the experiments are in agreement with the basis of the Dutch guidelines for the use of sewage sludge in agriculture.

1. INTRODUCTION

Sewage sludge has some value in agriculture because of its contents of organic matter and plant nutrients. In the long term, frequent use of sewage sludge may alter the nutrient status of soils and the composition of crops. This was investigated from 1976-1986 in two field experiments with repeated applications of large amounts of sewage sludge. The results of the investigation were tested against the Dutch (suggested) maximum permissible levels of heavy metals in crops and soils.

2. MATERIALS AND METHODS

During ten years field experiments were conducted on a reclaimed peat soil and a clay soil. The organic matter content, the percentage soil particles < 16 μm and pH-KCl were 14.7%, 4.1% and 5.0 for the reclaimed peat, and 3.9%, 26.0% and 7.2 for the clay soil. Liquid or dewatered sewage sludge was applied to sugar beets and potatoes in the preceding autumn. This resulted in eight and five applications of sewage sludge on the peat and clay soil, respectively. The sewage sludge treatments in quadruplicate were 0, 10 and 20 tonnes dry matter per ha. The mean composition of the sewage sludge is given in table 1. Inorganic fertilizers were applied to ensure optimum crop growth. Sugar beets, potatoes and wheat were grown as test crops.

The top layers of the treatments (peat 0-20 cm, clay 0-25 cm) were sampled in 1976, 1982 and 1986 and analysed for pH-KCl, total contents of organic matter, N, P, K, Ca, Mg, S, Fe, As, Cd, Cr, Cu, Hg, Mn, Mo, Ni,

Pb, Zn and extractable contents of Mg (0.5 N NaCl), P (water), K (0.1 N HCl), B (water) and Co (2.5% HAc). All marketable products of the cultivated crops were washed with deionised water (except wheat grain) before determining the total Cd, Cu, Mn, Ni and Zn contents.

Nutrient contents in the soils of the 10 and 20 tonnes sewage sludge treatments were compared with the contents in the soil of the 0 tonnes treatment both in 1982 and in 1986. Increase or decrease in both years were recorded and statistically tested (sign test). The test was executed for each separate soil and for both soils together. The same was done with the elemental composition of the crops cultivated during the ten years of the experiments.

3. RESULTS AND DISCUSSION

Repeated use of sewage sludge significantly ($P = 0.05$) increased the content of organic matter, total contents of N, P, Cd, Cu, Zn, Fe, Pb, Hg and extractable amounts of P, K, Mg and B of the soils. The 1986 soil contents are given in table 2. Heavy-metal contents did not exceed the Dutch suggested "safety" levels for agricultural soils (1). For reclaimed peat and clay soils these levels are: Cd 0.5 and 1.0, Cu 50 and 200, Hg 2 and 2, Pb 100 and 200 and Zn 100 and 350 (mg/kg), respectively. Sewage sludge raised the Cd, Cu and Zn contents of sugar-beet leaves, Zn in sugar beets, Cu and Zn in potatoes (tubers) and Cd and Cu in wheat grain. The Dutch Cd limit for fodder of 1.14 mg/kg d.m. (2) was not exceeded in the experiments. The maximum permissible Cd levels for food are 0.10 and 0.15 mg/kg fresh weight of potatoes (tubers) and wheat grain, respectively (4). This level was exceeded in two wheat grain samples; this was not related to the sewage sludge treatments.

In the Netherlands guidelines have been set for the use of sewage sludge in agriculture (3) to ensure a safe agricultural crop production on all agricultural soils. Two tonnes sewage sludge (dry matter) per year during 100 years is permitted. Then the load of the soil with Zn, Cu, Pb, Cr, Ni, Cd, Hg and As is equal to or lower than 400, 120, 100, 100, 20, 1, 1 and 2 (kg/ha), respectively. In the experiments total loads of these elements in the highest sewage sludge treatment of the reclaimed peat and clay soil were 151 and 115, 88 and 25, 37 and 19, 11 and 5.5, 3.5 and 2.2, 0.6 and 0.3, 0.3 and 0.2 and 0.8 and 1.1 (kg/ha), respectively. The moderate changes in the composition of the topsoil and the crops, and the fact that the (suggested) maximum permissible levels for heavy metals in soils and crops were not exceeded are in agreement with the basis of the guidelines for the use of sewage sludge in agriculture.

Acknowledgement

The investigation was initiated and subsidized by The Foundation for Applied Wastewater Research, Rijswijk. Detailed results were published (in Dutch) in Verslag no. 98, April 1990, of PAGV at Lelystad by P.J. van Erp, P. van Lune and A. de Jong. The title is: Zuiveringsslib in de akkerbouw. a) Landbouwkundige waarde van zuiveringsslib (A. de Jong, PAGV); b) Invloed van het gebruik van zuiveringsslib op de bodem- en gewassamenstelling (P.J. van Erp en P. van Lune, IB). Part b is presented in the poster.

4. LITERATURE

(1) Landbouw Advies Commissie Milieukritische stoffen (1986) Signaalwaarden voor de gehalten van milieukritische stoffen in grond met het oog op landbouwkundige gebruiksmogelijkheden van verontreinigde bodems. LAC-rapport nr. 86.1.
(2) Produktschap Veevoeder (1975) Verordening voor ongewenste stoffen en produkten.
(3) Unie van Waterschappen (1985) Richtlijn voor de afzet van zuiveringsslib ten behoeve van gebruik op bouw- en grasland. Den Haag, 1985.
(4) Warenwet (1985) Ned. Staatscourant, nr. 85.

Table 1. Mean composition of sewage sludge (on a dry-matter basis) used on the reclaimed peat and clay soil.

		Reclaimed peat soil	Clay soil
Organic matter	%	45.8	36.9*
N (total)	%	4.00	2.01
N (mineral)	%	1.22	0.12
P_2O_5 (total)	%	3.44	4.11
K_2O "	%	0.32	0.19
CaO "	%	2.69	4.44
MgO "	%	0.38	0.50
Na_2O "	%	0.17	0.08
Cl "	%	0.16	0.08
S "	%	0.82	0.84
As "	mg/kg	5.0	11.3
B "	"	167	41
Cd "	"	4.02	3.44
Co "	"	2.5	1.0
Cr "	"	68	55
Cu "	"	549	250
Fe "	"	13509	26770
Hg "	"	1.7	1.7
Mn "	"	204	1201
Mo "	"	3.8	5.3
Ni "	"	22	22
Pb "	"	230	192
Zn "	"	942	1149

* Mean concentration in the first three sewage sludges applied

Table 2. Composition of the topsoil (on a dry matter-basis) in 1986 of the treatments with 0, 10 and 20 tonnes sewage sludge per ha.

		Reclaimed peat soil			Clay soil		
		0	10	20	0	10	20
Org. matter	%	11.3	11.3	11.5	2.7	2.8	2.8
Part. < 16 μm	%	4.1	4.1	3.9	27.3	28.1	27.3
pH-KCl		4.8	4.7	4.7	7.4	7.4	7.4
N (total)	%	0.25	0.27	0.26	0.13	0.13	0.14
P_2O_5 (total)	%	0.08	0.13	0.16	0.16	0.19	0.22
P_2O_5 (water)	mg/l	43	92	117	17	23	30
K_2O (total)	%	0.10	0.10	0.10	0.72	0.71	0.72
K_2O (0.1 N HCl)	mg/100g	6	10	7	17	17	18
CaO (total)	%	0.48	0.51	0.52	6.87	6.53	6.78
MgO (total)	%	0.06	0.07	0.07	1.12	1.12	1.12
MgO (0.5 N NaCl)	mg/kg	161	186	183	63	67	73
S (total)	%	0.07	0.08	0.06	0.17	0.17	0.20
Fe "	%	0.28	0.31	0.32	2.15	2.21	2.16
Cd "	mg/kg	0.22	0.29	0.33	0.30	0.38	0.35
Cr "	"	21.7	19.8	14.4	70.1	72.9	68.5
Cu "	"	9.8	23.8	27.9	24.9	27.9	34.0
Mn "	"	160	167	151	522	529	560
Ni "	"	3.6	4.7	3.9	20.3	20.7	19.7
Pb "	"	31	37	33	18	19	23
Zn "	"	22.7	40.6	47.2	60.6	76.2	84.1
Mo "	"	0.48	0.41	0.34	0.81	0.88	0.66
As "	"	2.6	2.1	2.2	19.1	22.0	19.1
Hg "	"	0.05	0.07	0.10	0.05	0.07	0.09
B (water)	"	0.35	0.48	0.52	0.88	1.02	1.06
Co (2.5% HAc)	"	0.14	0.11	0.14	1.17	0.89	0.95

Presentation of a Research Project on Sludge Consistency

P. GEUZENS
Centre for Nuclear Research (SCK/CEN), Mol (Belgium)
W. DIELTJENS
Centre for Water Research (SVW), Antwerp (Belgium)

Summary

A generally accepted standard for the consistency of sludge-like waste materials, to be dumped on controlled landfills, would provide a legal security which is since long looked forward to. Based on a preceding literature study, the Belgian Centre for Water Research started a research project in october '89 in order to gather the necessary information on which to base such a standard. This project comprises a circumstantial sampling and characterization of sludges deriving from waste water treatment and drinking-water production plants, and an inventory of the actually applied dumping techniques for sludges on sanitary landfills and problems encountered with them.

Introduction

In the Belgian Flemish region a great part of the sludges emerging from sewage treatment and drinking-water production are disposed of at sanitary landfills.

Sewage treatment in the Flemish Region is undertaken by one public company, the VMZ. Three companies are active in the production of drinking-water, namely AWW, PIDPA and VMW. Both activities lead to the production of considerable amounts of sludge.

The sanitary landfills in the Flemish Region are distinguished into different classes. Drinking-water sludge and sewage sludge mostly may be dumped on class 2 landfills. These are primarily intended for the disposal of municipal refuse. Sometimes however the contamination of the sludges with heavy metals will make it necessary to dump on class 1 landfills. These are provided for the disposal of industrial, non-toxic wastes.

Flemish legislation requires that waste materials dumped on sanitary landfills are "congealed". The term "congealed" however is imprecisely defined. This gap in the law has already led to difficult juridicial conflicts.

The governmental organisation OVAM controls the disposal and treatment of waste materials.

VMZ, the Flemish drinking-water producers and OVAM have joined forces in the Centre for Water Research for their common scientific and practically oriented research.

In october 1989 this research centre started a project on the consistency of sludge, with special regards to the drinking-water and the sewage sludges. The project could lean on the results of a preceding literature study on determination methods for geomechanical properties of sludges (1).

1. Aims and Methods of the Research Project

A definition and numerical interpretation of the term "congealed" is important for different parties. The producer of sludge may develop a purposive research and management of the sludge dewatering process. Those responsible for the dumping sites will be able to more easily decide on the acceptance or refusal of certain sludges. The controlling authorities may better define and regulate the exploitment conditions.

The aims of the project are therefore :

- to conduct a circumstantial experimental research on the mechanical characteristics of the sludges, and on their suitability for landfilling. The influence of the origin of the sludge, the dewatering technique and added materials on the consistency of the sludge will have to be evaluated;
- to propose a regulation for the concepts of sludge-consistency, possibility to override and stability of the landfill. The consequences of such a regulation should also be evaluated technically, economically and ecologically.

As to establishing a standard for the mechanical strength - as part of the landfilling requirements that must be met by sludge-like waste materials - different routes may be followed:

- the empirical route will correlate mechanical characteristics with problems as experienced by the landfilling exploitation;
- a theoretical route may be relied on to estimate stabilities and safety factors met in the application of sludges to landfills;
- looking at presently used or available dewatering techniques and available adding materials, the feasibility of any proposed criterium will have to be established;
- an eye must be kept on research and regulation in neighbouring countries and in the European context.

We hope to be able to follow these four routes in order to propose a standard that satisfies the technical, ecological as well as economical requirements imposed on landfilling of sludges. If such a standard would be generally accepted in the near future, it could take away much of the confusion that now reigns in this field

2. Measuring the Consistency of Sludges

38 sludge production places, provided of an installation for mechanical dewatering or of sludge drying beds, were chosen for a systematic and periodical sampling of the dewatered drinking-water or sewage sludge. Recently 7 industrial waste water treatment plants, spread over different industrial activities, were also taken into account.

The following fysico-chemical analyses are performed on these sludges :

- dry matter content
- glow loss
- fermenting material (EAWAG-procedure)
- carbonate content
- sand content (inert materials)
- Fe
- Mg
- Ca
- Al

Two mechanical tests are used to determine the consistency of the sludges :

- the laboratory vane shear test
- a fall cone test

The laboratory vane shear test has been described repeatedly, also in the COST 681 context (see for example Otte-Witte in (2), p.309). The most recently published proposal for execution of the test is followed (3). The vane shear strength (symbol: Cv) is expressed in kPa (= kN/m^2) and is obtained as the average of three measurements on a sludge sample, which is pre-compacted in three layers using the so-called little Proctor hammer (see (3) for full details).

Figure 1 shows the results obtained until now on some 100 samples of drinking-water and sewage sludge. 90% of the measured shear strengths lay between 1 and 60 kPa. Once more it is seen that the dry solids content is of hardly any value to predict the consistency. Therefore direct measurements of the consistency are required.

In view of the research done until now on the suitability of sludges for landfilling (see for example (5) and (6)) a standard for the consistency of sludges will most probably be expressed in terms of vane shear strength. Gay and co-workers (5) proposed a value of 10 kPa. This value was often cited. Recently Loll and co-workers (6), on the basis of experiments with sewage sludge and household refuse proposed a value of 15 à 20 kPa.

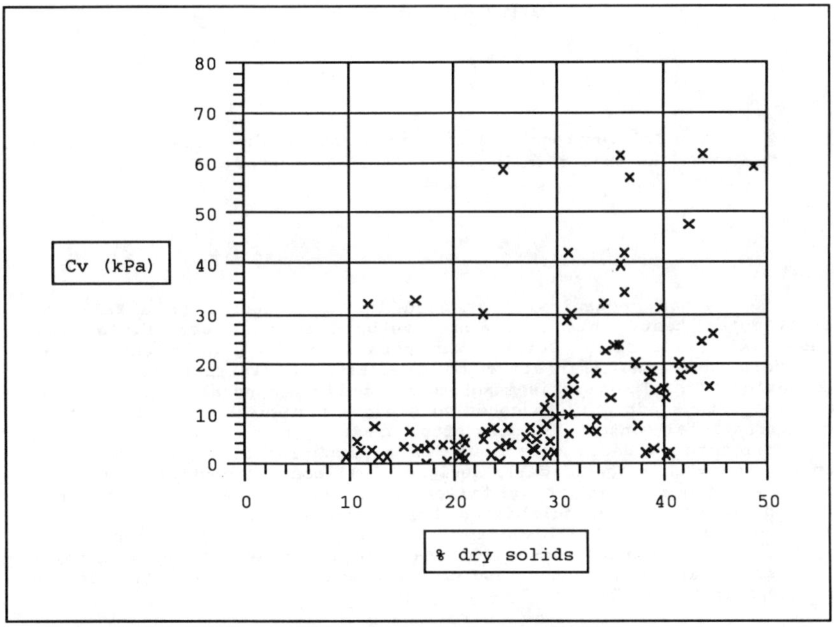

__Figure 1__ : Vane shear strength (Cv) versus dry solids content.

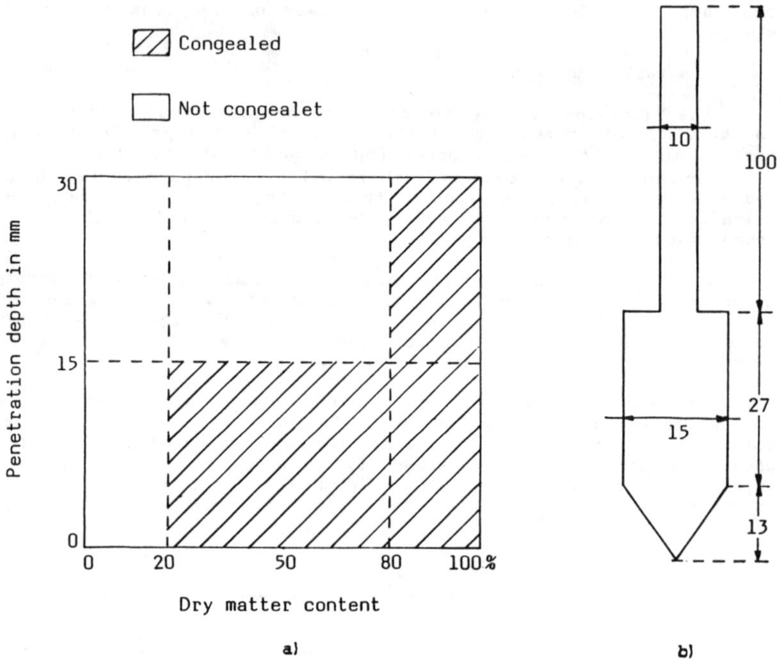

Figure 2 : The Indaver test : a) criterion and b) dimensions of the fall cone in mm. The fall cone is made of stainless steel and weighs 105 grams.

The second mechanical test used in our research is a fall cone penetrometer test. This test was developed by the responsible of an industrial landfill in Belgium who perceived the gap in the law and therefore took the initiative to propose a definition of the term "congealed" (4). The test is meant to be applicable to all kinds of waste materials, thus not only to cohesive sludges for which the vane test, in an international context, is more often used.

The test is based on the use of a high precision penetrometer in conjunction with a specifically designed fall cone. A container of heigth ±5 cm and diameter ±5 cm is filled with the sludge which is gently pressed by hand. The calibrated fall cone is placed on top of the equalised surface and allowed to fall for 10 seconds. The penetration depth is then accurately measured. As with the vane test an average of three trials is taken as the result. The dimensions of the fall cone used are given in figure 2.

Along with the test went a criterion for the definition of "congealed" materials which was formulated as follows :

- when the dry solids content is >80% the material always is congealed;
- when the dry solids content is <20% the material never is congealed;
- when the dry solids content is between 20 and 80% the consistency is determined by the results of the penetrometertest : a penetration depth <15mm gives a congealed material, a penetration depth >15mm gives a non-congealed material.

This criterion was developed on the basis of the practical experiences with different sludges on the landfill. These practical findings unfortunately were never published.

Figure 3 shows the results of some 50 simultaneous measurements with the fall cone test and the vane shear test. On the basis of these simultaneous measurements it was found that the above mentioned criterion (penetration depth 15 mm) accords with a vane shear strength of about 2 kPa.

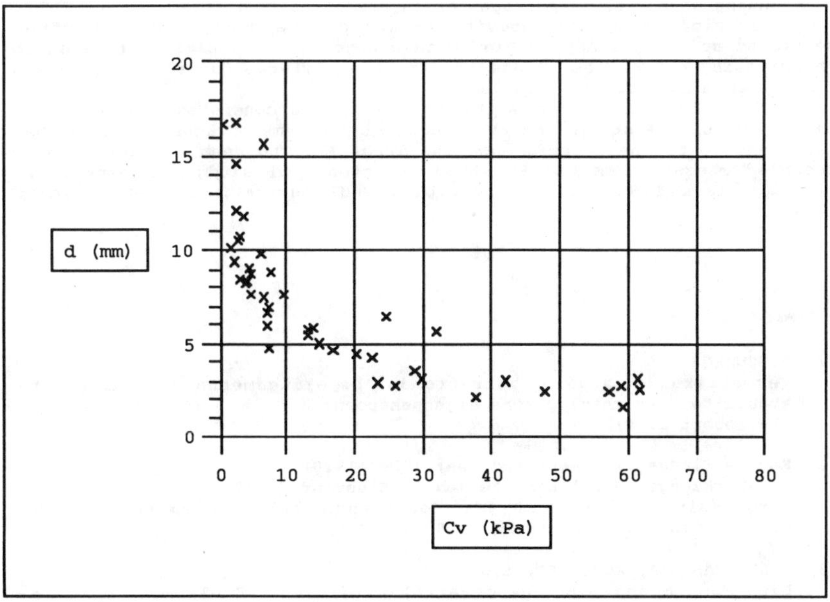

Figure 3 : Fall cone test (penetration depth d in mm) versus vane shear test (shear strength Cv in kPa).

3. Sludge at Sanitary Landfills

At present 13 public class 2 landfills and 4 public class 1 landfills are active in the Flemish region. 4 class 2 landfills do not accept sludge.

Based on literature and preliminary conversations with dumping site responsibles a questionnaire was made which asks for

- general information regarding the dumping site
- dumping techniques and way of treating sludge
- problems on the dumping site with sludge
- criteria used by the dumping site responsibles to accept or reject sludges.

This questionnaire was presented at the exploitants of all sludge-treating and public class 1 and class 2 landfills in the Flemish region.

The most important conclusions of this research are :

- The relative amount of sludge supplied at class 2 landfills mostly is less than 5 % (with one exception). On class 1 landfills this relative amount may be more than half the total amount of waste supplied. A great part of this class 1 sludge however is solidified prior to dumping (added materials are fly ash, cement and lime).
- Depending on the relative amount of sludge supplied, the sludge is build in either in thin layers which leads to a thorough mixing with the surrounding waste, or in packages or layers of up to 2 meters high.
- Slipping of waste-supplying trucks on the sludge is most often mentioned as the primary problem with sludge. This is closely related to the pressing out of the sludge towards the surface. Odour emissions are also often mentioned as a problem.
- Class 2 landfill responsibles evaluate the consistency of certain, well known sludge streams on the practical problems or possibilities they experience with these streams on the landfill. The need for a generally accepted measuring method for the consistency of sludge is especially felt by class 1 landfill responsibles and undertakings that solidify sludge.

References

(1) P.GEUZENS;
Vergelijkende studie betreffende bepalingsmethoden van slib-stabiliteit en rheologische eigenschappen;
SVW report 1989/1.

(2) Sewage Sludge Treatment and Use; (EUR 11918)
A.H.Dirkzwager & P.l'Hermite (ed.); Elsevier, 1989.
(Proceedings of a COST 681 Conference held in Amsterdam, The Netherlands, 19-23 September 1988).

(3) LOLL, KASSNER, KÖHLHOFF, a.o.;
Die Bestimmung der Deponierfähigkeit von Schlämmen mit der Referenzmethode "Laborflügelscherfestigkeit"; in : Korrespondenz Abwasser, 8/89, p.903-908.

(4) B.VANDERBORGHT & R.ANSOMS;
Steekvastheid van industriële afvalstoffen; in : Technivisie, nr.116, 1988.

(5) GAY, CH.G., HENKE, K.F., RETTENBERGER, G., TABASARAN, O., Standsicherheit von Deponien für Hausmüll und Klärschlamm, Stuttgarter Berichte zur Abfallwirtschaft, Band 14, E.Schmidt Verlag, Bielefeld 1981.

(6) MÖLLER, U., GAY, CH.G., KASSNER, W., KÖHLHOFF, D., LOLL, U., OTTE-WITTE, R., Neudefinition der Deponierfähigkeit von Abwasserschlämmen, Teil 1, Schriftenreihe Siedlungswasserwirtschaft Bochum, Band 6, Bochum 1985.

THE USE OF SEWAGE SLUDGE FROM THE CENTER OF BIOLOGICAL CLEANING, IN METAMORPHOSIS ATTIKI, FOR THE IMPROVEMENT OF SOIL FERTILITY AND THE PROTECTION OF THE ENVIRONMENT

G. CHARDAS, M. KARAYIANNI-CHRISTOU,
A. KOLLIAS, T. NIKOLAOU and N. PAPADOPOULOS
Soil Science Institute of Athens 14123 Lycovrissi, Greece

Summary

In order to research the possibility of using the sewage sludge from the center of biological cleaning of Metamorphosis Attiki, in agriculture we took samples every two weeks, in which we studied: 1. Some of its physicochemical parameters (humidity, organic matter, PH, water soluble salts, plant nutrition elements (N,P,K,Mg), and heavy metals (Zn,Cd,Ni,Cu,Cr,Pb). 2. The degree of its maturity, using the Lepidium Sativum method. 3. The effect of its mixing with the soil, on the growth of corn and wheat cultivated in pots. The results were very encouraging. Using sewage sludge as soil conditioner we can increase soil fertility and at the same time contribute to the protection of the environment.

1. INTRODUCTION

The disposal of the residues from agriculture, the urban areas and industry is of critical importance for human survival in the developed countries. Among these residues is the sludge from the biological claaning centers of the cities. A recent study (5) in the European Community (E.E.C) showed that forty-five percent of the sludge produced in these centers is used on soils. The alternative solution of burning is used basically in the U.S.A. and in Japan. The utilization of sludge in agriculture has been widespread in recent years, although inorganic and organic substances, like heavy metals, pesticides, PCV etc. may restrict its use in the fields. In Greece the first biological cleaning center for sewage sludge to receive the sludge of the northern suburbs of Athens, is located in Metamorphosis, Attiki. It started operating in 1986. From the center, we took samples of sludge in order to study: 1. its physical and chemical parameters, 2. its degree of maturity, and, 3. the effect of sludge application on soil fertility.

2. MATERIAL AND METHODS

The sludge samples from the center of biological cleaning center were taken every 15 days for a period of ten months. The samples were analysed for:

organic matter content, total nitrogen content (method of Kjeldahl), total phosphorus and potassium (dry ashing) and the heavy metals Zn, Ni, Cd, Cr, Cu, and Pb (wet ashing by nitric acid). The degree of maturity of the sludge was determined in the pressure of extract (250 atm), by putting a certain number of seeds of Lepidium Sativum to germinate in water/extract ratios and it was expressed by the Germination Index (G. I) (2,3):

$$G.I = \frac{G}{G_c} \times \frac{L}{L_c} \times 100$$ where:

G: is the number of seeds which germinated in a treatment,
G_c: is the number of germinated seeds in the control,
L: is the total length of roots in the same treatment, and
L_c: is the total length of roots in the control.

In the pot experiments two different soils were used with the characteristics shown in table 1.

TABLE 1

		Soil 1	Soil 2
Texture		SL	L
PH		7.7	7.4
Organic matter %		1.8	1.7
Electrical Conductivity	(mmhos/cm)	<3	<3
P (ppm, Olsen)		12	7
Exchangable K	(meq/100gr)	0,84	0,24

Plants were grown in the greenhouse in pots of 15L in percentage of sludge to soil by volume: 0%, 5%, 10%, 20%, 25% and 40%. Each treatment was performed simultaneously six times. The seeds were sown ten days after the mixing of soil and sludge in the pots. Corn (hybrid ARIS) was sown in the autumn of 1987. After 45 days the crops were cut, washed and dried. Corn roots were removed and in the same soil in the pots, wheat was sown (Var. Capeiti). Wheat plants were cut in 45 days after the sowing. Finally, corn was sown in the spring of 1988 (hybrid ARIS), which was cut three months later.

3. RESULTS

Picture 1 shows the concentration of total N, P and Mg in the sludge at the successive samplings, while Picture 2 shows the concentration of the heavy metals Cd, Ni, Cr, Pb and Zn, in the successive samplings of the sludge. Finally in Picture 3, the change of the Germination Index in the different dilutions of the sludge extract with water, from three successive samplings, is presented. In Table 2 the average amounts of nutrients taken up by the second culture of corn per pot are included, along with the respective regression equations.

TABLE 2

Average amounts of nutrients taken up by corn (gr/pot)

Nutrient	Sludge added to the soil(%)						Regression equations
	0	5	10	20	25	40	
N	0,88	3,60	5,10	11,23	12,28	13,86	$Y_N=0,41+0,67X-0,008X^2$ $R^2=0,98$
P	0,10	0,63	0,93	1,83	2,09	2,26	$Y_P=0,05+0,11X-0,0015X^2$ $R^2=0,99$
K	2,47	6,21	7,70	11,37	11,26	10,48	$Y_K= 2,76+0,63X-0,001X^2$ $R^2=0,98$
Mg	0,24	1,16	1,57	3,55	3,71	4,66	$Y_{mg}=0,54+0,19X-0,002X^2$ $R^2=0,98$

Y*: Y(N,P,K,Mg) measures the absorption of N,P,K and Mg by corn plants due to the application of increasing amounts of sludge to the soil.

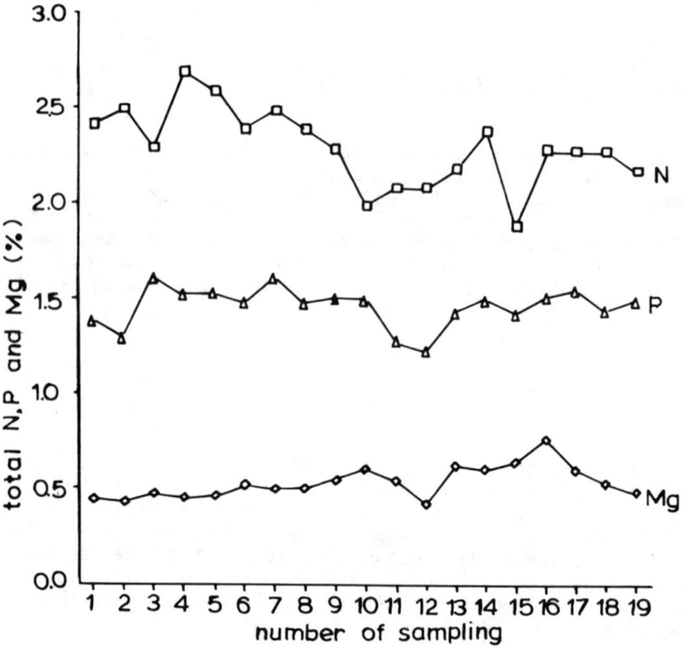

Fig 1 - Total N,P and Mg concentration in the sludge at the successive samplings

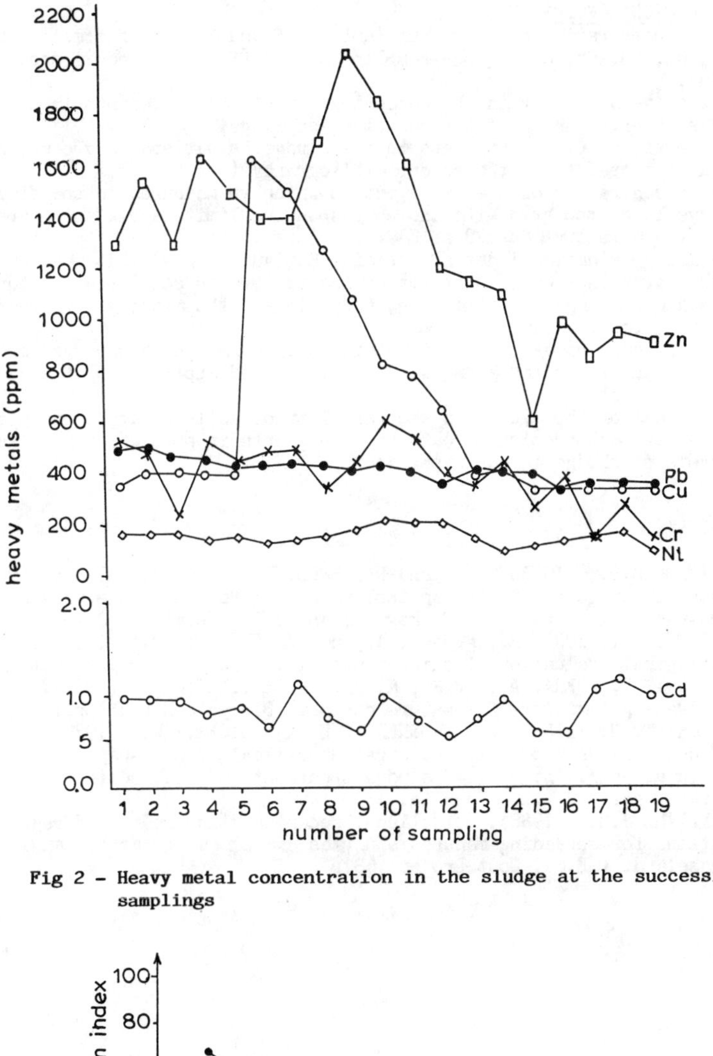

Fig 2 - Heavy metal concentration in the sludge at the successive samplings

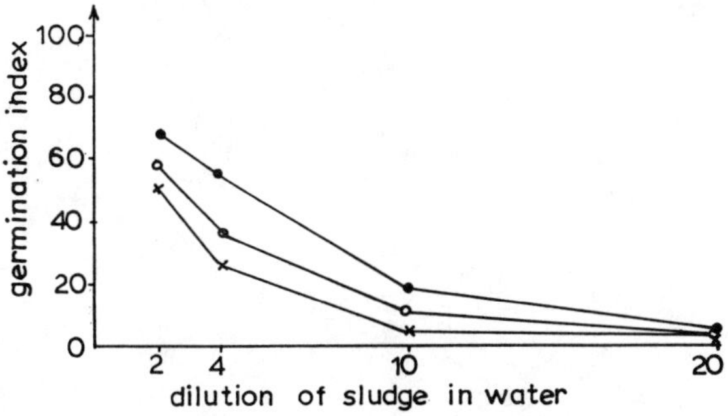

Fig 3 - Germination index with Cress in three samples of sludge

4. DISCUSSION CONCLUSIONS

The concentration of plant nutrients N, P and Mg and of some heavy metals Cd, Ni, Zn, Cr, during the sampling period (Fig: 1 and 2) shows the following:

a. A relative stability in the concentration of these elements in the sludge is observed during the successive samplings.

b. The levels of plant nutrients in the sludge is satisfactory compared with that referred to in the recent bibliography (1,4,5,).

c. The concentration of the heavy metals which we measured in the sludge, was always lower compared with the corresponding limit levels which are given by the directive 86/278 of EEC.

d. The low germination index of Lepidium Sativum (Cress) that was found suggests that this sludge as it leaves the factory is still too immature to be used as an organic fertilizer, according to the findings of other researchers, too (2,3).

e. A significant uptake of plant nutrients by corn, which was used as an indicator plant was noted along with the increased application of sludge in the soil substrate.

f. The effect of the sludge on the fertility of soil substrate was proved by the better macroscopical growth of corn plants in pots which received high levels of sludge in our experiment.

REFERENCES

(1) BAYERS, C.D., VIGERUST, E. and PAULSRUD, B. (1989). Alternative uses of sludge other than agricultural. In: Sewage sludge treatment and use, edited by A. H. Dirkzwager and P.L. Hermite. 85-99.

(2) ZUCCONI, F., FORTE M., MONACO, A. and DE BERTOLDI, M. (1981). Biological evaluation of compost maturity. Biocycle July/August. 27-29

(3) ZUCCONI, F., PERA, A.,FORTE , M. and DE BERTOLDI, M. (1981). Evaluating toxicity of immature compost. Biocycle March/April. 54-57.

(4) SIEGRIST, H., ALDER, A., BRUNNER, P.H. and GIGER,: W. (1989). Pathway analysis of selected organic chemicals from sewage to agricultural soil. In: sewage sludge treatment and use, edited by A. H. Dirk zwager and P.L. Hermite. 133-144.

(5) WILLIAMS J.H. (1988). Guidelines, recommendation rules and regulations for spreading manures slurries and sludge on arable and grassland. C.E.C., Cost Project 681.

ECOLOGICAL AND ECONOMICAL UTILIZATION OF WASTE-WATER FROM OLIVE-OIL PRODUCTION WITH PHYSIOCHEMICAL AND BIOTECHNOLOGICAL METHODS.

N. INIOTAKIS[1)], C.B. V.D. DECKEN[1)], C.J. ISRAILIDES[2)],

K. KATSABOXAKIS[2)], P. MICHAILIDES[3)], D. OIKONOMOU[2)] and D. PAPANICOLAOU[2)]

1. KFA-IEV - Jülich, 2. Institute of Technology of Agricultural Products, Athen, 3. University of Crete, Rethymno

Summary

On the basis of a combination of physicochemical and biotechnological methods an ecologically effective and economically attractive solution has been proposed for the problem of waste-waters yield by the olive-oil production (katsigaros). Our proposal leads to elimination of the toxicity for plants and the about 100 % utilisation of the ingredients of these waste-waters. Economotechnical studies have been performed. The results indicate that the expected revenues from the economical utilisation of the valuable substances of katsigaros overcompensate the total disposal cost.

1. INTRODUCTION

The production of olive oil yields a toxic waste water (Katsigaros), which cannot be purified with biological methods /1/. These waste waters are produced between November and February. An average olive oil factory produces up to about 2000 m^3 of Katsigaros per year. In the area of Iraklion, Crete, for exemple 120.000 m^3, in total Crete about 490.000 m^3, in Greece about 1.500.000 m^3 of those waste waters have to be treated annualy. The ratio between Katsigaros and olive oil is about 6 : 1.

As an economically acceptable treatment of the katsigaros in accordance with ecological requirements does not exist, nearly all olive oil factories, at least those in the Mediterranean countries, discharge their wast waters, without any treatment, into the nearest sea, river, or creck. This causes an acute ecological problem; surface, soil and ground water are polluted. This results in a significant reduction of drinking water supplies in the southern countries. At Iraklion for example, this reduction amount to 25 %. Therefore, the problem of Katsigaros is a significant ecological problem for all Mediterranean countries which certainly needs to be solved in an ecologically correct and economically acceptable way.

The Katsigaros contains, not only toxic substances (phenols, organic acids, taniles), but also valuable substances which can be used for the improvement and fertilizing of soils like potassium salts, ammonia salts and also nitrates and phosphates (about 80 kg/m^3). Moreover, the katsigaros contains in average about 4 - 30 kg/m^3 of olive oil, about 15 kg/m^3 of sugar, about 10 kg/m^3 Polyalcohols, colour substances etc. The recovery and the economic use of those valuable substances and of the residual oil constitutes a considerable and economic potential.

A solution for the treatment and disposal of the katsigaros problem, therefore, should not only consider the environmental aspects, but also the economic use of the contained valuable substances and of the residual oil. In this case, the cost for the treatment and disposal of katsigaros can be overcompensated. In this proposal we propose a new method for the about 100 % recycling of katsigaros, leading to a radical, ecologically effective and economically attractive solution of this problem.
In the following chapters the technical method to be applied and the economic potential will be presented.

2. OBJECTIVE OF THE PROPOSAL

2.1 Aims of the proposal

On the basis of a combination of physicochemical and biotechnological methods an ecologically effective and economically attractive solution to the problem of katsigaros as been proposed. The function of those methods has been already verified and partially demonstrated on a laboratory scale (30 l/h). The main results are:

1. Elimination of high toxicity for plants.
2. Recovery of high quality water.
3. Recovery and production of valuable substances, contained in katsigaros, for example:
   a) Fertilizer about 80 kg/m^3-katsigaros, with organic and inorganic ingredients.
   b) Residual oil between 5 - 30 kg/m^3-katsigaros
   c) Colour substances (anthocyanines) between 2 - 3 kg/m^3-katsigaros.
   d) Pullulan and/or ethanol in quantities 10 kg/m^3 or 12 kg/m^3-katsigaros respectively.

Moreover those methods work independent from each other when arranged in an adequate way.
The aims of our proposal are:

- verification of the laboratory results,
- optimisation of the whole process,
- demonstration of the applicability and effectiveness of the method on a technically intersesting scale,
- demonstration that the elimination of the toxicity for plants and the about 100 % utilisation of the ingredients of katsigaros, is economically attractive solution to the related environmental problem. Essentially if this method will be proved applicable, the problem of katsigaros will not occur at all.

Furthermore the marketing of the products will be considered.

2.2 Selection of the technical methods to be applied

As mentioned in the introduction, the point at issue is an ecological and economic disposal of katsigaros including a simultaneous recovery of the water and of the valuable substances contained in the katsigaros e.g. residual oil, fertilizer, colour substances, etc. These aims can be achieved in a process employing a combination of physicochemical and biotechnological methods, which eliminates the toxicity and separates water and valuable substances of the katsigaros in the sense of a total recycling and 100 % utilisation of its ingredients.

Realisation of this proposal involves combined application of the following techniques:

a) For the conversion of sugar to ethanol and/or Pullulan the biotechnological methods of Fermentation can be applied. First experiments in the National Foundation of Agricultural Research in Athens Greece have shown, that a rate of conversion of ethanol or Pullulan of 12 kg/m^3-, respectively 10 kg/m^3-katsigaros, can be expected /2. 3/.

b) For the recovering of the residual oil from katsigaros ceramic membranes have been developed in collaboration of the Research Center of Jülich (KFA) and the company ATP-Ltd, Iraklion, Creete. As an alternative to that, a physical extraction process for olive oil developed by the company ATP-Ltd, can be used. First experiments using both of those techniques give similarly good results. Accordingly, more than 99 % of the oil could be recovered. It quality depends mainly on the storing time of katsigaros. Using those membranes we achieve very cheap recovery although we believe that our membranes are capable of further optimation. The same applies for the physical extraction process developed by ATP-ltd.

c) For the recovery of the colour substances ion-exchange techniques can be used. According to experiments carried out at the National Foundation of Agricultural Research in Athens Greece, an amount of 2 - 3 kg/m^3-katsigaros of colour substances can be expected /4./.

d) For the elimination of the toxicity and the recovery of the organic and inorganic salts disolved in katsigaros and their use as fertilizers, we can apply the AZUR-process /5/. This process is based on the hydrolysis under simultaneous oxidation and total evaporation of katsigaros. As end-product we have here a sludge containing 30 - 70 % moisture. The energy supplied for the necessary water evaporation can up to 98 % be recovered through recondensation of the water vapor. This process has been developed by the Company AZUR GmbH, Aachen, F.R.G., in colaboration with the Research Center of Jülich (KFA) and the Municipal Works of Iraklion (DEYAH) for treating water-wastes containing organic as well as inorganic ingredients.

A research program performed by DEYAH, which aimed at sea water desalination with salt recovery, showed that with this process water of highest quality could be obtained and salt could be successfully separated. A test facility with a capacity of 40 l condensate per hour has been constructed at Iraklion and is operated there since 1986.

After termination of the programm for sea water desalination, katsigaros was treated on trial in the test facility at Iraklion. Starting in November 1988, first tests were carried out for two months.

The main results can be summarized as follows /5/:

- The katsigaros could be separated into reusable water of very good quality, and economically applicable sludge (moisture about 30 to 70 %).
- The volume could be reduced to 7 - 10 % of the waste water volume.
- All tests on toxic effects against plants were negative for condensate as well as for sludge diluted into the water (suldge concentration 10 %).

On the basis of these results it can be concluded that this process is well suited for the disposal of katsigaros in the sense of the aims of our proposal.

2.3 Description of the AZUR-PROCESS

The process is described indepth in /5, 6/, and in the patent /7/. In Fig. 1 the process is shown schematically. In the following, we present a short description of the process.

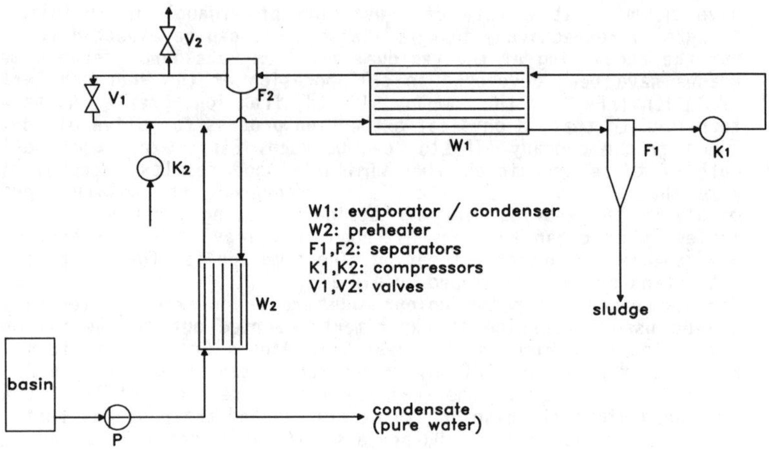

Fig 1 - Scheme of evaporation unit (AZUR-process)

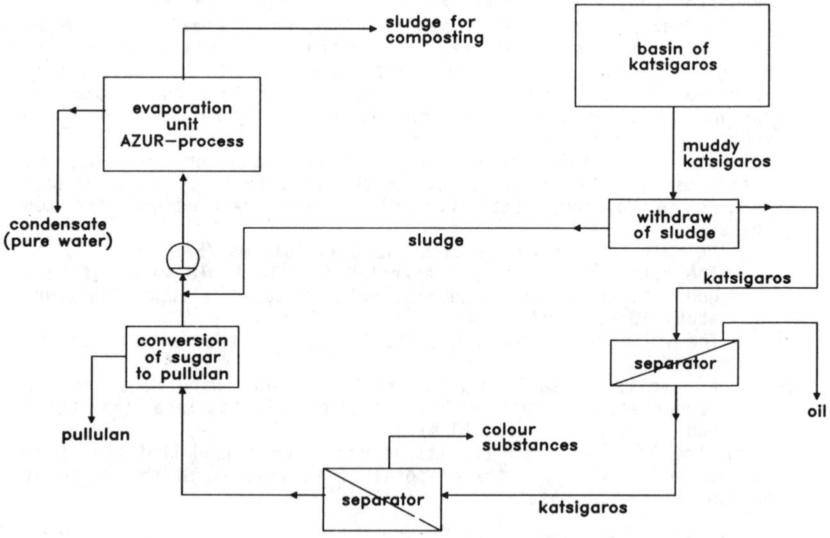

Fig 2 - Scheme of ecological and economical waste-water disposal at olive-oil production with recycling of valuables

In a first stage, the waste water is introduced and evaporated continuously within an entrainment gas circulating in a closed circuit. The evaporation takes place continuously on the evaporator side of a heat exchanger. Depending on the operating pressure, and on the ratio between the partial pressures of water and the entrainment gas, temperatures of up to 160 °C can be obtained on the evaporation side. The substances contained in the waste water will precipitate in crystalline form or as a sludge, depending on their nature and on the degree of dehydration. The water content of the salts is controllable. In the case of sludge, its water content can be chosen between 5 an 70 %. In a subsequent cyclone, salt or sludge are separated from the gas phase. Salt/sludge are removed from the process. Thereafter, the gas phase, i.e. the entrainment gas and the water vapor, are compressed, and led to the condensator side of the heat exchanger. The water vapor condenses, and the condensation energy is used for the evaporation of the waste water. The condensate is removed from the entrainment gas current in the form of pure water. The entrainment gas is recycled to the evaporation side of the heat exchanger in a closed circuit and mixed with the preheated waste water. The preheating of the waste water is effected by the hot condensate.

In the case of oxidable (organic) substances contained in the waste water, and if these should be oxidized, an entrainment gas with an oxidation potential is chosen, e.g. air, or air enriched with O_2. The tests in Iraklion were carried out with air.

Colour substances, Pullulan and the residual oil, cannot however be recovered by this process, because they would thermally disindegrate. Those substances should therefore be separated before applying the AZUR-PROCESS. A overview of the arrangement of the different steps necessery for the complete recycling of katsigaros is schematically shown in Fig. 2.

3. ECONOMIC POTENTIAL

According to the experiences with a pilot plant for the treatment of galvanic waste-waters, the energy consuption depends essentially on the elevation of the boiling point of the saturated concentrate. For the katsigaros an elevation of the boiling point of 2 °C has been measured. Therefore, depending on the layout of the heat exchanger, a total energy demand of less than 25 kwh/m^3 can be expected. The process does not require much labour. Based on the plant concept, requirements for material and maintenance will be very low. This must however be proved by longterm operation.

If the sludge is conditioned in an optimal way, revennus of at least 0,15 DM/kg (DM = German Mark) are reasonable. For the residual olive oil, the price should be 3,6 DM/kg, for colour substances 4,3 DM/kg and for pullulan 27,0 DM/kg.

Therefore revenues per m^3 of katsigaros can be expected as follows:

-	12,0	DM	from compost
-	18,0 - 108,0	DM	from residual oil
-	8,6 - 12,9	DM	from colour substances
-	270,0	DM	from pullulan
Total	308,6 - 402,9	DM	

The first economic calculations for the standardised plant show specific total cost for the disposal of katsigaros, which depend on the plant capacity and amount between 19 DM/m^3 (big plants) and 65 DM/m^3 (small plants). In our calculations we have assumed a seasonal operation of 90 days (or 2160 hours) during the winter period under simultaneous

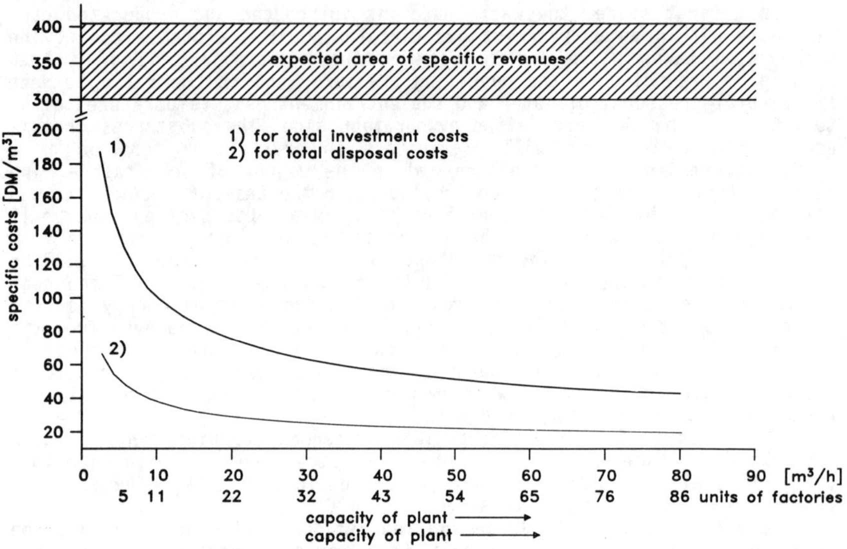

Fig 3 - Specific total costs of investment and disposal as a function of capacity of the plant service of several oil factories. The katsigaros is assumed to be transported to the plant by tank-lorries. The plant will be built in Crete, respectively Greece. For the employee-costs we assume a salary of 3.500,-- DM monthly and for the transportation, depending on the distance, a cost of 3 - 5 DM/m^3 is assumed. For the finances we did not take into account any subsidies and accepted an amortisation time of 10 years by an average year interest of 22 %.

Fig. 3 shows the results of our calculations. In more detail, it shows the total investment and disposal costs in DM/m^3 as function of plant capacity given in m^3/h or in units of oil factories served. For comparison we also show in fig. 3 the regime of expected revennes.

As it can be easily recognized an economic operation books very resonable because the revenues from the valuable substances, i.e. from compost, redidual oil, colour substances and pullulan exceed not only the disposal but also the investment costs. Thus the disposal of katsigaros can be solved radically not only ecologically effective but also economically very attractive. This is the innovative aspect of this proposal.

4. CONCLUSIONS AND ENVIRONMENTAL PROSPECTS

Our proposal leads to elimination of the toxicity for plants and the about 100 % utilisation of the ingredients of katsigaros. The expected revenues from the economical utilisation of the valuable substances of katsigaros overcompensate the total disposal cost. This provides an ecologically effective and ecnomically attractive solution to the related environmental problem. Essentially, if this methode is applied, the problem wil not occur at all.

That methodology and especially the AZUR-PROCESS can also be applied for recycling and evaluation of other toxic organic and inorganic waste waters, for example disposal of vinasse, waste-waters of galvanic-, paper-, leather-, colour-industries etc. It can therefore decisively contribute to economically solution of several enviromental problems.

REFERENCES

(1) MICHELAKIS, N., and KOUTSAUTAKIS, A. (1989). The problem of waste of olive oil factories. Dificulties and posibiliteis for its treatment. Proc. of the meeting of the Geotechnikon Epimelitirion Ellados, parartima Kritis, for disposal of waste from olive oil factories. Iraklion Crete. March 31, 1989, p. 9 - 25, (A) (in Greek language)

(2) ISRAILIDES, C.J. and A.W. ANDERSON (1980). Aureobasidium pullulans: A potential microorganism for single cell protein. Proc. at the 13th meeting of the Federation of European biochemical societes (FEBS). Jerusalem, Israel, Aug. 24 - 29, p. 272

(3) ISRAILIDES, C.J., private communication.

(4) PAPANICOLAOU, D., private communication.

(5) INIOTAKIS, N. et. al., (1989), Purification of katsigaros by simultaneous use of valuable substances. Published in (A). p. 81 - 84.

(6) INIOTAKIS, N., FRÖHLING, W. (1984). Description of a Water Desalination Process combined with the Extraction and Seperation of inorganic Salts. KFA-IRB-IB-9/84, Oktober 1984

(7) INIOTAKIS, N. et. al., (1988), Apparatus for waste water purification, United States Patent Nr. 4,767,527, Aug. 30. 1988

HEAVY METALS IN URBAN SEWAGE SLUDGES: CHEMICAL FORMS AND POSSIBLE AVAILABILITY

M. ANGELIDIS* and R.J. GIBBS**

* Environmental Studies, Athens, Greece

** Center for Colloidal Science, College of Marine Studies, University of Delaware, USA.

Summary

Urban sewage sludge contains various amounts of metals that could be harmful to the environment, depending on their chemical associations in the sludge. The dissolved or easily available metals, can be uptaken by organisms living in the sludge dumping area. The use of an extraction technique on digested urban sludge (New York City), revealed that organic matter and sulfides are the most important carriers of metals in the sludge. More than 85% of Cu, Pb and Zn, and more than 60% of Cd and Cr, were found in this phase. The acid-reducible phase was important only for Fe (26-40%). The dissolved metal phase was important only for Cd (3-25% of the total Cd). The dissolved Cd may represent a threat to the environment in the areas of sludge dumping, and the oxidizable Cu, Pb and Zn, are also easily available for uptake by the organisms living in the sludge dumping area. Do to the availability of the greater part of the metals in the sludge, it is essential that the total metal concentration is known, before any use of the sludge in agricultural land.

1. INTRODUCTION

The metal concentration in sludge show important variations from one treatment plant to another, reflecting the differences in the wastewater inflow and the proportion and quality of industrial input. In most cases of urban effluents, there is a small industrial contribution, which increases the metal content of the sludge. However, the effluents of residential areas, contain important amounts of metals. Klein et al. (1) calculated that the residential areas contribute 49% of Cd, 47% of Cu, 28% of Cr, 25% of Ni and 42% of Zn, to the New York City wastewater treatment plants. In the same plants, the average percent industrial inflow was 4% (range 0-8%) of the total average daily inflow (2).

Most studies of metals in sludge deal with the total concentrations. However, in order to use the sludge in agricultural land or to dispose it off on the ground or into the sea, it is important to know the possible availability of the metals in it, which can be determined by studying the partitioning of the metals among different chemical phases (3, 4). The phases determined in the present study were (a) dissolved, (b) adsorbed, (c) acid-reducible, (d) oxidizable and (e) residual. The extraction technique used was based on similar extractions for soils and aquatic sediments (5, 6, 7). The metals in the dissolved phase and the adsorbed phase are obviously the most available to organisms, and the metals in the oxidizable phase (organic matter and sulfides) are also easily available either through the food chain or due to physicochemical changes (oxidation) after the disposal of the sludge. The metals in the residual phase do not represent any threat to the environment because they are strongly bound to the sludge particles and can not be released. Finally the metals in the acid-reducible phase could be partially released in acid conditions, but generally are much less available to the environment than the metals in phases (a), (b) and (d).

2. METHODOLOGY

Sludge samples were collected from four wastewater treatment plants in New York City (Ward's Island [WI], Bowery Bay [BB], Newtown Creek [NC] and Hunt's Point [HP]). The samples were centrifuged at 13,000 rpm for 20 min. The supernatant liquid was considered as the dissolved phase of the metals in the sludge. The solid residue was then treated consecutively with 1M $MgCl_2$ (adsorbed phase), 0.1 M $NH_2OH.HCl$ in 25% HOAc (acid-reducible phase), 30% H_2O_2 in HNO_3 (oxidizable phase) and finally with HF, $HClO_4$ and HNO_3 (residual phase) (5, 6, 7). The dissolved phase and the extracted phases were analysed for Cd, Cr, Cu, Fe. Pb and Zn using a Perkin-Elmer 5000 Atomic Absorption Spectrometer (8).

3. RESULTS AND DISCUSSION

The total and dissolved metal concentrations in the sludges examined are presented in Table 1.

Table 1. Total and dissolved metal concentrations (mg/l)

Plant	Cd	Cr	Cu	Fe	Pb	Zn
WI Total	0.369	14.32	89.10	646.2	19.4	53.84
Dissolved	0.094	0.088	0.373	2.51	<0.50	0.215
(percent)	(25.5)	(0.62)	(0.42)	(0.39)		(0.40)
BB Total	0.540	44.37	74.64	475.4	12.3	165.60
Dissolved	0.058	0.165	0.123	2.30	<0.50	0.369
(percent)	(10.7)	(0.37)	(0.16)	(0.49)		(0.22)
NC Total	3.044	83.16	96.21	766.7	85.4	124.80
Dissolved	0.099	0.523	0.322	2.51	<0.50	0.280
(percent)	(3.3)	(0.63)	(0.33)	(0.33)		(0.22)
HP Total	0.690	10.06	44.89	512.0	9.9	53.15
Dissolved	0.067	0.105	0.269	3.83	<0.50	0.348
(percent)	(9.7)	(1.04)	(0.60)	(0.75)		(0.65)

All sludges were primarily domestic with 2-8% of industrial inflow. The sludge from Newtown Creek with 6% industrial inflow had the highest concentrations of Cd, Cr, Cu, Fe and Pb. In the other treatment plants, the industrial inflow was: 2% in Ward's Island, 5% in Hunt's Point and 8% in Bowery Bay (maximum concentration of Zn).

Among the metals examined, the dissolved phase was important only for Cd. The dissolved Cd represented 3.3 - 25.5 % of the total metal in the sludge. On the other hand all other dissolved metals represented less than 1.0% of the total metal concentration. It is therefore important to note that, with the exception of Cd, the particulate phase of the sludge carries almost the entire metal load.

In Figure 1 is presented the partitioning of metals among the chemi-

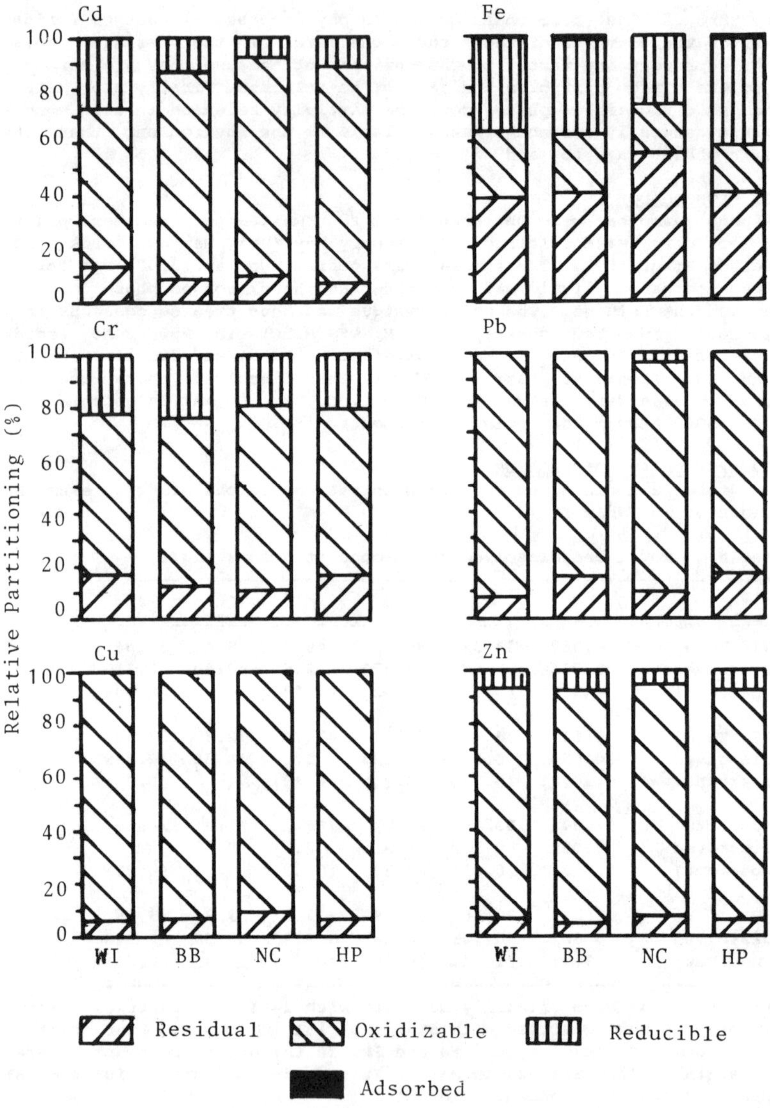

Figure 1. Relative partitioning (%) of Cd, Cr, Cu, Fe, Pb and Zn in sludges: Wards Island (WI), Bowery Bay (BB), Newtown Creek (NC), Hunts Point (HP).

cal phases examined, in the particulate matter of the sludges. We note that all sludges present similar relative (percent) chemical partitioning, although their absolute metal concentrations vary considerably (Table 1). The similar redox and bacterial conditions existing in all four anaerobically digested sludges, is probably the reason for this result.

The mean relative partitioning of the metals in the four anaerobically treated sludges of the New York City wastewater treatment plants is presented in Table 2.

Table 2: The mean percent of Cd, Cr, Cu, Fe, Pb and Zn in the adsorbed, acid-reducible, oxidizable and residual phases of sludges from New York City plants.

	Cd %	Cr %	Cu %	Fe %	Pb %	Zn %
Adsorbed	-	-	-	0.77 ±0.47	-	-
Reducible	14.8 ± 8.5	20.8 ± 2.0	0.57 ± 0.48	32.7 ± 5.1	0 - 3.2	8.0 ± 0.6
Oxidizable	75.2 ± 10.6	64.0 ± 3.8	91.8 ± 1.2	19.4 ± 2.8	85.4 ± 3.6	85.0 ± 0.3
Residual	10.0 ± 2.6	14.8 ± 2.9	7.6 ± 1.5	44.5 ± 7.5	13.8 ± 3.9	6.7 ± 0.9

The adsorbed phase does not carry important amounts of metals. Possibly, the $MgCl_2$ is not able to extract the strongly bound adsorbed metals due to the presence of carboxylic groups in the surface of the sludge particles and to the strong complexation by organic compounds (8). Therefore no important metal amounts can be released to the environment by ion exchange, after the disposal of the sludge on land. On the other hand, the oxidizable phase is a very important metal carrier. This phase consists of organic material and sulfides and carries 75.2% of Cd, 64.0% of Cr, 91.8% of Cu, 85.4% of Pb and 85.0% of Zn, present in the sludge particles. It is clear, that with the exception of Fe, all the other metals examined are massively transported by this phase. Once the sludge is disposed on land, oxidation may occur, and the metals present in the oxidizable phase can be released to the environment. Also the organic matter may be used by organisms for metabolic processes, and in this case, the metals enter the food chain of the ecosystem. Therefore, the metals found in this phase present a potential threat to the environment of the sludge disposal area.

The acid-reducible phase is not an important metal carrier, with the exception of Fe (32.7% in this phase), and in a lesser degree of Cd and Cr (14.8% and 20.8% respectively).

The metals in this phase, should not be considered as scavenged by Fe/Mn oxides, because it is not likely that there is any formation of oxides in the environment of the anaerobically digested sludges (8). It is suggested, that this phase consists primarily of acid soluble sulfides that are freshly precipitated on the particle surface. This phase may also contain some strongly adsorbed metals and carbonates (9).

Finally, the residual phase does not carry an important metal concentration, and it consists mainly of inorganic particles which are part of the inflow of the wastewater treatment plants.

4. CONCLUSIONS

The particulate phase carried almost the totality of Cr, Cu, Fe, Pb and Zn found in the sludges from four wastewater treatment plants of the New York City area. On the other hand, the dissolved Cd represented the 3.3 - 25.5 % of the metal in the sludges, indicating that any sludge with relatively high Cd content may represent a threat to the environment of its disposal area.

In the particulate part, the oxidizable phase was the most important carrier of most of the metals examined. It carried 75.2% of Cd, 64.0% of Cr, 91.8% of Cu, 85.4% of Pb and 85.0% of Zn (mean values). After its disposal on land, the sludge will undergo chemical changes due to oxidation, and it is possible that part of the oxidizable metals will be released to the environment. Additionally, because the oxidizable phase consists mostly of organic material, it will be possibly used by organisms for metabolic processes, resulting the introduction of some metals into the food chain of the local ecosystem. The above results indicate that the use of sludge on land should be done very carefully, because of the bioavailability of the greater part of the metals it contains.

The acid-reducible phase consist probably of freshly precipitated sulfides, carbonates and strongly adsorbed metals and it plays a less important role as metal carrier with the exception of Fe (32.7% in this phase). The metals found in this phase are not easily available to the environment, unless there is a lowering of the PH of the sludge. The residual phase of the sludge is not an important metal carrier, and the metals found there are not at all available to the environment of the sludge disposal site.

5. REFERENCES

(1) KLEIN, L.A., M. LANG, N. NASH and S.L. KIRSCHNER (1974). Sources of metals in New York City wastewater. J.W.P.C.F. 46, 2653-2662 .

(2) PAPAMICHAEL, F, C. YAPIJAKIS and L.A. KLEIN (1985). Heavy metals in New York City wastewater: Their source and impacts. Presented at "Hazardous Materials Conf., Baltimore, MD.

(3) LUOMA, S.N. and E.A. JENNE (1977). The availability of sediment bound cobalt, silver and zinc to a deposit-feeding clam. In Biological Implications of Metals in the Environment, ed. H.Drucker & R.E.Wildung, US NTIS, CONF-750929, Springfield, Va, 213-231.

(4) BLOOMFIELD, C. and G. PRUDEN (1975). The effects of aerobic and anaerobic incubation on the extractabilities of heavy metals in digested sewage sludge. Envir. Pollut. 8, 217.

(5) GIBBS, R.J. (1973). Mechanisms of trace metal transport in rivers. Science 156, 1734-1737.

(6) TESSIER, A., P.G.C.CAMPBELL and M. BISSON (1979). Sequential extraction procedure for the speciation of particulate trace metals. Analyt. Chem. 51, 844-850.

(7) SALOMONS, W. and U. FORSTNER (1979). Trace metal analysis on polluted sediments. Part II: Evaluation of environmental impact. Environ. Techn. Ltrs., 1, 506-517.

(8) ANGELIDIS, M and R.J. GIBBS (1989). Chemistry of metals in anaerobically treated sludges. Water Res. 23, 29-33.

(9) STOVER, B.C., L.E. SOMMERS and D.J. SILVIERA (1976). J.W.P.C.F., 48, 2165-2175.

SOURCES AND OCCURRENCE OF ORGANIC MICROPOLLUTANTS IN WASTEWATER AND SEWAGE SLUDGE IN GÖTEBORG, SWEDEN.

Paxéus Nicklas, Robinson Peter and Balmér Peter.
GRYAAB, S-417 22 Göteborg, Sweden.

Summary

Organic compounds in the influent wastewater and the raw sludge have been identified and quantified using GC-MS. The major components found both in the wastewater and in the sludge are of domestic origin. The content of micropollutants in the raw sludge was found to be low and comparable with that of low-levelled polluted sludges in Germany and the UK. As for the influent quantities of priority pollutants the results indicate a decline in the levels of monocyclic aromatics and halogenated solvents during 1984-1989, while levels of bicyclic aromatics, phenols and plasticisers have remained relatively unchanged.

INTRODUCTION

The Göteborg Regional Sewage Work, GRYAAB, serves a population of about 550,000 people and in addition treats 18 Mm^3/year of industrial effluents. At GRYAAB, approximately 120 Mm^3 of wastewater is treated annually with a corresponding sludge production of about 32,000 tons dry weight (DS). In modern society a great number of synthetic organic chemicals are used both in households and in industry. A major part of these chemicals can be expected to be present in the wastewater and concentrated in the sewage sludge. Some of these may be potentially harmful for the environment. In order to improve the sludge quality and to meet the demands for a sludge acceptable for long term agricultural disposal, a program to chart the occurence of organic micropollutants and their sources in the influent, effluent and sludge started in 1988. The selected parameters included the pollutants from the Norwegian priority pollutants list (1). In the present work the results obtained for the influent wastewater and the sewage sludge during 1988-1990 are reported.

EXPERIMENTAL

The 24 h flow-proportionally taken influent wastewater water samples (totally 11) and raw sludge samples (totally 4) were collected at GRYAAB between December 1988 and May 1990. Two sludge samples and 4 water samples were analysed at the Center for Industrial Research in Oslo (SI) according to the procedures described by Sporstøl et al.(1) and Ringstad et al.(2). The same concentration procedure for the influent water (1), namely extraction with dichloromethane (DCM), was used at GRYAAB. However, sludges were treated in a different manner according to the scheme outlined in Fig.1 and each fraction was analysed separately. The DCM extracts obtained were analysed using gas-chromatography-masspectrometry (GC-MS), partly as analysis of the target compounds and partly as a manual search and identification of other organic components present in the sample. The experimental conditions were as follows. GC-MS computerised system: INCOS 50 (Finnigan MAT). Injector and transfer line temperatures were held at 250° C and the column temperature was programmed from 5°C to 300°C at 4°C/min after an inital isothermal period of 5 min. A 30mx0.3mm i.d. fused silica column coated with 0.25μm chemically bound phase DB-5(J&W) was used. The ion source temperature was held at 150°C. The masspectrometer was operated in electron impact mode. Identification and quantification of the target compounds were performed using a modified version of Finnigan MAT's TCA software procedure. Volatile halogenated solvents were analysed as described by Sporstøl et al.(1).

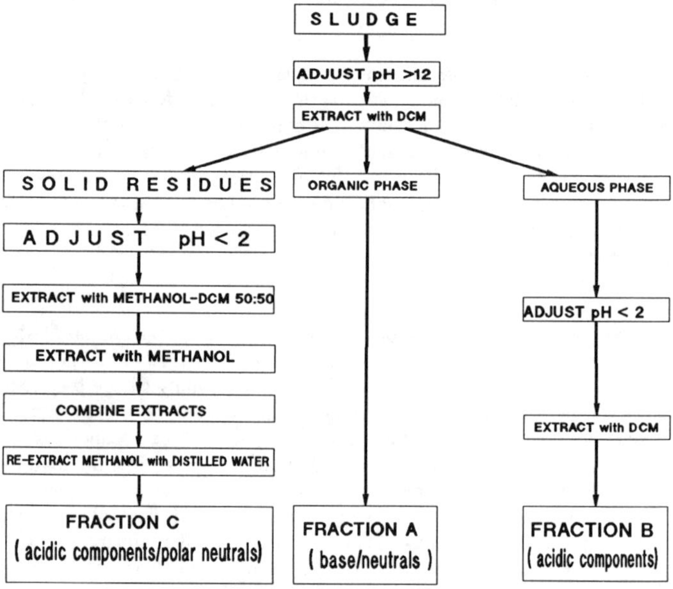

Figure 1. Analytical scheme for analysis of organic constituents in sludge.

RESULTS AND DISCUSSION

The results obtained from the analyses of the influent and sludge are shown in Table 1. As appears from Table 1, the major components of both the sludge and influent wastewater are long-chain carboxylic acids and their esters as well as sterols, which reflects the large proportion of domestic-type wastewater. Other compounds of domestic origin are aldehydes, formed through the decomposition of edible oils and fats, indoles, BHT (butylated hydroxytoluene), which is widely used as antioxidant in food (3,4) and terpenes - odorous compounds, which are often added to cleaning products. The small proportion of aldehydes in the sludge can be explained by their susceptibility to oxidation by the atmospheric oxygen to acids. Another large group of major components detected are aliphatic hydrocarbons, which are also known to be formed as a result of oxidation of edible oils and fats. However, the distribution pattern of these compounds shows that the major part of the aliphatic hydrocarbons found in the influent and sludge is of petrochemical origin.

Table 1. Organic micropollutants found in the wastewater and sludge and range of their concentrations.

COMPOUND	CONCENTRATION Water µg/l	Sludge mg/kg dry weight GRYAAB	Literature values
Aromatic hydrocarbons			
Benzene	0.1-5	0.1-2.0	0.02-20 *
Toluene	0.1-5	0.2-1.1	0.04-19200 *
Ethylbenzene	1-5	0.01-0.3	0.02-840 *
m/p-Xylene	1-16	0.01-1.2	
o-Xylene	1-10	0.01-0.5	
Isopropylbenzene	0.1-5	0.01-0.025	
m/p-Ethyltoluene	0.1-5	0.01-0.45	
o-Ethyltoluene	0.1-5	0.01-0.15	
1,3,5-trimethylbenzene	0.1-5	0.01-0.10	
p-Isopropyltoluene	0.1-5	0.01-0.16	
Tetramethylbenzenes	0.1-5	0.01-0.4	
Diethylbenzenes	0.1-5	0.01-0.35	
Butylbenzenes	0.1-5	0.01-0.40	
Linear Alkylbenzenes(LAB),			
C-5 to C-18 n-alkyl	0.1-10	0.1-0.94	
Linear Alkylbenzenes(LAB),			
Branched alkyl			
C-11	0.1-2	0.1-2.6	
C-12	0.1-2	0.2-0.7	
C-13	0.1-2	0.5-1.8	
Styrene	0.1-4	0.01-0.1	
Naphthalene	0.1-15	0.1-0.5	0.02-6610 *
1-methylnaphthalene	0.1-10	0.05-0.15	
2-methylnaphthalene	0.1-10	0.08-0.1	
2,3,5-trimethylnaphthalene	0.1-5	0.02-0.15	
Biphenyl	0.1-5	0.1-0.20	
Phenanthrene	0.1-0.5	0.05-0.9	
Pyrene	0.1-0.5	0.01-0.7	
Fluoranthene	0.1-0.5	0.01-0.7	0.5-60 *
Indeno(1,2,3-c,d) pyrene	0-0.1	0-0.005	0.2-8 *
Benzo(g,h,i) perylene	0-0.1	0-0.005	0.1-9 *
Dibenzofuran	0-0.6	0-0.05	

Table 1. cont.

Dichlorobenzene 1,4-	0-0.5	0.01-0.05	<0.2-33.9 #
Trichlorobenzene 1,2,4-	0-1.0	0.02-0.05	0.02-4.81 #
<u>Nitrogen-containing</u>			
Aniline	0.1-5	0.1-3.0	
Indoles	0.1-5	1.0- 2.0	
<u>Phthalates/ Adipates</u>			
Dimethylphthalate	0-10	0.01-2.0	0.1-940 *
Diethylphthalate	0-20	0.01-0.15	0.02-3780 *
Di-propylphthalate	0-4	0.01-0.02	
Di-n-butyl phthalate	1-100	0.01-14.5	0.02-3210 *
Butylbenzylphthalate	1-10	0.01-0.05	0.05-12,800 *
Di-(2-ethylhexylphthalate)	5-200	6.7-28	0.04-119 *
Di-n-octylphthalate	1-10	0.05-0.20	0.02-261 *
Di-(2-ethylhexyl)adipate	0.1-3	0.01-0.40	
<u>Phosphate esters</u>			
Tri-n-butyl phosphate	0.1-1	0.01-0.30	
<u>Phenols</u>			
Phenol	0.1-50	0.4-0.9	
o-Cresol	0.1-4	< 0.01	
m/p Cresol	0.1-50	0.01-0.3	
2,6-dimethylphenol	0.1-50	0.01-2.2	
p-Nonylphenol	10-700	4.3-260	10-2500 *
Substituted anisoles	0.1-5	0.01-2.6	
BHT	0.1-5	0.01-0.04	
<u>Aliphatic hydrocarbons</u>			
Octane	5-50	0.1-0.3	
Nonane	5-50	0.4-1.5	
Dekane	10-100	8-12.5	
Undecane	10-100	8-12.5	
Dodecane	10-100	7-21.5	
Tridecane	2-30	4-11.5	
Tetradecane	1-20	1-3.5	
Pentadecane	1-30	0.5-4.0	
Hexadecane	1-30	0.5-9.0	
n-Alcanes (C-17 - C-32)	50-300	15-46.5	
<u>Terpenes</u>			
Limonene	0-20	0.01-1.0	
3-Carene	0-5	0.01-0.05	
4-Carene	0-5	0.01-0.05	
<u>Sterols</u>			
Cholesterols	100-300	150-215	
Cholestanols	100-300	200-430	
<u>Aldehydes</u>			
(C-5 - C-14)	5-100	0.01-0.05	
Benzaldehyde	0-10	0.01-0.05	

Table 1. cont.

Acids and their esters		
Aliphatic (C-3 - C-6)	100-500	60-135
(C-7 - C-18)	3500-38000	8000-17000
Benzoic	0-40	0.05-0.2
Benzenacetic	0-40	0.6-2.0
Benzenepropanoic	0-40	0.6-10.5
<u>Miscelanious</u>		
Cyclosiloxanes*	0-100	1.5

* - BRD sludges (7) # - UK sludges (4)

Table 2. Estimation of influent quantities of organic priority pollutants at GRYAAB.

	Monocyclic aromatic compounds	Bicyclic aromatic compounds	PAH	Phenols	Phthalates +adipates	Halogenated solvents
1984-1986 average						
mg/cap. day	68	2.2	0.56	123	49	27
avg. kg/y	14,700	480	120	26,000	10,500	6,000
1988 average						
mg/cap. day	2.5	4.0	0.17	160	61	5.4
avg. kg/y	300	790	33	32,000	12,000	1,100
Household contribution %	30	28	40	14	>70	3
1989 average						
mg/cap. day	10	2.6	1.9	90	49	2.7
avg. kg/y	2000	520	380	18,000	9,700	530

A group of aromatic hydrocarbons found in the influent water and sludge contain a large number of environmentally potentially harmful compounds. Benzene and alkylbenzenes with the alkyl chain length between 1 and 4 carbons are constituents in many products such as gasoline, degreasants etc, which are widely used both in industry and households. A number of linear alkylbenzenes (LAB) found i the influent water and sludge is probably a result of the large-scale usage of linear alkylbenzene sulphonate (LAS) detergents which always contain unsulphonated hydrocarbon residues (6). As concerns toluene, ethylbenzene and xylenes the estimated household contribution accounts for approximately 20-50% of the total load (5). On the other hand, the discharge from car-washing facilities seems to contribute to a large extent to the load of (C-3 and C-4 alkyl)benzenes. Our analyses of wastewater from Göteborg's 29 different car-washing facilities equipped with obligatory oil-separators have revealed that (C-3 and C-4 alkyl)benzenes account for more than 98% of the total amount of benzenes in the oilseparator effluents. The use of degreasants is probably the main source of these compounds in the influent and sludge.

From a group of chlorinated compounds from the Norwegian priority pollutants list (1), including chlorinated benzenes, pesticides and phenols, only two were detected both in the influent water and sludge (Table 1), namely 1,4-dichloro- (p-DCB) and 1,2,4-trichlorobenzene (TCB). p-DCB is known to be a constituent in so called spool-blocks used in WC's. Concerning TCB main sources of this compound are known to be dielectric fluids, synthetic transformer oils and lubricants. Noteworthy is that the concentrations of both chlorinated benzenes in the sludge are low compared with the values reported for UK sludges (Table 1).

A range of di- and polyaromatic hydrocarbons was found in the influent water and sludge with naphthalenes predominating. However, the concentration of naphthalene in the sludge is low in comparison with the results of German sludges (see Table 1). Concentrations of the few PAH's found in the present study are also comparable with the lowest levels found in sewage sludges in the UK (0.08-11.4 mg/kg) and in Germany (0.09-4.39 mg/kg) (8,9).

The class of esters given in Table 1 encompasses a number of phthalates, an adipate and tributyl phosphate (TBP) with di-ethylhexylphthalate predominating. Compared with German sludges the found concentrations are low. Phthalates as well as phosphate esters are known to be constituents in many products widely used in industry and in households. As estimated by Robinson et al.(5) households account for between 10 and 50% of the total phthalates and more than 50% of total TBP load to GRYAAB.

Phenols detected in the influent and sludge are dominated by p-nonylphenol (NP), which arises from the degradation of polyethoxylated alkylphenols used as industrial surfactants. The concentration of NP in the influent water tends to be much lower in summer. While the household contribution of NP is estimated to be less than 10% (5) of the total NP load to GRYAAB, analyses of the effluents from car-washing facilities in Göteborg show that their discharge of NP, especially under cold and rainy periods of year, is significant.

The compounds given in Table 1 represent a major part of the organic matter extracted by the adopted method and amenable to analysis by GC-MS. However, the samples of influent and sludge contained a number of other organic components, which despite of their relatively high concentrations have not yet been identified. In many cases these components are chromatographically poorly separated thus giving rise to a number of overlapping peaks. One possible strategy to solve this problem is to use other, more specific concentration methods and/or the use of different clean-up procedures.

An estimation of influent quantities of the most frequently occuring groups of priority pollutants is outlined in Table 2. The corresponding estimations for the period of 1984 - 1986, based on previously unpublished results for 13 effluent wastewater samples analysed by SI, are included for comparison. The results from Table 2. indicate a decrease in the influent quantities of monocyclic aromatic hydrocarbons and halogenated solvents, while influent quantities of bicyclic aromatic hydrocarbons, PAH's, phenols and phthalates+adipates seem to be unchanged. The influent quantities of several groups of pollutants appear to be very low. For example, the average daily amount of chlorinated solvents is down to a few kilograms. This is surprising bearing in mind that these solvents are widely used an that GRYAAB is taking care of almost 10% of all wastewater in Sweden. Even for many other priority pollutants the concentrations are at such low level that it is motivated to study their "natural background levels".

In Göteborg an industrial effluent control program has been in effect for over 20 years. The results presented in this work (Table 2.) indicate that this program has been successful as shown by the progressively increasing dominance of domestic sewage as a principal source of many organic pollutants and especially for phthalates.

REFERENCES

(1) SPORSTØL, S., URDAL, K., DRANGSHOLT, H., and GJØS., N. (1985) Description of a method for automated determination of organic pollutants in water. Intern. J. Environ. Anal. Chem., 21,pp. 129-138.

(2) RINGSTAD, O., DRANGSHOLT; H., and ØRELD, F.. (1989) Analysing organic micrpollutants in sewage sludge using GC/MS Priority Pollutants scheme. In: Organic Contaminants in Waste Water, Sludge and Sediment: Occurence, Fate and Disposal. Published by Elsevier Applied Science pp.66-71.

(3) FRANKEL, E. (1985) Chemistry of autoxidation: Mechanism, Products and Flavor Significance. In: Flavor Chemistry of Fats and Oils. American Oil Chemists' Society. pp 1-39.

(4) CRATHORNE B., DONALDSON K., JAMES H.A., and ROGERS H.R. (1989) The determination of organic contaminants in UK sewage sludges. In: Organic Contaminants in Waste Water, Sludge and Sediment: Occurence, Fate and Disposal. Published by Elsevier Applied Science pp.45-65.

(5) ROBINSON P., AVERGÅRD I., and MATTSON J. (1989) Hushållsspillvatten, Rapport 1989:2, GRYAAB - VA-verket (in Swedish).

(6) TAKADA, H., and ISHIWATARI, R. (1987) Linear alkylbenzenes in urban riverine environments in Tokyo; distribution source and behaviour. Environ. Sci. Technol., 21, 9, pp 875-883.

(7) DRESCHER-KADEN, U., MATTIES, M., and BRüGGMANN R. (1989) Organische Schadstoffe in Klärschlammen. Wasser Abwasser, 130, 12, pp.614-620

(8) MCINTYRE, A.E., PERRY, R., and LESTER, J.N., (1981) Analysis of polynuclear aromatic hydrocarbons in sewage sludges. Anal. Lett. 14, pp 291-309.

(9) GRIMMER, G., BOHNKE, H., and BROWITZKY H. (1978) Profile analysis of polycyclic aromatic hydrocarbons in sewage sludge by gaschromatography. Fresenius' Z. Anal. Chem.., 289, pp 91-95.

BIOLOGICAL AND CHEMICAL INDEXES FOR EVALUATING NITROGEN AVAILABILITY OF SEWAGE SLUDGES

M.D. SERNA AND F. POMARES

Departament of Ecology, Instituto Valenciano de Investigaciones Agrarias. Apartado Oficial. 46113-Moncada (Valencia) Spain.

Summary

Biological and chemical methods for estimating N availability in sewage sludges were used. The two biological methods (maize plants grown in pots, and soil-sludges mixtures incubated at 2, 4, 6, 8, 12 and 16 weeks) and four chemical methods (autoclave, $KMnO_4$ 0.5N, pepsin and HCl 6) were compared to determine N availability in twelve sewage sludges in a given soil. In the mineralization test, the aerobically treated sewage sludges gave higher mineralization rates than the anaerobically treated wastes. The simple correlation between available N, estimated from the plant N uptake during 6 weeks and the N extracted by chemical methods showed that both HCl 6N and pepsin appeared to be the best single indexes. Prediction of N availability of sewage sludges by the plants in growth chamber improved if N mineralized and several chemical methods were included in a multiple regression analyses.

1.INTRODUCTION

Recycling organic wastes through agricultural land application is an old practice that has received renewed interest in recent years due to the increased cost of fertilizers and the need for low cost disposal of large quantities of municipal organic wastes. The amount of sewage sludge applied to agricultural land is often based upon the N content or N availability of the waste (18 ,19). Because most of the N in sludge is organic (10 ,14 ,15 ,19 ,21), information about the rate of N mineralization is necessary to predict N availability during a cropping season.

Numerous laboratory studies have been conducted to determine the rate and extent of sewage sludge mineralization in soil. The results change widely not only according to sewage sludge sources (1), treatment process (7, 8, 14, 24), and applied rates (17, 19), but also for laboratory procedure employed (16). Generally, procedures involving the determination of N mineralized during incubation are considered the most satisfactory of those methods currently used as N availability indexes (5, 12, 13, 16). However, incubation procedures are quite laborious and time-consuming, and therefore, many rapid and convenient chemical extraction procedures providing suitable indexes of N availability have been proposed (4, 13, 16). The possible utility of any chemical index depends on the degree to which it correlates with reliable biological measurements of N availability, e.g, N uptake by plants or N mineralizable.

The objectives of this study were to evaluate chemical methods of assessing N availability via regression analysis with total N uptake by maize (*Zea mays L.*) grown in growth chamber and with the N mineralized during incubation.

2. MATERIALS AND METHODS

2.1. Soil and sludges

Aerobic and anaerobically digested sewage sludges were obtained from different sewage treatment plants in Spain. The soil materials employed were taken from surface layer (0 - 15-cm). Soil and sewage-sludge samples were air dried and ground to pass through a 2-mm sieve.

2.2. Incubation procedure

Sludges were mixed with soil at the rate of 50 Mg/ha (28 g/Kg soil). The samples were incubated aerobically for a nonleached procedure. The incubation period was 16 weeks at 25°C; the moisture content was adjusted weekly to keep a constant moisture during the incubation period (about 2/3 of field capacity). The extractant employed was 2M KCl.

Mineralized N ($NH4^{+}$, $NO3^{-}$+ $NO2^{-}$) was determined following 2, 4, 6, 8, 12 and 16 weeks of incubation.

2.3. Experiment in growth chamber

Maize plants were grown during six weeks in a culture chamber at 22°C day and 16°C night temperature and 16 h photoperiod. Plants were grown in 1 Kg pots containing sludges at a equivalent rate of 25 Mg/ha (14 g waste/Kg soil) added with 700 Kg/ha of 0/24/12 fertilizer (0.38 g fertilizer/Kg soil). An unamended soil was included as control. Pots were irrigated twice weekly to maintain the moisture level during the experimental period. After a six weeks culture period, plants were cut, washed with distilled water and ground in a stainless steel mill.

2.4. Chemical analyses

Total N in plant, soil and sludge samples was determined by semi-micro Kjeldahl method which employed salicylic acid to include nitrates (2). NH4 and ($NO3^{-}$+ $NO2^{-}$) were determined by steam distillation with MgO and Devarda alloy reduction of $NO3^{-}$ and $NO2^{-}$ (3). Organic N was calculated by subtraction of $NH4^{+}$ from total-N. Organic C was determined by a Walkley & Black modified procedure.

The chemical test of N availability by an autoclave procedure was that of Smith & Stanford (1971). The acid permanganate (KMnO4) oxidation method was essentially that Stanford & Smith (1978). The HCl 6N method was that of Castellanos & Pratt (1981b). The N released by pepsin was determined by a procedure designed for the determination of protein digestibility in animal feeds (9).

2.5. Statistical analysis

Simple and multiple correlation coefficients were calculated to indicate the relationships between N uptake by plants and different indexes of N availability, using the computer program "statgraf".

3. RESULTS AND DISCUSSION

3.1. Crop N uptake

N absorbed during a cultivation period of 6 weeks ranged from 20 and 73 mg N/pot. The lowest value was obtained from the Castellon sludge (AN2), and the highest, from Burriana sludge (AE1) (table 1). In general, the aerobically digested sludges showed a higher ability for mineralization of organic N than anaerobically treated sludges. We therefore may conclude that the stabilization process was a crucial factor as to determine the rate of mineralization of a given amendment .

3.2. N mineralization

N mineralized in an incubation period of 16 weeks ranged from 74 and 428 mg N/Kg soil, referred to dry weight. This range was similar to those reported by Stark & Clapp (1980) and lower to those reported by Chae & Tabatabai (1986) (table 1). The aerobically treated sewage sludges gave higher mineralization rates than the anaerobically treated wastes (267 ± 115 and 131 ± 73 , respectively). In this study, the lowest value was obtained from the Castellon sludge (AN2), and the highest to the Olive sludge (AN2). These results are attributed to the higher stability of the organic N fraction in anaerobically digested sludges, owing to this fraction is comparatively more resistant to the biodegradation by microrganism than the aerobically treated sludges; this resistence may be originated by the stabilization process.

	mg N/Kg soil	
Sewage sludges	Maize	Incubation
AN 1	35	105
AN 2	20	74
AN 3	31	97
AN 4	50	262
AN 5	47	78
AN 6	49	174
AE 1	73	251
AE 2	44	223
AE 3	44	247
AE 4	66	359
AE 5	52	428
AE 6	34	95

Table 1. Sewage sludges N uptake by maize and N mineralized during a 16 weeks period of aerobic incubation

Relationships between maize N uptake and N mineralized during a six weeks period of aerobic incubation are shown in table 2 . Correlation coefficients for all incubation periods were statistically significant. However, the relationships for the 6th and 16th showed a somewhat closer association. The regression coefficient changed very little with increased incubation time, but the value of the intercept decreased. The overall predictive value did not improve significantly with an increase of the incubation time.

Time of incubation, weeks	Regression equation	Coefficient of regression,r
2	Y = 1879 + 10.4 X	0.670 *
4	Y = 1796 + 7.98 X	0.667 *
6	Y = 1933 + 7.53 X	0.671 *
8	Y = 2083 + 6.93 X	0.665 *
12	Y = 1924 + 5.99 X	0.638 *
16	Y = 1891 + 5.54 X	0.677 *

* Significant at the 0.05 level.

Table 2. Regression equations and coefficients of correlation for the relationships between available N by during 6-weeks cropping period and N mineralized during laboratory incubation.

3.3. Chemical indexes of N availability

Table 3 shows tha values of N extracted by different chemical extraction methods. The HCl 6N method provided the highest extraction ability, followed by pepsin, autoclave, and KMnO4 0.5N, in this order.

	mg N/Kg waste		* Min N/Org N (%)	
Chemical indexes	range	mean	range	mean
Autoclave	300-1820	1378±498	1.31-19.3	6.88±4.74
KMnO4 0.5N	322-1034	559±199	0.92-7.0	2.82±1.83
Pepsin	1160-3400	2330±624	6.89-23.4	10.7±4.53
HCl 6N	5800-30700	16300±7320	29.0-99.1	78.9±18.3

* Extracted N expressed as % of organic N

Table 3. Ranges and means of N extracted by several chemical indexes of N availability

3.4. Correlation between biological and chemical test

Correlations for the N extracted by several chemical methods, and the two biological indexes tested in this study, N mineralized in a 16 weeks period of incubation, or N uptake by maize have been proposed. Total N was also considered as a chemical index for N availability.

Table 4 shows the correlation coefficients obtained by several N extraction methods. The autoclave and the KMnO4 0.5N were not suitable for predicting N availability in these wastes.

	Autoclave	KMnO4 0.5N	Pepsin	HCl 6N
N uptake by maize	0.421	0.311	0.651 *	0.668 *
N mineralized during 16 weeks	0.131	-0.303	0.630 *	0.971 ***
Total N	0.232	-0.178	0.763 **	0.838 **

*,**,*** : Significant at the 0.05, 0.01 and 0.001 levels, respectively.

Table 4. Correlation coefficients between several chemical and biological indexes

The correlation coefficients obtained with the pepsin method allowed to conclude that this was a relatively suitable index to predict potentially N mineralizable in sludges, because the correlations obtained with the two biological indexes was significant at the 0.05 level. Conversely , the HCl 6N method proved highly efficient to estimate the potentially available N in this wastes , since it was adequately correlated with N mineralized during incubation (r = 0.971, significant at the 0.001 level), and although at a lower significance level, with the N uptake by maize (r = 0.668, significant at the 0.05 level).

Both the pepsine and the HCl 6N methods showed a good correlation with total N (r = 0.763 and 0.838, respectively).

Results of multiple regression analyses to select the best prediction of net N mineralized during a 6-week cropping period are presented in table 5. Considering maize N uptake as a dependent variable, and N extracted by the various chemical indexes and the relation C/N as independent variables, high correlation coefficients were obtained; this increase was more remarkable for autoclave and permanganate methods, if compared with the linear regression, although the highest degree of correlation was obtained with the pepsin method.

The multiple regression analysis, including all the studied variables, states that the highest correlation indexes were attained with the pepsin method, in all the incubation periods, particulary for two weeks, and with the $KMnO_4$ 0.5N and HCl 6N methods, in the 16 weeks period for both of them.

Indexes involved	Equation	r
I. Chemical indexes	Y = 4825 - 265 X1 + 0.55 X2	0.673 *
	Y = 6254 - 276 X1 - 0.98 X3	0.635 *
	Y = 3525 - 228 X1 + 0.73 X4	0.787 **
	Y = 4114 - 180 X1 + 39.9 X5	0.621 *
II. All indexes studied	Y = 3834 + 0.28 X6 - 129 X1 + 3.70 X7 - 1.31 X3	0.671 *
	Y = 3989 - 0.10 X6 - 294 X1 - 4.09 X8 + 0.93 X4	0.751 *
	Y = 4691 - 0.25 X6 - 221 X1 + 10.1 X7 - 118 X5	0.648 *

*, ** : Significant at 0.05 and 0.01 levels , respectively.

Y: N uptake by maize; X1: C/N ratio; X2: N released by autoclave (mg N/Kg waste); X3: N released by acid permanganate 0.5 N (mg N/Kg waste); X4: N released by pepsin (mg N/Kg waste); X5: N released by HCl 6N (g N/Kg waste); X6: Inorganic N; X7: N mineralized during 16 weeks of incubation; X8: N mineralized during 2 weeks of incubation.

Table 5. Prediction equations of the N mineralization rates in the sewage sludges calculated from chemical and biological indexes

REFERENCES

(1) BEAUCHAMP, E. G. et al. (1979). Nitrate production from chemically treated sewage sludges in soil. J. Environ. Qual. 8, 557-560.

(2) BREMNER, J. M. (1965a). Total nitrogen. In Methods of soil analysis. Part 2. Agronomy 9. Eds. C A Black et al. pp 1149-1178. Am. Soc. of Agronomy, Inc. Madison, Wis.

(3) BREMNER, J. M. (1965b). Inorganic forms of nitrogen. In Methods of soil analysis. Part 2. Agronomy 9. Eds. C A Black et al. pp 1179-1237. Am. Soc. of Agronomy, Inc. Madison, Wis.

(4) CASTELLANOS, J. Z. and PRATT, P. F. (1981a). Mineralization of manure nitrogen-correlation with laboratory indexes. Soil Sci. Soc. Am. J. 45, 354-356.

(5) CASTELLANOS, J. Z. and PRATT, P. F. (1981b). Nitrogen availability in animmal manures and crop yields. Agrochimica. 25, 443-356.

(6) CHAE, Y. M. and TABATABAI, M. A. (1986). Mineralization of nitrogen in soils amended with organic wastes. J. Environ. Qual. 15, 193-198.
(7) CHAUSSOD, R. and GERMON, J.C. (1977). Mineralization in the soil of nitrogen in the sludges from different sewage sludge works. C.R. Seances Acad. Agric. Fr. 63, 525-531.
(8) EPSTEIN, E. et al. (1978). Mineralization of nitrogen from sewage sludge and sludge compost. J. Environ. Qual. 7, 217-221.
(9) HORWITZ, W. (1970). Official methods of analysis 11th ed. pp 127-128. Association of Official Analytical Chemist, Washington, D. C.
(10) KING, L. D. and MORRIS, H. D. (1972). Land disposal of liquid sewage sludge: The effect on yield, in vitro digestibility and chemical composition of Coastal Bermudagrass (Cynodon dactylon L. Pers.). J. Environ. Qual. 1, 325-329.
(11) KING, L. D. (1984). Availability of nitrogen in municipal, industrial and animal wastes. J. Environ. Qual. 13, 609-612.
(12) LOEWEN-RUDGERS, L.A. et al. (1981). Incubation of pulverized household refuse with soil and sewage sludge, poultry manure or (NH4)2SO4. Can. J. Soil Sci. 61,109-121.
(13) MAGDOFF, F.R. and AMADON, F. F. (1980). Nitrogen availability from sewage sludges. J. Environ. Qual. 9, 451-455.
(14) MAGDOFF , F. R. and CHROMEC, F. W. (1977). Nitrogen mineralization from sewage sludges. J. Env. Sci. Health A12, 191-201.
(15) MITCHELL, M. J. et al. (1978). Effects of different sewage sludges on some chemical and biological characteristics of soil. J. Environ. Qual. 7, 551-559.
(16) PARKER, C. F. and SOMMERS, L.E. (1983). Mineralization of nitrogen in sewage sludges. J. Environ. Qual. 12, 150-156.
(17) PREMI, P. R. and CORNFIELD, A. H. (1969). Incubation study on nitrification of digested sewage sludges added to soil. Soil Biol. Biochem. 1, 1-4.
(18) PREMI, P. R. and CORNFIELD, A. H. (1971). Incubation study of nitrogen mineralization in soil treated with dried sewage sludges. Environ. Pollut. 2, 1-5.
(19) RYAN, J. A. et al. (1973). Nitrogen transformations and availability of an anaerobically digested sewage sludge in soil. J. Environ. Qual. 489-492.
(20) SMITH, S. J. and STANFORD, G. (1971). Evaluation of a chemical index of soil nitrogen availability. Soil Sci. 111, 228-232.
(21) SOMMERS, L. E. (1977). Chemical composition of sewage sludges and analysis of their potential use as fertilizers. J. Environ. Qual. 6, 225-232.
(22) STANFORD, G, and SMITH, S. J. (1978). Oxidative release of potentially mineralizable soil nitrogen by acid permanganate extraction. Soil Sci. 126, 210-218.
(23) STARK, S.A. and CLAPP, C.E. (1980). Residual nitrogen availability soils treated with sewage sludges in a field experiments. J. Environ. Qual. 9, 505-512.
(24) TESTER, C.F. et al. (1977). Decomposition of sewage sludge compost in soil : I Carbon and nitrogen transformation. J. Environ. Qual. 6; 459-463.

ANAEROBIC DIGESTION AND SLUDGE CONDITIONING AT THE "METAMORPHOSSIS" WASTEWATER TREATMENT PLANT

S. HOULAKI Ph.D
P. DIAMANTOPOULOS DEA

Kinopraxy for the Major Athens Area Wastewater Treatment Plant

13th Km National Road Athens - Lamia
Metamorphossis, Attiki

Summary

The "Metamorphossis" wastewater treatment plant treats septic and domestic effluents from the Major Athens Area. The sludge is treated by anaerobic digestion followed by dewatering on belt presses.
The plant control data as well as the quality of the dewatered sludge after anaerobic treatment show a satisfactory digestion efficiency.

1. THE "METAMORPHOSSIS" WASTEWATER TREATMENT PLANT.

The "Metamorphossis" Sewage Works in Athens treats sewage flows from two sources. Septic sludges are treated with Ferrous Sulphate and Lime prior to sedimentation. A conventional sewage system flow also undergoes sedimentation and the two effluents are mixed prior to activated sludge treatment and final settlement. The supernatant is chlorinated while the sludge is passed to anaerobic digestion plant and finally dewatered by belt press. The design inlet flow parameters are 24000 m^3/d of septic effluent and 20000 m^3/d of domestic effluent.

The treatment process is shown in Fig.1 and Fig.2 shows monthly variations of the incoming loads expressed in tons per day of suspended solids, COD and BOD_5 for both kinds of effluent (5).

2. ANAEROBIC SLUDGE DIGESTION

The two primary digesters of 7900 m^3 each are heated at 35° C by means of the biogas recycled into the installation. The secondary digester of 3000 m^3. is used as a sludge thickner and is not heated.

The daily digestion control in the plant includes the measurement of the pH of the digested sludge and the total alcalinity and volatile acids determination (4). Under normal operation of the digesters the sludge pH remains at 7.1 - 7.2 and the ratio volatile acids and total alcalinity expressed in meq CH_3COOH/meq $CaCO_3$ is always less than 0.2 . When a problem in the digestion occurs the sludge pH and the biogas production fall down, due to the high concentration of volatile acids which are not transformed to biogas after hydrolysis of the organic matter (3).

The parameters that influence the digestion efficiency for a given temperature are the sludge retention time and the percent content of the volatile matter in the raw sludge.

As can be seen from Fig.3 a sludge retention prolongation for more than 30 days -by thickening the sludge prior to digestion- will not influence the volatile matter reduction.

The percent reduction of the volatile matter was supposed to be at least 50% provided the ratio volatile solids/dry solids in the sludge feeding the primary digesters was 70% or more (1,2,3). This is true for the septic sludge entering the plant but not for the septic sludge prior to anaerobic digestion because

i) a part of reduction of the volatile matter has already taken place during pre-treatment and primary sedimentation.

ii) The addition of slaked lime into the effluent in order to enhance the flocculation in the settling tank, reduces the ratio volatile solids/dry solids. Thus, the raw sludge feeding the digesters contains less than 60% VS/DS (5) as can be seen from Table 1.

MONTH	SLUDGE RETENTION TIME (d)	VS/DS RAW SLUDGE(%)	VS RAW SLUDGE(t)	GAS (m^3)	l GAS/Kg VS RAW SLUDGE
AUG 86	48	56	368	137194	372
SEPT 86	44	57	392	155474	390
OCT 86	32	55	536	185127	340
NOV 86	32	56	540	185457	343
DEC 86	44	60	361	152756	420
JAN 87	50	59	322	125308	389
FEB 87	42	57	384	104718*	270*
MAR 87	39	61	382	44529*	110*
APR 87	30	62	519	246558	470
MAY 87	29	55	558	212724	380
JUN 87	29	55	521	197766	370
JUL 87	39	57	392	160026	410
AUG 87	44	59	283	128915	460
SEPT 87	48	56	300	120005	400
JAN 88	49	54	283	96527	341
FEB 88	47	50	341	132899	390
MAR 88	41	53	460	157421	342
APR 88	39	49	432	158864	360
MAY 88	46	57	409	165150	403
JUN 88	42	49	367	145144	395
JUL 88	46	53	318	9240*	30*
AUG 88	46	48	336	26340*	80*
SEPT 88	44	54	367	148730	400
OCT 88	44	48	339	120736	350
NOV 88	42	47	312	133307	420
DEC 88	32	51	353	118309	335

* Shock in the digesters

Table 1 : Average monthly parameters at the "Metamorphossis" anaerobic digestion installation.

International experience on wastewater treatment has shown that a mixture of primary sludge and secondary biological sludge, which is the case at the "Metamorphossis" Plant is more difficult to be digested than a purely primary sludge (1). The average biogas production of 381 l of biogas per kg of volatile solids feeding the digesters (see Table 1) indicates a satisfactory digestion efficiency compared to the typical production of 400 - 500 l of biogas per kg of organic matter in the raw sludge when a primary domestic sewage sludge is treated anaerobicaly (3).

Previous measurements for heavy metals concentrations in the sewage sludge from the Major Athens Area have shown that although they do not inhibit the anaerobic digestion they may be responsible for low gas productions (6).

3. SLUDGE DEWATERING.

The digested sludge is pumped from the bottom of the secondary digester onto 4 beltpresses with a capacity of 600 kg/h.m each, after being mixed with the appropriate dose of polymer solution.

Laboratory evaluations as well as plant trials have been carried out in order to find the most cost effective treatment. Various products with slight or medium cationicity have worked satisfactorily. The polymer consumption varies from 2 to 3.5 kg of dry product per ton of sludge dry solids and the final cake concentration is at least 30%. Fig.4 shows daily sludge quantities treated at the "Metamorphossis" dewatering installation as well as the polymer doses.

Experience from the "Metamorphossis" sludge dewatering has shown that the dewatering efficiency depends on

i) The previous good anaerobic digestion
ii) The digested sludge solids concentration. For the same polymer dose a sludge with a dry solids concentration of 50 gr/l gives a drier cake than a sludge with 40 or 45 gr/l. (the relation is not linear).
iii) The ambient temperature. Sludge dewatering is easier during summer.

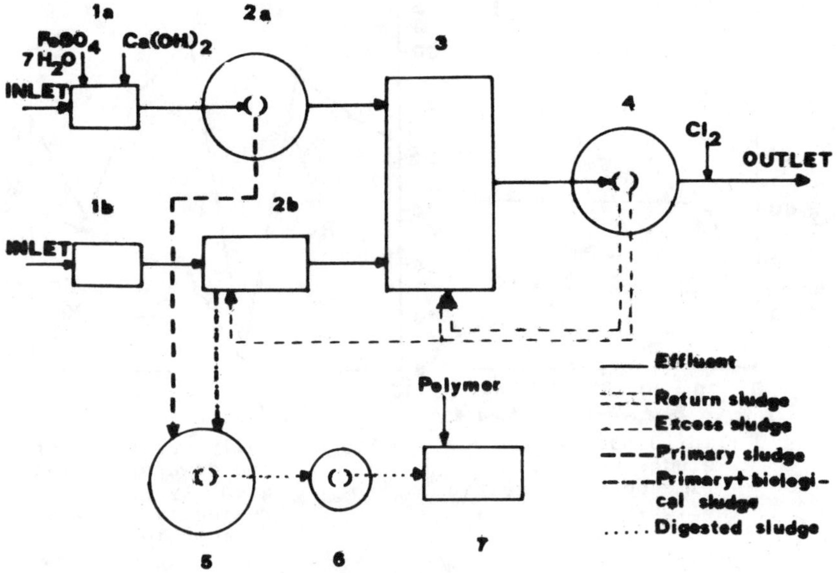

1a. Septic effluent pretreatment
1b. Domestic effluent pretreatment
2a,2b Primary sedimentation
3. Aeration tank
4. Secondary sedimentation
5. Primary digester
6. Secondary digester
7. Sludge dewatering

Fig.1 : Plant lay-out of the "Metamorphossis" Sewage Works.

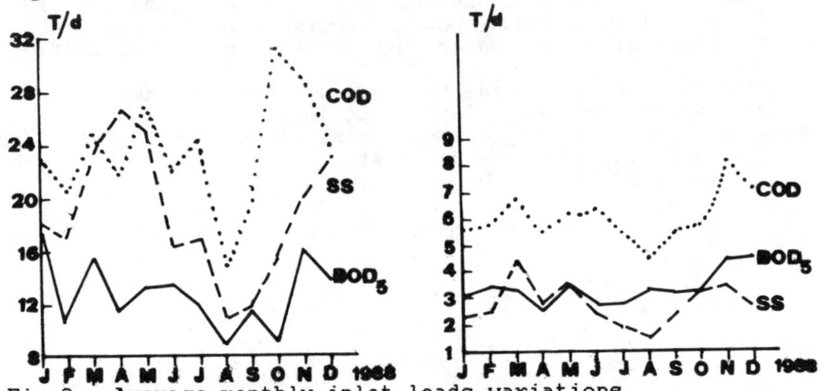

Fig.2 : Average monthly inlet loads variations
2a : Septic effluent 2b : Domestic effluent

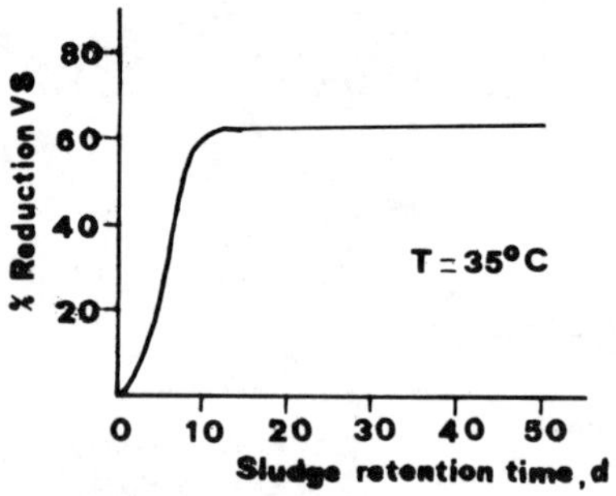

Fig.3 : Influence of the sludge retention time on the digestion efficiency.

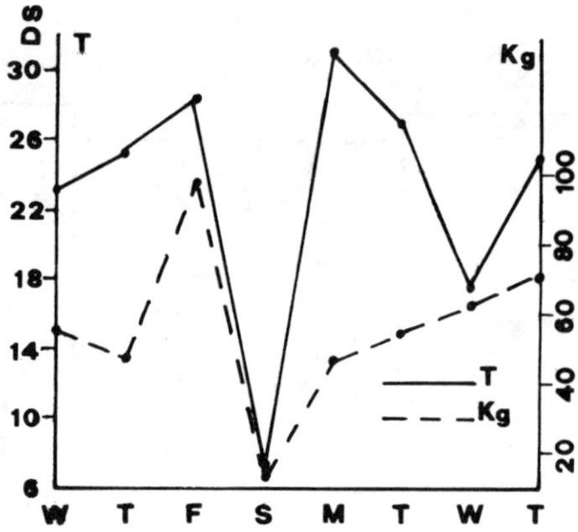

Fig.4 : Daily variation of the dewatered sludge quantity and the respective polymer consumption.

REFERENCES

1. US Environmental Protection Agency, "Process Design Manual for Sludge Treatment and Disposal", Technology Transfer, EPA 625/1-79-001, 1979.
2. American Society of Civil Engineers, "Wastewater Treatment Plant Design", Manuals and Reports on Engineering Practice No 36, N. York, 1977.
3. Degremont "Memento technique de l' eau" 8eme edition, 1978
4. American Public Health Association, "Standard Methods for the Examination of Water and Wastewater", 15th Edition, 1980.
5. Monthly Control Data Reports of the "Metamorphossis" Wastewater Treatment Plant (1986-1989)
6. N. Kartsonas, A. Kouzeli - Katsiri "Anaerobic Digestion of Sewage Sludge from the Major Athens Area", Tech. Chron.-A. Greece, 1987, Vol 7, No 4.

ENVIRONMENTAL IMPACT OF PIG UNITS IN PORTUGAL

E. Fernandes, L. Campos, J.Q. Santos

Department of Agricultural Chemistry
Instituto Superior de Agronomia
Tapada da Ajuda, 1399 Lisboa Codex
Portugal

Summary

As a part of a research strategy for pollution control, the Department of Chemistry of the Faculty of Agronomy at Lisbon implemented two pilot units consisting of two types of anaerobic digesters, with laboratory facility which includes a compreensive range of analytical equipment. The program of work comprises five projects on anaerobic digestion of piggery wastes: scale-up, start-up, stability, monitoring and control, and reutilization of the by-products.

1. INTRODUCTION

With the intensification of pig production and the tightening of environmental regulations (recently published, March 1990*), piggery wastes are becoming a severe ecological problem thoroughout Portugal.

In order to offer a solution for this problem, an integrative process has been developed in our department with the financial support of the Ministry of Environment.

2. FRAMEWORK

The plan of action for this process comprises:

i) An evaluation of pig units in Portugal (there are around 20,000 pig units, 400 of these units having an average of 7,000 animals**);

ii) An experimental and demonstration program involving two pilot scale biogas reactors - a completely mixed anaerobic reactor and a fixed film bioreactor.

iii) A follow-up program to determine the environmental, agricultural and energy-related advantages and disadvantages of large biogas plants (there are around 32 installed in Portugal***) in relation to other options.

iv) Organizational and promotional activities to ensure that experience obtained from the implementation of the plan is continuously exchanged among farmers, local and regional authorities involved in environmental and energy planning.

---

* Diário da República, I Série, nº 55 (March 7th (1990)

** Instituto Nacional de Estatística (National Institute of Statistics), April 1989, Lisboa, Portugal.

*** March 1990.

The results of three years of trials have revealed that composition of pig slurries in Portugal is relatively independent of the breeding ration and modes; undiluted daily volume of pig manure represents 9-12% of the body weight, 1.4 to 14.0 liters per day dilution being the major factor which affects the digesters performance.

The anaerobic treatment of pig manure is a valid method for removing considerable amounts of their organic load (around 50-60% in the completely mixed reactor and 70-80% in the fixed one) with lower energy costs.

The anaerobic digestion process can be applied effectively to the digestion of swine manure. The digestion produces large quantities of high quality (60-70% methane) biogas. The specific gas production rate during the monitoring program remained consistent at 0.50-0.70m^3 biogas/kg vs.fed/day.

However effluents from anaerobic treatment, which according to the present regulations still contain loads which are too polluting to be discharged into surface waters, require predischarge treatments.

Although spreading of effluents on land, which combines disposal and treatment, is without any doubt the cheapest system, it is not, however, free from difficulties, especially in the case of large volumes, and becomes even more of a problem when, owing to a lack of availability of land, it is impossible to carry out fertilizing irrigation and it is necessary to install a proper treatment plant. To overcome those problems other solutions has been implemented by our Department, namely an hydroponic pilot scale plant and a bench scale system to recover the protein fraction for incorporation in feedstuffs.

3. CONCLUSION

The results from this program should indicate the most suitable integrative treatment process for piggery wastes based on anaerobic digestion, as well as an operational strategy applicable to many other biomethane wastes.

Fortunately, stressing by governmental authorities on environmental quality favours the development of biogas technologies, since such application may accomplish concomitant improvements in waste treatment beyond the benefits of alternative energy production.

AGRICULTURAL USE OF POLLUTANT ORGANIC WASTES

F. Cabral, E. Vasconcelos, F.C. Pinto, J.Q. Santos

Department of Agricultural Chemistry
Instituto Superior de Agronomia
Tapada da Ajuda, 1399 Lisboa Codex
Portugal

Summary

Some experiments concerning the use as fertilizers of "pulp mill sludges" from the primary treatment of the pulp paper industry and piggery wastes before or after being submitted to anaerobic treatment, were carried out with the aim of deffining utilization limits of those byproducts to different crops and evaluate the real advantages of its incorporation in the soil.

Pig manure applications to the soil, if not strongly diluted, affects significantly seed germination and crop yields.

Levels higher than 50t/ha of pulp mill sludges affect yields, in variable degrees to different crops, though seed germination is not affected.

1. INTRODUCTION

The beneficial influence of organic matter on physical, chemical and biological fertility of the soils, and since it may avoid soil pollution caused by excessive or unbalanced use of fertilizers, its interest under an ecological point of view must be considered.

As far as organic matter is concerned, the majority of portuguese soils are defficient. As a matter of fact, the natural manure availabilities are becoming lower as a result of the modern methods of cattle raising. On the other hand, the low yields obtained in all cereal crops in this country, and the current trends in cereal prices determined by the Common Agriculture Policy, are leading to a reduction in organic residues from those crops.

Therefore, the possibility of using organic residues of several origins (municipal wastes, agricultural and forest residues etc.), that are pollutant in variable degrees, must be considered. Their utilization in agricultural and forest ecosystems, can replace or at least decrease the need of waste treatment, and will allow the incorporation of organic matter to the soil and in some cases of non-negligeable amounts of nutrients.

Furthermore, the generalized acidic characteristics of most portuguese soils, stresses the advantages of some of these residues which show strong liming potential.

Nevertheless, the use of those residues must be adapted in a convenient way, to different agro-climatic conditions, without what the risk of soil pollution problems may occur.

These kind of problems may become very dangerous, not only because a non-renewable resource is affected, but also because a polluted soil may eventually cause the pollution of both water and agricultural products.

Some experiments concerning the use as fertilizers of organic residues from different origins were carried out in the last three years at the Department of Agricultural Chemistry of the Instituto Superior de Agronomia, Lisboa, in order to obtain elements to be considered in a rational agricultural utilization of those byproducts.

The main purpose of the present work is to present a summula of the results obtained at the end of some of those experiments, concerning the study of the effects on seed germination and crop development, when "pig manure" before or after being submited to anaerobic digestion as well as "pulp mill sludges" proceeding from the primary treatment of the pulp paper industry, are applied to soil.

The main objectives of these studies are to deffine utilization limits of that kind of byproducts, to different crops, and evaluate the real advantages under an agricultural and ecological point of view of its incorporation in the soil.

2. GENERAL CONCLUSIONS

PIG MANURE

Effects on seed germination

According to the results obtained it is possible to conclude that pig manure if not strongly diluted when applied to the soil, affects significantly seed germination, particularly when legume crops are concerned (1).

The main causes that lead to the decrease of seed germination (%), may be due both to the specific ammonium (NH_4^+) effect and to the high salt concentration of pig manure, that motivates a strong increase of soil solution electric condutivity (EC).

The low germination rate observed when legume crops are concerned, is in perfect accordance with the higher susceptibility of those crops to soil salinity.

Effects on crop development

According to the results obtained it is possible to maintain that the irrigation of the plants using non diluted pig manure or different dilutions of this kind of effluent, lead to a significant decrease of the yields, as a result of the increase of soil solution EC, exceptuating the case when a 10% dilution was used, where the yields obtained were similar to those obtained to the blank (irrigation with demineralized water)(2,3).

When the application of pig manure to soil, has the aim of replacing nitrogen top dressing fertilization, a decrease of the yields occurs, unless its utilization is fractionated in several applications (4).

The yields obtained when pig manure after being submitted to anaerobic treatment is applied to soil, are lower than those obtained when the untreated effluent is used, probably as a consequence of the low content in some nutrients, namely phosphorus and calcium, presented by the effluent after the anaerobic treatment (2).

In every considered case the use of pig manure lead to an increase of soil ph values, as well as in some cases of the soil organic matter levels.

PULP MILL SLUDGES

Effects on seed germination

According to the results obtained, it is possible to conclude that the incorporation in the soil of pulp mill sludges do not affects in a significant way seed germination, even when very high amounts are used (160t ha^{-1})(5).

Effects on crop development

According to the results obtained it is possible to draw the following conclusions:

When the amounts of pulp mill sludges incorporated in the soil at planting, do not exceed 50t ha^{-1}, no depressive effects in the yields are observed. Higher amounts of that material lead to a decrease of the yields, resulting from an unbalanced plant nutrition, particulary in what concerns nitrogen, as a consequence of the high ratio C/N of the sludges (6,7,8).

As expected, the use of raising amounts of pulp mill sludges, lead to an increase of the soil ph, organic matter and exchangeable calcium values (6,7).

The incorporation in the soil of different amounts of pulp mill sludges in a convenient period before planting, leads to the complete disappearence of the depressive effects in the yields. This fact reflects the biodegradable characteristics of the sludges (9).

In the same way the depressive effect in the yields is completely eliminated, when pulp mill sludges instead incorporated in the soil, are applied at its superficial layer.

3. REFERENCES

(1) Cabral, F.M., Vasconcelos, E. and Monjardino, P. (1988). Impacto da utilização de chorume de suínos na germinação de sementes de algumas espécies forrageiras. IX Reunião de Primavera da Sociedade Portuguesa de Pastagens e Forragens, Monfortinho, Portugal.

(2) Santos, J.Q. et al (1988). Pollutant organic wastes as fertilizers. Proceedings of the Seminar on Storing, Handling and Spreading Manure and Municipal Wastes, Uppsala, Sweden.

(3) Vasconcelos, E. and Cabral, F.M. (1989). Caracterização e aproveitamento de resíduos orgânicos poluentes. III. Utilização como fertilizantes. Proceedings das 1ªs Jornadas sobre a Indústria e ambiente, Lisboa, Portugal.

(4) Vasconcelos, E., Cabral, F.M. and Monjardino, P. (1989). A utilização de chorume de suínos em milho para forragem. II Reunião Ibérica de Pastagens e Forragens, p.87, Elvas-- Badajóz, Portuga-España.

(5) Monjardino, P. (1988). Impacto de alguns correctivos orgânicos na germinação de sementes de milho e alface. Relatório de estágio do curso de Produção Agrícola, Universidade dos Açores, Angra do Heroísmo, Portugal.
(6) Santos, J.Q., Vasconcelos, E. and Cabral, F.M. (1990). Utilização das lamas celulósicas como fertilizante. IV Encontro Nacional de Saneamento Básico, Aveiro.
(7) Cabral, F.M. et al (1989). Critical application levels of pulp mill sludges to different crops. Proceedings of 5th Conference on Biomass for Energy and Industry, Fundação Gulbenkian, Lisboa, Portugal.
(8) Vasconcelos, E. et al (1989). Critical application levels of pulp mill sludges to yellow lupin crop. Proceedings of International Symposium on Fertilization and the Environment, Leuven, Bélgica.
(9) Cabral, F.M., Vasconcelos, E. and Monjardino, P. (1990). Biodegradabilidade no solo das lamas celulósicas. Proceedings da 2ª Conferência Nacional da Qualidade do Ambiente, vol. 1, p. E-1, Lisboa, Portugal.

IMPLEMENTATION OF REGULATIONS CONCERNING THE USE OF SEWAGE SLUDGE AND ORGANIC WASTE MATERIALS ON FARMLAND IN DENMARK.

E.H. JENSEN and PETER WEILE

Ministry of the Environment
National Agency of Environmental Protection
Strandgade 29, DK-1401 Copenhagen K.
Denmark

Summary

A new Danish regulation for application of organic waste materials to farmland has been prepared. This regulation restricts the amount of dry matter, heavy metals and nutrients which can be applied to farmland. In Denmark 130,000 tons of sewage sludge (dry solids) was produced in 1987. Approx. 37% was used for agricultural purposes and about 80% of the total sludge quantity could meet the limit values established for heavy metal content.

1. INTRODUCTION

Regulation of waste materials applied to farmland in Denmark has up till now taken place in many different ways. Separate regulations apply to the use of:

- Wastewater from e.g. processing of potato starch-including maximum limits for the amount of water applied per ha (1).

- Sewage sludge - including limits for application of heavy metals (2).

- Farmyard manure - including maximum limits for application of nutrients (3).

None of the abovementioned regulations specify any common limits concerning application of nutrients and heavy metals on farmland, which means that the protection of groundwater and farmland resources has been quite diffuse.

In Denmark the EEC Sludge Directive in 1986 was adopted in an atmosphere of increasing interest in recycling of organic waste products for agricultural purposes.

There was for instance an increasing interest in using compost made on the basis of source separation together with application of different sorts of organic waste from processing of vegetable and animal food.

Not only from an overall environmental point of view, but also from the point of view of the producer and consumer, joint regulation of the use of waste products for agricultural purposes was required. This regulation was implemented in 1989 and will be discussed in this symposium paper, and the Danish maximum limits for heavy metals in sludge will be compared to the EEC guidelines. Finally quantities and qualities of sewage sludge in Denmark will be discussed.

2. NEW REGULATION OF SLUDGE APPLICATION IN DENMARK

The new Statutory Order (4) concerning application of sewage sludge and organic waste materials on farmland entered into force in January 1990. The idea of this new regulation is to promote recycling of waste materials containing valuable plant nutrients, without harming the environment. So, the order deals with waste materials used on land to achieve plant production, which means it can be used for fertilization on farmland, in parks, gardening and horticulture etc.

To protect the groundwater from pollution by leaching of nitrate, maximum limits for application of nutrients have been established. These limits are 250 kg N and 40 kg P per year. And whereas phosphorus can be applied as a 5 year average (200 kg P in one application every 5 years), no more than 250 kg of nitrogen can be applied in one year. For nitrogen there are further restrictions on liquid waste materials such as waste water, which means that no more than 100 kg N can be applied in the period from the 1st of September to the 15th of February. Application in this period has to be on land covered with vegetation and no more than 1,000 m^3 per ha can be applied. For the whole year 3,000 m^3 waste water per ha can be applied. A maximum limit for application of waste water in autumn and winter has been laid down to avoid that nutrients remaining in the soil are washed from the rootzone and down into the groundwater.

The amount of waste materials which can be applied to farmland is not only restricted by the amount of nutrients and m^3 of liquid waste products, but also by tons of dry matter added per year. On a 10 year average 10 tons of dry matter can be added per year (max. application in one year is 100 tons of dry matter).

As regards the regulation of heavy metals in Denmark, two sets of maximum limits have been given. Of these limits either one can be followed. The difference between the two limits is that one relates to dry matter (and will mainly be used for compost and similar waste materials), and the other relates to phosphorus, since the amount of heavy metals in fertilizer is usually related to phosphorus. The phosphorus-related values will mainly be used for sewage sludge and other waste materials with a considerable content of nutrients - such as phosphorus. The idea of relating maximum limits of heavy metals in waste materials to the phosphorus content is first of all that only a certain amount of phosphorus should be added per ha. Secondly the contribution of heavy metals per kg phosphorus in waste materials used for fertilization should be comparable to the amount of heavy metals following each kg of phosphorus fertilizer.

In table 1 the maximum limits for heavy metals in waste materials are listed. In order to show the development of regulations dealing with heavy metals in Denmark old limits set up in 1984 and coming limits which will be put into force in 1995 are listed - together with the limit values in EEC (5) and Sweden (6).

	DENMARK					EEC	SWEDEN
	From 1984	From 1990		After 1995		From 1986	From 1989
	g/tDM	g/tDM	g/tP	g/tDM	g/tP	g/tDM	g/tDM
Cadmium	8	1.2	320	0.8	200	20-40	2
Mercury	6	1.2	320	0.8	200	16-25	2,5
Lead	400	120	15000	120	10000	750-1200	100
Nickel	50	45	4000	30	2500	300-400	100
Chromium	100	100	-	100	-	100-200	150
Zinc	4000	4000	-	4000	-	2500-4000	1500
Copper	1000	1000	-	1000	-	1000-1750	200

(g/tDM=gram/tons dry matter - g/tP=gram/tons Phosphorus).

Table 1: Maximum limits for heavy metals in waste materials which can be applied to farmland.

As will appear from table 1, the Danish limit values are comparable to the Swedish values and much more restrictive than the values given for the EEC as a whole.

Concerning risks to human and animal health it is necessary for some of the waste materials to undergo treatment before being used in agriculture. Moreover, a certain period must elapse between application of sludge and putting stock out to pasture.

3. SYSTEM OF ADMINISTRATION

Treater´s obligations:

The administration of the Order is made as simple as possible in an attempt to encourage and promote the use of waste products on farmland. However, one has to keep in mind that no regulation can be implemented without a certain amount of paper work.

When dealing with waste products on farmland typically 4 persons/bodies are involved:

- The treater (producer) of the waste product
- The user of the waste product
- The local council, and
- The regional council

In some cases another person will act as intermediary between the treater and the user e.g. transporter of the waste product.

The treater of the waste product is defined as the person/party who produces, processes or imports a waste product intended for transfer for agricultural purposes.

It will come naturally that it is the treater of the waste product who informs the user of the contents of the waste product, informs the local council of the use of a waste product and, where required, informs the regional authorities of the amount of this production.

The treater shall:

- every six months take out representative samples of the finished product
- have the samples analyzed for specified parametres
- send the results of the analyses to the regional council
- work out a declaration on the waste product
- forward the declaration to the regional and the local council (where required to the user)
- inform the local council of the amount delivered together with name and address of the user
- every year report to the regional council the quantities of each product supplied for use on farm land in the preceding year.

In annex 1 you will find a graphical description of the relations between the different parts of a waste product distribution system.

<u>User´s obligation:</u>

The user has no obligations in reporting to the authorities. His obligations are limited to legitimate storage and use of the product.

<u>How to apply for use of a waste product:</u>

There are two ways in which waste products can be applied on farmland in Denmark: One requires that the waste product is listed on a "positive list", and the other requires an application.

In the first case the waste product can be used without permission for agricultural purposes in accordance with certain provisions outlined in the Statutory Order. (See section 2 of this paper). If the product is not on the list, the user will have to send an application to the regional council in order to obtain an authorization to use the waste product. The regional council is obliged to follow the requirements set out in the Statutory Order. It is, however, allowed to lay down more strigent conditions, if necessary.

The positive list contains products which are regarded as harmless to the environment provided the conditions of the Statutory Order are followed.

The main part of waste products is expected to fall in this category. Other products which are less well known and are therefore somewhat more complicated, have to be analysed and estimated more carefully before use.

It should be noted that the regional council can only authorize the use of a waste product within the borders of the county. The treater of a waste product who wants to distribute his product for use in several counties have to file an application with all the relevant regional authorities.

Supervision:

The local council shall see to it that waste products are stored and used in accordance with the provisions of the Statutory Order wheras the regional council supervises that the provisions on control and product requirements and transfer are observed. This means that the user is supervised by the local council, the treater by the regional council.

4. SEWAGE SLUDGE QUANTITIES AND QUALITY IN DENMARK

In 1987 a total of 130,000 tons of sludge (dry solids) was produced in Denmark, corresponding to 1,000.000 tons of wet sludge (7). In table 2 the amount of nutrients in this sludge is compared to the amount of nutrients supplied by farm yard manure and fertilizers in 1987 (8).

	Tons N	Tons P
Fertilizer used	381,300	46,800
Farm yard manure	174,100	70,100
Sewage sludge	5,200	2,500

Table 2: Nutrients in fertilizer, farm yard manure and sewage sludge - 1987.

From table 2 appears that the nutrients in sludge account for approximately 1% of the total N-consumption and approximately 2% of the total P-consumption in Denmark.

Not all sludge was used in agriculture. Final disposal methods used in 1987 were:

Agriculture	37%
Landfill	28%
Incineration	27%
Composting	2%
Other methods	6%

"Other methods" include sludge farms and sludge depots, and a small portion spread in orchards and forest areas.

The quality of sludge in terms of cadmium, mercury, lead and nickel appears from table 3. The table lists the maximum concentrations of substances occurring in sludge in a given fraction. Example: In 1987 50% of the sludge quantity contained cadmium in concentrations not exceeding 117 g/tons phosphorus.

% of sludge Quantity	Heavy metal concentration (g/t phosphorus)			
	Cd	Hg	Pb	Ni
5	20	20	1,683	471
10	34	29	1,912	609
20	56	42	2,526	800
30	78	53	3,458	1,000
40	100	67	4,375	1,182
50	117	86	5,714	1,438
60	144	125	6,714	1,786
70	176	145	8,545	2,200
80	200	200	10,222	2,867
90	314	318	14,786	4,000
95	600	550	19,333	6,545

Table 3: Concentrations of heavy metals observed by fractiles of sludge quantities (7).

As will appear the heavy metal values have been related to the content of phosphorus in sludge in order to be able to calculate the sludge quantities which can meet the required limit values.

When the concentrations of heavy metals in table 3 are compared to the required quality in table 1, it has been calculated that 80% of the total sludge quantities can observe the requirements today. In 1995 new limit values for sludge will be introduced (see table 1), and it has been calculated that 55% of today's sludge can observe these requirements. Moreover, the content of phosphorus in sludge is expected to increase when waste water treatment plants are extended and equipped with phosphorus reduction.

REFERENCES

(1) MINISTRY OF THE ENVIRONMENT (1980). Statutury Order No. 139 of April 15th on Dispersion of waste water to the soil.

(2) MINISTRY OF THE ENVIRONMENT (1984). Statutury Order No. 574 of November 14th 1984 on Use of sewage sludge in agriculture.

(3) MINISTRY OF THE ENVIRONMENT (1988). Statutury Order No. 568 of September 22nd 1988 on Farm yard manure and silage etc.

(4) MINISTRY OF THE ENVIRONMENT (1989). Statutury Order No. 736 of October 26th 1989 on application of sludge, sewage and compost etc. for agricultural purposes.

(5) THE COUNCIL OF THE EUROPEAN COMMUNITIES (1986). Council Directive of June 12th 1986 on the Protection of

the environment and in particular of the soil, when sewage sludge is used in agriculture.
(6) NATURVÅRDSVERKET (1990). Allmänna Råd 90:xx. Slam från kommunala avloppsreningsverk.
(7) NATIONAL AGENCY OF ENVIRONMENTAL PROTECTION (1989). Guidelines No. 10, 1989 on Sewage sludge from municipal waste water treatment plants in 1987.
(8) AGRICULTURAL STATISTICS (1987).

ANNEX 1

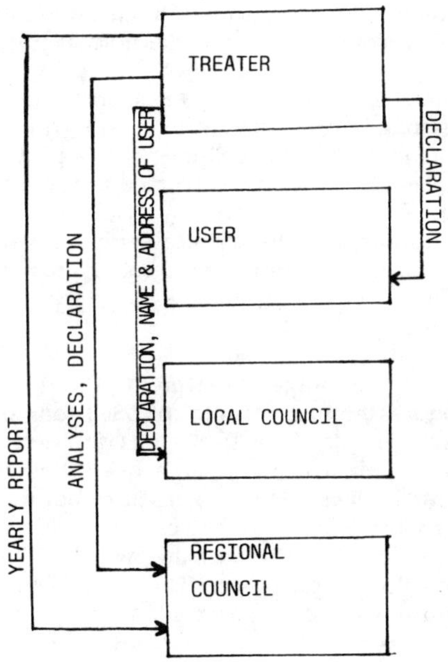

TREATER'S OBLIGATIONS:

1) REPRESENTATIVE SAMPLES
2) ANALYSES
3) DECLARATION
4) AMOUNT, NAME & ADDRESS OF USER
5) YEARLY REPORT ON SPECIFIED AMOUNTS PRODUCED

UTILIZATION OF COMPOSTED MUNICIPAL SLUDGE AS GROWTH MEDIUM OF SOME ORNAMENTAL POT FOLIAGE PLANTS

G. PETROPOULOS[1], G, CHARDAS[2], A. FATOUROS[3] and K. CHRISTOULAS

1: Agricultural Research Center of Athens
2: Soil Research Institute of Athens
3: Institute of Agricultural Science
4: Aliartos Agricultural Research Station

1. S. Venizelou, 141 23 LYCOVRYSSI, Attiki-Greece

Summary

Municipal wastewater and Septange Sludge, that have been composted prior to their use as medium was tested raw or in different mixtures with peat and perlite without any fertilization as substrate of growing in flower pots seedlings of the two species of the ornamental genus of **Ficus**, one salt sensentive and the other salt tolerant in comparison with the normal used substrate.

The addition of sludge even up to 100% has not any detrimental effect in the growth response. There were strong indications that the plant growth response was better as compared to the mixture D (50% Sludge + 25% peat + 25% perlite) especially with the salt tolerant plants, that is seemed to provide a recommended growth medium. This superiority is discussed in combination with the chemical composition and the physical properties of the applied mixtures. The **F. benjamina** plants presented a better growth development.

1. INTRODUCTION

The Municipal Wastewater and Septage Treatment Plant at "Metamorphosis" site of Athens produces annually a total of 5500 tons of sludge calculated as dry matter corresponding to about 15.000 tons dewatered sludge. Sludge from a waterpurification plant contains high quantities of organic matter (1). These plant nutrients in the sludge are a resource but this can cause environment impact at careless handling, that can be reduced through the choise of suitable way of final use (5). So being worth to be investigated on its value for agricultural purposes. The dominating environment factor regarding sludge use is the cadmium content that is the risk for health effects following the uptake of cadmium in crops (2,5). For this reason sludge should not be used on ley or pastures and in gardens for products aimed for direct consumption (2,5).

In many countries have felt it worth sludge to be tested out as horticultural substrate with different ornamental plants (1,6) as the risk of cadmium can be avoided and the demand for peat as a substrate in floriculture in the last decades had increased condiniously while its availability is decreasing and becoming more and more expensive, can be insured as these sludges were introduced as peat substitute in container media after proper composting (3).

As the quality of sludge depend on the kind of industries and the other conditions under which the purified waste water had produced (4,5) we have felt it worth to carry out this experiment in flower pots to find the behaviour of this sludge in ornamental foliage plants.

2. MATERIALS AND METHODS

Sludge from the above Municipal Plant that have been composted prior to their use as medium was tested raw or in different mixtures with peat and perlite as substrate in comparison with the normal used substate and a standard of peat and perlite in the following six treatments.

A = 60% peat + 20% garden soil + 20% manure (normal used substate)
B = 75% peat + 25% perlite (standard)
C = 50% peat + 25% sludge + 25% perlite
D = 25% peat + 50% sludge + 25% perlite
E = 75% sludge + 25% perlite
F = 100% sludge

The determaned chemical composition and the physical properties of the above six mixtures are tabulated below:

Mixtures	$CaCO_3$ %	Organic matter %	Available P OLSEN p.p.m	Available K MEQ/100GR.
A	6,9	11,5	102	2,84
B	0,5	58,0	-	0,38
C	10,8	43,5	254	0,24
D	21,1	38,0	266	0,22
E	32,5	30,0	295	0,40
F	35,1	27,0	305	0,38

Table 1: Chemical composition of the applied six mixtures

Mitxures	PH	Moisture equivalent %
A	7,4	46
B	5,0	165
C	7,0	112
D	7,1	98
E	7,0	68
F	7,1	70

Table 2: Physical properties of the applied six mixtures

Seedlings of the two species of **Ficus** namely **F. benjamina** (salt sensetive) and **F. microcarpa** (salt tolerant) were planted on November 1988 in plastic pots 20 cm diam. under Greenhouse conditions growing without any fertilization.

The experimental design was complete randomized blocks in four replications and the length of growth increase of the main and secondary shoots was determining every three months (six periods of growth).

3. RESULTS AND DISCUSSION

The average length of the growth increase of the experimental plants is presented below in Table 3.

		Mixtures						Differences between		
Period of growth	Species	A	B	C	D	E	F	Treatments	Species	Interaction
1st	1*	10	7	14	12	13	14	ns	ns	***
	2	6	9	15	26	7	6	***		
2nd	1	46	63	73	64	65	38	ns	ns	ns
	2	58	78	67	102	41	48	ns		
3rd	1	135	169	193	180	157	127	ns	*	ns
	2	96	118	118	194	96	137	ns		
4th	1	154	192	216	211	164	151	ns	*	ns
	2	115	132	130	214	128	145	ns		
5th	1	165	205	239	227	184	178	ns	*	ns
	2	122	145	141	238	156	176	ns		
6th	1	227	360	389	364	265	273	ns	***	ns
	2	188	216	193	301	229	271	ns		

* 1 = Ficus benjamina
2 = Ficus microcarpa

Table 3: Average length of growth increase in cm

As the analysis of variance of the original length of the seedlings has shown no significant difference any consideration for analysis of covariance on the experimental results is not necessary.

The data of Table 3 was analysed separately for the six periods of growth (mesurments) conducted after three, six nine, twelve, fifteen and eighteen months from the establishment of the experiment and the results are presented also in the same table.

According to the above results the addition of sludge even up to 100% has not any detrimental effect in the growth response of the experimental plants. There are strong indications that the plant growth response is better as compared to the mixture D especially with the plants of **F. microcarpa** that are characterized as salt tolerant with a high adaptability. Although the reasons for this mixture's D superiority at least over the normal used substrate are not fully understood, however the following possible mechanisms are suggested in relation with the chemical composition and the physical properties of the mixtures: (i) The high levels of nutrients ($CaCO_3$ included) and mainly of humic substances which are potential growth stimulators (8,9). (ii) The high activity of the microbial population because of the sludge content, may suggest increased growth response due to rhizosphere microorganisms (3). (iii) Mixture D offers proper watering of the plants because of its higher moisture equivalent which guarantees a higher available moisture content in combination with the presence of the perlite that helps the better aeration of substrate and it is very important (7), although the hydraulic conductivity is not determined in order to estimate the permeability of the substrate.

The **F. benjamina** plants were proved with a better growth development and were confermed its reputation as a fast active cell division plant, while the **F. microcarpa** plants showed a higher adaptability to unfavourable conditions of growth.

No odour (real or perceived) problems due to sludge were appeared.

The conclusion drawn up from this experimentation is that the studied sludge can be mixed up from 25% to 100% with the best results at 50% and it provides inexpensive and high quality substrate for growing ornamental pot foliage plants as well a solution for environmental problems of waste disposal.

REFERENCES

(1) CHEN, Y., HARAR, Y. and INBAR, Y. (1985), The use of composted Agricultural Waste as peat substitute in Horticulture, Ed. CEC, Procceedings of the fourth Intern. Symp., Held in Rome, Italy, 8-11 October 1985, page 389-392.

(2) COMMISSION OF EUROPEAN COMMUNITIES, Guidelines, Recommendation Rules and Regulations for spreading manures, slurries and sludge on Arable and Grassland, Cost Project 681, page 15-30

(3) DE BOODT, M. and VERDONCK, O. (1972), The physical properties of the Substrates used in horticulture, Act. Hort. 26: 37-44.
(4) LEE, V.S. and BARTLETT, R.J. (1976). Stimulation of plant growth by humic substances, J. Soil. Sci. Soc. Am., 40 : 876-879.
(5) MANRIQUE, A., ARROYD, I., SANZ, A.M., GARCIA DE BUSTOS, J. M. and NEBREDA, A.M. (1985), A pot experience with a high level copper sewage sludge, Ed. CEC, Proceedings of the Fourth Intern. Symp. Held in Rome, Italy, 8-11 October 1985, page 460-464.
(6) RECAMEY, P. (1985), Sludge composting and utilization in West Switzerland, Processing and use of organic sludge and linquit agricultural wastes, Ed. CEC, Proceedings of the Fourth Intern. Symp., Held in Rome, Italy, 8-11 October 1985, page 431-435.
(7) SCHNITZER, M. and POAPST, P.A. (1967). Effect of soil humic compounds on root initiation, NATURE (Lond.) 213:548-599.
(8) SWEDISH ENVIRONMENT PROTECTION BOARD, (1988), Handling of sludge from Municipal Sewage Works, Ed. CEC, SAFE AND EFFICIENT SLURRY UTILIZATION, Edited by H. VETTER, G. STEFFENS and P. L'HERMITE, page 285-299.
(9) VERDONCK, O, DE VLEESCHAUWER, D and DE BOODT, M. (1979). The use of sludge in Horticulture and Agriculture, Sym. Eur. sur la caractérisation, le conditionnement et l'utilization des boues résiduaires, Cadarache, 13-15, Fevr. 1979.

PLANT UPTAKE AND DISTRIBUTION OF EIGHT HEAVY METALS IN SOILS TREATED WITH SEWAGE SLUDGES.

H. DOMINGUES, J.M.VIEIRA E SILVA and E.MENEZES SEQUEIRA

Estação Agronómica Nacional, Departamento Pedologia,
Quinta do Marquês, 2780 Oeiras,
Portugal

Summary

In a pot trial, two soil samples derived from schists and alluvium were mixed with urban sewage sludges which presented high levels of zinc and copper. Ryegrass, the test plant cropped in these mixtures, did not presented phytotoxicity symptoms although had showed high levels of heavy metals. The results of the soil samples analysed by sequential extraction showed that after the mixture with the sludges the Cu and Zn levels increased too much in some fractions; the available forms, "exchangeable"and "organic matter", increased twice, but in the unavailable forms, "oxides" and residual fractions, the increase was greater. Other heavy metals supplied by the sludge in small amounts were distributed into different fractions of the soils. Some specific accumulation of Cu in Evora soil and Zn in Lamaçais soil were found.

1. INTRODUCTION

Arable land pollution caused by trace metals from the sewage sludges can be a problem because, more and more, the soils are amended with sludges and so a large amount of metals in soils may saturate the "specific sites" and still leave an excess of more reactive and potentially toxic forms. For that reason, to land protection it is necessary to know the forms of combination of trace metals in the sludges and their transformations in the arable soil.

Sewage sludge is a complex mixture of organic and inorganic materials and it is not easy the distint fractions to be isolated in order to localise their trace metals. Therefore, some workers used the sludges incorporated into the arable soils in experiments to study the plant uptake and they also used some "specific extractants" to verify the heavy metals distribution in different soil fractions.

In Portugal, the application of sludges in arable land is at the beginning, and so these studies are most suitable to advise the farmers about the rates of sewage sludges to be applied in different soils.

The aim of this paper is to know about the plant uptake, the distribution and the accumulation of heavy metals in two arable soils treated with sludges high in Zn and Cu contents.

2. MATERIAL AND METHODS

Two composite soil samples derived from red shists (Evora) and alluvium (Lamaçais-Belmonte) were taken at 0-25 cm depth and they were air-dried and sieved at 2 mm.

A pot experiment was laid out in a greenhouse with the soil samples mixed with three rates of different urban sewage sludges. The quantities of sludges were calculated from the Zn equivalent (Eq = Zn + 2Cu + 8Ni), and they were 0 (S0), 39,5(S1) and 69 (S2) t/ha for Evora sludge and 0 (S0), 39(S1) and 78(S2) t/ha for Belmont sludge.

The soil samples taken at the begining of the experiment were physically and chemically (1) analyzed. The clay fraction was analyzed by X-ray diffraction (2).

The soil samples taken at the end of the experiment were analized by chemical sequential extraction (3) to determine the distribution of heavy metals in the different soil fractions: "exch"(exchangeable),"O.M."(organic matter), "MnOx" (manganese oxides), "AFeOx"(amorphous iron oxides) and "CFeOx" (crystalline iron oxides).

The amount of heavy metals in the sludges and in the soil samples mixed with sludges were determined by aqua regia extraction with reflux (4) and measured by atomic absorption.

In the pot experiment ryegrass was used as the test plant and it was submited for three cuts when it reached 20 cm height. The dry matter was determined and the plants were analyzed for heavy metals content (1).

3. RESULTS

3.1 - The results of the soil samples analyses (Table 1) show that:

	Lamaçais soil	Evora soil
Sand (%)	67.3	62.0
Silt (%)	21.4	13.9
Clay (%)	11.3	24.1
pH (H2O)	4.9	5.1
O.M.(%)	1.7	0.9
CEC (me/100g)	10.2	11.0
Total Fe (%)	2.7	4.6
" Cd (mg/Kg)	2.0	3.0
" Co "	9.0	13.5
" Cr "	29.0	88.5
" Cu "	12.0	86.0
" Mn "	400.0	520.0
" Ni "	6.5	20.0
" Pb "	46.0	44.0
" Zn "	83.0	58.5

Table 1: Some physical and chemical soil characteristics

- they are acid soils (pH≈5), the soil from Lamaçais has the highest organic matter content (1.7%) and the soil from Evora exhibits the highest percentage of clay (24,1%);

- in Evora soil, the total amount of some heavy metals (Cu-86.0, Cr-88.5 and Ni-20.0 mg/Kg) are higher than in Lamaçais soil, while Zn content (83.0 mg/Kg) is higher in Lamaçais soil;the amounts of other heavy metals are similar in soil samples.

3.2 - The results of the sewage sludges analyses (Table 2) show that Evora sludge contains higher amounts of total Cu-975 mg/Kg, Cr-67.5 mg/Kg and Ni-116 mg/Kg, whereas Belmonte sludge contains higher total Zn 2375 mg/Kg and Pb-260mg/Kg.

Elements		Belmonte sludge	Evora sludge
Fe	(mg/Kg)	1300.0	1.2
Cd	"	4.0	3.0
Co	"	4.5	6.0
Cr	"	46.5	67.5
Cu	"	249.5	975.0
Mn	"	176.5	182.0
Ni	"	27.0	116.0
Pb	"	260.0	240.0
Zn	"	2375.0	850.0

Table 2: Total heavy metals content in the sewage sludges

3.3 - From the results of the analysis of variance of the effect of the treatments in the plants uptake (Table 3) we may conclude that:

	Lamaçais soil			Evora soil		
	Sludge rates			Sludge rates		
Elements	S0	S1	S2	S0	S1	S2
Co	0.68	7.48	6.78	1.85	1.91	3.33
Cr	0.33 b	6.40 a	9.49 a	0.96 b	3.18 a	2.95 a
Cu	15.69 c	63.73 b	98.25 a	16.26 c	49.78 b	76.60 a
Fe	410 b	1298 a	1192 a	735 b	1232 a	1158 a
Mn	365 b	493 a	345 b	352 b	635 a	582 a
Ni	5.08 b	35.74 a	43.15 a	6.98 b	24.49 a	39.94 a
Pb	16.55	34.89	60.71	20.19	25.50	27.38
Zn	71.00 c	1480 b	2389 a	67.30 c	497 b	933 a

Note: The values with different letters mean that are statistical differents.

Table 3: Some heavy metals plant uptake (μg/pot)

- the two rates of sludges (S1 and S2) affected the uptake of Cu and Zn, in the two soils, but did not influence significcately the uptake of Co and Pb probably due to the low amounts of sludges added to the soils, to the low number of replicates in the experiment and to the analytical errors.

- either in Belmonte or Evora soils the plants uptake of Cr, Ni, Fe and Mn were affected by the first rate of sludges (S1), but Cr, Ni and Fe were not statistically different in both S1 and S2 levels of sludges; in Lamaçais soil there was no statistical difference between the Mn uptake in both S0 and S2 treatments and in Evora soil there was no difference between the Mn uptake in both S1 and S2 treatments .

3.4 - The results of the heavy metals extracted by aqua regia in the soil samples at the end of the experiment (Table 4) show that Cu, Zn and Ni are accumulated in the soils, mainly Cu in Evora soil and Zn in Lamaçais soil when higher rates of sludges were applied.

	Lamaçais soil			Evora soil		
	Sludge rates			Sludge rates		
Elements	S0	S1	S2	S0	S1	S2
Co(mg/Kg)	4.5	5.0	5.0	12.0	11.0	8.0
Cr "	15.0	15.5	14.5	33.0	34.0	36.0
Cu "	11.0	13.0	16.0	23.0	33.0	42.0
Fe "	23000	22900	21000	40700	41800	40900
Mn "	285	280	290	398	381	380
Ni "	9.0	9.5	11.0	17.0	17.0	22.0
Pb "	31.5	28.5	31.0	30.0	30.0	40.0
Zn	81.0	97.5	121.0	51.0	52.0	60.0

Table 4: Heavy metals content extracted by aqua regia

3.5 - The results of some heavy metals separated by sequential extraction are presented one by one.

Copper - The results (Table 5) show the following:

		Sequential Fractions					
Soils	Sludge rates	Exch	O.M.	MnOx	AFeOx	CFeOx	Sum
		mg/Kg					
	S0	0.24	1.40	2.50	3.50	3.40	11.0
Évora	S1	0.65	3.50	5.60	7.40	4.50	21.6
	S2	0.75	5.60	6.30	9.00	5.00	26.6
	S0	0.20	3.40	1.75	3.90	2.20	11.5
Lamaçais	S1	0.25	4.50	1.80	4.10	2.10	12.7
	S2	0.35	6.70	2.80	4.50	2.15	16.5

Table 5: The copper content in the soil fractions

- in the soil samples not treated (S0), Cu occured mainly in "AFeOx,"CFeOx", "MnOx" and "O.M."fractions of both samples;

- by the treatment of the soil samples with sludges, two rates of Cu were added [Evora: 11.9 (S1) and 23.8 (S2) mg/Kg; Lamaçais 3.4 (S1) and 6.8 (S2) mg/Kg];

- in the soil samples treated (S1) and (S2), Cu added by sludges occured mainly in the "O.M.", "MnOx" and "AFeOx" fractions of both soil samples.

Zinc - The results (Table 6) show that:

Soils	Sludge rates	Exch	O.M.	MnOx	AFeOx	CFeOx	Sum
		Sequential Fractions mg/Kg					
	S0	0.24	0.07	0.40	0.70	1.35	2.8
Evora	S1	4.10	1.00	2.10	1.85	1.90	11.0
	S2	6.90	2.30	4.50	3.00	1.90	18.6
	S0	0.70	1.70	0.85	1.80	3.20	8.3
Lamaçais	S1	9.00	12.50	4.10	4.30	4.00	33.9
	S2	24.30	30.00	9.10	8.20	5.00	76.6

Table 6: The zinc content in the soil fractions

- in the soil samples not treated, Zn occured mainly in the "CFeOx" and "AFeOx" fractions of both soil samples and also in the "O.M." fraction of Lamaçais soil;

- by the treatment of the soil samples with sludges, two rates of Zn were added [Evora: 10.4 (S1) and 20.8 (S2) mg/Kg; Lamaçais: 32.3 (S1) and 64.6 (S2) mg/Kg];

- in the soil samples treated, Zn added by sludges showed a remarkable increase in "Exch","O.M." and "MnOx" fractions of both soil samples, but more considerable in Lamaçais soil.

Iron - The results of this metal (Table 7) show that:

Soils	Sludge rates	Exch	O.M.	MnOx	AFeOx	CFeOx	Sum
		Sequential Fractions mg/Kg					
	S_0	0.15	3.80	40.0	3570	7980	11594
Evora	S_1	0.15	3.70	40.5	3700	8400	12144
	S_2	0.16	3.90	40.0	3700	8350	12094
	S_0	0.50	9.00	73.5	3800	2650	6533
Lamaçais	S_1	0.70	10.00	75.0	4000	2650	6736
	S_2	0.80	14.00	80.0	3980	2750	6825

Table 7: The iron content in the soil fractions

- in soil samples not treated Fe occured mainly in the "CFeOx" and "AFeOx" fractions of both soil samples;

- by the treatment of the soil samples with sludges two rates of Fe were added [Evora: 149 (S1) and 298 (S2) mg/Kg; Lamaçais: 179 (S1) and 358 (S2) mg/Kg];

- in the soil samples treated (S1) and (S2), Fe added by sludges occured mainly in the "AFeOx" and "CFeOx" fractions of both soil samples and in the "O.M." fraction of Lamaçais soil.

Manganese - The results of this metal (Table 8) show the following:

Soils	Sludge rates	Exch	O.M.	MnOx	AFeOx	CFeOx	Sum
		Sequential Fractions mg/Kg					
Evora	S_0	20.0	11.0	59.0	29.0	6.8	125.8
	S_1	19.0	16.0	57.0	28.0	7.0	117.0
	S_2	19.0	16.0	56.0	28.0	7.8	126.8
Lamaçais	S_0	4.0	10.0	11.0	130.0	10.0	165.0
	S_1	5.5	11.0	10.0	131.0	10.0	167.5
	S_2	4.5	11.5	10.5	125.0	9.5	161.0

Table 8: The manganese content in the soil fractions

- in the soil samples not treated Mn occured mainly in the "MnOx" and "AFeOx" fractions of Evora and in the "AFeOx" fraction of Lamaçais soil;

- by the treatment of the soil samples with sludges two rates of Mn were added [Evora: 2.2 (S1) and 4.4 (S2) mg/Kg; Lamaçais 2.4 (S1) and 4.8 (S2) mg/Kg];

- important changes in the sequential fractions did not appear at the soil samples treated with sludges.

Lead - The results of this metal (Table 9) show that:

Soils	Sludge rates	Exch	O.M.	MnOx	AFeOx	CFeOx	Sum
		Sequential Fractions mg/Kg					
Evora	S_0	n.d.	4.50	1.00	4.20	2.80	12.5
	S_1	n.d.	4.80	1.00	4.20	3.80	13.8
	S_2	n.d.	4.80	1.50	3.80	4.00	14.1
Lamaçais	S_0	n.d.	6.10	1.00	4.20	2.80	14.1
	S_1	n.d.	7.30	1.00	4.40	3.00	15.7
	S_2	n.d.	7.30	1.00	4.30	2.50	15.1

Table 9: The lead content in the soil fractions

- in the soil samples not treated Pb occured mainly in the "O.M." and "AFeOx" fractions of both soils;

- by the treatment of the soil samples with sludges two rates of Pb were added [Evora: 2.9 (S1) and 5.9 (S2) mg/Kg; Lamaçais 3.5 (S1) and 7.0 (S2) mg/Kg];

- in the soil samples treated (S1) and (S2) the Pb added by sludges did not occured in the "Exch" fraction, but was lightly associated with the"O.M." fraction of Lamaçais soil.

Niquel - The distribution of this metal show (Table 10) the following:

		Sequential Fractions					
Soils	Sludge rates	Exch	O.M.	MnOx	AFeOx	CFeOx	Sum
		mg/Kg					
	S_0	0.25	n.d.	0.15	n.d.	n.d.	0.40
Evora	S_1	1.00	0.20	0.40	n.d.	n.d.	1.60
	S_2	1.40	0.50	0.60	n.d.	n.d.	2.50
	S_0	0.10	0.12	0.12	n.d.	0.30	0.64
Lamaçais	S_1	0.15	0.15	0.12	n.d.	0.30	0.72
	S_2	0.25	0.15	0.20	n.d.	0.30	0.90

Table 10: The niquel content in the soil fractions

- in the soil samples not treated Ni occured mainly in the "Exch" and "MnOx" fractions of Evora soil and in the "CFeOx", "Exch", "O.M." and "MnOx" fractions of Lamaçais soil;

- by the treatment of the soil samples with sludges two rates of Ni were added in each soil [Evora: 1.4 (S1) and 2.8 (S2) mg/Kg; Lamaçais 0.4 (S1) and 0.7 (S2) mg/Kg];

- in the soil samples treated (S1) and (S2) the Ni added occured preferentially in the "Exch", "O.M." and "MnOx" fractions in both soil samples.

Cobalt and Cromium - In soil samples not treated Co occured mainly in the "MnOx" fraction of Evora soil(1.20 mg/Kg) and "AFeOx" fraction of Lamaçais soil (1.00 mg/Kg) and Cr occured mainly in the "CFeOx" fraction of both soils (Evora- 8.00 mg/Kg and Lamaçais-1.50 mg/Kg). In soil samples treated with sludges low rates of Cr were added [Evora: 0.82 and 1.44 mg/Kg; Lamaçais: 0.7 and 1.3 mg/Kg] and traces of Co. No significante increase in Co and Cr contents in the soil fractions was observed.

4. DISCUSSION

The sewage sludges only present high levels of Zn and Cu and the reygrass fertilized with them does not exhibit phytotoxicity symptoms although shows a high level of heavy metals.

The sum of the metals contained in the different sequential fractions comparing with the total contents extracted by aqua regia (Table 4) shows that both values are similar in Lamaçais soil, indicating that all Cu (Table 5) in Lamaçais soil (including Cu added by sludges), occured in the different sequential fractions analized, unlike the Evora soil where this metal is occured mainly in the residual fractions, sand, silt and clay. (See Cu values in Table 1)

Compairing the Cu content in both soils we verified that the Cu added by sludges was lower in Lamaçais soil than in Evora soil, but the plant uptake (Table 3) was higher in Lamaçais soil confirming the easiest availability of Cu in this soil. This can be due to the different forms of combination of Cu in both soils and sludges and also to the content (Table 1) and mineralogical composition of clay fraction in both soils [Evora: kaolinite (40-45%), illite (20-25%), montmorillonite (15-20%); Lamaçais: kaolinite (35-40), illite (15-20), vermiculite (15-20%)].

Comparing the total heavy metals content (Table 4) and the heavy metals sum in the different sequential fractions in soils treated with sludges we can conclude that there was Zn and Cu specific adsorption in Lamaçais and Evora soils, respectivelly.

At moment, the remaining heavy metals seems do not present great problems, although a light accumulation of Ni in both soils, and of Cr and Pb in Évora soil has been observed.

5 - CONCLUSIONS

The high levels of Cu and Zn in the sludges did not cause phytotoxicity symptoms.

The accumulation of heavy metals in the soils treated with sludges, saturing the "specific sites", did not exceed the maximum level of 86/278/CEE Directive (5), even in case of Evora soil with high heavy metals contents.

These sludges can be used as fertilizers in adequate rates, but it is necessary to supervise periodically the metal levels in the sludges and in the soil samples using the sequential extraction to identify the forms of combination of the metals in soils.

6 - ACKNOWLEDGEMENTS

The autors appreciate the laboratory assistence provide by R.Monteiro, M.Ema Faustino, M.Celeste Campos, M.Fernanda Freire and Ana M.Santos and the review of this paper by Corina Carranca.

7 - REFERENCES

(1) SILVA, A.A., ALVIM, A.J.S. and SANTOS, M.J.(1975). Métodos

de análise de solos, plantas e águas. Pedologia, Oeiras, 10: 153-544.

(2) ROBERT, M. and TESSIER, D. (1974). Méthode de préparation des argiles des sols pour les études minéralogiques. Ann. agron., 25: 859-882.

(3) SHUMAN, L.M. (1980). Fractionation method for soil micro-elements, Soil Sci, 140: 11-22.

(4) Din 38414 part 7 (1983). German standart methods for the examination of water, waste water, sludges and sediments. Deutsche Norm: 1-3.

(5) 86/278/CEE Directive - Directiva do Conselho relativa à protecção do ambiente, em especial solos, na utilização agrícola de lamas de depuração. Jornal Oficial das Comunidades Europeias Nº1, 181/6 de 86/7/4: 6-12.

ANAEROBIC -AEROBIC TREATMENT OF OLIVE-OIL PROCESSING WASTES

D. DALIS and K. ANAGNOSTIDIS

UNIVERSITY OF ATHENS, DEPARTMENT OF BIOLOGY
SECTION OF ECOLOGY AND SYSTEMATICS
Panepistimiopolis, 157 84 ATHENS, GREECE

Abstract

Greece, ranks third among the olive-oil producing countries in the world after Spain and Italy, yielding 230.000 - 280.000 tn of olive-oil per year depending to climatological conditions (15% of the world olive-oil production). Approximately 3.500 olive-oil mills are in operation in Greece which treat an amount of 1.275.000 tn of olives per year, discharging, at the same time, equal amount of liquid wastes

Concerning the organic load, the olive-oil mill wastes, together with Vinasses, are considered the most heavily polluted agricultural - industrial wastes. (It is indicative that the wastes of a medium potential olive-oil mill i.e. 50 m^3/per day, BOD = 40 g/l are equavalent with urban wastes of 30.000 inhabitants).

The waste treatment is becoming hard because of the physicochemical properties of the wastes (low pH values, high content in phenolic compounds, fat acids and pigments, phytotoxic properties), the technical - economical infrastructure of the oil-mills (of average economic efficiency in the private and/or cooperative, and agricultural-industrials units of seasonal operation) as well as their sparse geographic distribution.

A basic goal of the research programme which is conducted in the Section of Ecology and Systematics of the University of Athens is the development of methodology on the anaerobic/aerobic treatment of oil-olive wastes. Research will be focused on:

- Efficient reduction of the organic load in the olive-oil wastes, so that their discharge will not change the environmental status substantially.
- Exploitation of the "rich" elements included in the wastes and especially of the energy produced during the anaerobic treatment of methane.

The up to date results- in laboratory conditions - show that the anaerobic treatment of olive-oil wastes is obtainable. The original wastes, without any pretreatment (apart from the adjustment of the ration C/N) i.e. without additional CaO for the pH adjustment of removal of the oil layer etc., go directly to the anaerobic fermentation reservoir. After a rest of 5 days, 75-80% of the initial organic load is decomposed and 1,2m^3 of biogas/m^3 of establishment volume is produced (80% CH_4, 20% CO_2).

Part of the produced energy could be used for the needs of the establishment itself and the rest could be disposed for the operational needs of glass-houses, piggeries or other units adjacent to the anaerobic establishment. In this way, the developmental initiatives of the area could be enhanced and the technological equipment could be enriched. Parallel experiments have been initiated using a simple soil filter, through which a further reduction of the organic load (90-95%) has been achieved accompanied by both a significant decolourization of the liquid and an indication of fertilizing capabilities.

The above encouraging data would be testified in pilot-plant establishments which are being in operation from the olive-gathering period of 1989-90. In case of successful results, the applicability of the method is guaranteed.

ANAEROBIC DIGESTION OF MIXTURES OF SEPTAGE AND SEWAGE SLUDGES

A. D. ANDREADAKIS, V. KAGELARIS and P. PAPAKOSTA

Division of Water Resources, National Technical University of Athens

Summary

The Metamorphosis Treatment Plant receives 7,500 $m^3 d^{-1}$ of septage and 11000 $m^3 d^{-1}$ of domestic sewage contributed by approximately 700,000 people. The mixture of primary and secondary sludge is treated in anaerobic digestors operating under residence times in the range 25-30 days. Percentage organic matter destruction averaged at about 40% which is lower than the anticipated design figure of 50% at a residence time of 14 days. This deviation is attributed to the low organic content of the septage sludge (around 50%) which constitutes about 80% of the total sludge mass fed to the digestors. Gas production was on average 0.35 m^3/kg vs fed but the gas production per vs destroyed varied with the digestion efficiency. The need to maintain a high retention time in the digestors in order to achieve sludge stabilisation creates the need for efficient future thickening of the surplus biological sludge prior to anaerobic digestion.

1. INTRODUCTION

The Greater Athens area has a population of about 3,200,000, most of which is served by the existing sewerage system. A main sewer conveys approximately 0.7-0.8 x 10^6 $m^3 d^{-1}$ of sewage to the north of Pireaus and discharges it without treatment into the Saronicos Gulf at a site called Acrokeramos.

The existing sewerage system serves about 75% of the population. The remaining 25% uses cesspools and septic tanks which produce approximately 7,000-8,000 m^3 of septage per day. This septage is collected and subjected to treatment at the Metamorphosis Treatment Plant prior to discharge to a nearby torrent.

The Metamorphosis Treatment Plant is designed to treat 24,000 $m^3 d^{-1}$ of septage and 20,000 $m^3 d^{-1}$ of domestic sewage from the nearby areas. Current septage flow is about 7,500 $m^3 d^{-1}$ and the sewage flow is about 11,000 $m^3 d^{-1}$. Septage is subjected to preliminary treatment including chemically assisted sedimentation in circular tanks (with addition of lime) and subsequent biological treatment by activated sludge. Sewage treatment consists of screening, degritting, primary sedimentation in rectangular basins and subsequent biological treatment applied to both sewage and primary treated septage. Primary and surplus secondary sludge is subjected to anaerobic digestion in primary and secondary completely mixed digestors followed by dewatering in belt filter presses.

The plant was commissioned in August 1984 and for two years received only septage subjected to primary treatment. The biological treatment stage started in March 1986 when domestic sewage was first introduced to the plant. This paper attempts to evaluate the performance of the anaerobic digestion of the sludge over the last six years with emphasis on the data collected during the period 1987-1989.

2. DIGESTION PERFORMANCE

The sources of sludge fed to the digestors are the following:

a. Sludge from the septage clarifiers. This sludge is the result of chemically assisted sedimentation with addition of lime. The daily quantity of the sludge fed to the digestors is approximately 20 tn, which corresponds to a volume of about 280 $m^3.day^{-1}$ with a concentration of around 7%. The volatile matter portion of the sludge is of the order of 50%, which is lower than the organic solids portion of the septage suspended solids (about 55-60%) due to the addition of lime.

b. Primary and biological sludge from the sewage primary sedimentation tanks. This sludge consists of about 2.5 tn per day of primary sewage sludge and approximately 2.0-2.5 tn per day of surplus biological sludge which is transferred to the primary sedimentation tanks from the final clarifiers. The concentration of the mixture of primary and secondary sludge varies between 1.2%-2% depending on the SVI values of the secondary sludge. Thus the daily volume fed to the digestors lies in the range 200-400 m^3. The volatile matter portion of the sludge is between 70-80%.

The two primary anaerobic digestors have a total volume of 15800 m^3. The design ultimate sludge flow rate of 1,000 $m^3 d^{-1}$ corresponds to a retention time of around 16 days and the anticipated digestion efficiency with respect to volatile solids reduction was of the order of 50%. These design figures were based on experience from conventional sewage sludge digestors which in view of the operational data do not appear to be realistic.

The monthly average digestion efficiency values (in terms of % volatile solids reduction) over the period November 1984-September 1989 and the corresponding retention times are presented in Table 1. Figure 1 shows weekly average efficiencies over the period January 1987-September 1989.

The whole period can be divided into two distinct subperiods. From November 1984 until August 1986 only septage sludge was introduced into the digestors resulting in high retention times (on average about 50 days). The average digestion efficiency over this period was about 42%, clearly lower than the anticipated 50%. This can be attributed to the low initial organic matter of the septage solids (in the

Table 1: Retention time and digestion efficiency

Month	Ret. Time	% VS Reduction	Month	Ret. Time	% VS Reduction
Nov. 1984	29	35.0	May	22	40.9
Dec.	29	30.0	June	23	40.5
Jan. 1985	29	28.6	July	32	40.7
Feb.	31	33.0	Aug.	33	33.5
Mar.	38	38.0	Sept.	37	45.2
Apr.	50	50.1	Oct.	-	-
May	33	37.5	Nov.	-	-
June	32	35.4	Dec.	-	-
July	35	45.6	Jan. 1988	33	42.1
Aug.	44	51.4	Feb.	28	35.5
Sept.	51	50.0	Mar.	18	37.0
Oct.	51	41.0	April	24	-
Nov.	55	41.0	May	32	40.5
Dec.	59	40.8	June	23	37.5
Jan. 1986	56	38.0	July	-	-
Feb.	55	46.0	Aug.	29	35.2
Mar.	83	50.0	Sept.	33	42.2
April	83	50.0	Oct.	34	41.3
May	85	62.0	Nov.	35	37.5
June	94	40.5	Dec.	28	35.9
July	52	40.0	Jan. 1989	28	45.0
Aug.	39	30.0	Feb.	30	43.5
Sept.	38	17.0	Mar.	30	41.8
Oct.	28	30.0	April	29	38.9
Nov.	25	30.0	May	29	33.3
Dec.	35	43.0	June	28	33.8
Jan. 1987	39	39.1	July	27	43.4
Feb.	31	46.6	Aug.	31	43.5
Mar.	45	41.2	Sept.	31	42.5
Apr.	23	49.7			

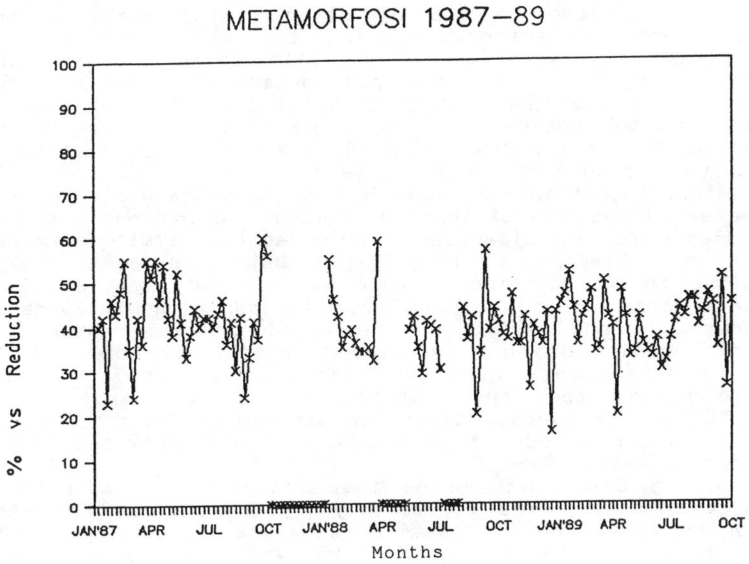

Fig 1 - Digestion efficiency vs time

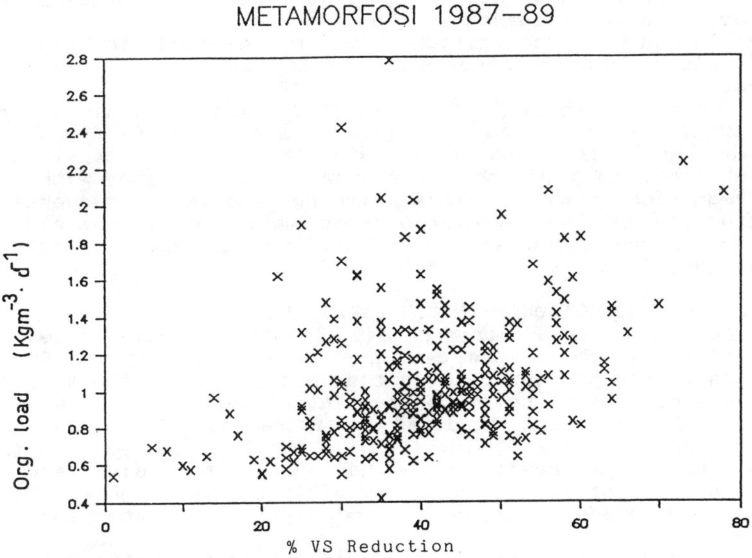

Fig 2 - Relationship between digestion efficiency and organic loading

range 55-60%) due to the digestion taking place in the cesspools or in the septic tanks. It is well known (4) that the percentage reduction of volatile solids by digestion depends in part upon the original content of volatile matter in the sludge and that for an original content of 55-60% the sludge can be considered to be digested after a percentage reduction of about 38-42% (5). In view of this the attained average efficiency of 42% is satisfactory.

Since August 1986 in addition to the septage sludge, the above mentioned quantitites of sewage and secondary sludge were fed into the digestors. This led to a small increment of the mass loading, by some 20-25%, but at the same time it resulted in a significant reduction of the retention time, which dropped, from about 50 days to approximately 30 days. Although the new sludge was richer in organic matter, the organic loading did not increase significantly (only by 25-30%) due to the small quantities (in terms of mass) that this sludge represented, and remained low in the range 1-2 kg vs.$m^{-3}d^{-1}$. The average digestion efficiency did not improve and in fact was reduced to around 40%, obviously due to the reduced retention time.

During this last period the efficiency varied between 33.50-49.70% in terms of monthly average values, and the range was even wider in terms of weekly average values. The main factors related to the observed efficiencies were loading variations and operation disturbances due to quantitative and qualitative shocks which resulted in high volatile acid concentrations and low pH values (2). Figure 2 shows clearly that higher efficiencies were attained at higher organic loadings while figure 3 shows that high volatile acid concentrations, which resulted in reduced efficiencies, correlated well with pH values in the digestors.

During the whole period, gas production was satisfactory and the production rate was equal to about 0.35m^3 of gas per kg vs fed. Gas production rate per kg vs destroyed was variable and as indicated by figure 4, it increased with low vs reduction rates. This finding implies a sequential destruction of the various organic materials in the sludge and is in agreement with similar findings reported in the literature (1,3).

3. FUTURE PROJECTIONS - RECOMMENDATIONS

Currently the Metamorphosis Treatment Plant receives approximately 45% of the design flow and only about 35% of the design organic load. Over the last six years the septage flow has been practically constant at around 7,500-8,000 m^3d^{-1} and it is not expected to increase. Therefore the design capacity of 24,000 m^3d^{-1} will not be realised. On the other hand the sewage flow has been increasing and is expected to continue to do so, possibly at even higher rates, due to the gradual completion and expansion of the sewerage system.

It is possible that in the near future the Metamorphosis

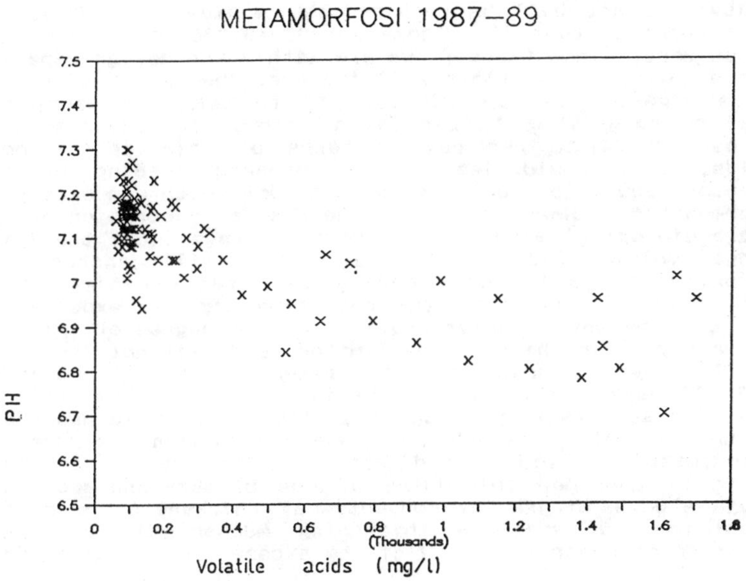

Fig 3 - Relationship between volatile acids concentration and PH in the digestor

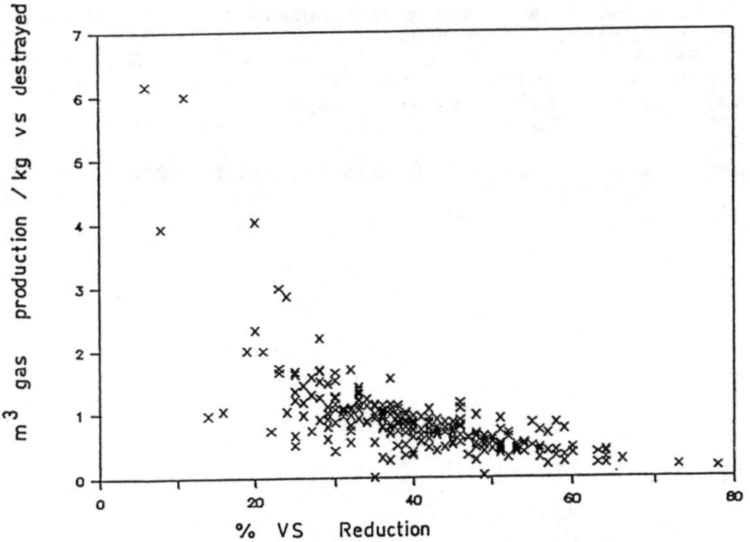

Fig 4 - Relationship between digestion efficiency and specific gas production rate

Treatment Plant will be required to receive, in addition to the present septage flow, some 35,000-40,000 m^3d^{-1} of sewage if not more, since these flows are within the design capacity of the plant. Such flows will increase the amount of sludge to be digested by approximately 10 tn per day reaching a total of roughly 35 tn per day in terms of total suspended solids and 21 tn per day in terms of volatile suspended solids. This would lead to an organic loading of the digestors equal to about 1.4 kg vs $m^{-3}d^{-1}$ which can easily be accommodated. However, given the low concentration of the sewage-biological sludge which is on average 1.5%, the daily sludge volume fed to the digestors will increase to approximatly 1,300 m^3d^{-1} leading to detention time of 12 days. This reduction of the detention time is expected to lead to substantial deterioration of the degree of volatile solids reduction which may be limited to 30% if not less.

In order to maintain a high retention time of the order of 25-30 days, which, on the basis of current observations, result in satisfactory sludge stabilisation, it is necessary to thicken the mixture of sewage/biological sludge to approximately 5% prior to digestion. This can be achieved either by combined thickening of the primary and secondary sewage sludge in gravity conventional thickeners, or by the adoption of a suitable thickening mechanisms (e.g. air flotation or centrifugation) of the excess biological sludge.

REFERENCES

(1) ANDREADAKIS, A.D., (1983). Anaerobic Digestion of Piggery Wastes. Sc. Papers N.T.U. Vol 7, pp. 15-29.

(2) KAGELARIS, V., PAPAKOSTA, P. (1990). Evaluation of Performance Data of the Metamorphosis Treatment Plant. Project Report (in Greek).

(3) PFEFFER, J.T., LEITER M., WORLAND, J.R., (1967). Population Dynamics in Anaerobic Digestion. J. Wat. Poll. Control Fed., Vol 39, p 1305.

(4) WPCF (1976). Operation of Wastewater Treatment Plants. MOP/11

(5) WPCF (1968). Anaerobic Sludge Digestion. MOP/16.

LIQUID PIGGERY WASTES APPLICATION ON COTTON CULTIVATION

J. AKRIVOS, Agricultural Research Station of Vardates Phthiotis, C. PASCHALIDIS, S. CHRISTIDOU, Athens Soil Institute

Summary

In a cotton fertilization experiment with liquid stabilized (under anaerobic physical conditions) piggery wastes, and in combination with anorganic chemical fertilizers, which has been realized at Vardates Agric. Research Station field, and for the fourth (4th) continuous year, period 89, firstly had been examined the arable land waste application potential and has been studied the effect of five fertilization levels (0, 05, 10, 15, and 20 $kgN/1000m^2$) on cotton yield, on N, P, K, Ca and Mg leaf concentration and soil nutrient enrichment. It has been studied also the waste residual effect.

A positive piggery waste application effect has been found on 1. cotton cultivation with yield increase and nitrogen leaf concentration increase and 2. on soil enrichement with nitrates, as the waste doses were increased.

Finally, the combination of lower waste doses with chemical fertilizers are maybe the most effective applications in contrast to the higher manure doses, where it has been noticed there is increased soil nitrate concentrations, with a hazard in the long term, because of its leaching; unfavourable environmental impact creation.

INTRODUCTION

In intensive agriculture heavy demands are made on the soil as a factor in plant and mineral production. Increasing mechanization, ample use of chemicals, increasingly narrow crop rotations, high yields, and increasing pig livestock densities with mostly inadequate waste treatment (environmental pollution hazard) require close attention to the maintenance of soil productivity, both in the short and the long term.

To improve and maintain soil fertility, among other things, various fertilizers are used. Organic manure, as liquid piggery wastes, increase crop yields when these are applied to soils that already received optimum amounts of fertilizers NPK. This "specific" manure effect on crop yields is partly chemical in nature and therefore it should be possible to produce the same results with more extensive fertilization, with more complete mineral fertilizers and the remainder effects finds its origin in physical, physiological, biological and synergistic processes (2). These factors make regular application of organic manure important in the maintenance of soil fertility, but the effect may vary strongly, depending on soil fertility, weather conditions and the care taken in choosing the time and method of manure application. Incorrect ways of application may lead to negative effects. Also, the "specific" effect occurs in the year of application as well as in subsequent years.

Liquid manure doesn't affect essentially the physical properties and organic matter of soil as its organic matter is only about 1-2%, but it contains macronutrients in sufficient amounts, and in the pore available form in comparison to other manure types.

Nitrogen with the greatest effect on cotton yields (6), is contained in ammonium form about 70-80% and only 20-30% in organic nitrogen. Similarly, Phosphorous with 50% availability is contained in organic and inorganic form and finally pig manure's potassium is in available for at about 90%.

The objective of this field study was the liquid pig wastes effect assessment combined with chemical fertilizers on cotton cultivation and to determine the optimum pig waste loading rate for maximum cotton yield without creating environmental problems.

METHODS AND MATERIALS

In 1986 permanent experimental cotton fertilization was installed on Vardates Research Station field on silty clay soil with 5-8% $CaCO^3$, 2,2% organic matter, PH 7,7 and sufficient productivity and fertility. Anaerobically digested (physically in basin) liquid piggery wastes were applied to plots in '86, '87, '88 and in '89 then were applied to half plots in order to examine the waste residual effect of the rest of the half plots. The experimental design was in factorial order where each plot had $40m^2$ area and was cultivated with the cotton variety "Sinthos 80".

The treatments which were studied were:
1. Martyr (without fertilization), 2. Fertilization with chemical fertilizers N10P8K4 (full dose), 3. Fertilization with chemical fertilizers N5P4K2 (1/2 dose), 4. 1/2 dose of Ch. fertilizers + manure - N5, 5. 1/2 dose of Ch. fertilizers + manure - N10, 6. Manure - N10, 7. Manure - N5, 8. Full dose of Ch. fertilizers + manure - N5, 9. Full dose of Ch. fertilizers + manure - N10.

Soil samples as with leaf samples were taken for lab analysis. Methodology was as below:
Mechanical analysis (BOUYOUKOU method). Organic matter (WALLKEY-BLACK method). T. Nitrogen (KJELDAHL method). Nitrate-nitrogen (Phenolodisoulphonic acid method). Phosphorous (OLSEN method). Phosphorous on leaves (CHAPMAN and PRATT method diverted by JACKSON). Potassium flame photometrically calcium and magnesium with Perkin & Elmer Atomic absorption.

RESULTS AND DISCUSSION

Figure 1 shows cotton yields related to Nitrogen level fertilization with chemical fertilizers and liquid piggery wastes respectively for '89 period (fourth continuous year), while Figure 2 shows the residual effect of a three-year application of liquid piggery wastes combined with chemical fertilizers, on cotton yields for the '89 period.

Table 1 gives the percentage of Nitrogen, phosphorous, potassium, calcium and magnesium cotton leaf concentration in dry weight matter per treatment, for '89 period, fifty days before cotton collection (end of August) on new, mature leaves (3rd or 4rth leaf from the top) and for both, fertilized and not fertilized experimental plots.

Liquid piggery waste components were as below:
pH: 7.5, E.C at 25 C: 4.6 mmhos/cm, total solids: 1-2%, $NO3^-$: 18ppm, $NH4^+$: 750 ppm, N: 0,12%, p: 35ppm, K^+: 220ppm, Ca^{++}: 140ppm, Mg^{++}: 55ppm, Na^+: 35ppm.

From the results it has been ascertained that the liquid pig waste had positive effects on cotton yields, which was higher as its doses were increased.

Of course, in the first two years there were no significant differentiations on cotton yields as there were in the last two years because the experimental field had adequate soil productivity and fertility. Also, manure application advantage has appeared in subsequent years because of its residential action. The pig manure specific effect of all levels gave a superiority of about 50 kg yield above chemical fertilizers with an ascending trend. Similar behaviour was noticed on unfertilized plots (Figure 2), where the manure superiority was about 30 kg above chemical fertilizers, but surely in lower yields related to fertilized plots (Figure 1). So, with liquid manure use, soil productivity and fertility is increased, as it is ascertained by the 3-3.25% Nitrogen leaf concentration in full manure doses in contrast to 2.50-2.80 of martyr and chemical fertilizers (Table 1).

Treatment	NITROGEN		PHOSPHOROUS		POTASSIUM Fertilization		CALCIUM		MAGNESIUM	
	Yes	No	Yes	No	Yes	No	Yes	No	Yes	No
1. ---	2,48	2,48	0.30	0.30	0.86	0.86	3.97	3.97	0.83	0.83
2. F*	2.75	2.56	0.20	0.25	0.80	0.87	4.20	4.23	1.15	0.89
3. 1/2F	2.87	2.64	0.25	0.25	0.67	0.88	3.33	4.60	0.85	1.08
4. 1/2F+1/2P**	3.07	2.57	0.23	0.23	0.75	0.79	3.77	4.17	0.88	1.00
5. 1/2F+P	3.25	2.67	0.22	0.25	0.84	0.69	3.57	4.07	0.87	0.96
6. P	3.24	2.82	0.22	0.25	0.88	0.85	3.83	4.23	0.85	0.98
7. 1/2P	2.79	2.36	0.21	0.32	0.81	0.85	3.97	4.40	0.93	0.87
8. F+1/2P	2.94	2.71	0.21	0.23	0.84	0.80	3.83	3.80	0.93	1.02
9. F+P	3.25	2.75	0.20	0.27	0.89	0.80	3.85	4.10	0.90	1.05

F* Chemical Fertilizer-N10P8K4 P** Liquid Piggery Wastes-N10P8K4

Table 1 Nitrogen, Phosphorous, Potassium, Calcium and Magnesium cotton leaf concentration percentage (%) per treatment on August 1989 in dry weight matter (4th cultivated year with and without fertilization).

Leaf nitrogen concentration also increased, but in a lower degree, about 2.70-2.82% found on unfertilized plots, which had been applied with pig manure in three previous years an 2.50-2.65% where martyr and other treatments (Table 1) were applied. These results, and in connection with corresponding cotton yields, point out clearly manure fertilizing superiority in the year of manure application as in subsequent years, because of its positive residual effect, in comparison with chemical fertilizer application. Phosphorous and potassium leaf concentration didn't present any significant differentiations per treatment, as soil analysis gave sufficient levels at rates of 5-20 ppm on labile phosphorous per Olsen and 100-200 ppm on exchange potassium in 0-30 cm soil depth.

Without any differentiation were also Calcium and Magnesium leaf concentration in fertilized and unfertilized plots (Table 1).

Manure Nitrogen is the main and restrictive factor on land applications for crop fertilization, where an increased manure dose, related to plant fertilizing requirement can lead to negative results, as soil nitrate accumulation, which with their great leaching, may pollute subsurface and surface water ecosystem (5).

Nitrate soil solution analysis gave high undesirable concentration levels at rates of 150-200 ppm for 0-30 cm soil depth and 100 ppm for 30-60 cm depth respectively on full manure fertilized plots, while from other treatments nitrate concentration were surely lower with descending trend (Table 2).

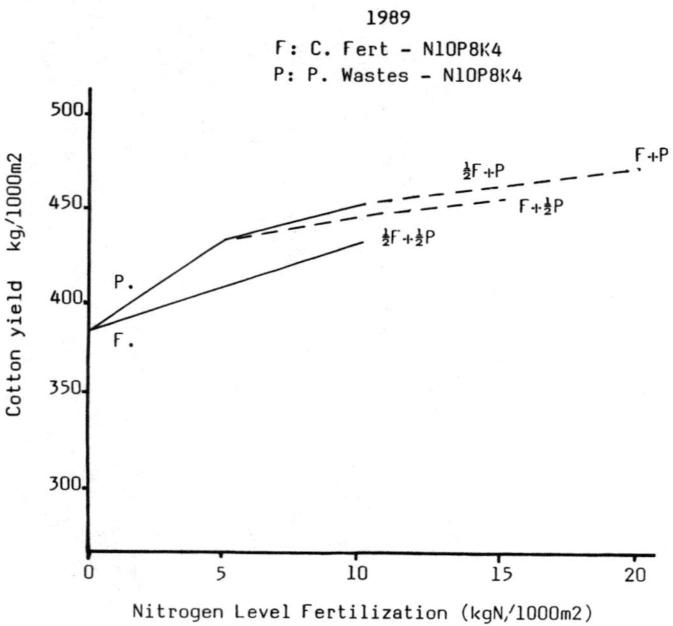

Fig. 1 – Cotton yields in fourth year, period '89 with piggery waste application combined with chemical fertilizers

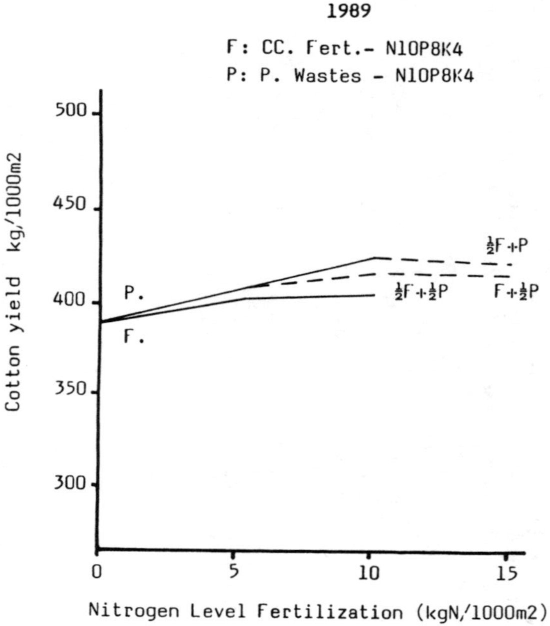

Fig. 2 – Cotton yields in 1989 without fertilization (residual effect) after three year application of liquid piggery wastes combined with chemical fertilizers.

Treatment	Soil Depth 0-30 cm Fertilization Yes	No	30-60 cm Yes	No
1. --	41	55	49	45
2. F*	134	45	61	54
3. 1/2F	109	40	52	53
4. 1/2F+1/2P**	109	53	77	45
5. 1/2F+P	134	57	125	35
6. P	167	60	92	81
7. 1/2P	94	43	53	38
8. F+1/2P	139	59	80	47
9. F+P	229	62	130	60

F* Chemical Fertilizer-N10P8K4
P** Liquid Piggery Wastes-N10P8K4

Table 2 Nitrate soil solution concentration in depths of 0-30 cm and 30-60 cm respectively at the end of the 4th cultivated year, period '89, with and without fertilization in ppm.

Concluding, it can be said that liquid piggery waste application on cotton cultivation is feasible and economic, without environmental hazard when the indispensable presuppositions are kept. Also, lower manure doses combined with chemical fertilizers, because of environment consideration are offered as a more effective fertilization method on cotton cultivation, where soil productive capacity is raised to a higher level than can be reached with mineral fertilizers alone.

REFERENCES

1. COOKE G. "Fertilizing for maximum yield." Grosby Lockwood London, 1972.
2. LANDE CREMER L.C.N de la. "Organic fertilizers for improvement and maintenance of soil fertility". 4th Intern ciec symposium, agricultural waste management and environmental protection Graunschweig, 14-17 May 1987.
3. PETERBURGSKI A. Agricultural Chemistry, Moscow, 1975.
4. SADRA MILAN. "Utilization of animal wastes, Czechoslovac, 1980.
5. SLUIJSMANS C.M.J. and KOLENBRANDER G.J. (1977). The significance of animal manure as a source of nitrogen and fertility management in intensive agric. Tokyo-Japan, 403-411.
6. HON PAUL LANDA L.M. "Environmental guidelines for intensive piggeries." State Pollution control commission, EGP-1, March 1979.

MICRO-POLLUTANTS IN SEWAGE SLUDGE

J.J. van den Berg,
Grontmij nv, consultant engineers, De Bilt, the Netherlands,
L.E. Duvoort van Engers,
National Institute of Public Health and Environmental Protection, the Netherlands

SUMMARY

In recent years the interest in the presence of micro-pollutants in sewage sludges has increased, in particular with relation to the thermical treatment of sewage sludge. The concentrations of some compounds, which are present in small quantities, and often below the detection limit, in sewage sludge will increase by treatment of sludge. On the other hand, degradation of some compounds can result in a higher toxicity level of the residues. This paper gives a survey on the actual data of organic compounds in sewage sludge and the presence of such compounds in the residues of sludge incineration. The data presented are the result of investigations on wet oxidation of sewage sludge by the Vertech-process.

1 INTRODUCTION

In recent years there is an increasing interest in micro-pollutants present in sewage sludge in the Netherlands. Micro-pollutants play an important role in the assessment of the effects of treatment of sewage sludge.
At this moment the oxidation of sewage sludge by incineration is considered to be the principal solution for the treatment (and disposal) of sewage sludge. A disadvantage of incineration is the emission of micro-pollutants. This requires an extended method of treatment of the off-gases. However, with other techniques of oxidation such as wet oxidation, emission of micro-pollutants can be avoided.
In this paper a brief survey will be given of the presence and occurrence of organic micro-pollutants in sewage sludge and the emissions due to incineration.
The paper will in particular describe the expected emission pattern of a planned treatment plant for the wet oxidation of sewage sludge in the Netherlands.

Experiments are being conducted with the treatment of the off-gases and effluents from this plant.
Based on sampling and analysis of the emissions, mass balance studies on the wet oxidation of sewage sludge are reviewed.

2 PRESENCE AND SOURCES OF MICRO-POLLUTANTS IN SEWAGE SLUDGE

Organic pollutants such as polychlorodibenzo-p-dioxines (PCDD) and polychlorodibenzo-furanes (PCDF) occur as the undesired products of incineration and the byproducts of some chemical compounds such as pentachlorophenol. As pentachlorophenol always is present in waste water, the water treatment plants will receive a load of PCDDs and PCDFs as well. These compounds will mainly be found in the sewage sludge.

Hagenmaier c.s. (1) found an average of 0,044 ppm PCDD and of 0,03 ppm PCDF for sewage sludges in Germany. The corresponding minimum-maximum values were 0,008-0,281 (PCDD) and 0,0001-0,019 PCDF) ppm.

Pols (2) reported values of 0,015-0,030 ppm PCDD for Dutch sewage sludge.

In both cases the concentrations are expressed as 2,3,7,8 TCDD-toxic equivalents.

In Germany and the Netherlands, as well as in other countries research on the occurence of dioxines and furanes is still in progress.

3 INCINERATION OF SEWAGE SLUDGE

One of the possible options for sewage sludge treatment is incineration. The organic compounds present in the sewage sludge will be fully or partly decomposed by the dry oxidation process of incineration. The reaction products will be released as (part of) the off-gases or will be adsorbed on the incineration ash or the fly ash.

Otherwise, it can not be ruled out that during the process of incineration and the after-treatment of the off-gases, compounds will be formed with a higher toxicity level than the original compounds. This is not only the case for the more conventional incineration process but can be expected to occur for each process of oxidation.

There are not many emission data on PCDD and PCDF available from sewage sludge incinerators. Williams (3) recently gave a review of TCDD an TCDF emissions from municipal waste incinerators: In the UK emission concentrations in the off-gas range from 0,81 to 204 ng/m^3 for TCDD and from 7,6 to 282 ng/m^3 for TCDF.

A fluidized-bed municipal waste incinerator in Sweden showed total PCDD emissions of 80 ng/m^3 of PCDD and 760 ng/m^3 of PCDF.

The latest Dutch data on PCDD and PCDF emissions from municipal waste incinerators are published by Sein e.a. (4). Mean values were found of 230 ng/m^3 of PCDD and 510 ng/m^3 of PCDF.

Woodfield (5) has compared the PCDD and PCDF concentrations for isomer groups in electrostatic seperator ash from incinators throughout the world, and in ash from other combustion sources.

Table 1: TCDD and TCDF in electrostatic precipitator ash from waste incineration

	TCDD		TCDF	
	ng/g ash	mg/tonne	ng/g ash	mg/tonne
UK	55	-	26	-
Amsterdam	9,4	23,6	38,8	97,6
Zürich	2,0	5,0	1,0	2,5
Rotterdam	18,0	42,2	70	176
Unspecified incinerator	110	276	-	-
US (1)	10	-	54	-
US (2)	85	-	209	-

mg/tonne = emission/tonne of waste incinerated

Difficulties arise in comparing plants since they have different capacities, are of different design, and may be operated under different conditions.

Table 1 shows the concentration of the tetra isomers of PCDD and PCDF; the total PCDD and PCDF will be much higher and, cumulatively, the non-TCDD and non-TCDF isomers may have a more toxic effect than some of the tetra isomers.

The control of PCDD and PCDF emissions may be approached by restricting their formation and/or by clean-up of the flue gases after they have been formed.

Combustion control has concentrated on the destruction of PCDD and PCDF at high temperatures. Consequently, recommended conditions are: temperatures above 1000°C, retention period of >1s, and turbulence to ensure good mixing, with excess air.

Post-combustion control of PCDD and PCDF has concentrated on the efficient collection of particules since PCDD and PCDF are shown to be mostly found on fly ash, either adsorbed or formed in-situ; they also exist at lower levels in the gas phase. Efficient collection of particules can be done by electrostatic precipitators and fabric filters, with the latter showing better retention for PCDD and PCDF.

Clean-up of the contaminated fly ash has been proposed by Hagenmaier et al (6), using reducing conditions in the presence of catalysts and temperature, to chemically reduce the PCDD and PCDF. Wet/dry scrubbers with lime slurry as the active scrubbing agent have been shown effective in the removal of PCDD and PCDF.

4 WET OXIDATION OF SEWAGE SLUDGE

Wet oxidation of sludge, a process which in some ways is similar to incineration, produces a substantially reduced volume of a material, simular to ash from incineration. In general wet oxidation of sewage sludge has a significantly reduced air pollution potential, but produces a liquid waste stream containing low moleculair weight organic materials.

In a wet oxidation process, such as the Vertech-process, organic contaminants are oxidized with oxygen, at elevated temperature and pressure. During the wet oxidation process several intermediates may be formed as the result of oxidation reactions, as well as hydrolysis and pyrolysis reactions.

The reaction with oxygen will generally cause small-molecular compounds with simple molecule structures to be transformed into end products which will not react further. With regards to high-molecular compounds, often with complicated molecular structures one reaction will not suffice to transform them into end products. These compounds will first be "split" into smaller pieces by means of one or more reactions, after which the pieces will react further with, for instance, oxygen, to form end products.

In this respect it is important that the substances which will be formed in the reactor during the oxidation process can have properties different from those of the substances which were present in the initial material. Thus, substances originally easily soluble in water may be transformed into substances that are hardly or not at all soluble in water. Substances of that kind may adhere to the solid matter present during the reaction. Others can be transformed into gaseous components which will be leaving the reactor along with the gaseous effluent, while others in their turn will be transformed into components soluble in water, leaving the reactor in the liquid effluent.

The most important question with respect to the reaction mechanism is whether it can be predicted how the reactions will occur, what products will be formed in the process and what their properties are in relation to further processing or final destination.
However, the reaction mechanism and its kinetics are not completely known.

5 INVESTIGATIONS

Investigations of the influence of wet oxidation on persistant organic compounds carried out by VerTech Treatment Systems have resulted in indications on the (bio)degradability of some organic compounds which are present in municipal waste water.

In Dordrecht in the Netherlands a test facility is available for the wet oxidation of (small quantities of) sewage sludge on pilot plant scale.

The results of tests are used for the design of a vertical tube reactor.

The test facility is used for the investigations on the wet oxidation of sewage sludge delivered by Zuiveringsschap Veluwe, who are in charge of the treatment of waste water of the region Veluwe.

The pilot plant has a capacity of 130 l/h., the several process conditions such as temperature, pressure and retention time, can be varied.

Tests have been carried out resulting in a COD-reduction of the sludge to be treated of 70%. The input (sewage sludge) as well as the output (liquid, solids and off-gases) are sampled. The samples have been analyzed on the "classic" chemical process parameters (P, SO_4, EOX, anorganic nitrogen, heavy metals) and on the presence of organic pollutants such as PCB's, organochloropesticides, chorophenols, dioxines and furanes.

6 RESULTS OF ANALYSIS

The results of the different analyses are indicated in tables 2 and 3. The results show that sludge loss on calcining is reduced by about 90% during its presence in the reactor. The Kjeldahl nitrogen content does not change significantly, while the ammonium nitrogen fraction increases by about 40%. About 90% of the EOX is decomposed. The amount of calcium in the liquid reactor effluent is about 30% lower than in the influent. This might be due to reactor scaling.

The analysis results show that 99% of the PCB's - except for fluorene - are decomposed.

Without any doubt the accuracy of the analyses and the detection limits are very important in the evaluation of the analysis results of heavy metals and organic micro-pollutions. As the amount of dry solids is reduced, substances in the sludge, in concentrations below the detecting limit, may be detectable in the effluent ash due to the concentration effect of the aqueous phase oxidations process.

Evidently, the conclusion cannot immediately be drawn that forming of such a substance has occurred. Concentrations of substances in the effluent are established after the relevant sample has been concentrated by evaporation. This means that all components originally present in the aqueous phase are included in the result: when the suspended solids content of the dry solids fraction is 10%, the concentration in the aqueous phase (in mg/kg) must amount to ten times the concentration in the suspended solids (in mg/l). This may result in incorrect interpretation of the analysis results. This can be illustrated by the example of 4,4 DDD. (see table 6, excerpt of table 2).

able 6 Analysis and mass balance for 4,4 DDD (excerpt from table 2)

1. INFLUENT SLUDGE		
Dry solids	6,5	%
4,4 DDD concentration	< 10	µg/kg d.s.
detection limit	10	µg/kg d.s.
2. EFFLUENT WATER (75% of effluent volume)		
dry solids	1902	mg/l
4,4 DDD concentration	2,25	µg/l
3. EFFLUENT WATER AND ASH (25% of effluent volume)		
dry solids	10	%
4,4 DDD concentration	15	µg/kg d.s.
detection limit	10	µg/kg d.s.
4. CALCULATED 4,4 DDD CONCENTRATION		
Influent	0	µg/l
Effluent	2,06	µg/l

The conclusion in part 4 of table 6 is that the concentration in the reactor increases from 0 to 2,06 µg/l. If, however, the separate analysis are evaluated, the concentration in the influent proves tot be below the detection limit of 10 µg/kg. On the other hand, the concentration in the dry solids effluent (table 6-3) is above this limit (15 µg/kg). In balance 15 µg/kg in 25% of the effluent volume with 10% d.s. would equal 5.8 µg/kg in the influent with 6% d.s., if no 4,4 DDD is formed or destroyed. However, 5.8 µg/kg is below the detection limit of 10 µg/kg d.s. This detection limit in the influent corresponds with a concentration in the effluent of 26 µg/kg d.s.
The concentration in the effluent water (table 6-2) amounts to 2,25 µg/l. This does not correspond with the concentration in the effluent water and ash mixture (table 6-3), since after evaporation in the concentration in the dry solids should be 22,5 µg/kg. Although at first a significant increase of 4,4 DDD seems to occur, the increase is not actually proven by the analysis results. It must be concluded for most of the organochloride pesticides that no formations of such substances can be established.

The issues related to the detection limits in the establishment of chloride phenols is identical to the above. The analysis results show that most of these compounds are mainly decomposed as a result of aqueous phase oxidation. No significant increase has been established for the other components, except 3,4 dichlorophenole.

It was (still) impossible to establish a conclusive mass balance for heavy metals in the pilot sampling process. On the basis of the limited number of analysis it may be concluded that heavy metals are "left behind" in the reactor. This corresponds with results obtained from the pilot installation in Lowry (USA), where it appeared that some metals were left behind in the reactor in the scaling on reactor walls. Future research will have to provide deeper insight in to this matter.

7 DIOXINS AND DIBENZOFURANS

At present, dioxins and dibenzofurans are rather topical issues. Most attention is given to the group of the (chlorinated) dioxins because of their presence in the environment and the toxicity of the number of isomers. About ten of the chlorinated dioxins are considered toxic, of which 2,3,7,8-tetrachlorodibenzo-p-dioxin (2,3,7,8-TCDD) is the most important. This compound is not very reactive with other substances and is chemically very stable. An important property with regard to the aqueous phase oxidation process is the low degree of solubility of this isomer in water. Dioxins will only be formed in a aqueous environment if sufficient pressure is provided. In case of atmospheric pressure, the boiling point of most discharge products is lower than the temperature required for dioxin formation. These discharge products are released in the aqueous phase before reaction to dioxin can occur. It is assumed that the temperature required for dioxin formation is to be found in the trajectory ranging from 180 to 400 degrees Celsius.

According to the relevant literature, decomposition occurs at temperatures above 500 degrees Celsius. Although testing has proved that complete decomposition can be reached at a temperature of 800 degrees Celsius, it has been proved that the basic form of dioxins remains intact up to temperatures of 1150 degrees Celsius, if the dioxins have been incorporated in dry solids. When thermally treated particles are cooled, (chlorinated) dioxins may recur in the critical temperature trajectory ("de novo" synthesis).

The toxicity of (chlorinated) dioxins and furans is expressed with reference to the most toxic 2,3,7,8-TCDD. The so called "2,3,7,8-TCDD Toxicity Equivalents Factor (TEF)" is commonly used to express the toxicity of a group of dioxins and furans.

The results of analyses for dioxins and furans of the sewage sludge used in the investigation and of the dry solids residue after aqueous phase oxidation are indicated in tables 4 and 5. The sludge has been analyzed for 17 <u>congeners</u> of polychlorodibenzo-p-dioxins and polychlorodibenzo-p-furans. Congeners are molecules with the same basic structure but a varying number of chlorine atoms on different substitutes. The congeners were selected in accordance with the advice of the Study Group of Toxicity Equivalent Factors". The study group assigned a 0 TEF value to the other congeners ½7¼.

The following conclusions may be drawn from the results:

- dioxins and furans are decomposed; the total content of dioxins and furans in the dry solids residue is lower than the content found in the sewage sludge;
- the concentration of dioxins in the dry solids increases because the suspended solids content is reduced by about 60%;
- the higher chlorinated congeners are transformed into tetra and pentachlordibenzo(p)dioxins and furans.

No direct conclusions can be drawn from the results with regard to the toxicity. Just as in the case of other organic micro-pollutions, the concentration effect on the concentration of the dioxins as a result of the aqueous phase oxidation process is very important. Several congeners which are below the detection limit in the influent sludge still have been found in the effluent sludge. Because a concentration factor of 2,6 has been reached, a detection limit of 0,02 mg/kg in the influent implies a limit value of 0,052 mg/kg in the effluent. This means that for all congeners with a concentration below 0,052 mg/kg in the effluent it cannot be established with certainty that formation of these congeners has occurred (see * in table 4).
The marked congeners add a total of 30 mg/kg to the toxicity equivalent. Moreover, it has been proved that a number of congeners exist in the effluent in concentrations below 0,10 mg/kg, which had not been proved to exist in the influent. Once again it is assumed that they are also present in the influent, though in concentrations below the detection limit. On the basis of these considerations the conclusion can be drawn that there might be some increase in toxicity, but that this increase is relatively small judging from these analysis results and falls within the accuracy of the analyses.

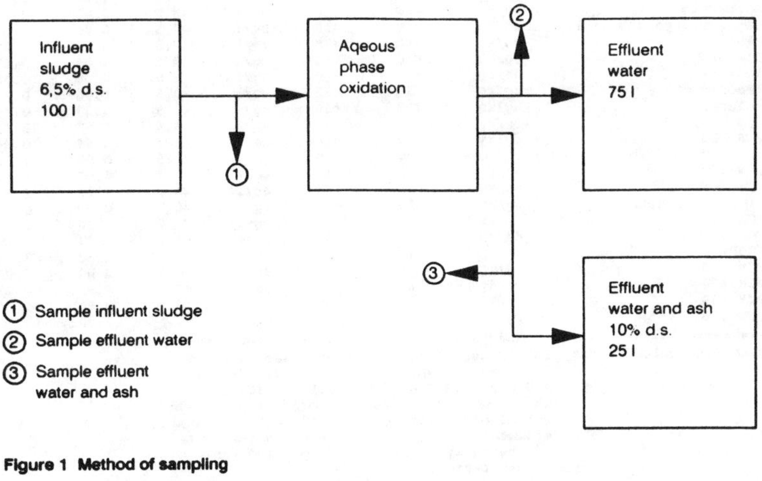

Figure 1 Method of sampling

Table 2. Wet oxidation of sewage sludge, analysis and mass balance

Volume				
influent	460 l	ratio	1	l
effluent ash	114 l		0.25	l
effluent water	346 l		0.75	l

Compound	unit d.s	sludge sample	ash sample	unit liquids	water sample	influent calculated	effluent results
D.S.	%	6.5%	10.0%	mg/l	1902	6.5%	2.5%
CLASSIC CHEMICAL ANALYSIS							
calcining loss	g/kg	42	10	mg/l	0	2726	248
N-Kj	g/kg	40	31	mg/l	2300	2596	2497
P	g/kg	21	52	mg/l	22	1332	1314
dissolved SO4	g/kg	2.7	16.3	mg/l	520	175	795
EOX	mg/kg	27	3	ug/l	120	1752	165
Anorg N	g/kg	18	18	mg/l	1600	1168	1649
Ca	g/kg	29	42	mg/l	135	1882	1141
POLYCYCLIC AROMATIC HYDROCARBONS							
Naftalene	mg/kg	<0.3	<0.05	ug/l	-		
Acenafthylene	mg/kg	<0.2	<0.05	ug/l	-		
Acenaftene	mg/kg	0.35	<0.05	ug/l	0.5	23	0.38
Fluorene	mg/kg	0.25	0.02	ug/l	1.6	16	1.70
Fenantrene	mg/kg	1.4	0.01	ug/l	0.1	91	0.32
Anthracene	mg/kg	0.15	<0.01	ug/l	-	10	0.00
Fluoranthene	mg/kg	2.4	<0.01	ug/l	0.01	156	0.01
Pyrene	mg/kg	2	<0.1	ug/l	0.02	130	0.02
Benzo(a)anthracene	mg/kg	0.8	<0.01	ug/l	<0.01	52	0.00
Chrysene	mg/kg	1.1	-	ug/l	-	71	0.00
Benzo(b)fluoranthene	mg/kg	1.4	0.01	ug/l	0.03	91	0.27
Benzo(k)fluoranthene	mg/kg	0.6	<0.01	ug/l	<0.01	39	0.00
Benzo(a)pyrene	mg/kg	0.7	0.01	ug/l	<0.01	45	0.25
Dibenz(ah)anthracene	mg/kg	0.2	<0.01	ug/l	<0.01	13	0.00
Benzo(ghi)perylene	mg/kg	0.7	0.02	ug/l	<0.01	45	0.50
Indeno(123cd)pyrene	mg/kg	0.7	0.01	ug/l	<0.01	45	0.25
Total 16 PCA	mg/kg	13	0.1	ug/l	2.2	844	4.13
Total 6 Borneff PCA	mg/kg	6.4	<0.1	ug/l	<0.1	415	0.00
ORGANO-CHLOROPESTICIDES							
alpha HCH	ug/kg	<10	40	ug/l	4.25	0.00	4.19
beta HCH	ug/kg	10	<10	ug/l	<0.5	0.65	0.00
gamma HCH	ug/kg	<10	<10	ug/l	<0.5	0.00	0.00
delta HCH	ug/kg	15	20	ug/l	3.5	0.97	3.13
Epsilon HCH	ug/kg	<10	<10	ug/l	0.8	0.00	0.60
HCB	ug/kg	15	30	ug/l	<0.5	0.97	0.74
Heptachlor	ug/kg	<10	<10	ug/l	<0.5	0.00	0.00
Heptachl.epoxide	ug/kg	<10	<10	ug/l	<0.5	0.00	0.00
Aldrin	ug/kg	<10	<10	ug/l	<0.5	0.00	0.00
Dieldrin	ug/kg	<10	15	ug/l	0.4	0.00	0.67
Endrin	ug/kg	<10	<10	ug/l	0.5	0.00	0.38
Isodrin	ug/kg	<10	25	ug/l	<0.5	0.00	0.62
Telodrin	ug/kg	<10	<10	ug/l	<0.5	0.00	0.00
4,4-DDE	ug/kg	15	25	ug/l	<0.5	0.97	0.62
2,4-DDD	ug/kg	10	<10	ug/l	1.25	0.65	0.94
4,4-DDD	ug/kg	<10	15	ug/l	2.25	0.00	2.06
2,4-DDT	ug/kg	<10	45	ug/l	<0.5	0.00	1.11
4,4-DDT	ug/kg	<10	100	ug/l	<0.5	0.00	2.48
alpha-endosulfan	ug/kg	<10	<10	ug/l	<0.5	0.00	0.00
PCB 28	ug/kg	<10	<10	ug/l	<0.5	0.00	0.00
PCB 52	ug/kg	15	150	ug/l	<0.5	0.97	3.71
PCB 101	ug/kg	20	85	ug/l	<0.5	1.30	2.10
PCB 118	ug/kg	<10	<10	ug/l	<0.5	0.00	0.00
PCB 138	ug/kg	15	25	ug/l	<0.5	0.97	0.62
PCB 153	ug/kg	10	30	ug/l	<0.5	0.65	0.74
PCB 180	ug/kg	<10	15	ug/l	<0.5	0.00	0.37

Method of calculation: The influent sludge has been sampled. The effluent is separated into an ash fraction and a liquid fraction (ratio 25:75) after which both fractions have been sampled. The actual concentrations in influent and total effluent have been calculated on the basis of the analyzed concentrations and known volume ratios.

Table 2 (Continued).

Compound	unit d.s	sludge sample	ash sample	unit liquids	water sample	influent calculated	effluent results
(CHLORO)PHENOLES							
2 Chlorophenole	ug/kg	220	<20	ug/l	<1	14.28	0.00
3 Chlorophenole	ug/kg	250	<20	ug/l	<1	16.23	0.00
4 Chlorophenole	ug/kg	<20	61	ug/l	<1	0.00	1.51
2,4/2,5 Dichlorophenole	ug/kg	740	<20	ug/l	<0.1	48.03	0.00
2,6 Dichlorophenole	ug/kg	110	<20	ug/l	<0.1	7.14	0.00
2,3 Dichlorophenole	ug/kg	<20	75	ug/l	<0.1	0.00	1.86
3,4 Dichlorophenole	ug/kg	70	178	ug/l	11	4.54	12.68
3,5 Dichlorophenole	ug/kg	<20	<20	ug/l	0.9	0.00	0.68
2,3,5 Trichlorophenole	ug/kg	<20	<20	ug/l	<0.1	0.00	0.00
2,3,6 Trichlorophenole	ug/kg	<20	<20	ug/l	<0.1	0.00	0.00
2,3,4 Trichlorophenole	ug/kg	71	23	ug/l	7	4.61	5.83
2,4,5 Trichlorophenole	ug/kg	<20	89	ug/l	0.6	0.00	2.65
2,4,6 Trichlorophenole	ug/kg	<20	<20	ug/l	0.6	0.00	0.45
3,4,5 Trichlorophenole	ug/kg	75	25	ug/l	0.7	4.87	1.15
2,3,4,5 Tetrachloropheno	ug/kg	<20	<20	ug/l	<0.1	0.00	0.00
2,3,4,6 Tetrachloropheno	ug/kg	<20	38	ug/l	1	0.00	1.69
2,3,5,6 Tetrachloropheno	ug/kg	51	55	ug/l	2.2	3.31	3.02
Pentachlorophenole	ug/kg	120	29	ug/l	<0.1	7.79	0.72
4 chloro-8 methylphenole	ug/kg	<20	<20	ug/l	<1.0	0.00	0.00

Table 3 Wet oxidation of sewage sludge, heavy metals mass balance

Compound	unit d.s	sludge sample	ash sample	unit liquids	water sample	influent calculated	effluent results
Cd	mg/kg	1	3	ug/l	0.1	65	74
Cr	mg/kg	76	150	ug/l	170	4932	3842
Cu	mg/kg	250	265	ug/l	21	16225	6577
Hg	mg/kg	1.6	2.7	ug/l	13	104	77
Ni	mg/kg	35	81	ug/l	12	2272	2014
Pb	mg/kg	70	200	ug/l	<1	4543	4952
Zn	mg/kg	770	1350	ug/l	53	49973	33463
As	mg/kg	5	12	ug/l	86	325	362

Table 4 Wet oxidation of sewage sludge
results of analysis, toxic congeneres of dioxins and furanes

	Sludge influent ug/kg	Ash effluent ug/kg	Water effluent ng/l	TEF	Influent TEQ ng/kg	Effluent TEQ ng/kg
2,3,7,8-TCDD	< 0.02	< 0.02	<10	1	0	0
1,2,3,7,8-PCDD	< 0.02	< 0.02	<10	0.5	0	0
1,2,3,4,7,8-HxCDD	< 0.02	< 0.02	<10	0.1	0	0
1,2,3,6,7,8-HxCDD	0.02	0.08	<10	0.1	2	8
1,2,3,7,8,9-HxCDD	< 0.02	0.04	<10	0.1	0	4 *
1,2,3,4,6,7,8-HpCDD	1	2.6	<10	0.01	10	26
OCDD	10	10	<10	0.001	10	10
2,3,7,8-TCDF	< 0.02	0.02	<10	0.1	0	2 *
1,2,3,7,8-PCDF	< 0.02	0.07	<10	0.05	0	3.5
2,3,4,7,8-PCDF	< 0.02	0.04	<10	0.5	0	20 *
1,2,3,4,7,8-HxCDF	< 0.02	0.09	<10	0.1	0	9
1,2,3,6,7,8-HxCDF	< 0.02	0.08	<10	0.1	0	8
1,2,3,7,8,9-HxCDF	< 0.02	< 0.02	<10	0.1	0	0
2,3,4,6,7,8-HxCDF	< 0.02	0.04	<10	0.1	0	4 *
1,2,3,4,6,7,8HpCDF	0.12	0.24	<10	0.01	1.2	2.4
1,2,3,4,7,8,9-HpCDF	< 0.02	< 0.02	<10	0.01	0	0
OCDF	0.43	0.6	<10	0.001	0.43	0.6
Dry solids 6.5%	Total TEQ at this d.s. contents:				24	
2.5%	Total TEQ at this d.s. contents:				62	98

TEF: Toxicity Equivalency Factor with respect to 2,3,7,8 TCDD

Table 5 Wet oxidation of sewage sludge concentrations of toxic congeneres of dioxins and furanes

	Concentration influent ng/l	effluent ng/l	TEF	Influent ng/l TEQ	Effluent ng/l TEQ	
2,3,7,8-TCDD	0.0	0.0	1	0.000	0.000	
1,2,3,7,8-PCDD	0.0	0.0	0.5	0.000	0.000	
1,2,3,4,7,8-HxCDD	0.0	0.0	0.1	0.000	0.000	
1,2,3,6,7,8-HxCDD	1.3	2.2	0.1	0.127	0.224	
1,2,3,7,8,9-HxCDD	0.0	1.1	0.1	0.000	0.112	*
1,2,3,4,6,7,8-HpCDD	63.7	72.8	0.01	0.637	0.728	
OCDD	637.0	280.0	0.001	0.637	0.280	
2,3,7,8-TCDF	0.0	0.6	0.1	0.000	0.056	*
1,2,3,7,8-PCDF	0.0	2.0	0.05	0.000	0.098	
2,3,4,7,8-PCDF	0.0	1.1	0.5	0.000	0.560	*
1,2,3,4,7,8-HxCDF	0.0	2.5	0.1	0.000	0.252	
1,2,3,6,7,8-HxCDF	0.0	2.2	0.1	0.000	0.224	
1,2,3,7,8,9-HxCDF	0.0	0.0	0.1	0.000	0.000	
2,3,4,6,7,8-HxCDF	0.0	1.1	0.1	0.000	0.112	*
1,2,3,4,6,7,8HpCDF	7.6	6.7	0.01	0.076	0.067	
1,2,3,4,7,8,9-HpCDF	0.0	0.0	0.01	0.000	0.000	
OCDF	27.4	16.8	0.001	0.027	0.017	
TOTAAL	737.0	389.2	Tot. TEQ	1.505	2.730	

Literature

(1) Hagenmaier, H., Brunner, H., Knapp, W. and Weberruss, U.; Untersuchungen der Gehalten an polychlorierten Dibenzodioxinen, polychlorierten Dibenzofuranen und ausgewählten Chlorkohlenwasserstoffen in Klärschlämmen UBA-forschungsbericht FKZ 10303305 (1988).

(2) Pols, H.B. Het voorkomen van gechloreerde dioxines en dibenzofuranen op rioolwaterzuiveringsinstallaties. DBW/RIZA nota nr. 88004 (1988).

(3) William, P.T. A review of pollution from waste incineration. J. JWEM 4-1990.

(4) Sein, A.A., Sluijmers, J.J., Verhagen, E.J.H. Samenvatting en conclusies van het eindrapport "Onderzoek emissies afvalverbrandingsinstallaties". RIVM nr. 738473010, 1989.

(5) Woodfield, M. The environmental impact of refuse incineration in the U.K. Warren Spring laboratory. HMSO, 1987.

(6) Hagenmaier, H., Kraft. M., Haag, R. and Brunner H. Energy recovery through Waste Combustion. Elsevier Applied Science. Essex, 1988.

(7) Advies van de werkgroep Toxiciteitsequivalentiefactoren, maart 1988. Voorstel tot een methode voor de beoordeling van de toxiciteit van mengsels van gehalogeneerde dibenzo-p-dioxinen en dibenzofuranen.

FOR A TECHNICO-ECONOMICAL APPROACH OF SEWAGE SLUDGE AND REFUSES REUSE : A FRENCH R + D EXPERIMENT

Maurice CASIMIR

BIOTECHNA - 10 Av de Boisbaudran - F 13015 MARSEILLE

The paper deals with an original R + D program now in progress in the South of France, on its Mediterranean border, conducted simultaneously with a full size operational phase, both dedicated to a technico-economical approach of sewage sludge and refuses reuse.

It is well know for a long time to mix sewage and cellulosic compounds to prepare via a process of fermentation - maturation - sieving a final compost. This product has found its main applications in agriculture till now.

The approach is quite different in the case of these pilot site of ENSUES-LA-REDONNE, thirty kilometers west of MARSEILLES, in a karstic, rocky, and semi-arid area with a rapid urban growth of family-housing type.

The aims of the research and subsequent development are multiple :

1. Testing a wide range of raw products and cellulosic compounds, these last ones preferably cheap and easy to found in the Mediterranean Region.

As raw products, the tests are conducted only with sewage treatment plants sludges and more originally, greases. It has been confirmed that the process is available with sludges produced in physico-cheminal plants, and is then even more performant, due to the higher siccity. We have to remind that the process accepts sludges with 18 to 25 % siccity, avoiding an expensive deshydratation phase of the sludges. The tests begin with greases and we have at the moment no significative data available.

Due to the rather low added value of the final product, it is indispensable to look for cheap and easy to found cellulosic compounds. On an other side, these products must allow a good oxygen inside circulation of the mixture. Fortunately, our mediterranean region is well doted in tree species and it has been possible to plan the checking of olive and pine wood with a predetermined chip size. The chips are produced in a mixed form by the surface cleaning of underwoods.

It is also possible, and done, to use refuses from sawmills and wholesale markets, these last one numerous in the Mediterranean Basin for fruits and vegetable processing and trade.

2. Certification of the final product according to the future European and national regulations at both levels of

- agronomical properties
- public health safety (especially long term inocuity)

We intend to work only with sludges and greases produced by rural and semi-rural sewage treatment plants. We hope to avoid the presence of heavy metals in the final product, and, according to an increase of temperature up to 70 - 72 Celsius, we insure a pasteurisation of the product.

Nevertheless, the test being conducted with the goal of economical feasibility, we want to be sure that the final product will receive all agreements and label of national Anthorities according to its final inocuity. Therefore have we to proceed to special investigations and analysis in three main fields : parasitology, virology and heavy metals. This will be done jointly with University laboratories.

To get a health safe product is necessary but not sufficient. We have also to be sure that it will compete with agronomic specifications making itself attractive for possible buyers, such as those listed below. A specific study of the agronomical properties will be conducted with the French specialized Institute in this field.

3. Assessment of an optimized process at both levels of engineering and cost.

Till now, the greater part of the investigations conducted all over the world are mainly dedicated to the feasibility from a biochemical point of view. We want to get a better knowledge of the process in terms of engineering, the only way for us to develop a real cost analysis study and to find the way of optimization.

This explains the contruction of a pilot plant able to receive a lot of samples of different raw products and cellulosic compounds and to produce a range of final products before the construction of plants dedicated to a specific product.

4. Market investigation enlarged to nurseries - public leisure and recreation areas - domestic uses, either for the final product in itself or as an enriched soil.

Many times, the mediterranean shore is mainly dedicated to housing and tourism. We have less and less agricultural activity and therefore if we want to develop the reuse as briefly described above, we need to find a market for the final product. We started from the idea of looking for clients needing rather an enriched soil than a fertilizer.

For in many places all around our sea, we have a lot of rocky spaces, highly eroded by automn rains and forest fires. Then, when an urbanization il decided, the future unhabitants are frequently confronted to the problem of creating green spaces and they need to bring soil for this - So we could check that there is effectively a market for the product which is of interest for nurseries, public leisure and recreation areas, including golfs and domestic uses - And this is a good new, because needs are lower than for agriculture, which fits with a production linked to the possibilities of rural and semi-rural areas.

5. Elaboration of a Department Master Plan for sewage sludges and refuses reuse, which could be enlarged to a regional, then to a national size.

Actually, the environmental policy developed by all riparian countries leads to the construction of a great lot of sewage treatment plant, especially in the seaside resorts. It is indispensable to manage and coordonate the sludges reuse in view of optimizing the number of process sites and the global cost for the community.

This can be done only by the setting of a master plan, envolving local and regional anthorities in a first step, and integrating itself in a national, and may be tomorrow, mediterranean master plan. Preliminary discussions with our regional anthorities have shown the pertinency of this approach.

6. Cost - benefit analysis dealing with the following parameters :
   - cost of sludges disposal without reuse
   - cost of reuse process
   - selling price of the final product

We know all of us that the sludges and refuses reuse will never be an economically profitable system. What we have to take into account is the global financial balance between the environmental cost of final disposition and the marginal cost of reuse.

We know very well actually the cost of sludges drying and disposal, which differs of course not only from one country to another, but also from one site to another. We know also the price we can sell the final product. We have a lot of things to learn on the cost of the reuse process and this is a very important part for our actual investigation. We hope that precise data will be available at the date of the Athena's Symposium.

7. Feasibility of a technology tranfer towards the Mediterranean countries (plus PORTUGAL) in the frame of E.E.C. Mediterranean Policy (ENVIREG) and U.N.E.P. Mediterranean Action Plan (BARCELONA Convention).

What seems may be to us the more interesting in the R + D project now in development is that it is really transferable to any mediterranean country with very few accomodments. The process is easy to manage (we do not say to optimize) in an open-air area rather isolated, and moreover, it needs no skilled manpower other than a technician able to act as process supervisor. This technician will easily receive an on job training. And the final product will find a market according to what has been writed above. This is evidently a topic of prime interest for the E.E.C. Mediterranean Policy, which seems to us enter pefectly in the frame of the ENVIREG program. This is also perfectly fitting with the aims of the Mediterranean Action Plan, because is there a better illustration of the Mustafa TOLBA's words "Development without destruction" resuming the ambitions of United Nations Environment Program for its Regional Seas Program ?

Actually, the project enters an active phase. Il has been decided to intregrate it in a French interregional cooperation between Languedoc-Roussillon and Provence-Alpes-Côte d'Azur (Southern High Tech Route).

The applied research will be conducted under the scientific leadership of the Process Engineering Laboratory of Montpellier University (Doctorate Thesis on a three years period). Il deals with aims 1 to 5.

The development phase will be conducted in cooperation with the French Environment State Secretariat, the Regional Agency for Environment Protection, the Department of Bouches du Rhône and will have the objective to create new composting sites in the Department, then in the Region, may be later in foreign countries. Il deals with aims 6 and 7.

The final aim is of course coherent with UNEP and E.E.C. policies in matter of Mediterranean ecosystem protection.

ROLE OF MANURE IN THE DISTRIBUTION OF LISTERIA MONOCYTOGENES

B. VAN RENTERGHEM, F. HUYSMAN, R. RYGOLE AND W. VERSTRAETE

Laboratory of Microbial Ecology
Centre of Environmental Sanitation
State University of Gent, Faculty of Agriculture
Coupure L 653, B-9000 Gent, Belgium

Summary

The following biotopes were examined for the presence of *Listeria monocytogenes* : 25 samples of fresh pig faeces, 25 samples of fresh cattle faeces, ten samples of stored liquid manure, 17 manured soil samples and 15 ground water samples. Respectively 16 % and 20 % of the collected samples of fresh pig and cattle faeces contained *Listeria monocytogenes*. However, after three weeks of storage *Listeria monocytogenes* could only be detected in one of the initially nine positive samples. None of the liquid manure or manured soil samples contained detectable levels of *Listeria monocytogenes*. *Listeria monocytogenes* was detected in only one ground water sample. Two months after inoculation of stored liquid pig manure, stored liquid cattle manure and soil with *Listeria monocytogenes*, this bacterium could not be traced in any of these biotopes. Radishes (*Raphanus sativus*) sown in soil inoculated with *Listeria monocytogenes*, were gathered after three months and examined for the presence of *Listeria monocytogenes*. Three out of six radish samples were found to be positive.

1. INTRODUCTION

Listeria monocytogenes is a nonsporing gram-positive rod-shaped micro-organism (1) that has become an important foodborne pathogen during the last decade. Until recently listeriosis, a severe and often fatal illness with clinical manifestations that typically include neonatal sepsis, meningitis, puerperal sepsis and/or flu-like illness during pregnancy (2) was usually associated with consumption of contaminated dairy products. Fleming et al. (3) associated the consumption of pasteurized milk that had originated from farms having listerial bovine encephalitis epidemiologically with human listeriosis.

In the natural environment, *Listeria* spp. are reported to be isolated from soil, mud, surface water, silage and faeces of animals and man (1, 4, 5). Recently Heisick et al. (6) detected the presence of nonpathogenic *Listeria* species as well as *Listeria monocytogenes* on fresh vegetables, particularly root crops such as potatoes (25.8 % positive) and radishes (30.3 % positive). Since *Listeria monocytogenes* contamination seems to be associated with vegetables and since faecal excretion of *Listeria monocytogenes* is rather common in cattle (6, 8) manure could function as a potential source of *Listeria monocytogenes*. This study attempts to determine the importance of manure as a transmission vector of *Listeria monocytogenes*.

2. METHODS

The following biotopes were examined for the presence of *Listeria monocytogenes* : 25 samples of fresh pig faeces, 25 samples of fresh cattle faeces, ten samples of stored liquid manure, 17 soil samples of the ploughlayer of manured fields and 15 ground water samples. The samples of fresh pig and cattle faeces were collected at a slaughterhouse in Eeklo, Belgium.

To examine the survival capacity of *Listeria monocytogenes* in soil and manure the latter biotopes were inoculated with the micro-organism. Sandy-loam soil of the ploughlayer of an agricultural field was air dried and passed through a two mm sieve. The soil was remoistened to 3/4 of field capacity and prior to use stored for two weeks aerobically at 25 °C. Soil, stored liquid cattle and pig manure were mixed with a *Listeria monocytogenes* enrichment culture in such a way that a final concentration of 10^5 *Listeria monocytogenes*/ml manure or g soil was obtained. The survival of *Listeria monocytogenes* was also examined in soil where inoculated pig manure was applied in a dose of 100 ton/ha. During the experiment, all samples were stored at an ambient temperature of 15 °C and soil samples were kept at 3/4 field capacity. Every week, triplicate samples were taken and examined for the presence of *Listeria monocytogenes*.

At the same time it was investigated if cultivating radishes (*Raphanus sativus*) in soil contaminated with *Listeria monocytogenes* would cause an enrichment in the rhizosphere. Therefore radishes (*Raphanus sativus*) were sown in (i) soil inoculated by means of a *Listeria monocytogenes* enrichment culture (10^5 / g soil) and (ii) soil inoculated by means of *Listeria monocytogenes* inoculated pig manure (10^5 / ml manure) (see before). After three months the vegetables were harvested. Triplicate samples of soil, unwashed and washed (with water until all soil was removed) root crops were examined for the presence of *Listeria monocytogenes*.

For the detection of *Listeria monocytogenes* a modified USDA-procedure incorporating a non-selective pre-enrichment step was used (9). This two stage enrichment procedure involved TSBY (Trypton Soya Broth, OXOID CM129, with Yeast extract, OXOID L21) as pre-enrichment medium for the resuscitation of the *Listeria monocytogenes* cells. For the solid samples (faeces, liquid manure and soil) ten g were suspended in 90 ml TSBY. For the water samples, 100 ml was filtered over a Millipore filter (0.45 μm). The filter was subsequently brought in 100 ml of TSBY. After 24 h incubation at 30 °C, one ml was inoculated in ten ml LEB (*Listeria* Enrichment Broth Base, OXOID CM882), with *Listeria* Selective Enrichment Supplement (OXOID SR141). This supplement contains acriflavine, nalidixine acid and cycloheximide as selective agents. Afterwards the samples were incubated during 24 at 30 °C on a shaker (80 rpm).

Subsequently 0.1 ml enrichment culture was streaked onto OXOID *Listeria* Selective Medium, Oxford formulation (10). This medium was obtained by adding, after sterilisation, *Listeria* Selective Supplement (OXOID SR140) to *Listeria* Selective Agar Base (OXOID CM856). After 48 h incubation at 37 °C, typical *Listeria* colonies with a black halo and a sunken center were streaked onto Trypton Soya Agar (TSA, OXOID CM131) for separation and verification of purity. After incubation during 24 h at 37 °C, isolated colonies were used for further qualitative analyses.

The qualitative determination of *Listeria* spp. was done by LISTERIA-TEK (Organon Technika), an enzyme-linked immunosorbent assay (ELISA). Further identification for *Listeria monocytogenes* was done according to the procedure given by Vanrolleghem et al.(11). Following tests were performed : microscopy; elucidation of the fermentation pattern on L-rhamnose, D-xylose and mannitol; hemolysis test; CAMP-test; urease test; motility test at 20 °C and 37 °C.

3. RESULTS

Respectively 16 % and 20 % of the collected samples of fresh pig and cattle faeces contained *Listeria monocytogenes* (Table 1). After three weeks of storage at an ambient temperature of 15 °C *Listeria monocytogenes* could however only be detected in one of the initially nine positive samples. In only one ground water sample *Listeria monocytogenes* was detected.

TABLE 1 : Prevalence of *Listeria monocytogenes* in various environmental samples.

	Number of samples examined	Number of positive samples
pig faeces	25	4
cattle faeces	25	5
manure	10	0
soil	17	0
ground water	15	1

Although almost 20 % of the fresh faeces samples were found positive, none of the manure samples, obtained from storage basements, or manured soil samples contained *Listeria monocytogenes*. Therefore the capability of survival of *Listeria monocytogenes* in these biotopes was investigated. *Listeria monocytogenes* was inoculated in soil and manure and every week triplicate samples were examined for the presence of *Listeria monocytogenes*. Three to four weeks after inoculation *Listeria monocytogenes* couldn't be traced in any pig manure samples. After six weeks the presence of *Listeria monocytogenes* in inoculated soil and cattle manure samples became less clearly. After eight weeks it wasn't possible to detect *Listeria monocytogenes* in any sample.

Listeria monocytogenes seems to be uncapable of surviving for long periods in soil or liquid manure. However, *Listeria monocytogenes* has regularly been isolated from agronomic crops and vegetables (6, 12). It was therefore examined if *Listeria monocytogenes* could be isolated on radishes (*Raphanus sativus)* sown and grown in soil inoculated with *Listeria monocytogenes* (Table 2). Two samples of radishes grown in inoculated unmanured and one sample of radishes grown in manured soil, were found to be *Listeria monocytogenes* positive. In one sample of manured soil where previously radishes (*Raphanus sativus)* had been cultivated the presence of *Listeria monocytogenes* could be established.

4. DISCUSSION

Listeria monocytogenes was detected in respectively 16 and 20 % of the investigated samples of pig faeces and cattle faeces. Skovgaard and Morgen (7) and Skovgaard and Nørrung (13) found 1.7 and 52 % of the investigated samples of fresh pig faeces and fresh cattle faeces *Listeria monocytogenes* positive.

TABLE 2 : Prevalence of *Listeria monocytogenes* on radishes and soil three months after the moment the vegetables were sown in artificially *Listeria monocytogenes* contaminated soil. Washed radishes were rinsed with water until all soil was removed.

	Number of samples examined	Number of positive samples
Unwashed radishes	6	3
Washed radishes	6	0
Soil where previously the radishes have grown	6	1

The prevalence of *Listeria monocytogenes* in the faeces of animals seems to be correlated with the kind of food the animals are fed. *Listeria monocytogenes* is a saprophytic micro-organism that has been regularly selected from silage, fresh and decaying vegetables (4, 5, 6, 7). Animals that are fed silage or vegetables could consequently be exposed to high levels of *Listeria monocytogenes*, while animals fed dry feed, where the possibilities for propagation do not exist under normal storage conditions, will probably have a lower risk of exposure to *Listeria monocytogenes* via feed (13). In this study the eventual correlation between the forage the animals are fed and the presence of *Listeria monocytogenes* in the faeces samples couldn't be traced.

Listeria monocytogenes has regularly been isolated from radishes, cucumbers, cabbage and potatoes (6). Ho et al. (14) suggested that consumption of raw celery, tomatoes and lettuce may be linked to listeriosis in hospitalized, immunosuppressed patients. Linking the results of these investigators to the high percentage of *Listeria monocytogenes* positive faeces samples, manure is likely to function as a transmission vector of *Listeria monocytogenes*. However, *Listeria monocytogenes* couldn't be detected in any manure or manured soil sample. This absence of *Listeria monocytogenes* in manure treated biotopes could be due to the absence of saprophytic conditions in manure, soil and ground water.

In a following step soil and liquid manure were inoculated with a *Listeria monocytogenes* enrichment culture. Although these artificially contaminated soil and manure samples contain at the start 10^5 *Listeria monocytogenes* / ml manure or g soil, all samples were found to be *Listeria monocytogenes* negative after a period of two months. The required conditions for growth and survival of *Listeria monocytogenes* are probably absent in soil and manure and, as a consequence, soil and manure can't be regarded as the natural biotope of *Listeria monocytogenes*.

Since vegetables are found to be regularly contaminated with *Listeria monocytogenes* and since this bacterium can't survive for long periods in soil or manure it is more likely to consider the plant-soil rhizosphere as the natural biotope of *Listeria monocytogenes*. In our experiment 66 % of the unwashed radishes were found to be positive, while only 33 % of the soil samples were previously the radishes had been cultivated contained *Listeria monocytogenes*. The presence of *Listeria monocytogenes* in the plant-soil ecosystem combined with the psychrotrophic nature of *Listeria monocytogenes* could envolve a risk for the public health : consumption of commercially prepared, ready-to-eat, fresh salad vegetables has increased considerably in recent years. Although these vegetables, whether offered for sale in retail groceries, restaurant salad bars or in fast food operations, are kept under refrigeration to maintain sensory qualities, the potential risk for growth of *Listeria monocytogenes* exists.

Manuring practices are probably not directly causal for the contamination of vegetables by *Listeria monocytogenes* but can be responsible for introducing the bacterium into the rhizosphere of the plants. The ecological characteristics of the organism, its ability to multiply at low temperatures and over a broad pH range as well as at a low temperature and under micro-aerobic conditions (1) enables it to multiply readily in the natural environment and in feed and refrigerated foods as well. Animals can become contaminated by grazing or eating vegetables, vegetables can become contaminated by *Listeria monocytogenes* positive faecal matter of these animals.

5. ACKNOWLEDGEMENT

This work was supported by the I.W.O.N.L., Instituut ter aanmoediging van Wetenschappelijk Onderzoek in Nijverheid en Landbouw.

6. REFERENCES

(1) SEELIGER, H.P.R. & JONES, D. (1986). Genus *Listeria* Pirie 1940. In *Bergley's Manual of Systematic Bacteriology* ed. Holt, J.C., Sharpe, M.E., Mair, N.S. & Sneath, P.H.A., pp. 1235-1245. Baltimore : The Williams & Wilkins Co.

(2) DOYLE, M.P. & SCHOENI, J.L. (1986). Selective-enrichment procedure for isolation of *Listeria monocytogenes* from fecal and biologic specimens. *Applied and Environmental Microbiology* **51**, 1127-1129.

(3) FLEMING, D.W., COCHI, S.L., MACDONALD, K.L., BRONDUM, J., HAYES, P.S., PLIKAYTIS, B.D., HOLMES, M.B., AUDUDIER, A., BROOME, C.V. & REINGOLD, A.L. (1985). Pasteurized milk as a vehicle of infection in an outbreak of listeriosis. *The New England Journal of Medicine* **312**, 404-407.

(4) LOVETT, J. (1989). *Listeria monocytogenes*. In *Foodborne Bacterial Pathogens* ed. Doyle, M.P. New York : Marcel Dekker Inc.

(5) WOOLFORD, M.K. (1990). A review : The detrimental effects of air on silage. *Journal of Applied Bacteriology* **68**, 101-116.

(6) HEISICK, J.E., WAGNER, D.E., NIERMAN, M.L. & PEELER, J.T. (1989). *Listeria* spp. found on fresh market produce. *Applied and Environmental Microbiology* **55**, 1925-1927.

(7) SKOVGAARD, N. & MORGEN, C. (1988). Detection of *Listeria*spp. in faeces from animals, in feeds, and in raw foods of animal origin. *International Journal of Food Microbiology* **6**, 229-242.

(8) COLBURN, G.K., KAYSNER, C.A., ABEYTA, C.JR. & WEKELL, M.M.(1990).*Listeria* species in a California estuarine environment. *Applied and Environmental Microbiology* **56**, 2007-2011.

(9) LAMMERDING, A.M. & DOYLE, M.P. (1989). Evaluation of enrichment procedures for recovering *Listeria monocytogenes* from dairy products. *International Journal of Food Microbiology* **9**, 249-268.

(10) CURTIS, G.D.W., MITCHELL, R.G., KING, A.F. & EMMA, J. (1989) A selective differential medium for the isolation of *Listeria monocytogenes*. *Letters in Applied Microbiology* **8**, 95-99.

(11) VANROLLEGHEM, P., PATIGNY, X., RICHARD, J.-F., FACHE, J. & VAN DIJCK (1989). A comparative study of different enrichment and direct plating methods for isolation and enumeration of pathogenic *Listeria* from

foods and other materials. *Proceedings of the Third Forum for Applied Biotechnology, 28 september 1989*, 1393-1400.
(12) WEIS, J. & SEELIGER, H.P.R. (1975). Incidence of *Listeria monocytogenes* in nature. *Applied Microbiology* **30**, 29-30.
(13) SKOVGAARD, N. & NØRRUNG, B. (1989). The incidence of *Listeria* spp. in faeces of Danish pigs and in minced pork meat. *International Journal of Food Microbiology* **8**, 59-63.
(14) HO, J.L., SHANDS, K.N., FRIEDLAND, G., ECKIND, P. & FRASER, D.W. (1986). An outbreak of type 4b *Listeria monocytogenes* infection involving patients from eight Boston hospitals. *Archive of Internal Medicine* **146**, 520-524.

DISTRIBUTION OF ANTIBIOTIC RESISTANT C. PERFRINGENS IN THE ENVIRONMENT

F. HUYSMAN, B. VAN RENTERGHEM and W. VERSTRAETE

Laboratory of microbial ecology
Centre of environmental sanitation
State University of Gent, Faculty of Agriculture
Coupure L 653, B-9000 Gent, Belgium

Summary

Clostridium perfringens resistant to either tetracyclin or oleandomycin were detected in all the investigated manure samples. They were present in high concentrations in the soil of manured fields and virtually absent in the soil of unmanured fields. Antibiotic resistant *C. perfringens* were also detected in the soil at a depth of 100 cm and in the groundwater of unmanured fields, while they were absent in the groundwater of unmanured fields. Antibiotic resistant *C. perfringens* were furthermore detected on the stem of maize, grown on manured fields. It is suggested that manure is a major source for the distribution of antibiotic, resistant *C. perfringens* in the environment.

1. INTRODUCTION

In Belgium large amounts of piggery manure (up to 200 tonnes/ha.year) are commonly applied on agricultural land as organic fertiliser. The effect of high doses of piggery manure on the quality of the environment is questioned. Therefore an intensive study to the contribution of manuring on the distribution of fecal bacteria in the environment was set up.

Antibiotics are largely used in animal husbandry. Various antibiotics (avoparcin, virginiamycin...) are added as feed additives in the mixed fodder of pigs in order to stimulate growth and performance. A series of antibiotics (tetracyclin, oleandomycin, penicillin...) are used for medical purposes. These antibiotics are different from those, used for as feed additives. Several authors have reported a higher incidence of antibiotic resistance among the bacterial populations in the manure of domestic animals in comparison with the bacterial population of unpolluted areas (6,7,4). The antibiotic resistance characteristics had a high stability and antibiotic resistant bacteria were detected in the manure several years after cessation of the addition of antibiotics (5).

The presence of antibiotic resistant bacteria in piggery manure and in soils which have received high doses of manure for several years is investigated. The distribution of antibiotic resistant bacteria in the environment is examined.

2. MATERIALS AND METHODS

Field and groundwater sampling procedure

Soil samples were collected from the A1 horizon of a series of manured and unmanured fields. The samples were stored at 4¦C and analysed within 14 days.

Four manured and three unmanured fields were selected. The manured fields received 80 to 210 ton piggery manure/ha. year. The fields were sandy to sandy loam soils. In each field, three PVC tubes (4 cm diameter and 3 m length) were installed in the soil profile. The tubes were placed in the field on one line at a distance of 10-20 m. The bottom 2.5 m of the tubes was perforated and wrapped with a nylon cloth to prevent the entrance of sand particles. The upper part of the tubes was closed and only opened to sample the groundwater. This was done monthly over a period of two years. Sampling was done by inserting in the pipes a small sterile tube to a depth of 15 cm beneath the water surface. The tubes were connected with a sterile flask of 1 l. The air was sucked out of the sampling flask with a hand pump and the flask filled with groundwater in an aseptic way (10). The first 400 ml was discarded and 500 ml was sampled for analysis. The flasks were stored at 4¦C and analysed within three days.

For bacterial analysis of the stem of maize, maize was harvested in autumn. The maize was collected from a field which in previous years regularly received manure. The lower 10 - 20 cm of the stem of maize was sampled, cut into smaller pieces, shaked for 1 h into fysiological solution, and the number of total and antibiotic resistant *C. perfringens* determined.

Enumeration of *C. perfringens*

An aliquot of 150 ml groundwater was filtered through a paper filter (S & S, 595 1/2) and subsequently centrifuged during 10 min at 10444 g. The pellet was resuspended in 16 ml 0.1 % proteose peptone and 5 ml was plated out on tryptose sulfite cycloserine agar (TSC), TSC with 10 ppm tetracyclin and TSC with 100 ppm oleandomycin using pouring plates. To 1 l TSC agar base (Merck) 5 ml Na_2SO_3 10% and 400 mg D-cycloserine (Oxoid) were added after sterilization. In this way, the selectivity of the TSC medium for *Clostridium perfringens* increased (3,1). The plates were anaerobically incubated at 37¦C for 48 hour. *C. perfringens* spores were selected by heat treatment of the sample at 60¦C for 15 min.

For the analysis of soil and manure, 10 g (wet weight) was mixed with 90 ml of physiological solution and shaken during 1 h. A dilution series was made in 0.1 % proteose peptone and the appropriate dilutions were plated out on TSC amended with the selective supplements using pouring plates.

The resistance against antibiotics was determined by plate counting on TSC and TSC with antibiotics. Following antibiotics were tested (data between brackets represent the concentration in ppm) : chloramphenicol (Cm, 50), ampicillin (Ap, 50), kanamycin (Km, 50), novobiocin (Nb,

50), spiramycin (Sp, 50), olaquindox (Oq, 50), erythromycin (Em, 50), penicillin (Pc, 50), tetracyclin (Tc, 10) and oleandomycin (Om, 100).

3. RESULTS

The degree of resistance against a series of antibiotics of the C. perfringens population of piggery manure and of a manured and unmanured soil was investigated by plate counting on TSC agar and TSC agar supplemented with different antibiotics (Table 1). An overall higher degree of antibiotic resistance of the C. perfringens isolated from piggery manure and manured soil was observed for penicillin, spiramycin, oleandomycin and tetracyclin. Tetracyclin respectively oleandomycin resistant C. perfringens were selected as indicator bacteria because oleandomycin and tetracyclin are commonly used in veterinary practice both for pigs and cattle and because C. perfringens is known to adapt to these antibiotics (9).

	Cm	AP	Km	Nb	Sp	Oq	Em	Pc	Tc	Om
pig manure	3	8	1	100	100	0	7	8	100	22
manured soil	0	1	20	1	100	0	20	100	57	59
unmanured soil	0	2	3	2	8	0	0	3	0	0

Table 1 : Degree of resistance (%) of the C. perfringens population.

Oleandomycin respectively tetracyclin resistant C. perfringens were detected in all the examined manure samples (Table 2) ; they were present in piggery manure as well as in cow dung. The amount of total and antibotic resistant clostridia in manure varied strongly from 2 to 5 log CFU/g.

		C. perfringens	
	Total	Oleandomycin^R	Tetracyclin^R
Pig manure			
P1	5.2	5.2	4.7
P2	2.0	2.3	2.3
P3	4.9	5.2	3.6
P4	3.2	3.2	2.5
Cow dung			
C1	4.7	3.8	4.8
C2	4.3	3.0	2.3
C3	5.2	4.5	3.9

Table 2 : Presence of tetracyclin and oleandomycin resistant C. perfringens in piggery manure and in cow ding (log CFU/g)

In all the manured fields, high numbers of antibiotic resistant C. perfringens were present in the soil (Table 3). In the soil of the unmanured fields low background levels of tetracyclin resistant clostridia were observed. Oleandomycin resistant clostridia were even less numerous; they were only detected in three of the investigated soils. Their densities in these three soils varied from 1 to 1.3 log CFU/g. These fields are currently not manured but may have been manured several years ago. The results indicate that C. perfringens, isolated on TSC with 10 ppm oleandomycin are virtually absent in normal soils and that their presence is a strong indication of manuring. No difference in the total number of C. perfringens was noticed between manured and unmanured soils.

	C. perfringens		
	Total	Oleandomycin[R]	Tetracyclin[R]
Manured fields			
number of fields	13	13	13
log CFU/g	3.4 ± 0.5	2.8 ± 0.5	2.7 ± 0.6
Unmanured fields			
number of fields	14	14	10
log CFU/g	2.7 ± 1.1	0.2 ± 0.4	0.9 ± 0.9

Table 3 : Presence of total, oleandomycin and tetracyclin resistant C. perfringens in the soil of manured and unmanured fields.

Oleandomycin respectively tetracyclin resistant C. perfringens were detected in the groundwater of the manured fields in respectively 42 and 60 % of the samples. The average number was 19 ± 47 respectively 62 ± 171 CFU in 100 ml. Oleandomycin resistant C. perfringens were detected in three of the 63 investigated groundwater samples obtained from the examined unmanured fields. Their number in those three samples was less than 2 CFU in 100 ml. Tetracyclin resistant C. perfringens were detected in 24 % of the examined samples of the groundwater of the unmanured fields in numbers varying from a few to more than 300 CFU in 100 ml.

The distribution of total, tetracyclin and oleandomycin resistant C. perfringens in the soil profile was examined. From four different fields with a long history of manuring soil samples from the plough layer and from a depth of 40, 60, 80 and 100 cm depth were analysed. The average number of total, tetracyclin and oleandomycin resistant C. perfringens in the plough layer of the four examined soil was 3.64, 2.83 and 2.63 respectively. The number of total, tetracyclin and oleandomycin resistant C. perfringens decreased linearly with the depth. The average number at a depth of 100 cm was 1.83, 0.7 and 1.15 respectively. The regression equations for total, tetracyclin and oleandomycin resistant C. perfringens for one soil were y = 4.07 - 0.024 * x, y = 3.16 - 0.023 * x, and y = 2.81 - 0.016 * x respectively with y the number of C. perfringens (log CFU/g) and x the depth in the soil profile (cm). The correlation coefficients (r^2) were 0.76, 0.95 and 0.90 respectively. Similar results were obtained for the other soils.

The stem of maize, grown on soil whigh has been manured for several years was examined. High numbers of C. perfringens were detected on the lower part of the stem. Appromimately 2 and 0.5 % of the C. perfringens bacteria isolated from the stem were resistant to tetracyclin and oleandomycin respectively. These percentages roughly correspond with the percentages antibiotic resistant C. perfringens determined in the soil.

4. DISCUSSION

Total and antibiotic resistant C. perfringens were detected in the whole soil profile and their concentration decreased liniarly with the depth. The results indicate that in soils under normal agricultural use, altochtonous bacteria can migrate vertically for considerable distances. This is important since transport of bacteria in the soil has been mentioned as an important factor for environmental risk assessment of the deliberate release of genetic engineered bacteria (13).

Antibiotic resistant C. perfringens were also detected in the groundwater of manured fields and on the surface of plants, grown on manured fields. By the consumption of groundwater which is not propely desinfected or of plants which are not properly washed antibiotic resistant clostridia can be taken up by man. Domestic animals thus act as producers of antibiotic resistant bacteria and the generalised addition of antibiotics at non-therapeutic levels to animal feed can

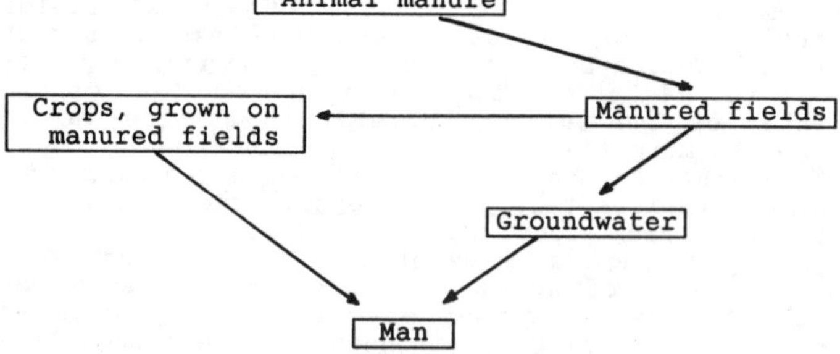

Figure 1 : Distribution of antibiotic resistant C. perfringens in the environment.

increase the dissemination of the antibiotic resistance genes in the environment. The different pathways by which antibiotic resistant clostridia can be consumed by man are illustrated in Figure 1. The distribution of antibiotic resistant bacteria and the possible public health risks are currently attracting increased interest (11). C. perfringens is very resistant to the classical water treatment systemes. Fujioka and Shizumura (2) reported decreases in the concentration of C. perfringens spores of only 1 log scale by a chlorination desinfection process in a waste treatment plant. Similar results have been reported by Sorensen et al. (12). Furthermore Rood (8) reported the conjugative transfer of tetracyclin resistance genes between C. perfringens species. Hence the presence of antibiotic resistant C. perfringens in manure and the distribution in the environmnet warrants further research with respect to longterm public health hazards.

ACKNOWLEDGMENTS

We acknowledge the excellent technical assistance of H. Van Hove and R. Rygole. This work was supported by the Institute of Scientific Research, Belgium (Instituut voor wetenschappelijk onderzoek in nijverheid en landbouw, IWONL).

REFERENCES

(1) EMSWILER, B.S., PIERSON, C.J. and KOTULA, A.W. (1977) Comparative study of two methods for detection of Clostridium perfringens in ground beef. Applied and Environmental Microbiology 33, 735-737.

(2) FUJIOKA, R.S. and SHIZUMURA L.K. (1985) Clostridium perfringens, a reliable indicator of stream water quality. Journal of Water Pollution Control Federation 57, 986-992.

(3) HAUSCHILD, A.H.W. and HILSHEIMER, R. (1974) Enumeration of foodborne Clostridium perfringens in egg yolk-free tryptose-sulfite-cycloserine agar. Applied Microbiology 27, 521-526.

(4) HINTON, M., KAUKAS, A. and LINTON, A.H. (1986) The ecology of drug resistance in enteric bacteria. Journal of Applied Bacteriology Symposium Supplement 77S-92S.

(5) LANGLOIS, B.E., CROMWELL, G.L., STAHLY, T.S., DAWSON, K.A. and HAYS, V.W. (1983) Antibiotic resistance of fecal coliforms after long-term withdrawal of therapeutic and subtherapeutic antibiotic use in a swine herd. Applied and Environmental Microbiology 46, 1433-1434.

(6) MERCER, H.D., PORCURULL, D., GAINES, S., WILSON, S. and BENNET, J.V. (1971) Characteristics of antimicrobial resistance of Escherichia coli from animals: relationschip to veterinary and management uses of antimicrobial agents. Applied Microbiology

22, 700-705.

(7) MINTON, N.P., MARSH, J. and ATKINSON T. (1983) The R-factors of multiple antibiotic resistant fecal coliforms isolated from a domestic dog. Journal of Applied Bacterialogy 55, 445-452.

(8) ROOD, J.I. (1983) Transferable tetracycine resistance in Clostridium perfringens strains from porcine feces. Antimicrobiology Agents Chemotherapy 13, 871-880.

(9) ROOD, J.I., MAHER, E.A., SOMERS, E.B., CAMPOS, E. and DUNCAN, C.L. (1978) Isolation and characterisation of multiple antibiotic resistance Clostridium perfringens strains from porcine feces. Antimicrobiology Agents Chemotherapy 13, 871-880.

(10) SCHAUB, S.A. and SORBER C.A. (1977) Virus and bacteria removal from wastewater by rapid infiltration through soil. Applied and Environmental Microbiology 33, 609-619.

(11) SOKARI, T.G., IBIELE, D.D. and OTTIH, R.M. (1988) Antibiotic resistance among coliforms and Pseudomonas spp. from bodies of water around Port Harcourt, Nigeria. Journal of Applied Bacteriology 64, 355-359.

(12) SORENSEN, D.L., EBERL, S.G. and DICKSA R.A. (1989) Clostridium perfringens as a point source in non point polluted streams. Water Research 23, 191-197.

(13) TIEDJE, J.M., COLWELL, R.K., GROSSMAN, Y.L., HODSON, R.E., LENSKI, R.E., MACK, R.N. and REGAL, P.J. (1989) The planned introduction of genetically engineered organisms: ecological considerations and recommendations. Ecology 70, 298-315.

PHOSPHORUS AND HEAVY METAL LOADS IN SEWAGE AND IN MAIN DRAINAGE CHANNELS BASED ON THE RESPECTIVE ELEMENT LOADS IN SEWAGE SLUDGE

Candinas, T., Chassot, G., Besson, J.M., Lischer, P.
Swiss Federal Research Station for Agricultural Chemistry and Environmental Hygiene, CH-3097 Liebefeld-Bern, Switzerland

Summary

In Switzerland a large proportion of the population is connected to sewage treatment plants (STP). Sludge quality is well known. Sludge quantities from every STP are measured. On the basis of these data, it was possible to determine, with great precision, the resulting loads of phosphorus, lead, and copper in sewage sludge. This formed the basis for the calculation of the corresponding total element loads in waste water as well as the remaining loads after purification. Total efficiency of waste water purification came to between 50% and 60% depending on the element. Element loads in sewage fed to main drainage systems are still very significant for the aquatic environment in spite of improvement in sewage treatment. With respect to lead and copper, the loads are much higher than has been previously assumed.

1. Introduction

Sewage sludge is, in many respects, a very interesting and important material:
- In all countries with well-developed sewage treatment, sewage sludge is a quantitatively **important** type of **waste** and requires special care with respect to safe disposal.
- Sewage sludge is a useful **multi-component organic fertilizer** in agriculture.
- Sewage sludge is also a very **interesting substance for environmental monitoring.**

Three requirements must be met in order to use sewage sludge for environmental monitoring:
- A major part of waste water produced is fed to sewage treatment plants.
- Because of intensive monitoring, the quality of sewage sludge is well-known.
- The quantity of sewage sludge produced in each treatment plant is documented.

If these requirements are met, then the loads of environmentally relevant substances contained in sewage sludge from waste water can be estimated. The result obtained is very reliable because the substances analyzed in sewage sludge occur in relatively high concentrations and are not difficult to analyze, and the fluctuations in the concentrations - as compared to those in waste water for example - are moderate. Element loads in sewage sludge are also extremely suitable for estimating the total quantities of substances which are supplied to the sewage system in defined catchment areas and which are delivered to the main drainage channels together with the purified waste water.

In Switzerland, the first two requirements are met to a sufficient degree. Since the beginning of the 80's, over 80% of the waste water from private households and industry has been delivered to over 900 sewage treatment plants. Since 1971, the quality of the sludges has been monitored with increasing intensity. The data are stored in a comprehensive data bank.

There was less information available on the quantities of sludge produced. An estimate of production based on the number of residents connected to sewage treatment plants and the population equivalent values, or the processed quantity of waste water is in no way sufficient, as shown by our studies. The investigation and evaluation of these data required a great deal of our time.

In Switzerland, initial attempts were made in 1982 to carry out load calculations for heavy metals in sewage sludge (1). In 1986, a continuing study by Nolte (2) was published, and in 1988 a project was initiated at FAC with the support of COST 681 called "Sewage sludge quality and sewage sludge utilization". To date, two reports have resulted from this project which applied this model to some degree (3,4).

In this report it is attempted to estimate - on the basis of element loads in sewage sludge - loads of phosphorus, lead and copper in Switzerland, which are delivered via the sewage system to the main trans-boundary drainage channels. We see this report as a critical contribution to the European Action Program "Rhine" (5), which, among other things, assigns every adjoining country the task of undertaking an inventory of the actual pollutant inputs to the Rhine river.

2. Materials and Methods

By the end of 1989, there were 924 central sewage treatment plants (STP) operating in Switzerland, which processed municipal as well as industrial and commercial sewage (mostly after pre-purification at the factory: **indirect contributors**) in various mixtures, together with rain and extraneous water. In addition, an unknown number of purely industrial STPs exist, which feed their waste water directly into the main drainage channel (**direct contributors**). Our work does not take the strictly industrial STPs into account. Thus, total loads are higher than those calculated here.

Figure 1 shows population development and the number of residents connected to sewage treatment plants between 1960 and 1990. The presentation indicates that the development of technical water protection is quite advanced in Switzerland. **Figure 2** shows that monitoring of sewage sludge quality has existed since 1974 and has been systematically expanded through the years.

Normally, estimates of sludge loads are based on empirical values. For example, it can be calculated that per resident connected to the sewage system (p) and population equivalent (pe) about 65 g of dry matter (d.m.) per day is produced after mechanical, biological and chemical purification and subsequent anaerobic sludge stabilization. This is equivalent to approximately 25 tons of dry matter per 1000 p+pe and year.

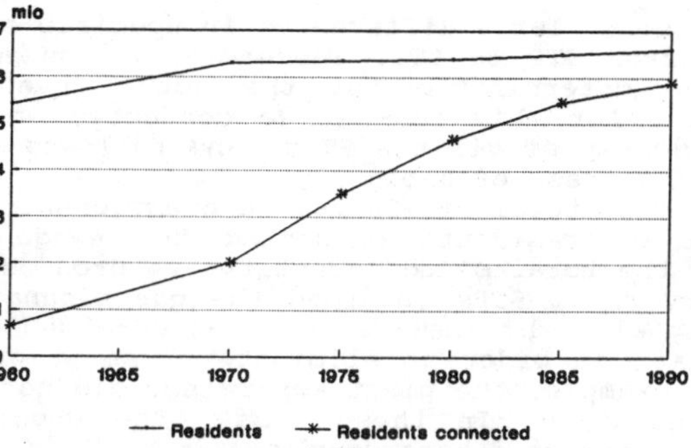

Fig. 1: Number of residents and residents connected to STPs in Switzerland since 1960 (7).

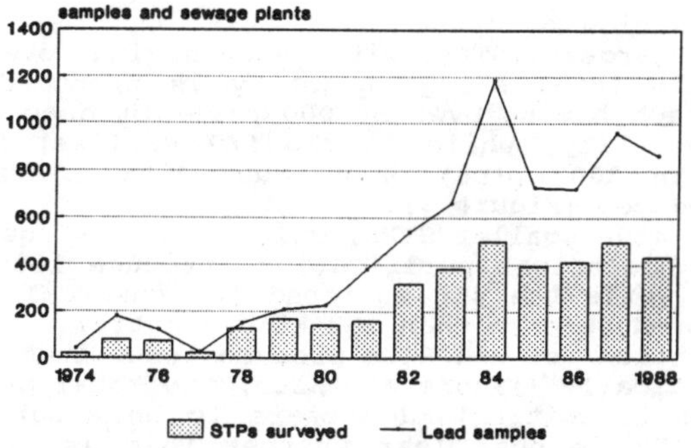

Fig. 2: Number of STPs investigated and number of sewage sludge samples since 1974.

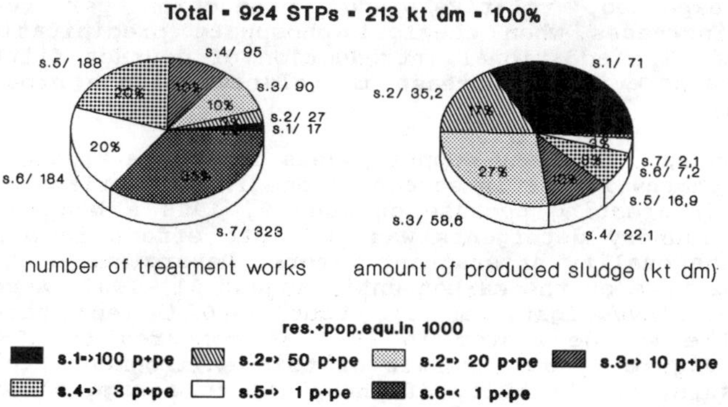

Fig. 3: Number of sewage treatment plants and sludge production in STPs of different size categories in 1989.

For various reasons, large differences in specific sludge production exist from STP to STP. Therefore, essential to this report was the determination of the actual quantities of sludge produced. For this reason, we conducted an extensive questionnaire survey of all the STPs, and followed with telephone calls in the case of ambiguity.

We received a total of 723 usable questionnaires. Based on the number of residents connected to sewage treatment plants, 97% of the total sludge quantity are produced in these STPs. Only very small STPs returned the questionnaire incomplete or not at all. The answers received were subjected to a plausibility test in order to eliminate those which were not usable if, for example, the plant supervisor did not correctly understand a question. In these cases, the questions were followed-up and corrected accordingly.

3. Results and Discussion

3.1 Sludge production

In the 17 largest STPs, with each having over 100,000 p+pe, 1/3 of the total sludge quantity is produced. The 44 largest STPs (each having over 50,000 p+pe) produce about 50% of the total quantity, while in the 507 smallest STPs (all having less than 3000 p+pe) only about 4.5% of the sludge quantity is produced (**Figure 3**).

A total of 449 smaller STPs, which are only equipped for mechanical and biological purification, produce less than 50 kt dry matter, while the 436 equipped for chemical phosphate precipitation produce more than 150 kt dry matter.

A large portion of all STPs produce less sludge than was expected theoretically (**Figure 4**). Also, the total quantity of sludge produced in Switzerland appears to be about 1/5 less than expected. It is not clear whether this is due to the possibility that the purification capacity of the STPs is not as good as expected, or because less solid material is produced by p+pe than was assumed.

As expected, relative sludge production per resident clearly increases when chemical phosphate precipitation is introduced. The additional introduction of contact filtration thus shows no evident increase in relative sludge production (**Figure 5**).

3.2 Phosphorus, lead and copper levels in sewage sludge

Phosphorus levels in sewage sludge in Switzerland are especially interesting because on June 9, 1986 a ban on phosphate in laundry detergents was put into effect in order to improve the quality of stagnant waters. Detergents containing phosphates were on the market until August 31, 1987. According to earlier investigations (3) about 4500 t less phosphate reached the sewage system in 1987 as compared to 1983. As shown in **Figure 6**, as a result of this environmental protection measure, the quantity of phosphate in sewage sludge has clearly decreased.

In STPs without chemical phosphate precipitation, the weighted mean concentration of phosphorus in sewage sludge in 1989 was still 80% of the level found in 1984, in STPs with

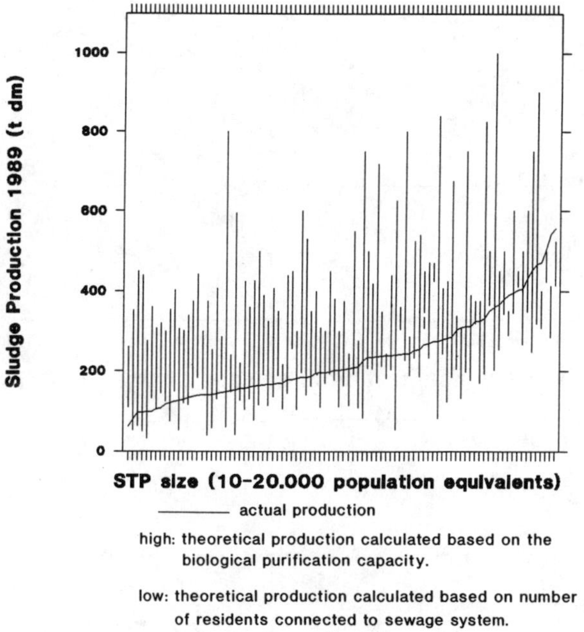

Fig. 4: Comparison between actual and theoretical sludge production (t dry matter) of some STPs.

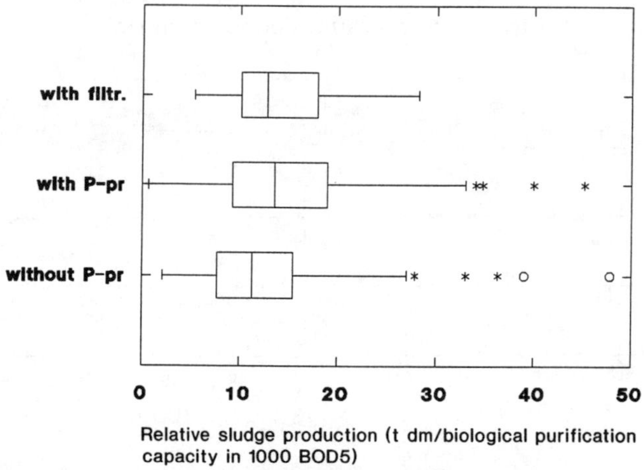

Fig. 5: Sludge production in sewage treatment plants relative to the number of residents connected to STPs and population equivalents, separated according to P-treatment of waste water (for 1989, only stabilized sludges).

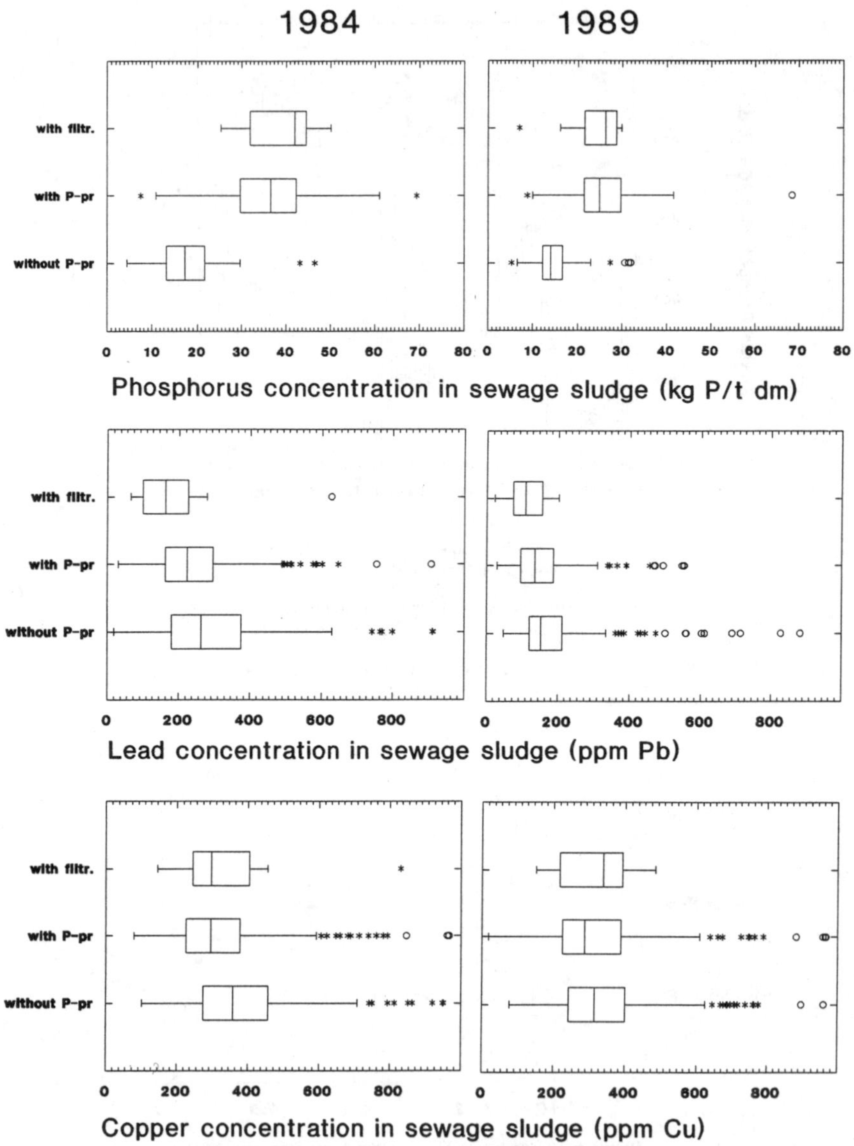

Fig. 6: Content of phosphorus, lead and copper in sewage sludge in 1984 and 1989, separated according to P-treatment of waste water.

phosphate precipitation 66%, and in STPs with additional contact filtration 68%. All STPs which underwent a change in the phosphorus treatment in sewage during the investigation were eliminated from these calculations.

Large changes in the levels of **lead** in sewage sludge between 1984 and 1989 are to be expected, as total lead emissions have decreased drastically since 1984 according to the Federal Office for the Environment (6). According to this source, 510 t Pb was emitted from traffic, 4.8 t Pb from households, and 170 t Pb from industry and commercial sources during 1984. For 1990 expected emissions are 290 t Pb (traffic), 4 t Pb (households) and 150 t Pb (industrial and commercial sources). In 1990, therefore, input of lead to the environment was reduced by 240 t (including the unknown quantity to aquatic ecosystems) as compared to 1984. The effect this reduction had on lead levels in sewage sludge can also be seen in **Figure 6**.

When comparing the two years, it becomes apparent that the weighted mean lead concentrations have decreased by about 40% (1984: 374 ppm Pb, 1989: 225 ppm Pb) and that the variation in concentrations have also clearly decreased. It is true for both years, that with the extended purification lead concentrations decrease, which we attribute to the fact that lead is only soluble in sewage water in very small quantities.

Copper is an environmentally relevant element, whose occurrence in the aquatic environment has not been considerably restricted. With copper, the decrease in the mean weighted concentration is much less spectacular, as shown in **Figure 6** (1984: 434 ppm Cu, 1989: 375 ppm Cu).

In the case of copper, the decrease in concentration with increasing purification efficiency at the STP is not apparent. In 1989, the concentration at STPs with contact filtration was even the highest. It should be noted, however, that this observation is based on only a comparatively small number of STPs (16 STPs).

Based on the results obtained it can be said that substances utilized for chemical phosphate precipitation have no considerable effect on the inputs of lead and copper to sewage sludge.

3.3 Phosphorus, lead and copper loads in sewage sludge

The element loads for each STP are calculated by multiplying the concentration times the dry matter production. The total load of all STPs can be calculated very exactly as the sum of all individual loads. In this way the following element loads in sewage sludge have been determined:

Element		Load in 1984	Load in 1989
Total phosphorus	(t P/year)	5563	4906
without P-precipitation	(t P/year)	1387	774
with P-precipitation	(t P/year)	3822	3761
with contact filtration	(t P/year)	354	371
Lead	(t Pb/year)	66	48
Copper	(t Cu/year)	77	80

When looking at this comparison of loads, it should be taken into account that the percentage of residents connected to STPs increased from about 84% to 89% between 1984 and 1989, and that the removal of sludge from sewage has been improved.

3.4 Phosphorus, lead and copper loads in sewage before and after treatment

Element loads in sewage sludge become especially important if they can be used as a basis for the calculation of mass fluxes through the entire sewage system. In Switzerland, results from the calculation of element loads in sewage sludge are very reliable due to favorable circumstances. We suggest that this should be taken advantage of in order to obtain useful estimates of element loads in waste water and main drainage channels.

This can be done approximately by applying the following simple formula:

$$L(SEW) = L(MDC) + L(SLU)$$

L(MDC) = Load in main drainage channel
L(SEW) = Load in sewage system
L(SLU) = Load in sewage sludge

$$L(MDC) = L(SLU)\ [100\ eff^{-1}\ (1 + x)\ (1 + y) - 1]$$

eff = Efficiency of the STP (different depending on technological equipment) relative to the element separation from sewage
x = Factor, accounts for the overflow of sewage from the sewage channel system during rainstorms
y = Factor, accounts for the additional element load in the sewage system from residents who are not connected to STPs

Because the efficiency of separation reported in the international literature varies to a large degree, these calculations hold a certain degree of uncertainty. Therefore, a variation of about +/- 20% for the following loads is to be expected. For our calculation, we use an average efficiency according to the Swiss Federal Institute for Water Resources and Water Pollution Control EAWAG (8) (adopted efficiencies: P, in STPs without chemical P-precipitation: 40% (1984), 50% (1989), with chemical P-precipitation: 77% (1984 and 1989), with contact filtration: 90% (1984 and 1989); Pb, in all STPs 70%; Cu, in all STPs 65%. Values for 1984 are calculated with a percentage of residents connected to STPs of 83% and a loss through overflow of 13%, those for 1989 with a percentage of residents connected to STPs of 88% and a loss through overflow of 10%):

```
L(MDC) 1984 =        3415 t P (STPs without P-precipitation)
                     3053 t P (STPs with P-precipitation)
                      191 t P (STPs with contact filtration)
            Total = 6659 t P         (L(SEW) = 12,222   t P )
                  =   64.9 t Pb      (L(SEW) =    152.8 t Pb)
                  =   86.7 t Cu      (L(SEW) =    163.3 t Cu)

L(MDC) 1989  =       1180 t P (STPs without P-precipitation)
                     2403 t P (STPs with P-precipitation)
                      149 t P (STPs with contact filtration)
            Total = 3732 t P         (L(SEW) =  8,638   t P )
                  =   38.7 t Pb      (L(SEW) =     86.9 t Pb)
                  =   75.6 t Cu      (L(SEW) =    155.9 t Cu)

L(MDC) 1984-1989  =-2927 t P         (L(SEW) = - 3,584   t P )
                  =-  25.9 t P       (L(SEW) = -    65.9 t Pb)
                  =-  11.1 t Cu      (L(SEW) = -     7.4 t Cu)
```

4. Conclusions

The results from these load calculations show that:

- the ban on phosphate in laundry detergents has had a clearly positive effect on the P-inputs to the main drainage channel;
- in spite of all the recent increased efforts in order to reduce P-pollution of the main drainage channels, the P-loads resulting from sewage are still very important for the surface waters;
- the measures which have been taken in order to reduce lead emissions have also led to a clear decrease in lead concentrations in sewage sludge and in lead inputs to main drainage channels;
- the P-load entering the Rhine river acting as as main drainage channel via waste water compares very well with the data collected in a completely different manner as part of the Action Program "Rhine" (5);
- the lead and copper loads entering the Rhine river via waste water seem to be very much higher than was estimated by the Action Program "Rhine" (5);
- sewage sludge is a very good indicator for the pollution of the aquatic environment within defined catchment areas, provided the following criteria are met: high percentage of residents connected to STPs, availability of a comprehensive data bank relative to sewage sludge quality, and availability of detailed information about the sludge quantity produced by each STP.

We are aware of the fact that our calculations may be subject to criticism because of some uncertainties. This concerns mainly the separation efficiency for materials from waste water which differs among sewage treatment plants, and the losses caused by overflow constructions. The model presented here, however, can still be refined. We believe that this effort is worthwhile.

5. Literature

1 Furrer, O.J., Candinas, T. and Lischer, P.: Schwermetallgehalt der Klärschlämme in der Schweiz. Landw. Forsch. SH 39, 1982.

2 Nolte, R.F.: Herkunft und Stoffbilanz der sieben Schwermetalle aller 562 hessischen Kläranlagen. Z. Gewässerschutz, Wasser, Abwasser 85, Aachen 1986.

3 Candinas, T., Gupta, S.K., Zaugg, W., Lischer, P. and Besson, J.M.: 15 Jahre Schwermetalle im Klärschlamm. Schweiz. Landw. Fo. 28 (3/4), 1989.

4 Candinas, T.: Ban on phosphorus in detergents: the effects on the phosphorus contents of Swiss sewage sludge and on the efficiency of phosphorus elimination by sewage treatment plants. Proceedings of the Commission of the European Communities, Sewage Sludge Treatment and Use, Amsterdam, 1988.

5 Internationale Kommission zum Schutze des Rheins gegen Verunreinigung: Aktionsprogramm "Rhein", Brüssel, 1989.

6 Bundesamt für Umweltschutz: Vom Menschen verursachte Schadstoff-Emissionen in der Schweiz, 1950-2010, Bern, 1987.

7 Gujer, W.: Die Entwicklung des Klärschlammanfalles in der Schweiz. Mitteilungen der Eidg. Anstalt für Wasserversorgung, Abwasserreinigung und Gewässerschutz 28, Dübendorf, 1989.

8 Heuseler G., Siegrist, H.: Personal communication.

THE STABILISATION AND DEWATERING OF MUNICIPAL SEWAGE SLUDGE AT LOW RATE TREATMENT PLANTS

V. CHERNISHOV, F. KARAGEZOV and E. CHERNISHOVA
Makeyevski Civil Engineering Institute

Summary

Investigations into the aerobic stabilisation of unthickened surplus activated sludge and primary sludge have shown that, when synthetic fibres are added to the stabilisation process, the time required to achieve stabilisation can be halved. A 45-54% reduction in sludge solids is achieved.

The process of stabilisation reduces the specific resistance to filtration (SRF) of a sludge, and the stabilised sludges can be dewatered under low vacuum pressure, without the use of conditioning chemicals, on drying beds. Filtration media have been selected which do not require regeneration. The stabilisation and dewatering techniques have been developed for application at low rate treatment plants.

1. INTRODUCTION

The main technique used for sludge treatment is stabilisation. At the low rate treatment plant stabilisation is mainly carried out under aerobic conditions. Surplus activated sludge, or co-settled primary and surplus activated are stabilised. However, a disbenefit of aerobic stabilisation is the 10-12 days retention time that is provided. Whilst there are a number of ways of intensifying this process, each results in additional costs being incurred (1,2,3). The second sludge treatment process is dewatering, to reduce sludge volume. Dewatering on drying beds using natural media bases requires the use of significantly large areas because of the low rate of filtration.

Mechanical dewatering techniques can be effective provided that sludge is preconditioned using chemicals. In some cases, however, the use of chemicals creates difficulties, requiring additional equipment and storage facilities.

In our view a contribution to solving or eliminating the above problems would be to reduce the retention time in the aerobic stabilisation process, through accelerating the rate of organic degradation, and cause an improvement in the dewatering properties of the stabilised sludge, such that chemical conditioners would not be required. We have undertaken research with the goal of resolving these problems and realising these objectives.

2. SLUDGE STABILISATION UNDER AEROBIC CONDITIONS USING SYNTHETIC FIBRE FILLERS

As the aerobic stabilisation of sludge proceeds, a number of processes that change the physical and chemical properties of the sludge take place. One of the properties that undergoes change is the dewatering characteristic of sludge as measured by its specific resistance to filtration (SRF). Initially, the unstabilised sludge has a high SRF value which may be ascribed to a high concentration of colloidal particles, perhaps resulting from the activated sludge. As stabilisation proceeds the colloidal structure breaks down, leading to a reduction in SRF. In agreement with other studies (4), our results indicate that a minimum SRF value is obtained following an aeration time of 2 to 4 days. At a retention time of this duration, however, full stabilisation is not achieved, and operation at longer retention times results in an increase in the SRF value. This is demonstrated particularly by data for the thickened stabilised sludge at a solids content of 25-30 g/l.

Typically, the reduction in solids content as the sludge stabilises in accompanied by a reduction in the pH value, and by an increasing BOD and nitrate content of the sludge liquor.

The research we have conducted has shown that reducing the retention time of aerobic stabilisation while increasing the rate of biodegradation is possible, employing nitrification and denitrification simultaneously. Both processes have been observed in an aerobic stabilisation reactor to which was added a known quantity of synthetic filler manufactured in our pilot plant.

The effects of adding the synthetic filler to the stabilisation reactor, in terms of the reduction in sludge volatile solids, are illustrated in Figure 1. Unthickened activated sludge with a total solids (TS) concentration of 6.7g/l was used in this experiment, conducted at a temperature of 18-20°C. These data demonstrate the considerable increase in volative solids biodegradation rate that can be achieved through the addition of synthetic fibre filler.

Organic, amoniacal and oxidised nitrogen, pH value, and oxygen consumption rate (OCR) were measured as a function of stabilisation time.

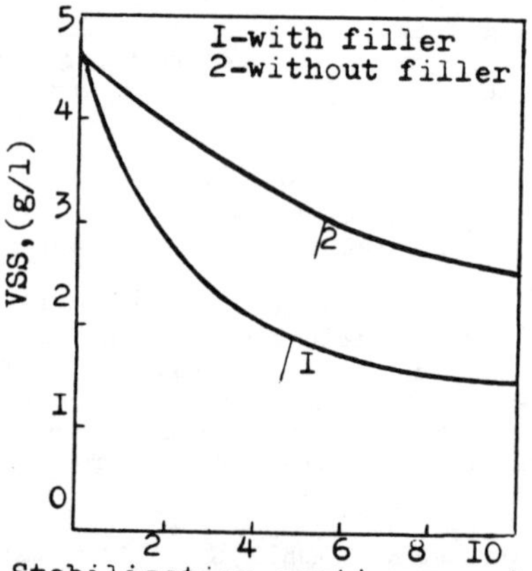

Fig.I VSS biodegradation

The data show that, where filler has been added, the concentration of total nitrogen decreases with an increasing period of stabilisation, and that this does not occur in the absence of filler. The reduction in total nitrogen concentration and increase in pH value as stabilisation proceeds in the presence of filler indicates that denitrification is taking place. The OCR value has been used as the criterion for sludge stability, and completion of the stabilisation process is considered to be achieved once the OCR decreases to a practically constant level.

It is also possible to deduce that sludge is stabilised more quickly as a result of the addition of filler, complete stabilisation being obtained after 3 days compared to 6 days in the absence of filler. With filler added, the total solids content of the sludge reduces from 6.7g/l initially to 3.2g/l after 3 days.

Numerous stabilisation experiments with unthickened sludge (TS less than 10g/l) to which filler has been added have confirmed a 45-54% reduction in total solids content after 3 days. At initial solids contents in excess of 10g/l the degree of enhancement is reduced until, at initial solids concentrations of 20-25g/l the filler has no beneficial effect and stabilisation occurs with or without the addition of filler.

The stabilisation of mixed unthickened activated sludge and primary sludge was also investigated, over retention times of 4-5 days. The degree of volatile solids reduction depended on the ratio of the two sludges in the mixture. At an activated to primary sludge solids ratio of 1:1 the degree of biodegradation observed was 50%.

The stabilisation of all sludges to which filler was added was accompanied by a reduction in specific resistance to filtration (SRF). Initially, sludges had an SRF value in the range of 1800-2500 $*10^{10}$ m/kg.

After stabilisation for 1 day the SRF reduced to 50-200$*$ 10^{10} m/kg, and remained close to that value thereafter.

3. INVESTIGATION OF STABILISED SLUDGE DEWATERING

The low values of SRF suggest the possibility that the stabilised sludge can be dewatered effectively without using chemicals.

Examination of well-known methods of dewatering indicated that a drying bed operated under vacuum could be used. One of the requirements of the filtration media is its ability to resist penetration by sludge particles. That makes it possible to avoid frequent regeneration or replacement of the filtration media and to remove the dewatered sludge simply through overturning the bed. In the course of this study it became possible to select a synthetic fabric having the desired characteristics. In order to maximise the sludge filtration rate, and throughput, it was observed that a (vacuum) under pressure of 0.04 to 0.05 MPa was required, that sludge should be applied to a depth of 0.10 to 0.15m, and that the applied sludge should have a solids content of 2%. Under these conditions the surface loading on the bed is 0.3 to 0.4m^3/m^2.h. In calculating this loading the time taken to discharge the sludge should be taken into consideration. The dewatered sludge has a solids content of about 13-15%. We have developed a design for the filtration system.

4. CONCLUSION

4.1 The addition of synthetic fibre filler to stabilisation reactors fed with surplus activated sludge and primary sludge of less than 10g/l TS results in a halving of the retention time necessary to achieve full stabilisation. Under these conditions a 45-54% reduction in total solids is achieved.

4.2 As sludges undergo stabilisation the specific resistance to filtration reduces from 1800-2500$*$ 10^{10} m/kg to 50-200 $*10^{10}$ m/kg over the first day. Little change occurs thereafter.

4.3 Stabilised sludge can be dewatered under a vacuum of 0.04-0.05 MPa without the use of conditioning chemicals. A filtration system has been developed which does not require regeneration of the filtration media. The design load is 0.3-0.4 m^3/m^2.h of the stabilised sludge at 2%TS.

PROCESS INTENSIFICATION FOR UPGRADING SEWAGE WORKS

N. ZOTOV, Makyevski Civil Engineering Institute

Summary

Over the life of a sewage works there will come a time when either the sewage flow into the plant exceeds the designed hydraulic loading, or the sewage strength increases to the point that the plant is no longer producing effluent that meets its consents, or both.

While the normal approach to increasing the capacity or treatment of a sewage works has been to expand the facilities, for example by putting in a larger primary settling tank or aeration tank, there is an alternative. This is to redesign the works to convert existing aeration tanks, percolating filters, and, if necessary, settling tanks, to an activated sludge plant where the aeration basin has foam packing added and the aeration systems are modernised to cope with the increased duty. This approach aims to create a treatment works that combines the best qualities of both activated sludge and percolating filter processes; as a bonus, the reliability and stability of the modified process is superior to both activated sludge and percolating filters.

The foam packing put into the aeration tank can increase the permitted tank loading by 10-20%; since the packing may be anaerobic at its centre, both aerobic and anaerobic reactions are taking place within the aeration basin, and this may lead to the degradation of a wider range of materials than a purely aerobic basin can achieve.

The mixed liquor suspended solids in the aeration basin, taking into account both the solids adhering to the packing and the free solids, can attain concentrations up to 4,800 mg/l. These high concentrations are responsible for the improved treatment of the modified process, and also for the increased stability against such effects as hydraulic shocks or toxins in the influent.

(The modifications described appear to be to convert sewage works to Ringlace-like activated sludge plants)

1. INTRODUCTION

Before building work on a sewage works can begin, some preliminary design work must be carried out. This requires determining the flowrate of sewage to the works, the concentration of BOD, suspended solids and amonia in the influent, the diurnal variations to be expected, as well as the treatment provided by existing processes on the site. From this can be developed the requirements for the new processes to be installed.

The use of mechanical screens should be considered as the first stage, and the cost of installing these screens compared against the reduction in solids that they may produce. Fine screens are useful not only for removing gross solids but also for removing floating solids, providing the screens have been designed to permit this.

While provision should be made to remove fats and oils floating on the surface of the sewage, primary settling tanks may not be necessary where the sewage uses packed activated sludge systems with subsequent aerobic sludge stabilisation. All the solids that would have been discharged in the sludge from a primary tank should be sent to the aeration tank for mineralisation of the organic fraction. The foam packing supplies a surface where both biomass and other solids can adhere; organic material trapped in the packings is used to feed the biomass.

2. BIOREACTOR DESIGN

The packed activated sludge tanks can be two or three stage systems, depending on the initial pollutants concentration and the availability of suitable tanks when upgrading the works. When calculating the effluent concentration from each of the stages, it should be remembered that the first stage treatment will solubilise some of the suspended solids, increasing the soluble fraction of the BOD, and this effect should be determined. For design purposes it is assumed that mineralisation only takes place in the second and subsequent stages.

The treatment to be expected from two to three stage trains depends on the oxidation rates of the organic material. The table gives typical results.

Sewate treatment from Packed Activated Sludge Plants

Three-stage train			Two-stage train	
Stage 1	Stage 2	Stage 3	Stage 1	Stage 2
BOD, mg/l				
Influent reduced to 50	Influent 50 effluent 15	Influent 15 effluent 3-5	Influent reduced to 15-30	Influent 15-30 effluent 3-5
Oxidation rate, mg 02/gh				
142	85	57.9	85	57.9

The rate at which biomass mineralises BOD depends on the concentrations of dissolved oxygen, BOD, and the biodegradability of the organic material; it is possible that inhibition may occur from the decay products of the biomass.

The mixed liquor suspended solids concentration can be controlled by choosing the packing density. This packing can support 20-40 g of active biomass in one metre, and up to 200m of packing can be installed per cubic metre of tank. This limits the MLSS to 4,800 mg/l, of which 2,100 mg/l is free biomass.

It is important that the aeration system can supply the oxygen demand of the biomass. While diffused air, jet or airlift aerators can be used, only airlift systems are suitable for the final stage.

3. EXAMPLES OF PACKED ACTIVATED SLUDGE PLANTS

Possible sewage works designs include

a) primary settling tanks, percolating filters, humus tanks;
b) extended aeration activated sludge plants;
c) primary settling tanks, activated sludge, secondary settling tanks.

3.1 Example 1

A treatment plant in Hkabarovsk was designed for 1,270m3/d; over time the sewage flow has increased, so that the plant now treats 4,00m3/d. The per capita flow had not changed from the expected 260 l/d. The plant comprises an activated sludge tank 30mx24mx3.1m and two secondary settling tanks, each 3m in diameter. The BOD flow to the plant is 1,155 kg/d, and suspended solids 1,000 kg/d.

By dividing the aeration tank into three sections, each filled with foam packing, the plant was able to satisfactorily treat the current flowrate. Only the first two stages have a settling tank with sludge recycle. The first two stages use diffused aeration systems to prevent solids deposition, while the third stage uses an airlift system; the third stage has virtually no free biomass, all the biomass being attached to the packing.

The retention time in the first stage is 1.55 h; in the second, 1.51h; and in the third 1.4 h. The air demands are 1,630 m3/h for the first two stages and the third stage requires 800 m3/h of air. The air demand required by the packing is 160 l/min/m3 for the first two stages and 120 l/min/m3 for the final stage. The effluent quality from the final stage is 5mg/l of BOD and 5 mg/l of suspended solids.

3.2 Example 2

Sewage with a BOD of 200 mg/l and a suspended solids content of 350 mg/l was entering the Balchisarai treatment works at a rate of 10,000 m3/d; unfortunately the works had only been designed to cope with 4,500 m3/d. The works comprised of gift removal, four primary settling tanks each of 12m diameter, four high-rate percolating filters, each 12mx12mx3m, and two humus tanks, each 9m in diameter.

To double the plant capacity, one primary settling tank was filled with the foam packing and operated anaerobically, while the other three primary tanks were filled with foam packing and operated aerobically, using fine-bubble aerators. These comprised the first stage where the initial BOD was reduced to 30 mg/l. A second aeration stage was created from the percolating filters. These were emptied of the old packing and foam packing installed in their place. The filters were modified to operate filled with liquid and an airlift aeration system installed.

465 km (28 tonnes) of foam packing were installed while uprating the plant, and the total air supply was 10,272 m3/h.

As can be seen from these two examples not only is the treatment capacity of the works improved, but also the effluent quality, by installing these packed aeration systems.

4. CONCLUSIONS

The use of packing with activated sludge allows plants to be uprated with a fairly low investment in the site. The annual savings could amount to 100-500 thousand roubles compared to other means of uprating these types of plant.

5. REFERENCES

1. Construction norms and rules. Sewerage. The outside mains and facilities. SNiP 2.04.03-85. - M.: Stroiizdat, 1985 - 97p.
2. Synthetic elements of type "Iorsh". TY 40.2070792.001-89.
3. Kulikov, N.I. Sewage water biological treatment with the communities of the free floating and adhesed microorganisms and hydrobionts. theses dr.techn.sc. - Makeyevka, 1987 - 32p.
4. Zotov, N.I. Sewage treatment with biodrums. theses cand. sc. (eng) - M.:1988 -32p.
5. Kulikov, N.I., Zotov, N.I. The sewage treatment plants of low overall dimension for the village. Water Supply and Sanitation Engineering, 1987 -N6.
6. Beletski, B.F., Zotov, N.I., Yaroslavski, L.V. The construction of water and sewage facilities. manual - M.: Stroiizdat, 1989, 218p.

FUNDAMENTALS FOR DERIVING REGIONAL SEWAGE SLUDGE MANAGEMENT CONCEPTS

Dr. U. STOLL and Dipl.-Ing. P. FINK
Lahmeyer International, Frankfurt/Main

Summary

Objectives in sewage sludge management are changing: new federal regulations can only be met by advanced technical and logistical standards. Thus, making an extensive structural change on current locally-oriented management concepts towards regionally-oriented concepts necessary. Especially thermal treatment of sewage sludge requires centralized facilities. A method to design regional sewage sludge management concepts, with special regard to the requirements of thermal sludge treatment, is presented. Goal-oriented model process chains have been developed. Their application to concrete concept design and cost effectivness evaluation is pointed out.

1. INITIAL SITUATION

The current situation in the field of waste management is determined by the following factors:

- increasing amounts of waste to be treated
- increasing requirements for disposal, e.g. pretreatment
- shortage of treatment capacities
- growing public criticism due to increasing concentrations of hazardous substances in the environment - the argumentative word being "dioxins".

However, in the field of municipal and hazardous waste management avoiding and reduction concepts might help to solve or mitigate the problems, this is unfeasible for sewage sludge. On the contrary, most meassures to improve waste water treatment result in larger amounts of sewage sludge to be treated. Utilization and dumping are the remaining treatment possibilities.

Utilization in sludge manamgement means either utilizing the material itself in agriculture or utilizing its energy potential in thermal processes.

In September 1988, analysis reports were published, documenting sewage sludge to be polluted with dioxins and other hazardous organic substances. Consequently, there is considerable aggravation towards agricultural sewage sludge utilization. During the last few months, considerable losses of agriculturally utilized sewage sludge have become evident. In this context, the amendment of the federal regulations of agricultural sewage sludge utilization is expected anxiously. These regulations are in the process of being updated and their commitments will have a considerable influence on the future development of agricultural sewage sludge utilization.

Utilizing the energy potential of sewage sludge means to incinerate its organic matter. Whether or not the energy balance of this process is positive depends to a large extent on the reference conditions. Undisputed is the fact that the direct, agricultural utilization of sewage sludge is the more cost-effective and the less equipement-intensitive path of sludge management.

To date, dumping is the most important disposal path. Approximatly 44 % of all sewage sludges are dumped, this practice is limited due to a lack of dumping-space, problems in handling, and increasing costs. Additionally, federal authorithies are developing technical guidelines, by which future dumping requirements will be regulated. Restrictions on the base of input criteria will be defined. It is expected that dumping of organic matter will be restricted and mineralization and immobilisation of the anorganic substances will be demanded. This would require a drasticly increased need for incineration capacities.

The above described situation shows the necessity of a new orientation in sewage sludge management. Whereas in the past local organisation and realization has been sufficient, increased requirements for future sludge management will demand for regional management concepts. The latter includes agricultural sludge utilisation, sludge incineration optional with previous sludge drying, disposal of the residual fraction on landfills, and logistical integration of all mentioned elements.

2. METHOD

In order to design a regional sewage sludge management concept first of all the region and the sewage treatment works have to be defined. Hereby it is recommended to adjust the concept region to already existing administration structures. This makes it easier to integrate sludge management in the whole complex of waste management.

Second, all relevant information related to sewage sludge need to be compiled in a data base. Including data for every clarification plant concerning the quantitiy of sewage sludge, specific sludge characteristics, as well as existence and capacity of sludge treatment techniques. Special interest has to be given to future development of this data. Therefore, all relevant planing activities must be taken into account. Especially an increase in sludge quantities, due to phosphate elimination or rain water treatment, must be considered in future oriented concepts.

Third, potential goals for sludge disposal of the concept region are to be selected and their specific requirements are to be definded. The latter will be done exemplarily for thermal sewage sludge treatment in the next chapter.

Fourth, goal-oriented pocess chains, connecting the concept regions clarification plants to the disposal goals need to be developed. It is recommended to prefer flexible concepts, which provide interfaces to different disposal goals rather than one dimensional solutions.

TABLE 1: Sewage sludge incineration processes

Type of incineration	Incineration of sewage sludge in lignite firing	Incineration of sewage sludge in coal melting chamber firing	Incineration of sewage sludge in waste firing		Separated incineration of sewage sludge (fluid column/double-deck-oven)		Incineration of sewage sludge in cement kiln
Solids of sewage sludge in % DS	approx. 25 %	> 90 %	< 40%	90%	40-50 %		minimum: 25-30 % optimum: > 75 %
Charge and kind of incineration of sewage sludge	combined with lignite (grinding dessication) coal dust firing	combined with coal (grinding) melting chamber firing	firing of waste mixed with sewage sludge	injection of dried and grinded sewage sludge over the grate	fluid column	deck	coal dust firing or simply by feeding shaft
Portion of sewage sludge	up to 10 % feasible, planned up to 2% of raw coal (referred to s.s. with 30 % DS)	up to 25 % feasible (referred to s.s. with 90% DS), 15% of the firing power of sewage sludge are planned	between 10 and 15% (sludge with 30 % DS)	up to 50 % feasible (s.s. with 90%DS)	100 %		up to 5 % of the clinker production (referred to s.s. with 100 % DS)
Residue management	deposition of ashes together with REA-products in tighted areas of exhausted surface mining	utilization of the granulat and REA-gypsum in construction industry	filter dusts will be dumped, slag can be used		waste dump		residuals are intercalated in clinker
Purification of exhaust gases	dedusting and de-sulphurization existant, NO_x-reduction by firing measures	dedusting, de-sulphurization, denox existant	flue gas purification existant, enlargement if necessary		flue gas purification existant		dedusting, NO_x-reduction by firing measures
Availability of plants	average utilization period per year 80 % (7000 hours)	average utilization period per year 50 % (4300 hours)	between 75 and 90 %		between 80 and 98 %		depending on construction industry

3. CONDITIONS FOR THERMAL SEWAGE SLUDGE TREATMENT

To date, the sewage sludge incineration practised has been considered as treatment method. In most cases pretreatment was not very well adapted to the incineration method applied. There was no utilization of the sludges' energy potential, in contrary these pocesses often required primary energy for support firing.

The specific conditions to be considered for an economically optimized thermal sewage sludge treatment for different incineration methods are presented in Table 1. In consequence, different requirements for pretreatment can be derived and have to be considered for regional sewage sludge management concepts by developing goal-orientated treatment or process chains.

4. DEVELOPMENT OF GOAL-ORIENTATED PROCESS CHAINS

Based on the specific requirements of the different incineration methods presented above, process chains can be developed as given in Figure 1. Since they take into account the different requirements of small-size, medium-size and large-scale clarification plants, these process chains have model charakter for sewage management concepts with thermal treatment as their primary goal, in general.

Starting-point of each process chain is sludge stabilization. Though the latter is not imperative when heading for sludge incineration, there are reasonable arguments for keeping this standard process:

- maintaining the interface to agricultural utilization
- reducing of sludge quantities
- improving dewatering effectiveness
- improving storage characteristics

The central process step towards sewage sludge incineration is sludge dewatering. Considering thermodynamic aspects of combustion processes as well as logistics (such as transportation, intermediate storage) and handling, dewatering is of major relevanz. In order to achieve higher solid contents dewatering is related to a drastically increasing demand for energy and augmentation of the demand for mechanical equipment. Therefore, an optimization of dewatering in each treatment phase is an important principle in developing process chains. Thus, capacities may be saved in subsequent process steps. Conventional dewatering will remain a decentral process step while advanced sludge dewatering and drying will be concentrated in one or more regional treatment centres. The incineration facility, however, has to be the most centralized proess step.

5. DETERMINATION OF CONCRETE CONCEPT VARIANTS AND COST EFFECTIVENESS EVALUATION

Working-out a sludge management concept for a region, it is necessary to determin concrete process variants based on the general process chains derived above . These variants must be adjusted to both, the specific requirements of the concrete regions' clarification plants as found out in the second workingstep and to the requirements of relevant incineration

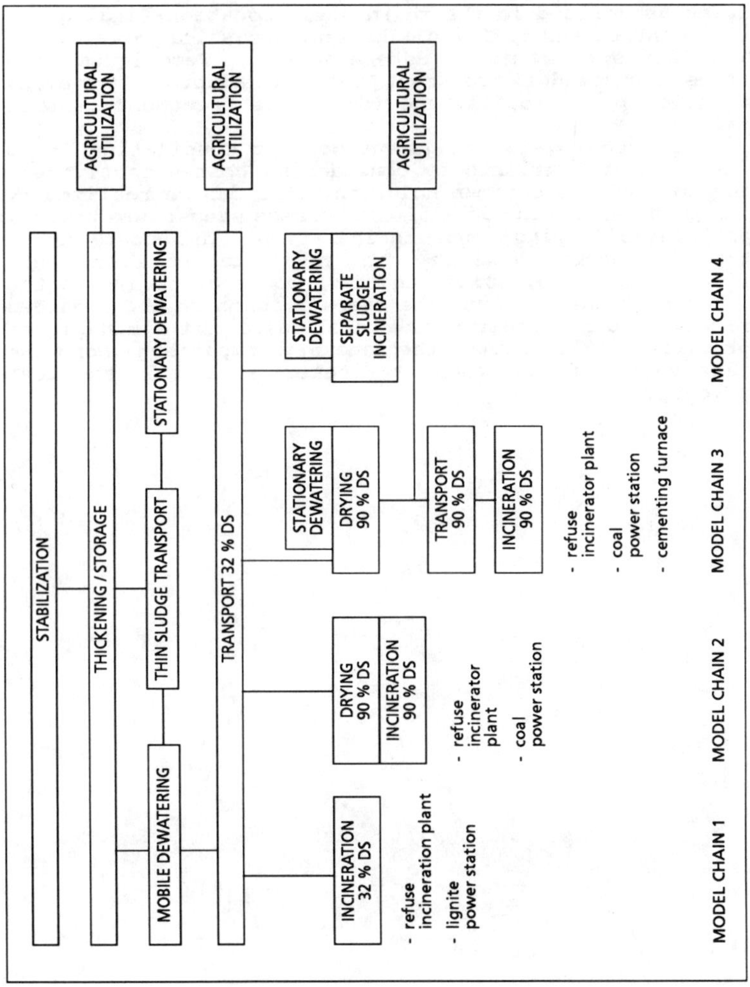

Figure 1: Diagram of model process chains

facilities as defined in the third step. Location-finding for treatment centres and defining catchment areas are parts of the concrete concept variant determination. Techniques and capacities for sludge treatment must be adapted for every clarification plant locally and for sludge treatment centres regionally.

The final proceeding of management concept compilation is to check the concept variants by considering cost-effectiveness in order to find the optimum solution. This can be realized by evaluating the specific costs of the process steps: dewatering, transportation, drying, and incineration. The costs of a management concept variant can be obtained for each clarification plant by combining these costs according to the process step of the variant. The overall costs of all regarded clarification works result in the total costs of the management concept variant. This allows the monetary comparison among the different variants and, thus, the determination of the most cost effective one.

CHANGES IN THE INORGANIC NITROGEN FRACTION DURING SEWAGE SLUDGE COMPOSTING

DIAZ-BURGOS M.A.*, CECCANTI B.** and POLO A.*
(*)Inst. Edafología y Biología Vegetal.CSIC. Madrid. Spain.
(**) Is. per la Chimica del Terreno.CNR. Pisa. Italy.

SUMMARY

The presence of NO3 and NH3 has been traditionally related to the maturity and toxicity indices of compost. Using these criteria as the bases for the experiment and with the purpose of gaining information about the maturity index, a study of the evolution of the inorganic nitrogen fraction during the sewage sludge process has been conducted.

The sewage sludges (municipal and industrial) mixed with grapevine branches and tree pruning waste were composted for 14 weeks under controlled conditions of temperature and humidity.

In general, NH3 content decreased and NO3 content increased during the process.

With regard to the maturity index, we were able to deduce optimum conditions favouring the non-toxicity of the compost: a gradual increase of NO3 and stabilization of ammonia content at minimum levels.

These criteria have been compared with the germination index proposed by Zucconi and with crop yields, from which limits of maturity are established. Maxima in these parameters correlate to the time when contents of oxidized forms of nitrogen are higher than the content of nitrogen soluble in a phosphate buffer.

1.- INTRODUCTION

Sewage sludge is an alternative to the use of traditional organic fertilizers, since elements necessary and beneficial to crop growth form a part in appreciable quantities.

However, it has been demonstrated that direct use of this sewage may have a negative effect due to potential toxicity generated by organic compounds and/or heavy metals (9,5,2) which also form part of the initial sewage sludge composition.

In order to reduce this negative effect, different alternatives have been sought, within which composting, prior to application, is one of them, since this process involves a stabilization of materials. Nevertheless, with the use of this technique, a second problem arises, centered on detecting the limit as from which it may be considered that the material does not involve risks for use.

The purpose of this paper is to concentrate on monitoring the inorganic forms of nitrogen throughout the composting process since the phytotoxicity of these materials may be due to the presence of reduced forms of nitrogen (ammonium, ammonia) (6,7).

2.- MATERIAL AND METHODS

Two sewage sludges were chosen with different characteristics as a function of their original composition. Both originated from treatment stations in the city of Madrid (Spain). Valdebebas is considered as being of an urban origin (Actual Density (g/cc) 1.46; Bulk Density (g/cc) 0.60; Porosity (%) 62.32; Solids Volume (%) 37.68; Water Ret. Cap. 270.0; pH 6.9; C.E. (mS/cm at 25ºC) 6700; C.E.C. (meq/100g) 57.8; %C 21.37; %N 2.13; %Na 1.35; % K 0.49; % Ca 3.54; %Mg 0.65; %P 1.09; %Fe 2.35; %Al 4.22; Cd (mg/kg) 14; Zn (mg/kg) 939. Butarque, with an industrial origin, displays the following features: Actual density (g/cc) 1.48; Bulk density (g/cc) 0.55; porosity (%) 62.83; Solids volume (%) 37.17; Water retaining capacity 261.8; pH 6.8; C.E. (mS/cm at 25ºC) 7200; C.E.C. (meq/100 g) 71.8; % C 23.71; %N 1.98; % Na 1.83; %K 0.31; %Ca 3.01; %Mg 0.53; %P 0.97; %Fe 2.84; %Al 2.58; Cd (mg/kg) 22; Zn (mg/Kg) 3567; Cu (Mg/Kg) 307; Ni (mg/Kg) 68; Pb (Mg/Kg) 810; Cr (mg/kg) 1239.

In the plant production experiment, soil classed as Typic Haploxeralf from Toledo (Spain) was used (3).

The composting process was undertaken for 14 weeks during which the mixtures were kept under controlled conditions of humidity (60%) and temperature (28ºC). The compost was periodically turned over in order to prevent anaerobiosis processes in the material.

This sludge was mixed in a 2:1 proportion by weight with chips from tree pruning in Madrid and with vine shoots in differentiated mixtures with the purpose of favouring the composting process's aeration conditions. Therefore, the mixtures used were: urban sludge with chips (VA), urban sludge with shoots (VS), industrial sludge with chips (BA) and industrial sludge with shoots (BS).

DETERMINATIONS: Nitrite and nitrate levels were determined in the water extract (1:5 proportion) by a Dionex Ion Chromatography 2000i/SP apparatus. The ammonium soluble in a phosphate buffer was carried out in the water extract with a selective electrode.

The germination index (Ig) was evaluated according to Zucconi et al. (11) in the water extract (1:10 proportion) using <u>Lepidium</u> <u>sativum</u> seeds.

The plant response experiment was undertaken using ray-grass plants as an indicator, sown in soil on which a 15 T/Ha dose of the samples was applied with different composting times. Production was cut off 4 weeks after the beginning of the experiment which was conducted in a hot-house.

3.- RESULTS AND DISCUSSION

During the sewage sludge composting process, changes in the nitrogenous fraction are very important. Within the large variety of forms in which this element is present, the inorganic fraction is particularly interesting, since it is closely linked to oxidizing processes which are the basis of composting.

Thus, the greater or lesser presence of oxidated forms of nitrogen determines how composting has evolved. That is to say, a greater presence of nitrites, and above all, nitrates, will mean that the initial forms of nitrogen (ammonium and ammonia) have evolved during the process. That is why the nitrate content has been used as a control parameter of the degree of maturity, whilst the presence of ammonia or ammonium ion has been directly related to the degree of toxicity.

Nitrite levels (table 1) appear in very low concentrations although they increase slightly at the end of the time considered.

TABLE 1
CHANGES IN THE INORGANIC FRACTION DURING SEWAGE SLUDGE COMPOSTING

COMPOST	WEEK	NH3*	NO2&	NO3&
VA	1	212	0.6	---
VA	2	138	0.8	---
VA	5	137	---	---
VA	10	92	0.1	9.1
VA	14	36	5.5	34.6
VS	1	184	---	---
VS	2	201	---	---
VS	5	81	---	4.1
VS	10	72	0.1	3.0
VS	14	11	0.4	38.5
BA	1	252	7.4	27.1
BA	2	252	1.3	7.4
BA	5	93	5.1	17.4
BA	10	34	4.3	12.8
BA	14	18	---	---
BS	1	113	10.9	11.7
BS	2	80	0.3	---
BS	5	19	6.2	26.0
BS	10	7	4.2	22.3
BS	14	5	6.5	22.9

* µmol/g.h & mg/100g

Nitrates are what display more differentiated behaviour: whilst the mixture of industrial sludge with shoots (BS) and the two of urban sludge (VA and VS) increase their content in these anions during the whole process, Butarque with chips (BA) shows a reduction with composting time and totally disappears after 14 weeks. It is obvious that in this latter case, a nitrification inhibition is occurring, which may be due to a strong presence of ammonium ion, a product of microbe activity (4).

Ammonia soluble in a phosphate buffer is used for

determining this compound's levels in the hydrosoluble extract. It furnishes important information since some of the enzyme activities are determined by the evaluation of these contents after hydrolisis of the compounds generated by their hydrolitic reactivity.

The concentration of this nitrogenous fraction decreases during the whole process, which indicates a gradual reduction in toxicity. There is no differentiation with regard to behaviour as a function of the type of additive added to the sludge, since a gradual reduction takes place in all cases, tending towards stabilization in residual values.

In the light of the results, it may be stated that the nitrifying activity is more intense in the sludge mixtures of an urban origin and in the industrial sludge with shoots, whilst in Butarque with chips, this activity is gradually inhibited during the process, which may be due partly to factors influencing the activity of bacteria responsible for these processes.

The germination index's evolution in the water extract was frequently used as a parameter enabling the evolution of phytotoxicity during composting to be followed (11,10,8).

TABLE 2
VARIATIONS OF THE GERMINATION INDEX DURING COMPOSTING

COMPOST	Ig	COMPOST	Ig
VA2	2.7	BA2	39.1
VA4	2.7	BA4	25.7
VA6	2.6	BA6	40.2
VA8	20.9	BA8	27.0
VA10	40.0	BA10	97.9
VA12	51.4	BA12	54.2
VA14	96.4	BA14	41.4
VS2	3.7	BS2	66.9
VS4	12.9	BS4	17.5
VS6	7.6	BS6	91.5
VS8	11.1	BS8	134.7
VS10	49.5	BS10	90.8
VS12	80.0	BS12	40.0
VS14	75.1	BS14	88.7

The mixtures used show a differentiated behaviour as a function of the type of sludge (table 2). For Valdebebas, composting time elapsed involves an obvious reduction of phytotoxicity, both in the mixture with chips and with shoots. In both cases, values between 2 and 3 change to levels near to the control level. The increase in Valdebebas with chips (VA) is more significant, whilst in Valdebebas with shoots (VS) two stages were differentiated (between the 2nd and 8th weeks and between the latter and the 14th), with the most notable increase occurring in the latter period.

In the case of industrial sludge, indices are much higher than in Valdebebas, i.e., nearer to the control and, in some cases, they exceed the latter's value. The mixture with chips shows a more or less constant evolution around the 40 index, although it shows its maximum value after 10 weeks. The Butarque sludge with shoots (BS) shows a maximum after 8 weeks.

By relating this parameter with the dynamics of the inorganic fractions of nitrogen, we see that in urban sludge (V), the crossing of the variation curves of nitrate and ammonia soluble in a phosphate buffer coincides with the greatest increase in the slope of the germination index curves, which may indicate that the greatest reduction in toxicity commences then. On the other hand, in industrial sludge mixed with shoots, this reduction in toxicity occurs after the crossing but at the end of the composting period it diminishes again, which may be attributed to interferences caused by the presence of heavy metals. In the case of this sludge mixed with chips, there is no predictable behaviour.

As regards plant yield, sludge of an urban origin is that showing the greatest increase, reaching values of up to 30% more than the witness under the most favourable conditions (table 3). The effect of composting time is different as a function of the type of material with which it has been composted. Thus, the VS mixture displays a gradual increase with time. On the other hand, the same sludge mixed with chips shows a production maximum after 5 weeks, whilst the remaining treatments keep around a 10% increase. This may mean the appearance of toxic compounds as from the 5th week which would mainly come from the degradation of the chips, since the same sludge mixed with shoots does not show this effect.

TABLE 3
YIELD OF RYE-GRASS GROWN
(15 T/Ha. Cut at 4 weeks.Control = 100)

TREATMENT	% D.W.*	TREATMENT	% D.W.*
VA1	116.3	BA1	106.0
VA2	117.9	BA2	103.3
VA5	129.9	BA5	105.6
VA10	123.4	BA10	101.1
VA12	114.9	BA12	103.5
VA14	113.6	BA14	101.5
VS1	121.1	BS1	109.3
VS2	123.6	BS2	100.7
VS5	119.3	BS5	100.7
VS10	124.1	BS10	101.3
VS12	127.7	BS12	96.1
VS14	134.3	BS14	101.5

(*)Dry weight

In the case of Butarque, the addition of compost makes yields increase with respect to the witness. However, the reduction in materials composted 10 weeks for BA and 10 and 12 for BS down to levels practically the same as the control level must be noted; nevertheless, recovery occurs later and the maximum values are reached in both cases when the material composted during 14 weeks has been used.

It must be pointed out, however, that in no case does a negative effect occur, i.e., values below the witness's yields as regards dry matter. This brings to mind the possibility of the difrerences being due not only to a phytotoxic effect, but also to the presence of compounds favouring plant growth in the Valdebebas sludge.

There does not seem to be a clear relationship between the variations in plant production and nitrogen's inorganic fraction, in view of the fact that this effect may be masked by other factors such as the specific sensitivity of the crop used to this type of toxic substance or theinfluences the soil used has.

CONCLUSIONS

In relation to the information supplied by monitoring the inorganic fraction of nitrogen during composting as regards the degree of maturity, optimum conditions that will favour the non toxicity of the composts can be deduced rather than specific values and, in the cases studied these would be a gradual increase in nitrate ion content and the stabilization of ammonia levels. Therefore, the cutting point between the graphs representing the evolution of nitrates and ammonia would mark the point as from which these conditions start being favourable, but this criterion can only be used in the case of sludge of an urban origin, since an overlapping of other polluting factors it would be necessary to evaluate occurs in those of an industrial origin.

REFERENCES

1-.CLAIRON, M., ZINSOU, C. and NAGOU, D. (1962). Etude des posibilités d'utilization agronomique des compost d'ordures ménagères en milieu tropical. I. Compostage des ordures ménagères. Agronomie, 2(3), 295-300.

2-.DeVLEESCHAUWER, D.,VERDONCK, O. and Van ASSCHE, P. (1981). Phytotoxicity of refuse compost. Biocycle, 22(1) 44-46.

3-.DIAZ-BURGOS, M.A. (1990). Compostaje de lodos residuales: aplicación agronómica y criterios de madurez. Tesis Docotoral. Facultad de Ciencias. Universidad Autónoma de Madrid.

4-.EPSTEIN, E., TAYLOR, J.M. and CHANEY, R.L. (1976). Effects of sewage sludge compost applied to soil on some soil physical and chemical properties. J. Environ. Qual., 5, 522-526.

5-.JUSTE, C. and SOLDA, P. (1986). Heavy metals availability in long term experiments. In factors influencing sludge utilization practices in Europa. Davis, Hoeni and l'Hermite. Elsev. Appl. Publish. Bruselas.

6-.KATAYAMA,A., KER,K.C., HIRAI, M., SHODA, M. and KUBOTA, H. (1986). Stabilization process of sewage sludge compost in soil. In: Compost: Production, quality and use. Eds. Bertoldi, Ferranti, l'Hermite and Zucconi. Elsev. Appl. Publish. Udine. 341-350.
7-.KATAYAMA,A., HIRAI, M., SHODA, M., KUBOTA, H. and MORI, S. (1985). Inhibitory factor of sewage sludge compost for growth of komatsuna Brassica campestris L var. rapiferafroug. Environ Pollut. 38 45-62.
8-.LOBO, M.C., DIAZ-BURGOS, M.A. y POLO, A. (1987). Evolución del grado de madurez de diferentes compost mediante el estudio electroforético de la fracción hidrosoluble. VII Congreso Nacional de Química. Química Agrícola y Alimentaria-2. Volumen II. Octubre 1987. Sevilla. 111-118.
9-.MANRIQUE, A., ARROYO, I., GARCIA DE BUSTOS, J.M., PORTAL, M.E., SANZ, A.M. y MORE, A. (1987). Los metales pesados como factor limitante a la aplicación agrícola de los lodos procedentes de estaciones depuradoras urbanas. Comunicación presentada a las III Jornadas sobre Higiene Ambiental. Octubre 1987. Madrid.
10-.SAVIOZZI, A., RIFFALDI, R. and LEVI-MENZI, R. (1986). Compost maturaty by water stract analyses. In: Compost: production, quality and use. Eds. Bertoldi, Ferranti, l'Hermite and Zucconi. Elsev. Appl. Sci. Publish. Udine. 533-542.
11-.ZUCCONI, F., PERA, A., FORTE, M. and De BERTOLDI (1981). Evaluating toxicity of inmature compost. Biocycle, 22(2), 54-57.

EFFECTS OF AEROBIC THERMOPHILIC PRETREATMENT ON ANAEROBIC SLUDGE STABILISATION AND SUBSEQUENT TREATMENT STEPS

U. BAIER, H. P. ZWIEFELHOFER, UTB

Summary

An evaluation of operating data aerobic thermophilic conditioning of sludges from sewage treatment plants has shown that a high degree of disinfection can be achieved with this process. It was also found that the thickening properties of the treated sludge are enhanced. The total decomposition of organic dry matter in the thermophilic preliminary treatment stage and in the mesophilic digestion chamber following this stage is higher than in single-stage anaerobic plants. The production of biogas in the digestion chamber can be increased by adding an aerobic thermophile preliminary treatment stage.

1. INTRODUCTION

During the last ten years, most European countries have been faced with constantly rising amounts of organic sludges procuced by an increasing number of sewage works serving an ever larger percentage of the population. Be it for landfill disposal of liquid or dewatered sludge, for incineration, or for agricultural use of stabilized sludge, most treatment works are now implementing optimisation plans for sludge treatment, with four primary objectives:

- constant quality of stabilized sludges with respect to further treatment.(control of stabilisation processes).
- destruction of potentially pathogenic organisms (e.g., worm eggs, salmonellae, viruses, bacteria) with respect to agricultural use of sludges.
- improvement of thickening properties of stabilized sludge with respect to storage volume, tankering costs and mechanical dewagering.
- improvement of dewatering characteristics of thickened sludge with respect to drying, incineration and disposal.

Anaerobic mesophilic digestion of primary and secondary sludges from sewage treatment has been common practice in Europe for more than 30 years, and still is the most widely used stabilisation method in European countries. Along with the beforementioned rising demands on sludge quality and operational control of treatment facilities, a number of mesophilic digesters can no longer meet treatment requirements, especially regarding hygienic properties of sludges to be used in agricultural land application.

In Europe this situation has primarily led to the development and technical application of processes for sludge hygienisation (1,2,3,4). Different treatment processes have proven to fulfill hygienisation criteria and have been in operation since the early 80's. Out of these processes,aerobic thermophilic pretreatment followed by anaerobic digestion (dual digestion) has been applied in most cases. This combination of treatment steps, although initially developed only for the purpose of pathogen reduction and weed seed removal and control, has shown its beneficial impacts on subsequent sludge treatment during the last ten years.

This paper will present data of different full scale operating dual treatment plants in Switzerland, Germany and Austria. The scope of this report is not only to show the long term pathogen inactivation and control potential of full scale operating plants, but also to clearly point out the beneficial impact of aerobic thermophilic pretreatment on anaerobic digestion and further treatment of sludge.

2. PROCESS DESCRIPTION

The dual sludge treatment process as investigated in this paper has been described in literature by different authors (5,6,7,8) and usually is a combination of the following treatment steps; prethickening of primary and waste activated sludge, either combined or separately; aerobic thermophilic pretreatment; anaerobic mesophile digestion; storage and consolidation.
(Figure 1)

Prethickening of combined sludge from primary and final clarifiers is performed in small tanks with short residence times of 10 to 20 hours (h). The tanks are not mixed and no additives are used. As an average, an icrease of raw sludge dry solids from 0.5 - 1.0% to 3-5% DS is achieved.

Aerobic thermophilic pretreatment of thickened raw sludge takes place in insulated cylindrical steel tanks with a heigh-to-diameter (h/d) ratio of 2 to 4. The vessels are closed and equipped with external recirculation by means of a pump. The recirculation line includes a jacketed heat exchanger and a venturi type gas injetor. Residence time in the thermophilic reactor is in the order of 18 to 24 h. Reactor temperature is controlled in the range of 60-68°C. Waste gas from the reactor is partially recirculated to the venturi injector to increase oxygen yield and to minimize vent gas heat losses. The thermophilic reactor is a CSTR (completely stirred tank reactor) with external loop an is run in a batch-fed mode, i.e., 5-20% of the reactor volume is withdrawn periodically and replaced by an equal volume of raw sludge.

For raw sludge heating, a sludge-sludge heat exchanger is included in the thermophilic pretreatment step. This recovers heat from hot sludge before introducing it into the mesophilic digester. Digester temperature is controlled through the final temperature of heat exchange. The temperature of sludge introduced into the digester is slightly higher than the digester temperature, to cover digester heat losses. The need for internal or external digester heating systems is eliminated.

The mesophilic digesters investigated are of conventional European design, which is based on cylindrical vessels with either a conical of flat bottom. The h/d ratio of digesters is found to be in the range of 1 to 2. The digester temperature is controlled in the range of 34-40°C depending on the site. Mixing of digester volume is achieved by means of either external pumps, internal draft tubes, internal propellers or biogas recirculation.

Biogas is either burned in a boiler for production of hot water or, more frequently, used to drive a gas engine equipped with electrical generator and direct/exhaust gas heat recovery (total energy module).

Final thickening of digested sludge before land application and further use takes place in relatively large consolidation and storage tanks equipped with mixing facilities. Mean hydraulic residence time is in the order of 1 week up to 6 months, the higher value applying where prolonged sludge storage is required during winter.

Consolidation /storage tanks are either run in a fill and draw mode or with continuous discharge of supernatant.

3. THERMOPHILIC PRETREATMENT FOR THE REMOVAL OF PATHOGENS

Agricultural use of stabilized sludge as an organic soil conditioner has been common practice in Europe for more than 20 years, and still is the most economical way to deal with the sludge problem. During the last ten years, however, there has been a growing awareness of potentially negative impacts of agricultural use of sludge on the environment. The danger of

TWO-STAGE SLUDGE TREATMENT

1. Stage Aerobic Thermophilic Pretreatment
Hydrolysis / Liquefaction / Decontamination
Thermal & Mechanical Treatment / flow equalization

2. Stage Anaerobic Mesophilic Digestion
Stabilization/Hygenization/Biogas Procuction
Consolidation/ev. dewatering

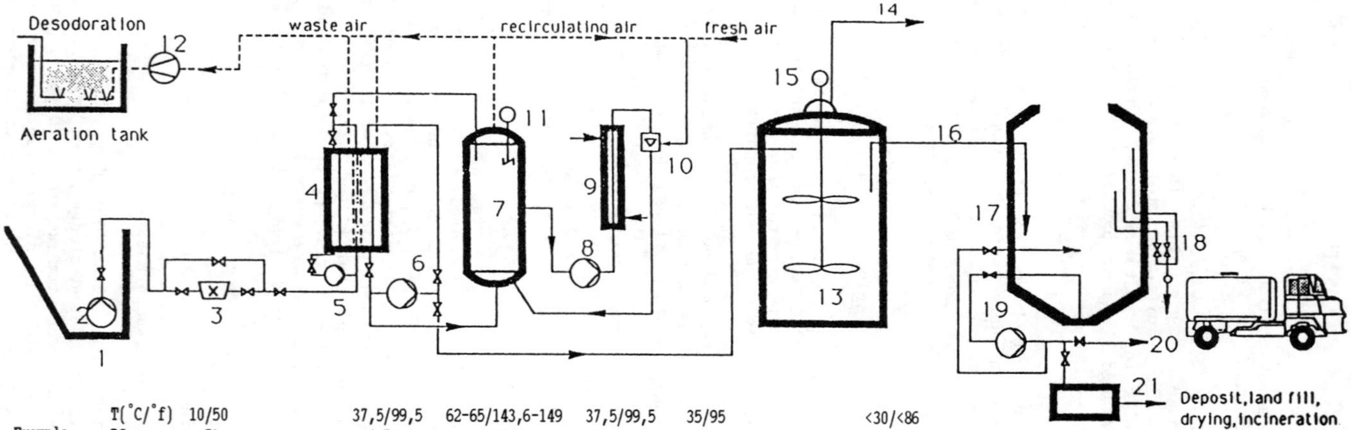

Example							
	T(°C/°f)	10/50	37,5/99,5	62-65/143,6-149	37,5/99,5	35/95	<30/<86
	DS	5%	4,5%	4,5%	4,5%	2.7-3%	>12%
	IIRT(d)	-	-	0,8-1	-	min. 15	-

ATS Pretreatment (1-12)

1 raw sludge sump
2 raw sludge pump
3 sieve/macerator
4 heat exchanger
5 warm sludge pump
6 hot sludge pump
7 reactor
8 heat exchanger
9 heat exchanger
10 air injector
11 foam cutter
12 air compressor

STABILIZATION (13-16)

13 anaerobic digestor
14 biogas
15 internal mixing
16 overflow

CONSOLIDATION/STORAGE (17-21)

17 storage/consolidation
18 supernatant withdrawal
19 sludge pump
20 sludge for agriculture
21 dewatering

Fig 1 - Scematic Drawing of Dual Sludge Treatment, System UTB-AEROTHERM

heavy metals and organic micropolluants accumulating in soil has been recognized, and the infectional risks of sludge appication have been demonstrated. As a consequence of this awareness, in 1981 the Swiss government enacted an ordinance on sewage sludge clearly regulating land application of sewage sludge (9). This ordinance demands effective sludge hugienisation in case of land application. In Germany, subsequently, an ordinance for sludge reuse was enacted in 1982 with an identical scope (10). In Europe, the regulations have led to the development and implementation, into existing treatment process schemes of decontamination and hygienisation of sewage sludge. A number of these processes have been discussed in other literature.

Conditions under which potentially pathogenic microorganisms are killed have been described by different authors (11,12) and allow the definition of a number of treatment processes for safe hugienisation (13,14), namely:

- Pre-pasteurisation of raw sludge
- aerobic thermophilic stabilisation
- aerobic thermophilic pretreatment followed by anaerobic digestion
- treatment with lime
- composting of sludge
- treatment with ionizing radiation

Any heat treatment of sewage or sludge will cause microorganisms opposed to high temperature to die. As heat inactivation is a process which is non-specific and irreversible, it has been used wherever there was a need to convert any sludge into a product free of potentially pathogenic cells. Swiss and German regulations have concentrated on three groups of organisms to be regarded as indicator organisms for hygienized sludge: entero-bacteriaceae, salmonellae and worm eggs (ascaridae, taenidae). The group of enterobacteriaceae is used most widely, since they are excreted by humans and animals altogether with pathogenic organisms and their presence is a good indicator for fecal pollution/contamination.

In Switzerland, enterobacteriaceae have evolved to the standard hygienisation parameter mainly due to ease of analysis. However, there is some concern regarding this parameter as enterobacteria can multiply in sludge whereas most pathogens cannot. The choice of enterobacteriaceae as hygienisation indicator organisms therefore has to be carefully balanced against other organisms.

Nevertheless, the majority of data available deals only with this group of microorganisms. Although the principle of heat inactivation of microorganisms has been known since Pasteur, its applicability for sludge treatment had to be shown first. Any heat treatment process has to fulfill two conditions: first, it must ensure a minimal residence time for each sludge particle at a given temperature; second, it is crucially for the treatment sludge to be introduced in an environment that does not allow reinfection nor regrowth of pathogenic organisms.

Aerobic thermophilic pretreatment will fulfill both of these requirements. The thermophilic reactor is operated in a semicontinous mode and mixed completely. By this mode of operation it is guaranteed that every sludge particle will be exposed during a minimal cycle time of 60 minutes to a temperature of 65°C. The pretreatment process, being made up of reactor and heat-exchanger, is constructed as a closed system. This allows thermophilic sludge to be transferred into a mesophilic digester directly, without any possibility for reinfection of pathogens. A completely mixed digester now has to provide a low level of readily degradable organic matter to ensure the level of pathogens eventually present remains low.

Up till now, more than 500 different samples of raw, thermophilic digested and stored sludge have been analysed with respect to enterobacteriaceae. These analyses clearly show that aerobic thermophilic pretreatment followed by anaerobic mesophilic digestion as sludge treatment is an effective method to meet the stringent Swiss legislative requirements, which allow a maximum number of 100 Enterobacteriaceae per gram of wet stabilized sludge (EBC/g).

Raw sludge usually contains between 10^6 and 10^7 counts of EBC/g. In the thermophilic reactor thery are reduced by a factor of about 10^7-10^8. As a result, we usually cannot find any Enterobacteriaceae in 10 g of pretreated sludge. In completely mixed digesters following thermophilic pretreatment, the level of EBC detected is found to be in the range of 1 - 50 EBC/g. As the regrowth potential in dual-stabilized sludge is practically zero, this concentration of EBC stays stable in sludge which is consolidated/cold stored during up to 1 year (Table 1).

Treatment Plant	Enterobacteriaceae (Counts per gr wet sludge) Raw Sludge	after aerobic thermophilic pretreatment	digested	to Farmland Average values
Alvaneu	-	<1	15	5
Balzers	880'000	<0.1	-	20
Broc	-	<0.1	0.1-10	10
Freienbach	-	<0.1	<0.1	30
Furthof	1'650'000	<1	0.1-40	40
Lachen	-	<0.1	0.1-10	5
Märstetten	-	<0.1	<0.1	15
Sorts	-	-	40	50
Thun	-	<0.1	-	25
Wartau	-	<0.1	5-350	30
Hofsteig	-	<0.1	-	50
Dillingen	15'000'000	<1	50	50
Ebersdorf	-	<1	10	30

Table 1 Level of Enterobacteriaceae in Dual Treatment Plants (aerobic-thermophilic Pretreatment followed by anaerobic-mesophilic stabilisation)

The concentration as well as the variety of salmonellae in raw sludge varies considerably. Therefore analysis of salmonella according to Swiss regulations is only qualitative. Independent of raw sludge concentration, however, the thermophilic reactor acts as a complete barrier for salmonellae. Whereas in mesophilic digesters without thermophilic pretreatment salmonellae were detected in the same order of magnitude as in raw sludge, salmonellae counts in dual-treated sludge lie below detection levels.

Salmonellae have been found in dual-treated sludge. Parasitological investigations of pretreated, as well as of digested, sludge has always failed to find any viable worm eggs (ascaris, taenia), even if present in of added to raw sludge.

4. HOW WELL IS SLUDGE STABILIZED?

Due to microbial activity organic material in either soluble or solid form is degraded during aerobic thermophilic pretreatment. At a given temperature in the completely mixed reactor, degradation rate is controlled by the availability of oxygen. By controlling the amount of oxygen introduced into the reactor, as well as the air circulation ratio, effective control of biological degradation can be achieved.

Degradation of organic material in mesophilic digesters is mainly dependent on temperature, hydraulic residence time, and mechanical energy input, i.e., complete mixing of tank volume.

Table 2 and Figure 2 show some results on degradation of dissolved solids vs. volatile solids (DS/VS) in a number of sludge treatment plants before and after aerobic thermophilic pretreatment was put into operation. Operational parameters of anaerobic digestion (temperature, retention time, mixing) were held constant. Without thermophilic pretreatment, all the investigated mesophile digesters showed a degradation of 25-30%/40-50% of fed DS/VS, respectively. No correlation between degradation and hydraulic retention time has been observed.

Treatment Plant	Retention Time		Biological Degradation DS/VS							
			No Pretreatment		With aerovic-thermophilic Pretreatment					
	HRT(d) Aerotherm	HRT(d) Digestor	δDS(%) Digestor	δVS(%) Digestor	δDS(%) Aerotherm	δDS(%) Aerotherm	δDS(%) Digestor	δDS(%) Digestor	δDS(%) total	δDS(%) total
Broc	1.4	80	n.a.	n.a.	6	10	26	47	33	57
Furthof	1.4	43	20	48	4	6	31	51	33	56
Lachen	1.4	28	n.a.	n.a.	7	12	25	45	32	56
Märstetten	2	30	n.a.	n.a.	18	24	20	31	30	48
Thun	1.6	27	29	47	10	14	29	45	36	53
Ebersdorf	1	25	n.a.	n.a.	7	10	63	81	65	83
Holzkirchen	2.1	23	n.a.	n.a.	10	n.a.	13	n.a.	22	n.a.
Schwarzach-gruppe	1.2	21	25	44	14	n.a.	39	n.a.	48	n.a.

Table 2 Degradation of DS and VS in Dual Treatment Plants (Values in % of sludge fed to the respective step) n.a. = no values available

HRT = Hydraulic Retention Time
DS = Dry Solids
VS = Volatile Solids

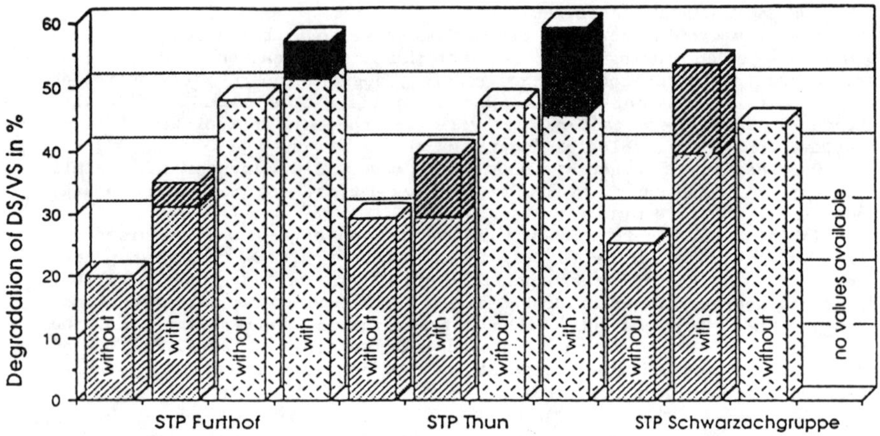

Fig 2 - Degradation of DS and VS with and without aerobic - thermophilic Pretreatment

Mesophilic digesters with aerobic thermophilic pretreatment showed degradation of solid material in the same order of magnitude as without pretreatment, i.e., 25-30/40-50% of pretreated DS/VS sludge. Again no correlation could be detected between hydraulic retention time and DS/VS degradation. On the plants investigated, the influence of mechanical energy input (mixing) on digester performance is of higher importance than the retention time. This can act as indication of digesters being far from completely mixed tank reactors.

About 3-10%/5-14% DS/VS, respectively, of raw sludge fed to the aerobic thermophilic reactor were degraded in this stage.

The combination of aerobic thermophilic pretreatment and anaerobic mesophilic digestion results in an enhanced overall degradation of 25% VS compared with mesophilic digestion alone. Dual treatment thus is producing a sludge with less organic material and a higher degree of stabilisation.

5. INFLUENCE ON BIOGAS PRODUCTION IN MESOPHILIC DIGESTERS

Microbial degradation of organic material in an anaerobic environment is linked with the production and release of biogas, a mixture of mainly carbon dioxide and methane. In mesophilic sludge digesters, biogas production is controlled by a number of different parameters. At a given temperature, crucial operational control parameters are hydraulic retention time, amount and composition of organic material fed into the digester (organic load), and mechanical energy input and distribution (mixing).

On most of the treatment plants investigated there was a general optimisation and upgrading of digester operation and equipment taking place concurrently with implementation of aerobic thermophilic pretreatment. These plants do not allow a comparison of before and after results, as biogas production is also generally enhanced due to optimized digester performance.

As a consequence, four plants have been chosen, on which mesophilic digesters were operated identically before and after introduction of aerobic thermophilic pretreatment into the sludge treatment scheme. To account nevertheless for hidden irregularities in mixing and/or load variations of digesters, specific biogas production has been calculated with respect to raw sludge, digester volume, and VS degraded (Table 3).

Treatment Plant	Raw Sludge Quantity (m3/d)	HRT ATS (D)	HRT Digestor (d)	Biogas Quantity (m3/d)	Specific Production of Biogas per Vol. Digestor (m3/m3 d)	per Vol Raw Sludge (m3/m3)	per kg VS in Raw Sludge (m3/kg VS)	per kg VS degraded in Digestor (M3/kG VSdeg)
Freienbach								
no Pretreatment, 86/87	52	-	26	530	0.41	10.1	-	-
with Pretreatment, 87/88	56	1.25	24	635	00.47	11.4	-	-
Furthof								
no Pretreatment, 1985	11	-	53	136	0.23	12	0.37	0.78
no Pretreatment, 1986	12	-	50	137	0.23	11.2	0.5	1.04
with Pretreatment, 1987	14	1.43	43	202	0.34	14.7	0.5	0.95
with Pretreatment, 1988	14	1.43	43	210	0.35	15.3	0.51	0.93
Savognin								
no Pretreatment, 1985	10	-	34	183	0.61	18	0.66	1.13
with Pretreatment, 1986	11	2.3	30	212	0.71	19.3	0.75	-
with Pretreatment, 1987	14	1.8	23	226	0.75	16.2	0.62	-
Thun								
no Pretreatment, 1985	208	-	24	3718	0.74	18.2	0.51	1
no Pretreatment, 1986	208	-	24	3876	0.78	18.7	0.55	1.16
with Pretreatment, 1988	183	1.57	28	3944	0.79	21.7	0.67	1.76

Table 3 Specific Production of Biogas in mesophilic Digestors with and without aerobic - thermophilic Pretreatment
HRT = Hydraulic Retention Time
ATS = aerobic thermophilic stage

The obvious general conlusion that, by degradation of organic material in the aerobic reactor biogas production in the digester will inevitably be affected negatively, has to be rejected. Although there was significant VS reduction in the thermophilic reactor, in none of the four investigated digesters was a reduction of the volumetric nor specific biogas production found. The amount of biogas produced was stable, or even enhanced, compared with digestion alone (Figure 3).

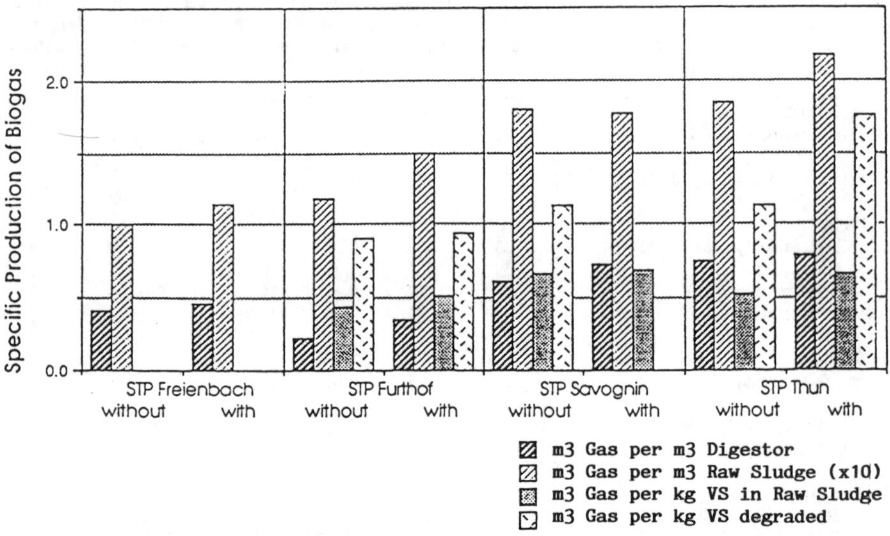

Fig 3 - Specific Production of Biogas in mesophilic Digestors of Dual Treatment Plants

Sludge pretreatment by aerobic thermophilic hydrolisation alters the amount, as well as the composition, of organic material fed into the digester. During pretreatment the activity of extracellular enzymes produced by thermophilic bacteria results in a reduction of insoluble, particulate and colloidal material. Volatile solids as one potential substrate for the anaerobic microbial community are degraded and, in their original form, no longer available for production of biogas. As a result of this thermophilic activity, soluble low-molecular products are formed (measured as DOC). Only part of these products are metabolized by thermophilic bacteria. The majority is left in the sludge to be an easily available substrate for anaerobic digester organisms (Figure 4).

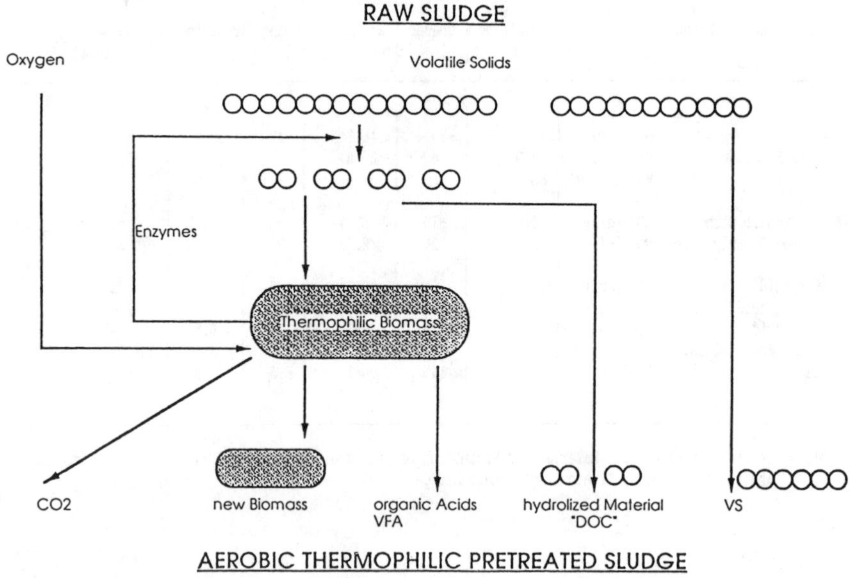

Fig 4 - Microbial Degradation of Volatile Solids and Production of dissolved organic Material in aerobic thermophilic Pretreatment

In the presence of carbohydrates, significant apounts of acetic acid are produced by the thermophilic microorganisms, which are directly available to mathanogens in the digester (15). The majority of VS hydrolyzed in the thermophilic reactor is present in pretreated sludge in the form of soluble intermediate products of aerobic metabolism (16).

Although these products, especially volatile fatty acids such as acetate and propionate, are not detected in a VS analysis, they nevertheless account for biogas production in the digester. Organic material as a whole, be it in a soluble or insoluble form, is not only partly degraded in the aerobic thermophilic reactor, but also hydrolyzed to give a substrate which is more readily available to the methanogenic organisms than would be raw sludge (Table 4, Figure 5).

Sewage Treatment Plant Thun		Raw Sludge	Change in %	after ATS	Change in %	digested	Change in % total
Dry Solids	DS (kg/m3)	51.2	-10	46.1	-29	32.6	-36
organic Fraction	VS (kg/m3)	33.9	-14	29.0	-45	16.0	-53
	VS (% d. DS)	66.3		63.0		49.0	
Total Organic Carbon	DOC (g/l)	18.6	-10	16.75	-51	8.27	-56
Dissolved Organic Carbon	DOC (g/l)	1.31	+105	2.68	-90	0.267	-80
Acetic Acid	HAc (mg/l)	1140	+191	3315	-90	320	-72
Propionic Acid	HPr (mg/l)	966	+105	1985	-100	4.6	-100
Butyric Acid	HBut (mg/l)	364	+115	781	-100	0.5	-100
Volatile Fatty Acids total	VFA (mg/l)	2470	+146	6081	-95	325.1	-87

Table 4. Degradation of organic Material and Production of intermediate Products during aerobic-thermophilic / anaerobic-mesophilic Treatment of Sewage Sludge

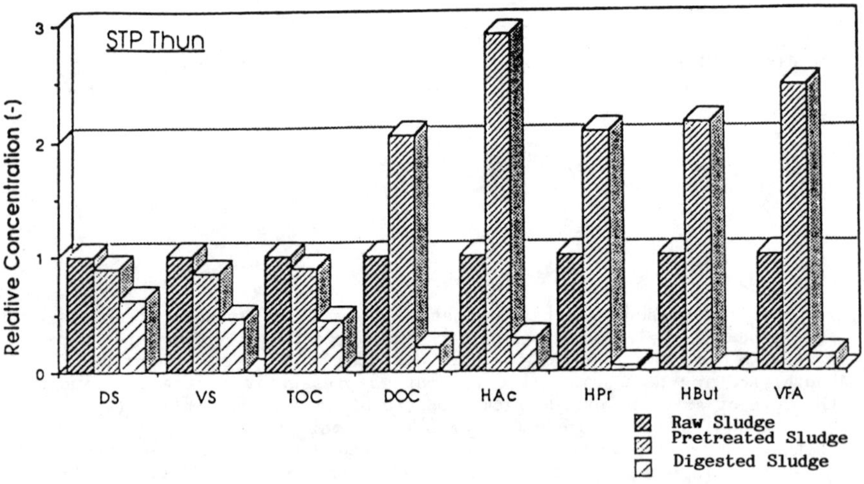

Fig 5 - Relative Concentration of organic parameters in Dual Treatment Plants

(Raw Sludge = 1)

Due to considerable variations in raw sludge quantity in some investigated plants without thermophilic pretreatment, biogas production was subject to corresponding fluctuations. To evaluate the stability of mesophilic digestion wiht thermophilic production of dual treatment plants under hydraulic load variations. It has been seen, that during periods of long HRT (>30 days) and of short HRT (15 days), no singificant fluctuation in specific biogas production could be detected. Digester performance of dual treatment plants was stable and gas production was high, even at prolonged periods of 15 days HRT in the digester.

6. WILL CONSOLIDATION AND DEWATERABILITY BE ENHANCED?

By definition, the aim of statical thickening or mechanical dewatering of sewage sludge is the separation of "water" from "solid material". In digested sludge, water (or better the liquid phase) is present in different forms: free water and intracellular water.

Free water, the majority of the liquid phase, can be quite easily separated from the solids present by taking advantage of the difference in specific weight (consolidation, centrifuge) or by compressing the solid phase and thus forcing free water to escape (belt filter, filter press). Main hindrances to the water escape are non-uniform particle size distribution and non-optimal operation of sludge thickeners. Operational problems have to be addressed by site specific solutions, whereas particle size distribution can be dealt with by process approaches. A sludge with a high standard deviation of the gaussian particle size distribution curve, say with a high amount of "large" and "very small" particles, will not release free water very efficiently, because inter-particle volume is high and water release channels are clogged easily. Any process that makes particle sized distribution uniform, either by eliminating small or breaking up big particles, will enhance free water release. Anaerobic digestion accomplishes much of this by hydrolyzing smaller particles. When it comes to breaking up "bigger" particulate material, anaerobic hydrolisation is slowed, and longer digester residence time is needed. Aerobic thermophilic pretreatment deals very efficiently with this problem; at elevated temperatures, the hydrolysis rate of aerobic organisms is an order of magnitude higher when compared to anaerobic mesophiles. By combining the very fast aerobic thermophilic hydrolysis with anaerobic stabilisation, a sludge with a very uniform particle size distribution and an enhanced free water release is obtained. It should be mentioned here, that with the chosen pretreatment reactor including external recirculation and air injection, the mechanical treatment of sludge by exposing it repeatedly to high shear forces effectively supports "big particle breakdown". The combined effect of high mechanical shear forces and elevated temperature results in death and lysis effects of microbial cells present in raw sludge. Cells exposed to high temperatures show a weakening of cell wall structure and an increase in membrane permeability. As a result, intracellular water is set free (17). The exposure of weakened cells to mechanical stress, as occurs in the venturi gas injector (low pressure), supports these effects.

A comparative study of consolidation properties of stabilized sewage sludge with and without aerobic thermophilic pretreatment is only possible under constant operating conditions of the thickeners during both periods. On most sites where investigations of aerobic thermophilic pretreatment were undertaken, operational changes in the digester and thickener have been implemented along with the AEROTHERM Installation. Out of 15 investigated plants, 10 showed a change in thickener operation and/or equipment. They are producing a stabilized sludge statically thickened to 9-13% DS.

On the remaining 5 plants with unaltered thickener operation, a comparative examination of the average DS concentration after implementation of aerobic thermophilic pretreatment (average weekly values of at least 8 months continuous operation, Figure 6). As an average, a 15-40% higher DS content could be achieved in well operated statical thickeners. Two other plants which showed no increase of DS clearly demonstrate the need of optimized thickener operation. These thickeners are run under non-optimal conditions either because of short-circuiting of non-stabilized sludge from the digester or by insufficient mixing/bubble release in the thickener. Along with enhanced thickening properties, dual treated sludge show an increased mechanical dewaterability on filter presses. A direct comparative study on the Hofsteig plant showed a reversible increase in DS in the dewater sludge (belt filter), from 27% without the aerobic thermophilic step to 37% with aerobic thermophilic pretreatment before anaerobic digestion. In other plants average DS concentration of dewatered sludge is found in the range of 35-42% DS. An increase of 8-10% is achieved by implementing aerobic thermophilic pretreatment compared with mesophilic digestion alone, with identical dosage of organic polymers. An average increase in DS content of 10% (35 vs 25%) will result in a reduction of water content of 40%. The amount of liquid separated will be 21% higher.

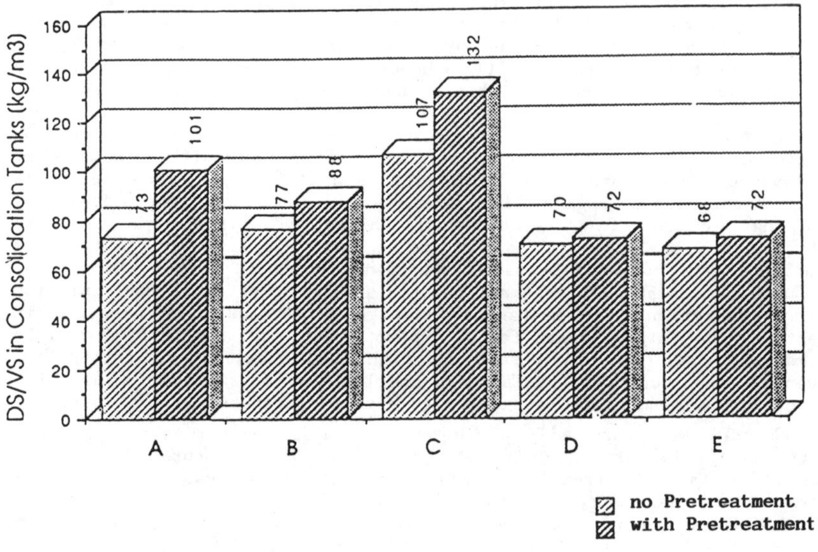

Fig 6 - DS/VS Concentration of consolidated Sludge with and without aerobic thermophilic Pretreatment

7. CONCLUSION

More than 70 plants with aerobic thermophilic pretreatment of raw sewage sludge (according to the UTB-AEROTHERM system) are at present in operation or under construction in Europe. Performance data from operation at periods of up 6 years is available from more than 30 plants (18). Based on this operational data the following conclusions can be drawn:

1) Aerobic thermophilic pretreatment followed by anaerobic mesophilic digestion (dual treatment) is an effective means to eliminate and control pathogenic organisms. The chosen combination of temperature (>60°C) and retention time (20-24 h, minimum 1 h) will guarantee an operation free of short circuiting. The process produces a sludge which is free of pathogenic organisms associated with human or animal disease, such as salmonellae, worm eggs, enterobacterial species and viruses. Reinfection potential of dual treated sludge is minimal.
2) Hydrolysis of organic material in the thermophilic reactor varies from 5-10% (VS) and can easily be controlled by variation of oxygen input.
3) The inevitable degradation of organic material in the thermophilic step will not occur at the expense of VS reduction in the digester. As a consequence, total VS reduction is enhanced in dual treatment.
4) As a consequence of the metabolic activity of aerobic thermophilic microorganisms, organic solids in raw sludge are hydrolyzed. Insoluble and complex compounds are transformed to become soluble by-products of bacterial metabolism (volatile fatty acids) or readily degradable intermediate products of less complex structure because of enzymatic breakdown. Pretreated sludge, therefore, is a substrate that is more amenable to anaerobic digestion than untreated raw sludge.
5) Biogas production in mesophilic digesters will not be affected negatively by thermophilic pretreatment. Due to altered substrate composition, volumetric as well as specific biogas production will be enhanced.
6) The specific parameter, "biogas volume per kg of VS degraded in the digester" can be higher than the theoretical value of 1.1. This is due to the fact that biogas is produced not only out of solid but out of soluble organics as well.
7) Thermophilic pretreatment will enhance statical thickening properties of digested sludge. In optimized statical thickeners, DS concentrations of 110-140 kg/m^3 can be achieved without any addition of conditioning agents. However, even the best of sludges will not thicken satisfactorily in a sub-optimal thickener tank.
8) Mechanical dewaterability of dual treated sludge is enhanced by 8-10% with a filter press as compared with anaerobic digestion alone.

All in all, the combinations of aerobic thermophilic pretreatment and anaerobic mesophilic stabilisation has proved to be a very effective and economical tool, not only to eliminate pathogen problems but also to deal with nonoptimal digester performance and insufficient thickening and dewatering of sewage sludge.

REFERENCES

(1) KASSNER W., "Verfahrnestechnische Moglichkelten der Klarschlammentseuchung", 1. Hohenheimer Seminar, 1986, 78-90, Universitat Hohenheim
(2) STACHOWSKE M., "Verfahrenstechnische Moglichkelten zur Klarschlammentseuchung", 5. Bochumer Workshop 1987, 43-79, Schriftenreihe Siedlungwasserwirtschaft Bochu, Bd. 10
(3) STRAUCH D (ed), Bericht des 2. Hohenheimer Seminars "Entseuchung von Klarschlamm",1988,Universitat Hohenheim

(4) HAVELAAR A.H., "Sludge Disinfection - an overview of methods and their effectiveness", 1984, Sewage Sludge Stabilisation and Disinfection, 48-60, A. Bruce (ed), Ellis Horwood, Chichester.
(5) ZWIEFELHOFER H.P., "Aerobic Thermophilic/Anaerobic-mesophilic Two Stage Sewage Sludge Treatment: Practical Experiences in Switzerland", 1985, Conservation & Recycling, 8/1,2/285-301.
(6) SCHAAD P., "Aerob-thermophile Schlamstabilislerung mlt anschliessender Faulung", 1988, 2. Hohnheimer Seminar, 114-133, Universitat Hohenheim.
(7) TEICHMANN H., GALLENT W. UND GORTLER R., "Die Aerob-thermophile Klarschlammhygienisierung mit anschliessender anaerob mesophiler Faulung - Fin Untersuchungsbericht". 1987. Korrespondenz Abwasser, 34/3.
(8) SIXT,H. und JUNGBAUER O., "Die aerob-thermophile Klarschlamm-Entseuchung und Konditionierung". 1986. GIT Supplement 5/27-33.
(9) SCHWEIZERISCHER BUNDESRAT, Klarschlammverordnung vom 8.4.1981, 814.225.23
(10) BUNDESMINISTER DES INNERN, "Verordnung uber das Aufbringen von Klarschlamm - Klarschlammverordnung (Abfklar V)", Bundrsgesetzblatt 1982, Teil I
(11) FEACHEM R.G., BRADLEY D.J., GARELICK H. and MARA D.P., "Sanitation and Disease - Health Aspects of Excreta and Wastewater Management", 1983, Wiley & Sons, Chichester.
(12) STRAUCH D., HAVELAAR A.H., L'HERMITE P. (ED)., "Inactivation of Microorganisms in Sewage Sludge by Stabilisation Processes", 1984, Elsevier, London.
(13) ATV, "Enteseuchung von Klarschlamm", 1988, Arbeitsberlcht der ATV/VKS - Arbeitsgruppe 3.2.2, Korrespondenz Abwasser, 35/1/71-74
(14) MOOLER U., "Entseuchung von Klarschlamm - Eine Stand-ortbestimmung 1987", 1987, Korrespondenz Abwasser, 35/1/24-30
(15) BAIER U., "Zur Physiologie thermophiler Bacilli", Dissertation Nr. 8423, 1987; ETH Zurich.
(16) HAMER G., "Aerobic thermophilic hygienisation - A supplement to anaerobic mesophilic waste sludge digestion"., 1986, Chem Eng res des, 64, 417-424.
(17) MASON C.A., "Microbial Death, Lysis and "Cryptic" Growth: Fundamental and Applied Aspects", Dissertation Nr. 8150, 1986; ETH Zurich.
(18) ZWIEFELHOFER H.P.,"Praxiserfahrung mit der aerob-thermophilen Vorbehandlung bei der Anpassung von Schlammfaulungsanlagen an die neuen Vorschriften bezuglich Hygiene und von Schlammfaulungsanlagen an die neuen Vorschriften bezuglich Hygiene und Stapelzeit. 1986, pro aqua - pro vita 86, Fachtagung II/Abwasser, Band 10 B.

METAL UPTAKE AND CROP GROWTH ON SEWAGE SLUDGE FIELD TRIALS WITH HEAVY METAL CONTENTS NEAR THE RECOMMENDED LIMIT VALUES

S. Lübben, E. Rietz and D. Sauerbeck
Institute of Plant Nutrition and Soil Science,
Federal Research Centre of Agriculture (FAL)

Summary

- Heavy metal contents of soils near the present limit values already retard plant growth and decrease crop yields.
- Sugar beets and potatoes as well as dicotyledoneous weed plants are much more sensitive than oats, wheat and maize.
- According to the metal contents and accumulation, the plant yield reductions must be attributed mainly to the phytotoxicity of Zn.
- Cd, Zn and Ni uptake was greater on the more acid soil, but the Cu, Pb and Cr uptake did not react on soil pH.
- The food guide value for Cd was frequently surpassed in the cereal grains.
- Cd and Pb contents were higher in leaves than in grains, tubers and beets, but Zn contents only in plants growing on the more highly polluted soils.
- Chromium contents were always small but higher in cereal grains than in straw. Both Cr and Pb uptake were hardly related to the degree of soil contamination.
- Cd, Zn, Ni and Cu accumulated in the plant roots, whereas Pb and Cr contents in the roots were lower than those in the soils.
- The present guide value for Zn in soils should be lowered for reasons of phytotoxicity, and the limit for Cd because of its transfer into the food chain.

1. Introduction:

Guide and limit values for heavy metals in sludge-treated soils have been issued in order to prevent them from excessive contamination. For Cd, Hg and Pb the main purpose is to minimize their transfer into the food chain. Other metals like Zn, Ni and Cu are known to be predominantly phytotoxic, and so their limitation was mainly thought to avoid crop growth depressions.

Most of these threshold values, however, have been derived only from experience with the individual metals concerned (e.g. Hofer and Schütz 1980), and little is known as yet about their possible interactions. It may well be that certain limit values are not safe any more if not only one but several metals exceed them simultaneously, especially as far as phytotoxicity is concerned.

2. Materials and Methods:

In order to test this possibility, a Luvisol (pH 6.5, Trial 1) and a Cambisol (pH. 5.5, Trial 2) have been treated since 1980 with 0 (SS0), 100 (SS1) and 300 (SS2) m^3/ha of liquid sewage sludge, corresponding to about 4 and 12 t d.w. each year. In one treatment (SSu) this sludge was used as such, in the other (SSc) the same sludge was brought to Cd, Zn, Ni, Cu, Pb and Cr contents of twice the permissible concentrations. This has resulted in increasing heavy metal contents of the two soils, which after the heaviest treatments (SS2c) are now near or even above the guide and limit values (table 1). A detailed description of the experimental layout has been given by Söchtig et al. (1983).

TABLE 1: German guide and limit values for heavy metals in soils and maximum metal contents obtained in the two field experiments (values for 1989 in mg/kg soil)

Element (Variant)	Cd	Zn	Ni	Cu	Pb	Cr
Offic. limit Value	3	300	200	100	100	100
Luvisol (SS2c)	2.8	360	23	102	101	95
Cambisol (SS2c)	3.7	387	24	111	115	105

The metal contents and plant yields since 1984 are discussed below. Since the results for wheat were similar in 1985 and 1988, only the latter data are shown. Each of the figures 1a - 1e represents one year and crop. The values for the acid Cambisol were frequently higher than for the more neutral Luvisol and are represented by the unshaded upper part of the columns. An asterisk indicates significant yield depressions in the high-sludge and high-metal plots (SS2c) as compared with the high-sludge but low-metal treatments (SS2u).

3. Results and Discussion:

3.1 Yield observations:

Already in 1984 the sugar beets o,in the treatment SS2c of the acid Cambisol suffered an almost complete yield failure, although the metal contents in the soil were still well below the limit values (Timmermann et al. 1985). A less pronounced but also significant yield loss was found for potatoes in 1986, in spite of the fact that none of the 6 heavy metals had already reached its limits for soils. In 1990, when potatoes grew again on the same plots which were now near the limit values (equal to fig. 1e), very pronounced yield reductions occured (analytical data not yet complete).

As a consequence of the gradually increasing heavy metal contents, both oats (1987) and wheat (1988) showed significant yield depressions, not only on the acid Cambisol but even on the SS2c treatment of the more neutral Luvisol. This has been found for both grain and straw. This heavy metal stress also caused a much more pronounced frost damage to the wheat plants on the SS2c plots as compared with the other treatments.

In contrast to wheat and oats, the maize in 1989 showed yield depressions only on the highly polluted SS2c Cambisol plot. This yield drop, however, was significant only for the corn stalks and not for the grain, although the initial development was strongly impeded. Another important observation was that the most highly polluted SS2c plots, which now had reached or exceeded the limit values, were almost free of dicotyledoneous weeds.

3.2 Cadmium contents:

Irrespective of the degree of pollution, the Cd uptake was almost consistently higher from the acid Cambisol than from the near neutral Luvisol. The Cd contents in plants increased more or less proportionately to the soil contents. The cereal straw was always higher in Cd than the grain, as were the leaves of potatoes and sugar beets compared with the tubers and beets. Similar findings have been reported e.g. by Diez and Rosopulo (1978) and Mitchell et al. (1978). The wheat and maize roots (fig. 1f) accumulated Cd to about 3 times the contents in soil.

The german limit value for cereal grain of 0.12 mg Cd/kg d.w. (1) was exceeded in most of the treatments, even if only the smaller sludge treatments (SS1) had been applied. On the acid Cambisol treatment SS2c resulted in up to 9 times excessive Cd contents in the grain, although the soil did not yet reach the 3 mg/kg limit for Cd. However, the experiments of Styperek et al. (1983), Oberländer and Roth (1987) and Kirleis et al. (1981, 1984) have shown that the majority of this Cd is removed during the milling process into the bran. On the other hand, the potato tubers still remained below the corresponding food guide value of about 0.5 mg Cd/kg d.w.. Nevertheless, the present limit value of 3 mg Cd/kg soil definitely needs to be reduced.

3.3 Zinc contents:

Just as for Cd, the Zn uptake was considerably higher from the acid Cambisol than from the neutral Luvisol, and the contents increased with increasing soil pollution. On the plots with low Zn contents (SS0 and SS1) the cereal grains contained more Zn than the straw, but this picture changed drastically with increasing Zn contamination in SS2c. This corroborates former similar findings by Mitchell et al. (1978), Brondi et al. (1984) and Lübben and Sauerbeck (1990). Zn contents of 180-500 mg/kg in the oat and of 500-1190 mg/kg d.w. in the wheat straw (depending on soil and pH) were sufficient to explain the observed yield depressions.

A similar picture was found for the sugar beet and potato leaves compared with tubers and beets. 330-500 mg Zn/kg in the sugar beet and 250-540 Zn/kg d.w. in the potato leaves were presumably sufficient to cause yield reductions, especially on the more acid Cambisol. On this soil the Zn contents in the wheat and especially in the maize roots (fig 1f) strongly exceeded the ones in the soil. These findings suggest that the present limit value for Zn of 300 mg /kg soil should be reduced.

3.3 Nickel contents:

As for Cd and Zn, the Ni contents in some plants showed a tendency to increase at the low pH values of the Cambisol. However, the contents differed considerably among the experimental plants, being relatively high in the oats but rather low in wheat and maize, and much higher in the sugar beets than in the potatoes. As formarly found by Hein and Sauerbeck (1988), oat and maize grains contained more Ni than the straw, but the sugar beet leaves and beets behaved just the other way.

In general, however, the Ni contents in all plants remained rather low, although the wheat and maize roots accumulated this element, especially if the pollution was high (SS2c) and the soil pH was low (fig 1f). All the shoot parts (except oat grain) contained less Ni than the soils, so that phytotoxic effects can be excluded. It should be noted, however, that the Ni pollution, compared with the limit value of 200 mg/kg for soils, was still rather low, because the Ni additions were relatively small.

3.4 Copper contents:

Similar as for Ni, almost all Cu contents in the plant shoots remained rather small, although the SS2c treatments gradually reached the limit value of 100 mg Cu/kg soil. In contrast to most other metals, the Cu contents in oat and wheat grain were higher than in the straw, whereas in sugar beets, potatoes and maize the leaves were higher in Cu than the other plant parts. Similar findings have been reported by Lübben and Sauerbeck (1990). Except for the potato leaves, the lower pH value of the Cambisol did not increase the Cu contents of the plants very much.

Only the Cu contents of the sugar beet and potato leaves on the highest SS2c treatments might have come near to phytotoxicity, although the limit values of 100 mg Cu/kg soil had not yet been reached. The Cu contents of the cereal plant shoots hardly reacted on differing soil treatments at all, even though the wheat and maize roots accumulated this element in all treatments up to twice the content of soils (fig. 1f). Accordingly, the limit value for Cu must be questioned only on acid sandy soils.

FIGURE 1a - f: Heavy metal uptake by different crop plants in a long term field experiment with metal polluted sewage sludge

a. Sugar-Beets
b. Potatoes
c. Oats
d. Wheat
e. Maize
f. Wheat and Maize Roots

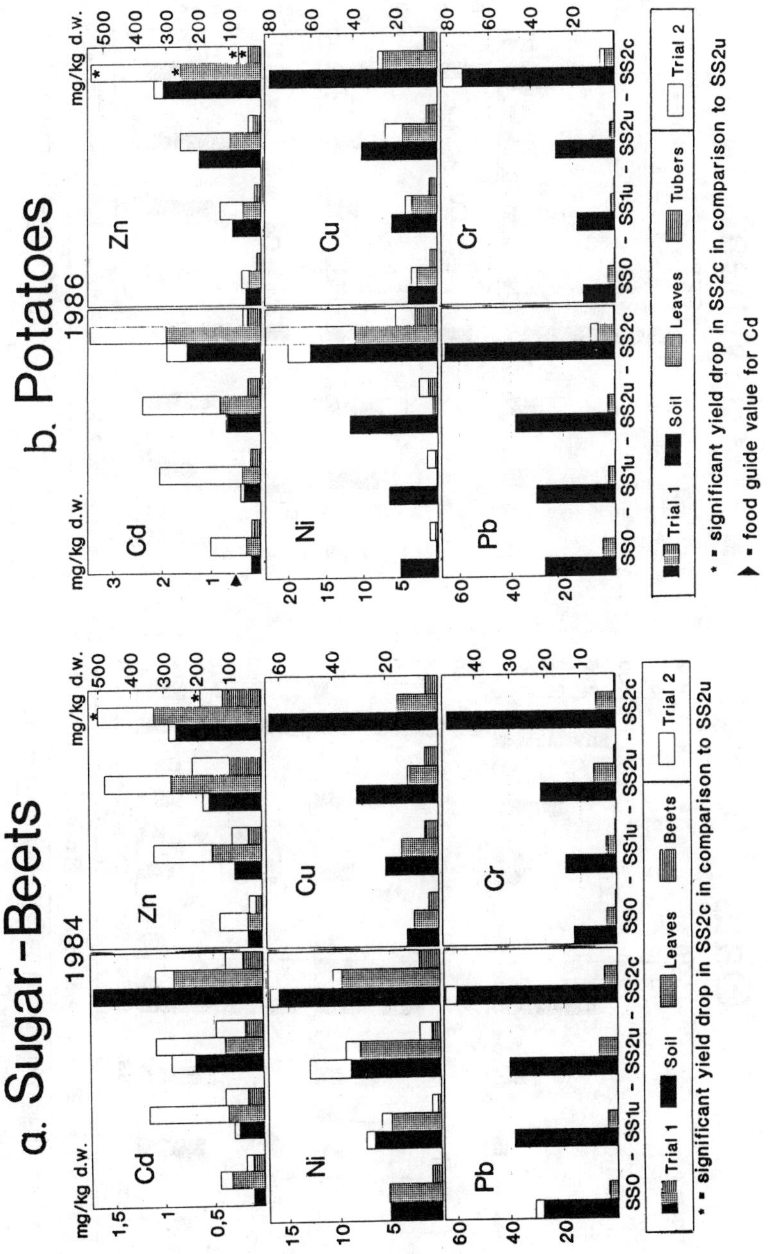
a. Sugar-Beets
1984
b. Potatoes
1986
mg/kg d.w.
Cd
Zn
Ni
Cu
Pb
Cr
SS0 - SS1u - SS2u - SS2c
Trial 1
Soil
Leaves
Beets
Tubers
Trial 2
* = significant yield drop in SS2c in comparison to SS2u
▶ = food guide value for Cd

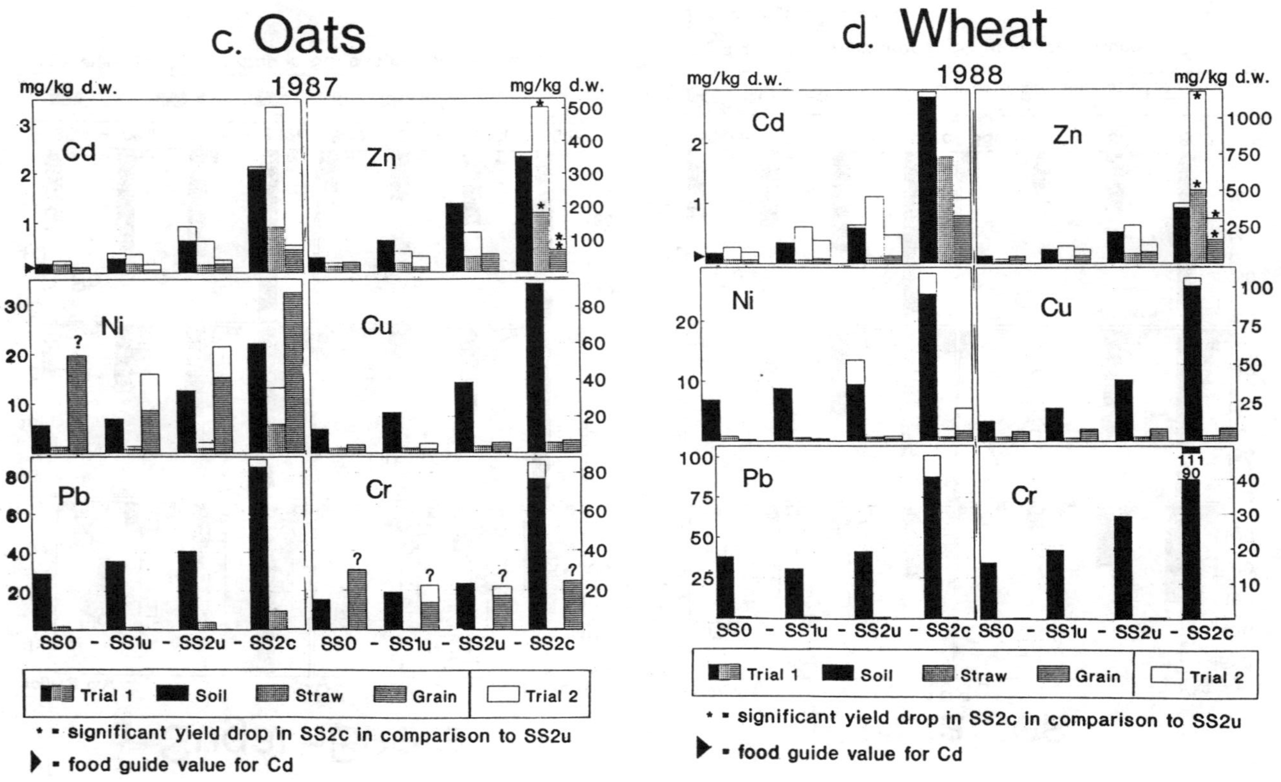
c. Oats
1987
mg/kg d.w.
mg/kg d.w.
Cd
Zn
Ni
Cu
Pb
Cr
SS0 - SS1u - SS2u - SS2c
SS0 - SS1u - SS2u - SS2c
Trial 1
Soil
Straw
Grain
Trial 2
* = significant yield drop in SS2c in comparison to SS2u
▶ = food guide value for Cd
d. Wheat
1988
mg/kg d.w.
mg/kg d.w.
Cd
Zn
Ni
Cu
Pb
Cr
111
90
SS0 - SS1u - SS2u - SS2c
SS0 - SS1u - SS2u - SS2c
Trial 1
Soil
Straw
Grain
Trial 2
* = significant yield drop in SS2c in comparison to SS2u
▶ = food guide value for Cd

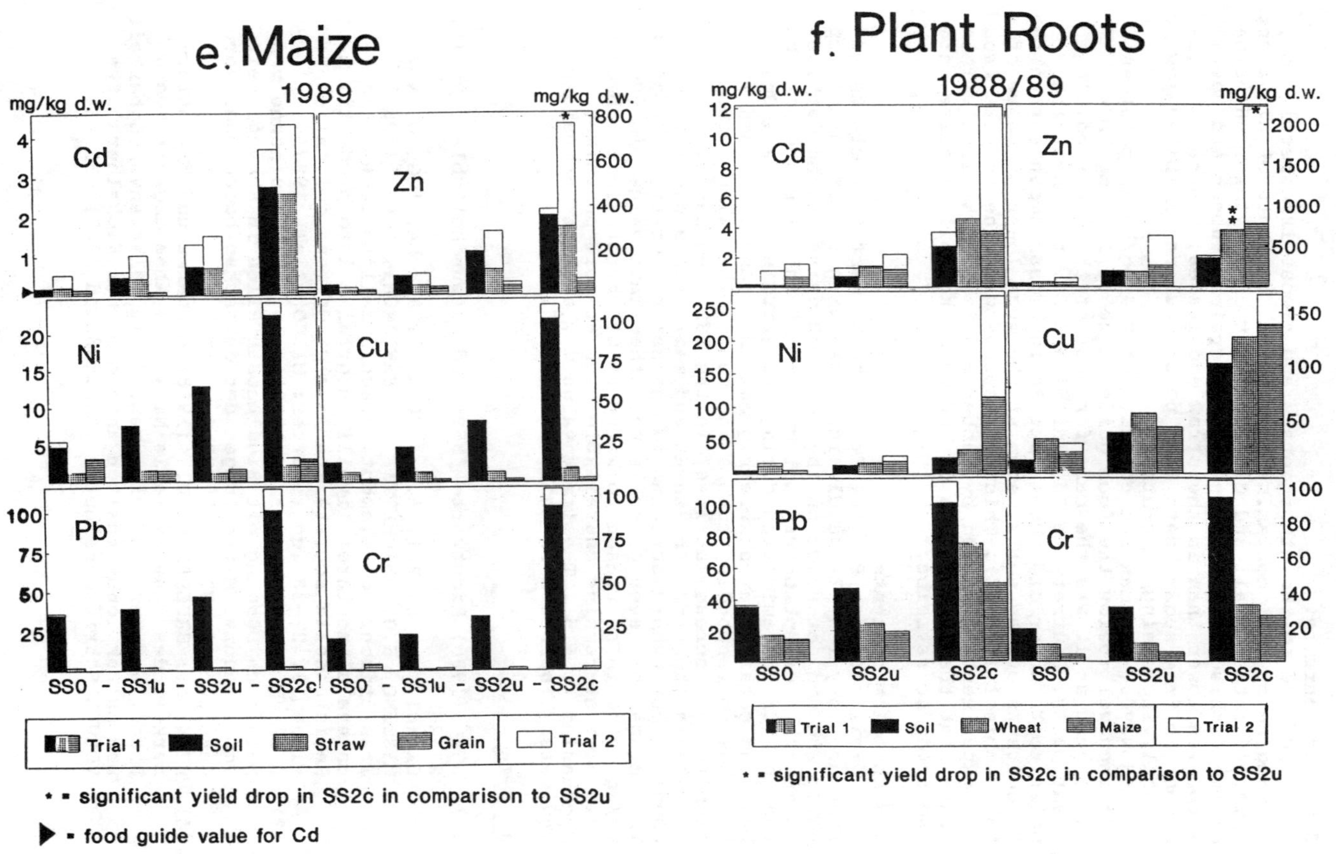

e. Maize
1989
mg/kg d.w.
mg/kg d.w.
Cd
Zn
Ni
Cu
Pb
Cr
SS0 - SS1u - SS2u - SS2c
SS0 - SS1u - SS2u - SS2c
Trial 1
Soil
Straw
Grain
Trial 2
* = significant yield drop in SS2c in comparison to SS2u
▶ = food guide value for Cd
f. Plant Roots
1988/89
mg/kg d.w.
mg/kg d.w.
Cd
Zn
Ni
Cu
Pb
Cr
SS0
SS2u
SS2c
SS0
SS2u
SS2c
Trial 1
Soil
Wheat
Maize
Trial 2
* = significant yield drop in SS2c in comparison to SS2u

3.5 Lead contents:

Opposite to the other metals and almost irrespective of the degree of soil contamination, the Pb contents of all plants remained rather small and did not react on the lower pH value of the Cambisol. The Pb contents in grains, tubers and beets were even lower than in the straw and leaves. Contrary to the almost Pb free sugar beets, the potato tubers exhibited low but slightly increasing Pb contents.

However, the maximum contents of 0.4 mg/kg in potatoes remained well below the food guide value of 1,25 mg Pb/kg d.w., just as it was the case for most of the other edible plant parts. Contrary to Cd, Zn, Ni and Cu, the wheat and maize roots never did accumulate Pb in higher amounts than the Pb contents of the soils, even at the much lower pH value of the Cambisol (fig 1f). This corresponds with the results from pot experiments reported by Lübben and Sauerbeck (1990). This means that the limit value of 100 mg Pb/kg soil is quite safe and needs no revision.

3.6 Chromium contents:

Similar as for Pb, the Cr uptake by most test plants remained very low. It hardly increased with the Cr content of soils and did not react on the lower pH value of the Cambisol. An apparent exception was oat grain, but this must be attributed to a Cr contamination during the grinding process. Sugar beet and potato leaves were highest in Cr and contained more than the corresponding tubers and beets. They also showed a significant increase at least in the SS2 treatments.

Again, however, the wheat and maize roots did not accumulate but even discriminated against the Cr from the soils. This means that even on soils near the limit value of 100 mg Cr/kg there is no reason to worry about phytotoxicity, because this element remains unavailable even at soil pH values decreasing to 5.5 . The limit value may even be reconsidered and safely increased if this is required.

4. References:

(1) BGA (1986) Richtwerte '86. Bundesgesundheitsblatt 29(1), 22-23

(2) BRONDI, M.; CIARDI, A.M.; CUBADDA, R.; DALL'OGLIO, M.; MASTINO, G.G.; MICHETTI, I.; SANTARONI, G.P. (1984) Trasferimento di elementi traccia dell'ambiente alla catena alimentare: livelli in grani e loro prodotti. La Revista della Società Italiana di Scienza dell'Alimentazione 13, 27-38 (quote after: MICCO et al. 1987)

(3) DIEZ, T.; ROSOPULO, A. (1978) Schwermetallaufnahme von Mais nach Düngung mit Siedlungsabfällen in Abhängigkeit vom Standort. Mitteilungen der deutschen bodenkundlichen Gesellschaft,27,, S.15-22

(4) HEIN, A.; SAUERBECK, D. (1988) Aufnahme und Extrahierbarkeit des Schwermetalls Nickel in Abhängigkeit von Boden, Herkunft und Pflanzenart. 1. Pflanzenverfügbarkeit und Extraktionsverhalten anthropogener Nickelherkünfte. VDLUFA-Schriftenreihe 23 (Kongreßband 1987), 327-341

(5) HOFER, H.; SCHÜTZ, E. (1980) Beitrag zur Ermittlung der pflanzenbaulichen Schadgrenze von Schwermetallen. Mitteilungen Schweiz.Landw.,28,(3) , S.66-77
(6) KIRLEIS, A.W.; SOMMERS, L.E.; NELSON, D.W. (1981) Heavy metal content of groats and hulls of oats grown on soil treated with sewage sludge. Cereal Chemistry,58,(6) S.530-533
(7) KIRLEIS, A.W.; SOMMERS, L.E.; NELSON, D.W. (1984) Yield, heavy metal content, and milling and baking properties of soft red winter wheat grown on soils amended with sewage sludge. Cereal Chemistry, 61,(6) , S.518-522
(8) LÜBBEN, S.; SAUERBECK, D. (1990) The uptake of heavy metals by spring wheat and their distribution in different plant parts. COST 681: "Alternative Uses for Sewage Sludge" - Intern. Workshop by the European Commission, University of York, UK, 05.-07.09.1989 (in press)
(9) MICCO, C.; ONORI, R.; MIRAGLIA, M.; GAMBELLI, L.; BRERA,C. (1987) Evaluation of lead, cadmium, chromium, copper and zinc by atomic absorption spectroscopy in durum wheat milling products in relation to the percentage of extraction. Food Additives and Contaminants 4(4), 429-435
(10) MITCHELL, G.A.; BINGHAM, F.T.; PAGE, A.L. (1978) Yield and metal compostition of lettuce and wheat grown on soils amended with sewage sludge enriched with cadmium, copper, nickel, and zinc. Journal of Environmental Quality 7(2), 165-171
(11) OBERLÄNDER, H.E.; ROTH, K. (1987) Aufnahme und Verteilung markierter, dem Boden zugesetzter Schwermetalle in Getreidepflanzen und deren Mahlprodukten. Die Bodenkultur 38 (4), 287-298
(12) SÖCHTIG, H.; TIMMERMANN, F.; KRAUSE, R. (1983) Wirkung einer Klärschlammdüngung auf Ertrag und Nährstoffgehalte von Pflanzen und Böden. Landwirtschaftliche Forschung, Kongreßband 1982, SH39, 262-275
(13) STYPEREK, P.; SAUERBECK, D.; TIMMERMANN, F. (1983) Cd-Verfügbarkeit in verschieden behandelten Böden in Abhängigkeit von Menge und Bindungsform. Landwirtschaftliche Forschung, SH39, 183-195
(14) TIMMERMANN, F.; SÖCHTIG, H.; KRAUSE, R. (1985) Wirkung mehrjähriger hoher Klärschlammdüngung auf Humus und Stickstoffhaushalt von zwei lehmigen Sandböden. VDLUFA-Schriftenreihe 16, 369-378

LIST OF PARTICIPANTS

H. AFRATEOS
N.T.U.A.
Patisson 142
GR - 11257 ATHENS

I. AGAPITIDIS
E.E.T.A.A.
19 Omirou
GR - 106 72 ATHENS

K. AICHBERGER
Bundesanstalt f. Agrarbiologie
Wieningerstr. 8
A - 4025 LINZ

J. AKRIVOS
Agricultural Research Station
of Vardates
351 00 N. KRIKKELO/LAMIA

T.A. ALBANIS
Department of Chemistry
University of Greece
GR - 453 32 IOANNINA

M. ALBERT
Université de Nancy I
Faculté de Pharmacie
Département de Microbiologie
5, rue Albert Lebrun
F - 54000 NANCY

G. AMANATIDIS
Ministry of Interior
Technical Service Dept.
27, Stadiou St.
GR - 101 83 ATHENS

P. AMMANN
Etat Genève
Dept. Travaux Pulics
Contrôle de Pollution
CH - 1211 GENEVE 8

A. ANDREADAKIS
National Technical Univ. of Athens
5 Iroon Polytechniou
GR - 157 73 ZOGRAPHOU-ATHENS

Th. ANGELIDES
Euromarket Ltd.
50, Efroniou Street
GR - 161 21 ATHENS

A. ANTIPARIOTOU
Air Force General Staff
Project's Office
48 Kafkassou
GR - 184 54 NIKEA

V. ANTONOPOULOS
Aristotle University of Thessaloniki
School of Agriculture
Dept. of Hydraulics
GR - 540 06 THESSALONIKI

Ms ARGHIRI
48, Kyprou Street
GR - 132 31 PETROUPOLIS ATTIKIS

G. BABOUNAKIS
Ministry of Agriculture
46, Chalkokondylis Str.
GR - 104 32 ATHENS

A. BALLAS
Ministry of Agriculture
46, Chalkokodylis Str.
GR - 104 32 ATHENS

P. BALMER
Chalmess Univ. of Technology
S - 41696 GÖTEBORG

D. BATZICOS
Ministry of Interior
Technical Service Department
27, Stadiou Street
GR - 101 83 ATHENS

E. BELIA
32, Ventouri Street
GR - 155 61 HOLARGOS - ATHENS

S. BELTSIOS
72, Kyprou Street
GR - 412 22 LARISSA

Ph. BON
Hoogheemraading West-Brabant
5, Boevingenlaan
NL - 4836 AA BREDA

A.D. BOSDOGIANNI
Ministry of Environment
Arkadias 4-8
GR - 115 26 ATHENS

A.M. BRUCE
Water Research Centre
Elder Way
UK - STEVEHAGE, HERTS SG1 1TH

F. CABRAL
Dept. of Agricultural Chemistry
Instituto Superiore de Agronomia
Tapada da Ajuda
P - 1399 LISBOA CODEX

V. CANDINAS
FAC
Swiss Federal Research Station
CH - 3097 LIEBEFELD

A. CAPSIS
L'Oréal Cosmetic company
6, Amorgou Street
GR - 112 56 ATHENS

D. CATSONI
Public Power Corporation
Navarinou Street 10
GR - 106 80 ATHENS

R.L. CHANEY
Soil Microb. Syst. Lab
U.S. Dept. of Agriculture
Building 318
Barc-East, Beltsville
USA - MARYLAND 20705

G. CHARDAS
Soil Science Institute of Athens
1, Sof. Venizelou Str.
Lycovrissi
GR - 141 23 ATHENS

G. CHATZICONSTANTINOU
National Technical University of Athens
5, Iroon Polytechniou Street
GR - 157 73 ATHENS

M. CHRISTOU
Ministry of Agriculture
4, Makariou Street
GR - 152 36 PENTELI

F. COLIN
Institut de Recherches Hydrologiques
11 bis, Rue Gabriel Péri
F - 54500 VANDOEUVRE LES NANCY

D.G. COUMOULOS
CASTOR Ltd.
5 Isavron Str.
GR - 114 71 ATHENS

J. DALY
Department of the Environment
Sanitary Services Section
Room 205
O'Connel Bridge House
IRL - DUBLIN 14

M. DE BERTOLDI
University of Udine
Institute of Industrial Microbiology
Via Marangoni 97
I - 33100 UDINE

P.H.A.M.J. DE BEKKER
Grontmij N.V.
De Holle Bilt 22
NL - 3732 HM DE BILT

K. DE BRABANDER
Instituut voor Hygiëne en
Epidemiologie
Juliette Wytsmanstraat 14
B - 1050 BRUSSEL

G. DIALINAS
Municipal Enterprises for Water and
Sewage
1, Vyronos Str.
GR - 712 02 IRAKLION - CRETE

W. DIELTJENS
Centre for Water Research (SVW)
Boeretang 200
B - 2400 MOL

Chr.A. DIMOWSKI
Institut f. Kommunalwirtschaft
1 Dobrila Str.
BG - 1606 SOFIA (BULGARIA)

A.H. DIRKZWAGER
Dienst Binnenwateren - RIZA
Maerlant 6
P.B. 17
NL - 8200 AA LELYSTAD

K. DONTAS
I.P.E.X.O.D.E.
57, Parnassidos Street
GR - 152 31 CHALANDRI - ATHENS

B. DRAKOS
Ministry of Agriculture
46 Chalkokondyli Str.
GR - 104 32 ATHENS

L.E. DUVOORT-VAN ENGERS
National Institute of Public Health
and Environmental Protection
Anthonie van Leeuwenhoeklaan 9
P.B. 1
NL - 3720 BA BILTHOVEN

J.F. EADES
The Agricultural Institute
Oak Park Research Centre
IRL - CARLOW

Ch. EFTHIMIATOU
Centre for Environmental Studies
39, Mavromichalalli Street
GR - 106 80 ATHENS

L. EISEN
Commission of the European Communities
DG XIII/C3
JMO B4/087B
Plateau du Kirchberg
L - 2920 LUXEMBOURG

F. ELLINGSEN
Norsk Hydro A.S.
Norsk Hydro. Research Centre
N - 3901 PORSGRUNN

L. EVRENOGLOU
School of Athens
196, Alexandras Street
GR - 115 21 ATHENS

G. FATOUROS
Ministry of Agriculture
Land Reclamation-Irrigation Projects
Chalkokonidily 46
GR - 104 32 ATHENS

E. FERNANDES
Dept. of Agricultural Chemistry
Instituto Superiore de Agronomia
Tapada da Ajuda
P - 1399 LISBOA CODEX

A. FINK
LAHMEYER INTERNATIONAL GmbH
Lyoner Strasse 22
D - 6000 FRANKFURT (Main) 71

G.A. FLEMING
"LORRAINE"
CARRIGLAN
IRL - WEXFORD

M. FOTOPOULOU
Ministry of Interior
Technical Service Dept.
27, Stadiou St.
GR - 101 83 ATHENS

I. FRANTZIS
E.S.D.K.N.A.
1, E. Makariou Street
GR - 161 21 ATHENS

I. GAVANOZIS
Douridos 27
GR - 116 34 ATHENS

A. GEOGARCOPOULOS
GSRT - Ministry of Industry
14-18, Messogion Avenue
GR - 115 27 ATHENS

G. GESSULIS
Water Supply and
Sewage Piping at Ioannina
Gen. Pindou Street 30
GR - 454 45 IOANNINA

H. GIGER
E.A.W.A.G.
CH - 8600 DUEBENDORF

A. GLEGG
Greenpeace UK
30-31 Islington Green
UK - LONDON N1 8XE

J. GRAMMATIKOU
Ministry of Interior
Technical Service Dept.
27, Stadiou St.
GR - 101 83 ATHENS

H. GRÜTTNER
Water Quality Institute
Agern Allé 11
DK - 2970 HORSHOLM

Ph. GUETTIER
Secrétariat d'Etat chargé de
l'Environnement
14, Bd du Général Leclerc
F - 92524 NEUILLY/SEINE

A. GUITONAS
Aristotle University
Dept. of Rural Engineering
GR - 540 06 THESSALONIKI

G. HADJIATHANASSIADIS
Ministry of Agriculture
6, Amassias Street
GR - 116 34 ATHENS

K. HADJIBIROS
National Technical University of Athens
5, Iroon Polytechniou Street
GR - 157 73 ZOGRAPHOU - ATHENS

P. HALAZONITIS
Ministry of Health
17, Aristotelous Street
GR - 101 87 ATHENS

J.E. HALL
WRC - Medmenham
Henley Road
Medmenham, P.B. 16
UK - MARLOW, BUCKS SL7 2HD

H. HAENI
Forschungsantalt f.
Agriculturchemie u. Umwelthyg.
Schwagenungstr. 155
CH - 3097 LIEBEFELD

J. HANNAN
Faculty of Veterinary Medicine
Ballsbridge
IRL - DUBLIN 4

J.A. HANSEN
Laboratories for Miljo/teknik
Aalborg Universitetscenter
Sohngaardsholmsvej 57
DK - 9000 AALBORG

N. HARKNESS
83 trent Vallay Road
UK - LICHFIELD WS13 6EZ

A.H. HAVELAAR
National Institute of Public Health
Anthonie van Leeuwenhoeklaan
NL - 3720 BA BILTHOVEN

J. HOJOVEC
University of Veterinary Science
Palackého 1-3
612 42 BRNO
CZECHOSLOVAKIA

S. HOULAKI
Karayianni street 10
GR - 71306 HERAKLION CRETE

F. HUYSMAN
R.U.G.
Laboratorium voor Algemene &
Toegepaste Microbiele Ecologie
Coupure Links 653
B - 9000 GENT

D. ICONOMOU
Ministry of Agriculture
1, Sofocli Venizelou Street
GR - 121 23 LYCOVRISSI - ATHENS

B. ILSOE
Dept. of Vet. Microbiology
Royal Vet. and Agr. University
Bülowsvej 13
DK - 1870 COPENHAGEN

N. INIOTAKIS
KFA - IEV Jülich
Postfach 19 13
D - 5170 JÜLICH

K. ISRAILIDIS
Institute of Technology
GR - ATHENS

C. JENSEN
Ministry of the Environment
National Agency of
Environmental Protection
Strandgade 29
DK - 1401 COPENHAGEN K

Jarl JOHANSSON
Swedish National Board
for Technical Development
Liljeholmsvägen 32
S - 100 72 STOCKHOLM

K.C. JONES
Lancaster University
Institute of Environmental
and Biological Sciences
UK - LANCASTER LA1 4YQ

V. KAGELARIS
National Technical University of Athens
76, Themistokleous Street
GR - 187 56 PIRAEUS

S. KAMARIOTIS
Design Office
A. Nika Street 13-15
GR - 341 00 CHALKIDA

G. KAMIZOULIS
Ministry of Health
Dept. of Sanitary Engineering
118 Aigaiou Str.
GR - 171 24 NEA SMYRNI ATHENS

E. KAPETANIOS
Assoc. of Munic. and Commun.
of Attika Region
2, Botassi Street
GR - 106 82 ATHENS

M. KAPSETAKI
3, Kononos Street
GR - 116 34 ATHENS

Th. KARADIMAS
Ministry of Agriculture
46, Chalkokondyli Street
GR - 104 32 ATHENS

N. KARAGEORGOS
CRYSTAL S.A.
102, 25th Martiou Street
GR - 502 00 PTOLEMAIS

F.G. KARAGEZOV
CRYSTAL S.A.
102, 25th Martiou Str.
GR - 50200 PTOLEMAIS

Th. KARAMANIS
O.A.E.D.
1, Miaouli and 28 Octovriou Str.
GR - 271 00 PYRGOS

Dimitris KASSIMIS
Olive Institute of Corfu
Derpfeld Street 14
GR - 491 00 CORFU

S. KATERGIANNAKIS
Municipal Enterprises for Water and Sewage
1, Byronos Str.
GR - 712 02 IRAKLION - CRETE

D. KATRIS
Anagnostopoulou 23
GR - 106 73 ATHENS

N. KATSIRIS
Athens School of Public Health
196 Alexandras Ave.
GR - 115 21 ATHENS

L.B. KEITH
Department of Agriculture & Fisheries for Scotland (D.A.F.S.)
Pentland House
47 Robb's Loan
UK - EDINBURGH, SCOTLAND EH14 1TW

A. KIESTRA
Water Board De Dammel
Molenpad 8
NL - 5281 JV BAXTEL

A. KLOZE
Commission of the European Communities
DG XII - G1
200, rue de la Loi
B - 1049 BRUSSELS

M. KOLLIA
28, William King Street
GR - 112 53 ATHENS

A. KOLIOU
D.E.Y.A.M.B.
161, Ghazi Street
GR - 382 21 VOLOS

K. KOMBOUGIAS
Ministry of Environment
23 Ath. Diakou Str.
GR - 262 25 PATRAS

N. KOMINATOS
Ministry of Interior
27 Stadiou Street
GR - 101 83 ATHENS

M. KONDYLIS
Bristol Hellas AEBE
11th Km National Road Athens-Lamia
GR - 144 51 METAMORPHOSIS ATTIKIS

D. KONSTAS
KONSTAS Engineering Ltd.
5, Pl. Victorias Street
GR - 104 34 ATHENS

A. KONTOS
President
M D S Marathon Data Systems
Management Resources
38, Kifissias Avenue
GR - 151 25 MAROUSSI

G. KOPSIDAS
Eleourgiki
Central Cooperative Union of Olive Products
4, Kyprou Street
GR - 157 73 ZOGRAPHOU - ATHENS

H. KORBER
Emschergenossenschaft
Lippeverband and
Main Committee of Detergents
Postfach 10 11 61
D - 4300 ESSEN 1

T.P. KORYALOS
CASTOR Ltd.
5 Isavron Str.
GR - 114 71 ATHENS

M. KOSMADAKIS
Greek Ministry of Defense
Hellenic Airforce General Staff
Kolokotroni Street 15
Ag. Paraskevi
GR - 153 42 ATHENS

F. KOSTAS
Ministry of Environment, Physical
Planning and Public Works
1, Levidou
GR - 302 00 MISSOLONGHI

V. KOSTIBAS
CRYSTAL S.A
102, 25th Martiou Street
GR - 502 00 PTOLEMAIS

I. KOTSAKA
Sandoz Hellas (SACI)
12th Km National Road
GR - 144 51 METAMORPHOSIS ATTIKIS

T. KOUIMTZIS
98, Tsimiski Street
GR - 546 22 THESSALONIKI

E. KOUKORINIS
Public Power Corporation
10, Navarinou Street
GR - 106 80 ATHENS

P. KOULOUMBIS
National Foundation of Agronomical
Research
Inst. of Soil Science
1, S. Venizelou Street
GR - 141 23 LYCOVRISSI - ATHENS

A. KOUTALOS
Consolt. Soil Scient.
3, V. Logothetidi Street
GR - 115 24 ATHENS

A. KOUZELI-KATSIRI
National Technical Univ. of Athens
5 Iroon Polytechniou
GR - 157 73 ZOGRAPHOU-ATHENS

A.B. LARSEN
Novo Nordisk Ltd.
Hallas Allé
DK - 4400 KALUNDBORG

H.I. LATOSTENMAA
National Board of Waters
and the Environment
PB 250
Finland - 00101 HELSINKI

Ch. LEMPESIS
Municipality of Kallithea
Technical Services
Matzagriotaki 76
GR - 176 72 KALLITHEA, ATHENS

R. LESCHBER
Institut f. Wasser-Boden-
u. Lufthygiene des
Bundesgesundheitsamtes
CorrSenplatz 1
D - 1000 BERLIN 33

R.S. LEVINEN
National Board of Waters
and the Environment
PB 250
Finland - 00101 HELSINKI

P. L'HERMITE
Commission of the European Communities
DG XII - E/1
200, rue de la Loi
B - 1049 BRUSSELS

S. LIAKOS
29 Miaouli Str.
GR - 151 21 PEFKI

A. LOUVROS
Olive Institute of Corfou
5, Figareto Street
GR - 491 00 CORFU

S. LUEBBEN
Federal Research Centre for Agriculture
Bundesallee 50
D - 3300 BRAUNSCHWEIG

Ms MALLIKOPOULOU
Athens Water and Sewage Corporation
156, Oropou Street
GR - 111 46 ATHENS

C. MANIOS
Technical Educational Institute
of Heraklion
GR - HERAKLION CRETE

R.G. MANOLIADIS
Technological Educational Institute
10 Athonasiou Filiou
GR - 115 24 ATHENS

A. MANRIQUE
Servicio Investigacion Agraria
Seccion de Analisis Ambiental
Delegacion Territorial de
Agricultura
avd Gral Vigon 23
E - 09071 BURGOS

P. MARAGOU
15, Kapetan Chrona Street
GR - 115 25 ATHENS

P. MARKATONATOS
56, Zan Moreas Street
GR - 152 31 HALANDRI - ATHENS

L. MELO
Estrada do Calhariz de Benfica
Lote 3, 30
Esquerdo
P - 1500 LISBOA

Y.L METAXAS
CASTOR Ltd.
5 Isavron Str.
GR - 114 71

M. MICHALAKOPOULOU
Athens School of Hygiene
196, Leoforos Alexandras St.
GR - 115 21 ATHENS

P. MICHOPOULOS
Forest Research Institute of Athens
Terma Alkmanos
GR - 115 28 ATHENS

C. MILIORITSAS
D.E.Y.A. Lamias
Sklivaniotou Street 8
GR - 351 00 LAMIA

N.D. MIMICOS
National Research Center for
Physical Science
Aghia Paraskevi Attiki
GR - 153 10 ATHENS

G. MININNI
C.N.R.
Water Research Institute
Via Reno 1
I - 00198 ROME

E. MITROPOULOU
11, Agathoupoleos Street
GR - 112 52 ATHENS

A. MOURTSIADIS
Ministry of Industry
Paradisou Street 37
GR - 171 23 NEA SMYRNI - ATHENS

P.J. NEWMAN
WRC Medmenham
Henley Rd.
P.O. Box 16, Medmenham
UK - MARLOW, BUCKS SL7 2HD

B. NIELSEN
Marketing Manager
Atlas Industries A/S
Balterpvej 160
DK - 2750 GALLENUP

J.B. NIELSEN
Enviroplan AS
Gladsaxevej 363
DK - 2860 SEBORG

V.C. NIELSEN
Agricultural Development and
Advisory Service
ADAS Unit, Wrest Park
Silsoe
UK - BEDFORD MK45 4MS

J. NICOLAOU
Ministry of Agriculture
1, S. Venezilou Street
GR - 141 23 ATHENS

P. OCKIER
V.M.Z.
Zandvoordestraat 375
B - 8900 OOSTENDE

H. OTT
Commission of the European Communities
DG XII - E/1
200, rue de la Loi
B - 1049 BRUSSELS

B.F. PAIN
AFRC Institute of Grassland
and Environmental Research
HURLEY
UK - MAIDENHEAD, BERKS SL6 5LR

O. PALM
National Environmental
Protection Board
S - 171 85 SOLNA

L.K PANAGIOTOPOULOS
Plant Protection Inst.
Ministry of Agriculture
Leoforos Amerifis
GR - 261 10 PATRAS

S. PANAGOULI
Ministry of Environment
Plant and Public Works
Varvaki Street 12
GR - 114 74 ATHENS

P.P. PAPABEIS
Greek Ministry of Defense
Hellenic Airforce General Staff
9 Pedelis Str.
Dafni
GR - 172 36 ATHENS

N. PAPACONSTANTINOU
Ministry of Agriculture
Direction of Environmental Protection
3-5 Ippokratous Street
GR - 101 64 ATHENS

A. PAPADAKI
Air Force Gneral Staff
Project's Office
Tzavela 53
GR - 152 32 HALANDRI - ATHENS

P. PAPADOPOULOS
Soil Science Instute
1, S. Venezilou Street
GR - 141 23 ATHENS

Th. PAPADOPOULOS
Environmental Office of Patras
Ministry for the Protection of the Environment
4, Iras street
GR - 264 41 PATRAS

S. PAPAGRIGORIOU
Ministry for Environment
17, Amaliados Street
GR - 115 23 ATHENS

P. PAPAKOSTA
National Technical University of Athens
73, Dodekanisou Street
GR - 142 35 ATHENS

E. PAPALEXIOU
9, Timandras Street
GR - 157 71 ZOGRAFOU - ATHENS

C. PAPAMATHEAKIS
Municipal Enterprise for Water and
Sewage of Iraklion
1, Byronos Str.
GR - 712 02 IRAKLION - CRETE

Ch. PAPAVASSILIOU
Ministry of Agriculture
6-8, Stassinou Street
GR - 116 35 ATHENS

G. PAPPA
Athens School of Hygiene
196, Leoforos Alexandras St.
GR - A15 21 ATHENS

P. PARASKEVAS
Aegean University
Environmental Dept.
E. Bostani Street 35
GR - 811 00 MYTILENE

J. PARTSALIDIS
D.E.Y.A. Katerinis
Kresnas Street 35
GR - 601 00 KATERINI

Ch. PASCHALIDIS
Institute of Soil Science
Ministry of Agriculture
1, S. Venezelou Street
GR - 141 23 ATHENS - LYKOVRISSI

Ch. PASCHALIDOU
Ministry of Agriculture
Direction of Environmental Protection
3-5 Ippokratous Street
GR - 101 64 ATHENS

B. PAULSRUD
Aquateam
P.O. Box 6326
Etterstad
N - 0604 OSLO 6

N. PAXEUS
GRYA AB
Karl IX's Väg
S - 417 22 GÖTEBORG

G. PETRAKAKIS
The Ministry of the Pulic Works
12, Vavraki Street
GR - 114 74 ATHENS

G. PETROPOULOS
Agricultural Research Centre of Athens
S. veniselou Street 1
GR - 141 23 LYCOVRYSSI-ATTIKIS

E.B. PIKE
Water Research Centre
Medmenham
P.O. Box 16
UK - MARLOW, BUCKS SL7 2HD

M. POALA
Greek Ministry of Defense
Hellenic Airforce General Staff
95, Kerkiras Street
GR - 113 63 ATHENS

A. POLO
Instituto de Edafologia y
Biologia Vegetal
Serrano 115 dpdo.
E - 28006 MADRID

F. POMARES
I.V.I.A.
Dep. of Ecology
Apartado Oficial
E - 461 13 MONCADA (VALENCIA)

B. POMMEL
INRA Laboratoire d'Agronomie
INA-PG Chaire d'Agronomie
F - 78850 THIEVERVAL GRIGNON

Y.V. PUOLANNE
National Board of Waters and
the Environment
P.B. 250
SF - 00101 HELSINKI

I.A. RAMOS
Comissao de Coordenaçao da
Regiao do Alentejo
Rua 24 de Julho, N° 1, 2°E
P - 7000 EVORA

R.C. RAMSAY
Dept. of the Environment
Romney House - room B4/52
43, Marsham Street
UK - LONDON SW1 3PY

B. ROGNERUD
Institute of Georesources and
Pollution Research
P.O. Box 9
N - 1432 AAS-NLH

A. SAABYE
Enviroplan A/S
Gladsaxevej 363
DK - 2860 SOEBORG

S.D. SAKELLARIADIS
Aristoteles Univ. Thessaloniki
GR - 540 06 THESSALONIKI

P. SANCHES DA GAMA
Direccao General Qualidade do Ambiante
Serviço de Residuos e Produtos Quimicos
Av. Gago Ceutinho 30, 6°
P - 1000 LISBOA

M.J. SANTANA
Comissao de Coordenaçao da Regiao
do Alentejo
Rua Z4 de Julho, 1, 2°E
P - 7000 EVORA

D. SCHEIDEGGER
Nordstrasse 314
CH - 8037 ZÜRICH

J. SCHWARTZBROD
Faculté de Pharmacie
Université de Nancy I
5, rue Albert Lebrun
F - 54000 NANCY

L. SCHWARTZBROD
Faculté de Pharmacie
Université de Nancy I
5, rue Albert Lebrun
F - 54000 NANCY

A. SGOUROPOULOS
Talos Engineering S.A.
6, Vas. Sophias Avenue
GR - 106 74 ATHENS

A. SIEGENTHALER
Swiss Federal Research Station
for Agricultural Chemistry and
Hygiene of Environment
CH - 3097 LIEBEFELD-BERN

H. SIEGRIST
EAWAG
Überlandstr. 133
CH - 8600 DÜBENDORF

A. SIMITZIS
Sokratous Street 29
GR - 145 61 KIFISIA - ATHENS

S. SOTIRACOPOULOS
Ministry of Environment, Phys. Plan
and Public Work (PERPA)
147, Patisson Street
GR - 112 51 ATHENS

E. SPIESS
Swiss Federal Research Station
for Agricultural Chemistry and
Hygiene of Environment
CH - 3097 LIEBEFELD-BERNE

L. SPINOSA
CNR-IRSA
Via de Blasio 5
I - 701 23 BARI

K. SPIROPOULOS
"Kykkos-Sferiki" S.A.
66, G. Geografou street
GR - 157 72 ZOGRAFOU - ATHENS

A.I. STAMOU
National Technical University of Athens
5, Iroon Polytechniou Street
GR - 157 73 ATHENS

N. STAVROPOULOU
Ministry of Agriculture
Direction of Environmental Protection
3-5 Ippocratous Str.
GR - 101 64 ATHENS

C. STERGIOU
Eleourgiki
Piraeus Street 37-39
GR - 10553 ATHENS

N. STRATIGEAS
Metsovou Street 29
GR - 106 83 ATHENS

P.J.W. TEN HAVE
Institute of Agricultural Engineering
IMAG
P.B. 43
NL - 6700 AA WAGENINGEN

I. THODOS
Ministry of Environment
Plant and Public Works
Varvaki Street 12
GR - 114 74 ATHENS

I. TOUBOUROU
Ministry of Public Works
12, Varvaki Street
GR - 114 74 ATHENS

E. TRIANTAFILLOU
Bitizaniou 13
GR - 152 32 HALANDRI/ATHENS

C. TSAFARAS
Ministry of Agriculture
Direction of Environmental Protection
3 Ippokratous Street
GR - 101 64 ATHENS

N.S. TSARAPATSANIS
18, Asimaki Fotila
GR - 11473 ATHENS

K. TZAVLAKI
Municipal Enterprises for Water and Sewage
1, Byronos Str.
GR - 712 02 IRAKLION - CRETE

P. TZIAS
Ministry of Environment
Plant and Public Works
Varvaki 12 Street
GR - 114 74 ATHENS

Ph. TZOUMERKAS
Athens Water and Sewage Corporation
156, Oropou Street
GR - 111 46 ATHENS

S. TZOVARIDIS
Ministry of Public Works
9, Fanarioton Street
GR - 101 78 ATHENS

R.J. UNWIN
A.D.A.S. Ministry of Agriculture
Burghill Rd.
UK - BRISTON BS1O 6NJ

D. VAFIDIS
Otme Consulting Engineers Ltd.
6 Alkmanos Str.
GR - 115 28 ATHENS

S. VAGIONAS
Ministry of Industry and Energy
23, Irodotou, Ano Poli
GR - 546 34 THESSALONIKI

I. VALI
Civil Service
Galatas
GR - 300 14 MESOLOGI

E. VALIANTZA
Creospectrum S.A.
Consulting Firm
56, Pindou Street
GR - 152 35 ATHENS VRILISSIA

N. VAN BREMEN
Dept. of Science and Geology
Agricultural University
Duivendaal 10
P.B. 37
NL - 6700 AA WAGENINGEN

J.J. VAN DEN BERG
GRONTMIJ N.V.
Postbox 203
NL - 3730 AE DE BILT

P. VAN LUNE
Institute for Soil Fertility Research
Oosterweg 92
P.B. 30003
NL - 9750 RA HAREN (GN)

G. VAVIZOS
2, Mavromataion Street
GR - 106 82 ATHENS

G.F. VERONESI
Ist. Genio Rurale
Univ. Bologna
Filippo Re.
I - 40126 BOLOGNA

P.A. VESILIND
Duke University
Dept. of Civil &
Envir. Engineering
USA - DURHAM NC 27706

E. VIGERUST
Agricultual University of Norway
Box 20
N - 1432 AS-NLH

J.H. VOORBURG
I.M.A.G.
P.B. 43
NL - 6700 AA WAGENINGEN

H.A. WEBERS
Witteveen & Bos
Consulting Engineers
99, Wilhelminasingel
P.O. Box 1080
NL - 6201 BB MAASTRICHT

G. WEIDENBACH
Commission of the European Communities
DG XII - E
200, rue de la Loi
B - 1049 BRUSSELS

P. WEILE
Ministry of the Environment
National Agency of Environmental Protection
Strandgade 29
DK - 1401 COPENHAGEN K

S.R. WILD
Lancaster University
UK - LANCASTER LH 4XQ

M. WIMAN
Swedish Environmental Protection
Agency
S - 171 85 SOLNA

H. WITTE
Universität GH-Siegen
Paul-Bonatz Str. 9-11
Postfach 101 240
D - 5900 SIEGEN

J.G. XANTHOPOULOU
Hydrodomiki Consultants
21, Theagenous Street
GR - 161 21 ATHENS

A. YLARANTA
National Board of Waters and
the Environment
BP 250
SF - 00101 HELSINKI

G. YOLARIS
9, Davaki Pindou street
GR - 157 73 ZOGRAFOU - ATHENS

M. ZACHARIAS
N.T.U.
60, Neapoleos Str.
Aghia Paraskevi
GR - ATHENS

E. ZAFIRI
Company Kruger Hellas
23, Tositsa Street
GR - 106 83 ATHENS

P. ZAGOREOU
Greek Ministry of Defense
Hellenic Airforce General Staff
23-25 Etolias Str.
Aghia Paraskevi
GR - 153 41 ATHENS

G.A. ZERVAS
Athens School of Public Health
196, Alexandras Avenue
GR - 115 21 ATHENS

A. ZIBIRIKAKIS
31, Ioulianou Street
GR - 104 33 ATHENS

A. ZIMBIRIKAKIS
Hospital Italian
Baskurt So. 29/3
Cihangir
TUR - ISTANBUL

N. ZOGRAFAKIS
Unversity of Crete
25, Atsipadon Street
GR - 713 05 HERAKLION CRETE

N. ZOGRAFOS
G. Apostolidou
GR - 341 00 CHALKIS

N. ZOTOV
CRYSTAL S.A.
102, 25th Martiou Str.
GR - 50200 PTOLEMAIS

A.I. ZOUBOULIS
Aristotelian University of Thessaloniki
Lab. of Gen. & Inorg. Chem. Technology
Dept. of Chemistry
GR - 54006 THESSALONIKI

INDEX OF AUTHORS